MICROBIAL BIOCHEMISTRY

MICROBIAL BIOCHEMISTRY

By

Dr. G.P. Garg

DISCOVERY PUBLISHING HOUSE PVT. LTD.
NEW DELHI-110 002

Published by:
Namit Wasan
DISCOVERY PUBLISHING HOUSE PVT. LTD.
4383/4B, Ansari Road, Darya Ganj
New Delhi-110 002 (India)
Phone : +91-11-23279245; 23253475; 43596065
E-mail : discoverybooksindia@gmail.com
discoverypublishinghouse@gmail.com
namitwasan9@gmail.com
web : www.discoverypublishinggroup.com

Edition: 2020

ISBN: 978-81-8356-625-4

Microbial Biochemistry

Printed at:
Infinity Imaging Systems
Delhi

Preface

Biochemistry today has made spectacular progress in unraveling the mysteries of animate nature. This progress has allowed us to gain deeper insight into the principles of vital activity and has to a very significant extent stimulated the development of applied disciplines, especially medicine. The present title *"Microbial Biochemistry"* is intended for those who wish to understand living organisms, especially man. Biochemistry is essential for this purpose, but it would be almost impossible for a student to survey on his own the massive body of existing knowledge, constantly augmented by a remarkable torrent of brilliat discoveries. The purpose of the book, then is to organize our knowledge into something that can be comprehended in a relatively short time and still convey a reasonable complete picture of the chemical structure and function of man. Readability without sacrifice of coverage has been a prime goal. An important device in gaining that goal is to keep attention constantly focused on function, with repeated use of rationalization to show that the chemical facts are not isolated, but part of a whole.

Biochemistry has two major goals as a fundamental science, it treats the vital functions from the standpoint of physical chemistry, and as an applied discipline, it points out practical applications for the wealth of scientific knowledge it has acquired. The dual purpose has been, as far as possible, take into account in the writing of this book. We have also tried to summarize our pedagogical experience in teaching biochemistry to students specializing in medicine and pharmacy. The material of this book has been organized according to the principle of functionally in order to trace the close relationship between the functions of the living organism and the structures of its constituents molecules as well as the chemical and physico-chemical process in which they are involved.

The aim of this book is to present a core of biochemical knowledge that is desirable for undergraduate and postgraduate students and also those involved in the field of medical, microbiology, biotechnology and pharmaceutical. Every attempt has been made to keep abrest of the advances in the subject and at the same time to include the fundamentals.

To make the work more comprehensive and informative, the author has consulted many authoritative books, research journals, abstracts, monographs etc. He is grateful to all those great scholars whose work are cited or substantially reproduced.

There can be no claim to originality except in the manner of treatment and much of the information has been obtained from the books and scientific journals available in the different libraries.

The author expresses his thanks to his friends and colleagues whose continue inspirations have initiated him to bring out this book.

The author expresses his gratitude to Mr. Wasan and Staff of M/s Discovery Publishing House Pvt. Ltd., for their whole hearted co-operation in the publication of this book.

In the mean time, the author will remain sincerely responsible for any shortcomings of the book and be grateful to the readers for their suggestions and constructive criticism for the continuous betterment of the book. He takes this opportunity to appeal to the readers to send their suggestions straightway to his publisher.

Author

CONTENTS

Structure—Initiation of Transcription—Bacterial RNA Polymerase—Promoter Seuences—Control of Stable RNA Synthesis—The Initiation Reaction—Elongation of RNA Transcripts—Control of Various Bacterial Operons—Many Repressors—Inducible Operons—Feedback Repression—Positive Control by Activator Proteins—Control by Looping—Bacterial Enhancers—Other Two-component Control Systems—Antisense RNA—Termination of Transcription in Bacteria—Attenuation—Rho and Other Termination Factors—Effects of Antibiotics—Processing of Bacterial RNA—Ribosomal RNA—Transfer RNA—Viral Rna in Prokaryotes—The Lysogenic State of Phage l—Replication of RNA Bacteriophages—Transcription In Eukaryotic Cells and in Archaea—Eukaryotic Nuclei and Transcription—RNA Polymerases —Transcriptional Units and Initiation of RNA Synthesis—Promoters, Transcription Factors, Enhancers, and Activators—The SV40 early promoter and transcription factor Sp1—The TATA Binding Protein and General Transcription Initiation Factors—Transcriptional Activators—Mediators and Coactivators—Enhancers—The Variety of DNA Binding Proteins—Leucine Zipper Transcription Factors—Control of Growth—Response Elements—Zinc-containing Transcription Factors—Winged Helix Transcription Factors—The NF-kB/Rel Proteins—HMG Proteins as Transcription Factors—The Variety of Transcriptional Responses—Nuclear Hormone Receptors—Nutrient Control—Stress Proteins—Light-induced Transcription—Homeotic Genes and Homeodomain Proteins—Transcription by RNA Polymerases I and III—Elongation, Modification, and Termination of Transcription—Conformational Properties of RNA—Processing Of Eukaryotic Rnas—Ribosomal RNA—Small Nuclear, and Nucleolar, and Cytoplasmic RNAs—Processing of Ribosomal RNA—Modification Guide RNAs—Transcription and Processing of snRNAs and snoRNAs—Processing of 5S RNA and tRNAs—Messenger RNA, Caps, and Polyadenylate Tails ——Splicing—Self-splicing RNA—Relationship to Viroids and Virusoids—Pre-tRNAs —The Spliceosome—Alternative Splicing Pathways—Trans Splicing—Modification and Editing of RNAs—Editing of RNA Transcripts—Finishing the Transcripts—Transcription of Mitochondrial, Chloroplast, and Vir Genes—Viral Transcription and Replication—Replication of RNA Viruses—Small RNA Viruses—Influenza Viruses—HIV-1 and Other Retroviruses—Accessory Regulatory Genes—Retrotransposons—The Antibiotics Rifamycin, Rifampicin, and Actinomycin D—Powerful Poisons from Mushrooms—Synthetic Antiviral Compounds.

CHAPTER

1 Introduction

The myriad different species of living cells, plant, animal, or protist, microbic or human, are obviously variations on a single basic theme—the cell theme. Some cells, like different makes of bicycles, have features not possessed by all, but all are obviously derived from the same fundamental pattern. All cells, however they may differ in physical and physiological characteristics, are composed of essentially identical elements combined in similar or identical compounds which, with small modifications from species to species, constitute similar or homologous cellular structures.

Chemical reactions of the same general (often identical) types underlie the processes by which foodstuffs are utilized in different species of cells as sources of energy and of cell substance. And, finally, it is now known that all of the variations in structure and physiology, including enzymic activities, that characterize different species can be ascribed to relatively small variations in the structure of certain complex molecular groups that are common to all cells; *i.e.*, the heredity-determining nucleic acids. Thus virtually all of the phenomena of life may be explained in terms of molecular biology and referred ultimately to the molecular structure of nucleic acids: deoxyribonucleic acid (DNA) and ribonucleic acid (RNA).

STRUCTURE OF ATOMS

From the foregoing it is obvious that a knowledge of molecular structure is essential to a modern appreciation of biology, especially microbiology. Therefore we shall, for the convenience of the reader, recapitulate some basic concepts concerning atoms, molecules, their structure and their roles in the phenomenon that we call life. It is assumed that the reader has at least a speaking acquaintance with physics and chemistry. Molecules, as the reader will doubtless remember, are made up of atoms.

Each atom, in turn, consists of a heavy, central nucleus surrounded by one or more negatively charged, weightless particles called *electrons* that encircle the nucleus in various inner and outer orbits. Both nucleus and electrons are minute, yet the distance between them is so vast, relative to their sizes, that the spatial relationship of nucleus to orbiting electrons suggests the relationship of the sun to the orbiting planets. The *atomic nucleus* consists of several kinds of particles, of which protons and neutrons are the most important in this discussion.

Each *proton* has a positive electrical charge. *Neutrons* have no electrical charge; they represent a combined proton and electron. Each uncombined proton and each uncombined neutron is assigned an arbitrary weight of 1 unit. The electron weighs about 1/1848 of this unit— not very heavy! Atoms of different substances have different numbers of protons and neutrons and of orbiting electrons. It is these differences, and especially the differences in number and arrangement of the electrons, that create the differences between substances. Under usual conditions the negative charges on the electrons of an atom are balanced by the positive charges of the protons in the nucleus of that atom: *i.e.,* the numbers of proton and of electrons are alike. Such atoms are alike. Such atoms are electrically neutral.

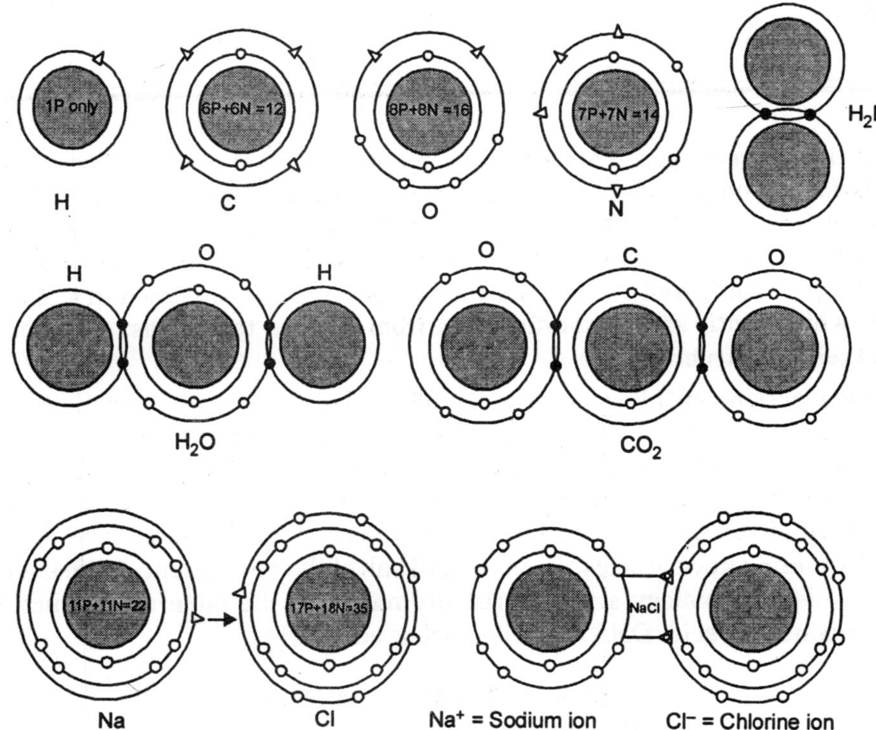

Fig. 1.1. Diagrammatic structures of atoms of hydrogen, carbon, oxygen and nitrogen, the hydrogen molecule and the molecules of H_2O and CO_2. In each diagram the atomic nucleus is shown as a relatively solid central portion with, protons and neutrons surrounded by electrons in various shells and orbitals. Electrons indicated as open circles are *paired;* those shown as open triangles are not paired, *i.e.,* they are *valence electrons;* those shown as closed circles are *shared, i.e.,* in *covalent bonds.* NaCl is shown as an *ionic pair* held together by weak *electrovalent bonds* (—◁), the single valence electron of Na having been taken over by the seven-electron, unsaturated outer shell of Cl.

The number of electrons (or of protons, since their numbers are equal) in any given type of atom is called the *atomic number* of that atom. The number of protons plus the number of neutrons is the *atomic weight* of that atom. Atomic weight is a relative quantity. It is usually expressed as the number of times heavier a given atom is than an atom of some other substance taken as a standard. This is now commonly carbon (6 protons + 6 electrons = at. wt. 12; 6

electrons = at. no. 6). The hydrogen atom is unique in having in its nucleus only one proton; atomic weight = 1.00^{+}.

A single electron, spinning on its own axis like the earth, orbits the nucleus with extreme rapidity, always at nearly the same distance from the nucleus. The centrifugal force of this whirling motion might cause the electron to fly off into space, were it not balanced by the constant attraction of the positively charged proton. So rapid is the motion and so strong and generalized is the influence of the electron around the proton that it forms a continuous spherical shell, or cloud, around the nucleus. The continuous zone of distribution of an electron around its nucleus is called a *shell* or *orbital.* Spherical orbitals are designated *s* orbitals; orbitals with patterns other than spherical (commonly figure-eight patterns) are called patterned or *p* orbitals. Depending on the element there may be one or several shells, the series shells from within outward being designated by the letters *k* to *p*. Carbon, for example, has two shells, an inner or *k* shell containing *paired* electrons in an *s* orbital close to the nucleus, and an outer or *l* shell with four electrons in modified (hybrid) *p* orbitals.

The six protons in the nucleus of the carbon atom balance the six electrons. The six electrons give it an atomic number of 6. The two electron shells give it a position in the periodic table of the elements in Period 2.

The Periodic Table of the Elements

This is a descriptive tabulation of all elements according to atomic structure. Arrangements vary, but in all tables the numbers of the *periods* (first column) represent the numbers of electron shells per atom. The *group* numbers represent numbers of electrons in the outermost shell. Of these outermost electrons all or, depending on patterns and arrangements of the orbitals, only certain outermost electrons can enter into the formation of compounds with other atoms. The outermost electrons available for the formation of bonds with other atoms or radicals are called *valence electrons* and their number is the valence number *(valence)* of that atom or element. Elements with one, two or three electrons in the outermost shell (Groups I, II and III of the table) tend to become stabilized by giving these electrons up, elements with five, six or seven outermost electrons (Groups V, VI and VII) tend to achieve stability by accepting electrons. Thus the elements of Groups I, II or III have a tendency to combine in particular with elements of Groups V, VI, or VII.

The readiness of an element to lose or gain electrons is expressed as *activity,* and the degree of activity is dependent in great part on the distance of the electrons from the nucleus. In general, those changes in electronic structure which achieve greater stability tend to occur and usually release energy. The outermost shells of elements in Group O are "saturated" with electrons; helium with two, the others with eight, the maximum for *any* outermost shell. This is a condition of complete stability; there is no tendency either to give up or to accept electrons. These are the *inert elements:* helium, neon, argon, krypton, xenon and radon. The transitional elements are subgroups of Periods 4, 5 and 6 and generally have two valence electrons, except Cu, Ag, Pt and Au, which have only one. The rare earth elements (lanthanide series, Period 6, and actinide series, Period 7) concern us but little in biology. Indeed, of over 100 known elements, only about 19 are of demonstrated importance in biology: Ca, C, Cl, Co, Cu, *H,* I, *Ft, Mg,* Mn, Mo, *N, O, P,* K, Na, Rb, S, Si. Those italicized are most abundant in living cells, although several of the others are absolutely essential in minute amounts. Some of those not listed may.

Valance and Compound

It is clear that the numbers of outermost electrons and their positions in regard to the nucleus are of critical importance in regard to combinations of any element with other elements, *i.e.*, in the formation of compounds. Electrons in the inner or *k* shell are generally very strongly held by the nucleus and, except for the hydrogen electron, are not involved in chemical reactions. Outer-shell structures are least stable because outer electrons are less strongly attracted to the nucleus. The strength of the attraction decreases in proportion to the distance of the electron from the nucleus; the outermost electrons are therefore more easily removed and the outermost shells are most readily added to. Although the outermost electrons of atoms are those farthest removed from the strong attraction of the nuclear protons and therefore most involved in β-combinations between atoms, as a general rule only those outer electrons that are *unpaired* tend to form bonds.

For example, oxygen has six outer electrons and is therefore in Group VI of the periodic table; but only two of these six electrons are unpaired, giving oxygen a valence of 2. Similarly, nitrogen has five outer electrons but only three are unpaired. Nitrogen therefore has a valence of 3. Under some circumstances it may exhibit a valence of 4 or 5. Some other elements show this variation of valency. In general, atoms tend to enter into combinations by which they can approach the stability that, for still obscure reasons, results from the presence of eight (four pairs) outer electrons. This is the structure of most inert elements of Group O. Hydrogen is an exception in that it has only one shell that requires only one more electron to saturate it. It then achieves the stability (and identity) of the inert helium atom with a pair of electrons. Because of its single electron, hydrogen is extremely active. It even reacts with itself to form H_2, a diatomic molecule.

The halogens, nitrogen and oxygen also generally occur as diatomic molecules. The capacity of each shell to contain electrons is limited, varying from the *k* shell (capacity two electrons) to the outermost shell (eight electrons). Intermediate shells may have up to 32 electrons. The outermost shell never contains more than eight electrons. A shell with eight electrons is a very stable structure and strongly resists alteration: *i.e.*, it is an inert element (Group O). Valence electrons cannot long be free or floating in the air, as it were; they tend to be pound as CH as a distinct and stable compound because three of the carbon valences would be left free; the simplest stable compound of carbon with hydrogen is CH_4 or methane, the principal constituent of marsh and sewer gas. Among the elements carbon, the basis of all organic compounds and therefore the key to life as we know it, is unique and most versatile.

In the first place, carbon atoms are ambivalent, *i.e.*, they can combine with positive or negative ions or groups simultaneously. They also have the rare property of combining with other carbon atoms. They thus can form almost indefinitely long chains of linked carbon atoms—straight, bent, branched or cyclic. This property is the basis of the large group of aliphatic compounds that underlie many structural parts of living cells. These will be mentioned later.

Ionization

We have already mentioned some properties of aqueous suspensions of large molecules (*e.g.*, proteins) or groups of molecules such as *colloids*. Colloids conduct relatively little electricity.

Table 1.1 Long form of the Perodic Table of the Elements*†

Period	IA		IIIB	IVB	VB	VIB	VIIB	VIII			IB	IIB	IIIA	IVA	VA	VIA	VIIA	O Inert Gases
			Transition Elements															
1	1 H 1.008																1 H 1.008	2 He 4.003
2	3 Li 6.939	4 Be 9.012											5 B 10.81	6 C 12.011	7 N 14.007	8 O 15.999	9 F 19.00	10 Ne 20.183
3	11 Na 22.990	12 Mg 24.31											13 Al 26.98	14 Si 28.088	15 P 30.974	16 S 35.453	17 Cl 35.453	18 Ar 39.948
4	19 K 39.102	20 Ca 40.08	21 Sc 64.956	22 Ti 47.90	23 V 50.94	24 Cr 51.996	25 Mn 54.94	26 Fe 55.85	27 Co 58.93	28 Ni 58.71	29 Cu 63.54	30 Zn 65.37	31 Ga 69.72	32 Ge 72.59	33 As 74.92	34 Se 78.96	35 Br 79.909	36 Kr 83.80
5	37 Rb 85.47	38 Sr 87.62	39 Y 88.91	40 Zr 91.22	41 Nb 92.91	42 Mo 95.94	43 Tc 99	44 Ru 101.07	45 Rh 102.91	46 Pd 106.4	47 Ag 107.870	48 Cd 112.40	49 In 114.82	50 Sn 118.69	51 Sb 121.75	52 Te 127.60	53 I 126.90	54 Xe 131.30
6	55 Cs 132.91	56 Ba 137.34	57-71 La-Lu Rare Earths	72 Hf 178.49	73 Ta 180.95	74 W 183.85	75 Re 186.2	76 Os 190.2	77 Ir 192.2	78 Pt 195.09	79 Ay 196.97	80 Hg 200.59	81 Tl 204.37	82 Pb 207.19	83 Bi 208.98	84 Po 210	85 At 210	86 Rn 222
7	87 Fr 223	88 Ra 226.05	89-103 Ac-Lw Actinides															
Rare Earths			57 La 138.91	58 Ce 140.12	59 Pr 140.91	60 Nd 144.24	61 Pm 145	62 Sm 150.35	63 Eu 151.96	64 Gd 157.25	65 Tb 158.92	66 Dy 162.50	67 Ho 164.93	68 Er 167.26	69 Tm 168.93	70 Yb 173.04	71 Lu 174.97	
Actinides			89 Ac 227	90 Th 232.04	91 Pa 231	92 U 238.03	93 Np 237	94 Pu 242	95 Am 243	96 Cm 247	97 Bk 249	98 Cf 251	99 Es 254	100 Fm 253	101 Md 256	102 No 253	103 Lw 257	

† Not that of all the elements only about 19 are of known significance in relation to microbiology. The darker shading indicates those elements that areof the most general importance as constituents of living cells, and the lighter shading those elements that are not so widely distributed among species or that occur in lesser amonts generally.

Electrolytes

Substances whose aqueous solutions readily conduct electricity are called *electrolytes.* If they are very active in this respect they are said to be strong electrolytes. Generally these are inorganic compounds: salts like sodium chloride, acids like sulfuric acid, bases like potassium hydroxide. Except carbon monoxide, carbon dioxide, carbonates and cyanides, *inorganic* compounds characteristically lack carbon, carbon-to-carbon or carbon-to-hydrogen bonds. These bonds are always found in *organic* compounds, which typically are *not* strong electrolytes. Substances whose aqueous solutions do not readily conduct electricity are called *nonelectrolytes.* They are generally organic compounds.

Organic colloids like protein and fat emulsions virtually nonelectrolytes. Organic acids like citric (of lemons), acetic (of vinegar) and lactic (of sour milk) are weak electrolytes, that is, their aqueous solutions conduct electricity to a slight degree. Electrolytes have several other properties that distinguish them from nonelectrolytes. Compared with nonelectrolytes, electrolytes markedly affect the colligative properties of water: *i.e.,* they (a) lower the freezing point, (b) raise the boiling point, (c) increase the osmotic pressure and (d) lower the vapor pressure of their aqueous solutions.

Dissociation

Electrolytes are electrically conductive because their molecules in aqueous solutions split *(dissociate)* into positively and negatively charged particles called *ions.* They has a high dielectric constant; *i.e.,* it has practically no conductance for electricity and (if pure!) is therefore a very good insulating material; (b) water molecules are *polar* because in each molecule the oxygen has attracted the electrons of the hydrogen unequally, giving it unbalanced negativity. These seemingly unrelated facts are easily integrated as follows.

If an electrolyte, for example HCl, (H^+Cl^-) is dissolved in water $\left(\begin{matrix}H^+\\H^+\end{matrix}\!\!>\!O^-\right)$ the positively charged hydrogen ion (H^+) (a *cation)* attracts around itself swarms of water molecules, each with its negative pole (oxygen) nearest the H^+. In a like manner, on electrolytic dissociation the negatively charged chlorine ion (Cl^-) (an *anion)* attracts other water molecules with their positive poles (hydrogen) nearest the Cl^-. The two ions (H^+ and Cl^-) are thus well insulated from each other by the dielectric action of the water molecules. The H^+ and Cl^- move about freely in the water.

If two oppositely charged, chemically inert electrodes from a battery (*e.g.,* platinum electrodes) are immersed in the solution at opposite sides of the vessel, any cations in the solution migrate to the negative electrode (the *cathode),* accepting electrons therefrom; any anions in the solution migrate to the positive electrode (the *anode),* yielding electrons thereto and thus demonstrating the conductivity of the solution. Each pair of H^+ ions, accepting electrons, becomes a hydrogen molecule, and passes off in hydrogen bubbles at the cathode. Each pair of Cl^- ions, yielding electrons, becomes a chlorine molecule, and passes off in bubbles at the anode: an example of *electrolysis.*

Various other reactions may occur at the electrodes, depending on the electrolyte in solution, the nature of the electrodes, the solvent, the voltage applied and other factors. The water itself is concomitantly decomposed into $2H_2$ and O_2. If the forces of attraction between ions in aqueous solution are great enough to overcome the insulating effect of the water molecules, then those ions will combine or react together.

Radicals (Groups)

A *radical* is a group of atoms held together by strong bonds but containing an excess (or deficiency) of electrons. The group commonly acts as an ion, the valence number of which depends on the net charge of the group; *e.g.*, ammonium, valence = +1 (NH_4^+); hydroxyl, valence = –1 (OH^-); nitrate, valence –1 (NO_3^-); sulfate, valence –2 (SO_4^{--}); phosphate, valence, – 3 (PO_4^{---}).

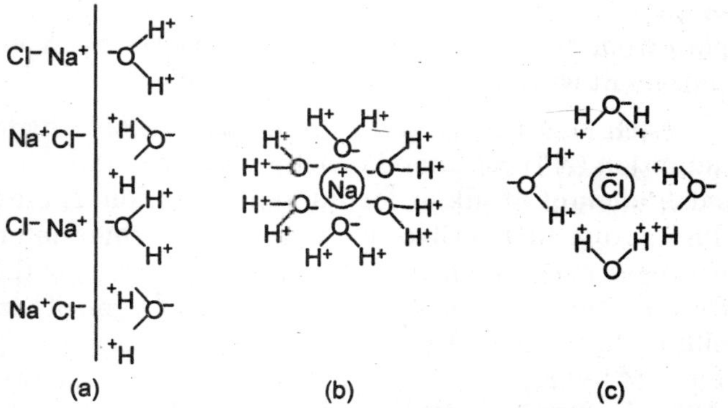

(a) (b) (c)

Fig. 1.2. Dissociation of sodium chloride in water.

Hydrogen and Hydroxyl Ions

Hydrogen and hydroxyl ions (H^+ and OH^-) are extremely important in biology because they are the constituents of water and the basis of the properties of acidity and alkalinity. These influence all aspects of cell life. Degree of *acidity* is commonly expressed as pH, with a number representing the concentration of hydrogen ions, since it is ionized hydrogen that determines the immediate acidic activity of any solution. *Alkalinity* of solutions may be expressed as concentration of hydroxyl ions (OH) (not hydroxyl groups attached to C). However, since concentrations of hydrogen ions and of hydroxyl ions are reciprocally related, as will be shown, degrees of either are commonly given in terms of pH only.

pH

Since H ions determine acidity, acids or alkalies may be strong or weak, depending on their degree of dissociation. This is always a fixed value for any given electrolyte and is generally expressed as the *dissociation constant.* Strong acids are those which, when dissolved in water, dissociate largely into positively charged hydrogen ions and negatively charged ions. For example, sulfuric acid dissociates into two hydrogen ions and a sulfate ion. Weak acid like acetic or citric also dissociate, but to a lesser degree.

The acidic activity of any acid solution depends upon the concentration of ions of hydrogen, and this is obviously dependent upon the ability of the acid to give them off into the solution or to dissociate. Thus, two acid solutions may be of the same concentration with respect to the total amounts of hydrogen available, yet have widely differing activities due to differences in the amount of active or ionized or dissociated hydrogen. Here we deal with a capacity effect, *i.e.*, total available (dis with an intensity or activity effect (dissociated acid or hydrogen ions alone. As an example, let us compare acetic acid and hydrochloric acid. A liter of a normal solution of each contains exactly 1 gm. of total available hydrogen, yet the activity of the N/1 Acetic acid is slight while that of the N/1 hydrochloric acid is great?

Of the gram of available hydrogen in the acetic acid solution only 1.36 per cent is in an ionized state, so that there is, in the liter of solution, only 0.0136 gm. of hydrogen ions. The gram of hydrogen in the liter of N/1 HCl solution is 91.4 per cent ionized, giving 0.914 gm. of hydrogen ions per liter. The N/1 HCl, therefore, is about 67 times as active or "strong" as the N/1 acetic acid. If one were to titrate the solutions, *i.e.,* add N/1 NaOH solution until each became neutral, the total amount of alkali required would be the same in each case. This is due to the fact that, as the alkali combines with the hydrogen ions, more hydrogen ions take their place from the undissociated acid, which strives to maintain a constant hydrogen ion concentration consistent with its dissociation constant.

Each acid finally gives up all its available hydrogen and, since each of the solutions by definition (N/l) contained exactly 1 gm. of available hydrogen to start with, each requires the same amount of alkali for its neutralization. A measurement of hydrogen ion concentration differs from such a titration, in that the former determines the actual concentration of ionized hydrogen at the moment, without calling out any of the reserve, undissociated acid. In acidimetry the term "normal" refers to the presence of 1 gm. of *total available,* hydrogen per liter (dissociated plus undissociated). By contrast, a solution normal only with respect to *ioni sed* hydrogen contains 1 gm. of hydrogen ions per liter. This implies the presence of 1 gm.-equivalent of a completely (100 per cent) dissociated acid; a N/10 solution would contain 0.1 gm.-equivalent of a completely dissoriatcd acid, and so on. As we were to express hydrogen ion concentrations or normality in terms of grams of hydrogen ions per liter we should have to deal with long words and long of zeros; a confusing and laborious system of nomenclature

In 1909 Sorensen devised a simpler system based on the fact that water is itself a very weak electrolyte. As noted previously the extent of dissociation of any electrolyte is a physical constant (K) for that electrolyte under standard conditions. A liter of pure, neutral water at 20° C, always contains 0.0000001 gm. (1×10^{-7} moles) of H^+ and, reciprocally, since water is actually $H^+ + OH^-$, 0.0000001 gm. (1×10^{-7} moles) of OH^+. In Sorensen's system, the term "grams of hydrogen ions per liter" is replaced by the symbol pH, while the number of moles of H^+ per liter (1×10^{-7} in the neutral water under discussion) is expressed as the logarithm of the reciprocal of the fraction, *i.e.,* the positive number 7.

The reaction of neutral water, and of any neutral solution, is therefore expressed as pH 7. Now the product of the concentration of H^+ and of OH^- in neutral water is always 10^{-14} $CH^+ \times OH^- = K_w = 10^{-7} \times 10^{-7} = 10^{-14}$). Since the product of the two is always the same (*i.e.,* since the two are reciprocally related), the term pH is commonly used to express either. For example, the pH of a solution containing 1 gm. of H^+ per liter (*i.e.,* normal [N/l] with respect to hydrogen ions or 1 gm.-equivalent of a completely dissociated acid) is 0 (log 1 = 0).

Reciprocally, this is also pOH 14; the smallest fraction of a gram of OH^- per liter possible on the Sorensen scale. Similarly, pH 6 implies pOH 8; pH 2 implies pOH 12, and so on. Since the number representing pH is derived from a fraction, the larger the fraction the smaller the pH number. Therefore pH numbers between 7 and 0 represent increasing degrees of acidity, and numbers between 7 and 14, increasing degrees of alkalinity. Unless one is familiar with the numbers they can at first be misleading. For example, a change in pH from 7 to 6 represents a 10-fold increase in concentration of hydrogen ions since the 7 and 6 are logarithms; a change from pH 7.0 to 7.3 represents a 50 per cent decrease in the concentration of hydrogen ions ($1/2 \times 10^{-7}$ = log 2 + 7 = 0.3 + 7 = pH 7.3).

ISOTOPES

All atoms of a given element have the same number of electrons and the same number of protons to balance those electrons. Atoms thus remain electrically neutral. But not all atoms of a given element necessarily have the same number of neutrons; *i.e.*, different atoms of a given element may have different numbers of neutrons and therefore different atomic weights. Such atoms are called *isotopes*. In any given sample of an element there is usually a mixture of isotopes.

Table 1.2. Relationships of hydrogen ion concentrations expressed in various ways

Reaction	*Fraction of Normality**	*Hydrogen Ions per Liter (grams)*	*Logarithms of H Ion Concentrations*	*Expressed as pH*
Acid...	N/1	1.0	−0	0.0
Acid...	N/10	0.1	−1	1.0
Acid...	N/100	0.01	−2	2.0
Acid...	N/1,000	0.001	−3	3.0
Acid...	N,10,000	0.000,1	−4	4.0
Acid...	N/100,000	0.000,01	−5	5.0
Acid...	N/1,000,000	0.000,001	−6	6.0
Neutral...	Pure water	0.000,000,1	−7	7.0
Alkaline...	N/1,000,000	0.000,000,01	−8	8.0
Alkaline...	N/100,000	0.000,000,001	−9	9.0
Alkaline...	N/10,000	0.000,000,000, 1	−10	10.0
Alkaline...	N/1,000	0.000,000,000,01	−11	11.0
Alkaline...	N/100	0.000,000,000,001	−12	12.0
Alkaline...	N/10	0.000,000,000,000,1	−13	13.0
Alkaline...	N/1	0.000,000,000,000,01	−14	14.0

*With respect to hydrogen or hydroxyl *ions*.

The international atomic weight of any element is listed as an average of the weights of all the various isotopes normally (commonly) present. For example, the weight of a single atom of normal (the most common isotope) chlorine is 35. But any considerable quantity of the element also contains about 25 per cent of heavy chlorine, Cl^{37}. The international atomic weight of chlorine as an element is therefore given as 35.457. Similarly, the atomic weight of normal hydrogen is 1 but the international atomic weight is 1.00814. Three isotopes of hydrogen are known: H^1, mass (at. wt.) 1.00814; deuterium (H^2), mass 2.01474; tritium (H^{3*}), mass 3.01701. Tritium is radioactive; it is so designated by the asterisk. There are isotopes of nearly every element; not all are radioactive. Chlorine has two principal (most common) isotopes with atomic weights of 35 and 37.

The atomic weight of an isotope is generally indicated by a superscript number (Cl^{35}, Cl^{37}). Lead has at least 16 isotopes, carbon five or six, and so on. In the periodic table all of the isotopes of an element are placed together, since the position of each element in the table depends only on the number of its *electrons* (at. no.), not its nuclear structure (at. wt.). Isotopes of an element are *chemically* similar since it is mainly the electrons that determine the chemical properties. However, since isotopes have different numbers of neutrons, they may differ markedly in *physical* properties because these are affected by their atomic weights. For example, of the four principal carbon isotopes C^{11*}, C^{12}, C^{13}, and C^{14*} one (C^{11*}) is lighter and two are heavier (C^{13} and C^{14*}) than the "normal" C^{12} carbon atom. The radioactive isotopes may be detected by means of a Geiger counter.

Radioactivity and Transmutations of Substance

Radioactivity is the emanation of various atomic particles and rays (energy) from certain elements that have unstable atomic structures (*e.g.*, isotopes). The emanations of radium (chiefly Ra^{226*}), for example, are: (1) alpha rays (particles) that are actually helium nuclei ($2P^{++}$ plus $2N^{00}$) traveling at about one fifth the speed of light; (2) beta rays or particles (actually streams of electrons) moving at nearly the speed of light; and (3) gamma rays. The last are nonparticulate radiant energy somewhat like hard (*i.e.*, short and penetrating) X-rays. They move at the speed of light (186,000 mi./sec.).

Radioactivity of isotopes is derived from instability of their nuclei only, and is fully manifested and detectable with a Geiger counter whether the isotopes are free or combined in compounds. Stability of an atomic nucleus is greatest when the numbers of protons and neutrons are equal; a ratio of 1:1. If too great a discrepancy exists in this ratio, in either direction, the nucleus is unstable and tends to adjust the ratio to greater stability by giving off energy in the form of various radioactive emanations. Adjustments are made toward greater stability by alterations in structure of the nucleus such as transformation of neutrons into protons by splitting off electrons from the neutrons.

For example, the unstable (radioactive) carbon isotope ${}_6C^{14}$, with 6 protons and 8 neutrons (a proton:neutron ratio of 3:4), loses an electron from one neutron and thus gains one proton, becoming the stable form of nitrogen (${}_7N^{14}$) with 7 each of protons and neutrons. The electron escapes with its energy; a beta particle. Transmutation of one element into another has occurred. Radioactive isotopes of a number of elements are commonly manufactured by shooting their nuclei with high-velocity atomic particles: neutrons, protons, alpha particles or gamma rays. For example, bombarding normal nitrogen atoms (at. wt. 14) with neutrons made to travel at extremely high speeds by means of an *accelerator (cyclotron* or *synchrotron)* splits the nitrogen nucleus, yielding radioactive heavy carbon (C^{14*}). C^{14*} is produced in considerable amounts in atomic bomb explosions. Because carbon enters into organic compounds, C^{14*} eventually gains entrance into the cells and tissue of living human beings.

As will be detailed later, radio-activity is an important cause of genetic mutations (altered inherited characteristics) and of cancer, hence isotopes in fallout are of intense interest not only to biologists but to everyone else. Since the nuclei of radioactive isotopes constantly give off nuclear particles and energy, the nuclei eventually become altered: *i.e.*, they decay. The element finally changes into another element of lower atomic number. For example, the nucleus of uranium (at. no. = 92; or ${}_{92}U^{238}$) becomes a thorium nucleus (at. no. = 90; *i.e.*, ${}_{90}Th^{232}$) by losing an alpha particle; the radium nucleus (at. no. = 88) becomes a radon nucleus (at. no. = 86) in the same way. When a beta particle is given off in radioactivity, a neutron (an electron combined with a proton) splits, permitting the electron to leave the decomposing nucleus.

The neutron residue remains in the nucleus as a proton. The atomic number (number of orbital electrons) is thus increased and the element changes into another; for example, uranium (at. no. 92) becomes first neptunium (at. no. 93) and then plutonium (at. no. 94). The emanation of gamma rays involves only loss of energy, which does not cause alteration in either atomic weight or number although it occurs during both alpha and beta radiations. All of these nuclear modifications result in the transmutation of substances—the mysterious change that medieval alchemists, wholly ignorant of atomic structure or the true nature of matter, tried so long and so hard to achieve with the aid of abracadabra and other incantations in order to make gold from lead! Many an ambitious prince of bygone centuries subsidized a court alchemist who never in his life came nearer to the secret of "the philosopher's stone" than the causing of weird lights and horrible odors! Often, instead of gaining a fortune, he lost his head.

Uses of Isotopes

Isotopes are of special interest to the biologist because he can use them to study many biological phenomena, such as determining exactly what a living cell does with any given element or compound that it takes in with its food; for example, carbon assimilation. By the use of carbon dioxide made with C^{14*} ($C^{14*}O_2$) it has been shown that, in photosynthesis, carbon dioxide is not combined directly with any organic compound in the cell. It is first reduced by transfer of electrons of hydrogen split from water. The carbon (C^{14*}) is then found in a number of compounds formed during the synthetic process: *e.g.*, phosphoglyceric acid:

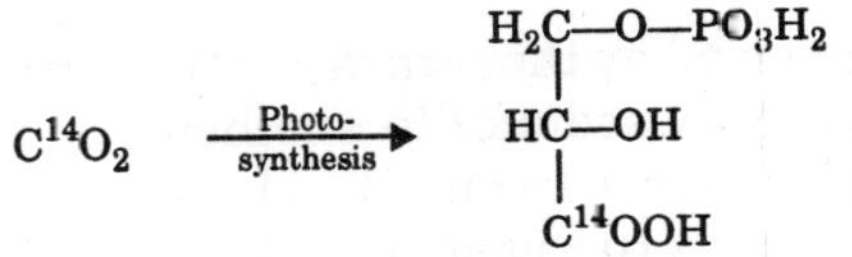

$$C^{14}O_2 \xrightarrow[\text{synthesis}]{\text{Photo-}} \begin{array}{l} H_2C\!-\!O\!-\!PO_3H_2 \\ \quad | \\ HC\!-\!OH \\ \quad | \\ C^{14}OOH \end{array}$$

3-Phosphoglyceric acid

If H_2O^{18} is provided in photosynthesis, all of the free oxygen given off is O_2^{18}. No O^{18} is found in the synthesized products. In many species of chemosynthetic bacteria carbon dioxide combines directly with already formed organic molecules. For example, Werkman, Wood, and their colleagues have "fed" carbon dioxide made with C^{14*} to propionic acid bacteria. The radioactive carbon can be followed, like a tracer bullet, through its course in the synthetic processes of the bacterial cell. For each molecule of C^{14*} one molecule of succinic acid or propionic acid is formed, probably from pyruvic acid via oxalacetic acid, or from glycerol:

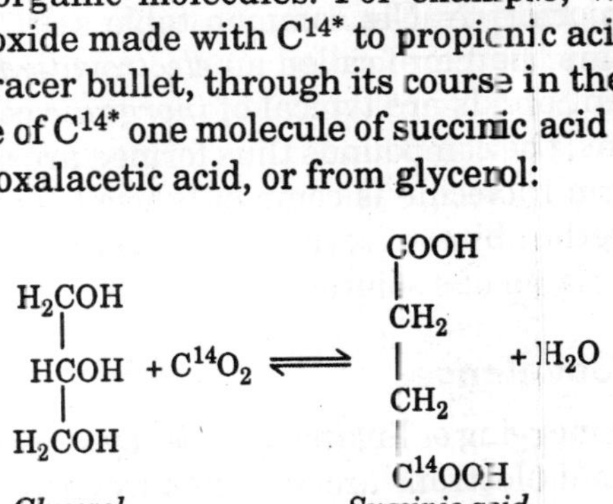

$$\begin{array}{c} H_2COH \\ | \\ HCOH \\ | \\ H_2COH \end{array} + C^{14}O_2 \rightleftharpoons \begin{array}{c} COOH \\ | \\ CH_2 \\ | \\ CH_2 \\ | \\ C^{14}OOH \end{array} + H_2O$$

Glycerol *Succinic acid*

A common soil saprophyte, *Aerobacter in-dologenes,* produces acetic, lactic and succinic acids as waste products of glucose utilization. When $C^{14*}O_2$ is added, C^{14*} is found in these acids. *Proteus vulgaris,* another common saprophyte of soil and water, produces lactic and succinic acids. The $C^{14*}O_2$ is found mainly in the carboxyl group of the acids. The pathway to the formation of these acids is via pyruvic acid according to the *Wood-Werkman reaction:*

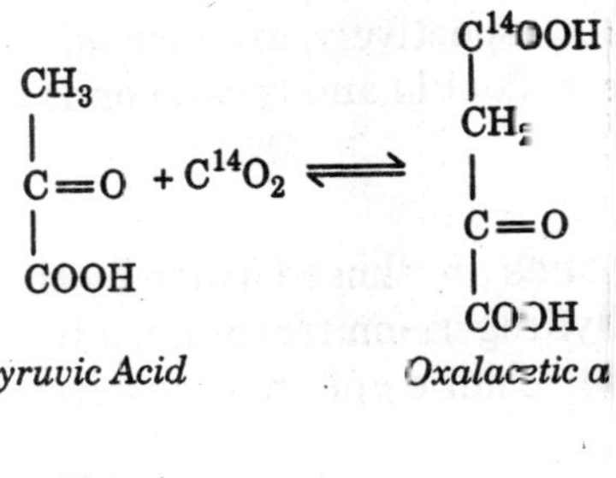

$$\begin{array}{c} CH_3 \\ | \\ C\!=\!O \\ | \\ COOH \end{array} + C^{14}O_2 \rightleftharpoons \begin{array}{c} C^{14}OOH \\ | \\ CH_2 \\ | \\ C\!=\!O \\ | \\ COOH \end{array}$$

Pyruvic Acid *Oxalacetic acid*

STRUCTURE OF COMPOUNDS

Compounds are substances consisting of atoms of two or more different elements in fixed proportions and steric relations, held together by chemical bonds. Molecules may be combinations of atoms of the same element (*e.g.*, H_2, Cl_2, O_2) or of different elements (*e.g.*, CH_4). A molecule is that minimal amount of a substance which, if further subdivided, ceases to be that substance.

CHEMICAL BONDS IN COMPOUNDS

The manner in which atoms are bonded together is of critical importance in all biochemical reactions. Three kinds of bond may be briefly considered here: electrovalent, covalent and hydrogen.

Inorganic Compounds: Electrovalence

An electrovalent bond is formed by the *transfer* (not sharing) of one or more electrons from the outer shell of one element, say sodium, to another, say chlorine. Sodium (periodic table Group I) has a single ("extra") outer electron; it lacks seven of the eight electrons needed for stability of the outer shell. Chlorine (periodic table Group VII) has seven outermost electrons; its outermost shell is one electron short of complete stability. Each atom would gain greater stability by combining with the other. The sodium tends to give up its single electron; the chlorine tends to saturate its outer electron shell by accepting the electron. Each becomes ionized thereby. The two ions therefore combine, the sodium transferring its electron to the chlorine, to form sodium chloride (NaCl), common table salt. The bond holding the two ions together is electrostatic and is therefore called an *electrovalent* or *ionic bond.* Such bonds are not very strong. Electrovalent bonds are typical of *inorganic* compounds. Since electrovalence results in aggregations of ions, the compounds thus formed are actually *ion aggregates,* not true molecules. However, the term molecule is commonly used for them. Unlike "true" molecules (see covalence) ions held together by electrovalence are readily separated; *i.e.*, such compounds readily ionize or dissociate in aqueous solutions.

Organic Compounds : Covalence

By *covalence* is meant the merging or linking together by *sharing* unpaired outermost electrons of two atoms so that the shared electrons are attracted by each nucleus. *Paired* inner electrons such as the two in the *k* (inner) shell of carbon are very strongly held by the nucleus and are not shared. It is the *unpaired* outer (valence) electrons that tend to be shared; unpaired electrons seek "mates." In the carbon atom there are four unpaired outer (valence) electrons; in hydrogen, one; in nitrogen, three; in oxygen, two. When electron pairs are shared (not transferred), the shared electrons are strongly held by each nucleus. Each pair of shared electrons is called a *covalent bond* or *electron-pair bond:* a strong and stable type of bond. Covalent bonds are most common in organic compounds. If two or three pairs of electrons are shared between two atoms, double and triple covalent bonds, respectively, are formed (see previous discussion of carbon). Instead of being stronger, however, double and triple bonds are progressively less stable.

Anhydro Bonds

Among the most important bonds are those formed by the removal of the elements of water from two adjacent molecules: a hydrogen ion from one, a hydroxyl ion from the other. The two ions combine to form water. The residues of the two altered molecules combine by what is sometimes called an *anhydro bond,* a type of covalent bond. The process is sometimes called *anhydrosynthesis.* For example:

$$H_3C-\overset{H_2}{C}-O\boxed{H \;+\; HO}-\overset{O}{\overset{\|}{C}}-CH_3 \;\xrightleftharpoons{H_2O}\; CH_3\overset{H_2}{C}-O-\overset{O}{\overset{\|}{C}}-CH_3$$

Ethyl alcohol — Acetic acid — Ester bond (an anhydro bond); Ethyl acetate

Anhydro bonds are extremely important in the process of *polymerization,* the mechanism by which complex *macromolecules* (proteins, polysaccharides, nucleic acids) containing thousands of smaller subunits (*e.g.,* glucose, amino acids, nucleotides), are put together by living cells. Anhydro bonds are important also because they can be readily made or disconnected by enzymic action. Anhydrosynthesis, polymers and their functions in living cells are more fully discussed farther on.

Some common and important anhydro bonds that will be mentioned again are here illustrated:

Ester Bonds

Between the carboxyl ($R-\overset{O}{\overset{\|}{C}}-OH$) group of an organic acid and the hydroxyl group (R—OH) of an alcohol, forming an ester (organic salt). These bonds occur in fats, oils and waxes and related compounds:

$$H_3C-\overset{H_2}{C}-O\boxed{H \;+\; HO}-\overset{O}{\overset{\|}{C}}-CH_3 \;\xrightleftharpoons{H_2O}\; CH_3\overset{H_2}{C}-O-\overset{O}{\overset{\|}{C}}-CH_3$$

Ethyl alcohol — Acetic acid — Ester bond (an anhydro bond); Ethyl acetate

Phosphate-ester Bonds

Between phosphoric acid and a hydroxyl group of a sugar, forming a phosphorylated sugar: these bonds are important in polymerization and the storage and transfer of energy in the cell:

$$R-\overset{H_2}{\overset{\|}{C}}-\boxed{OH \;+\; H}O\cdot P\cdot O_3H_2 \;\xrightleftharpoons{H_2O}\; R-\overset{H_2}{\overset{\|}{C}}-O\cdot P\cdot O_3H_2$$

Phosphate ester bond

Thioester Bonds

Between the carboxyl group of an organic acid and a sulfhydryl (⁻SH) group in a sulfhydryl-containing compound; important in the storage and transfer of energy in the cell:

$$R-\overset{O_2}{\overset{\|}{C}}-\boxed{OH \;+\; H}S-R \;\xrightleftharpoons{H_2O}\; R-\overset{O}{\overset{\|}{C}}-S-R$$

Thioester bond

Peptide Bonds

Between the carboxyl group of one amino acid and the amino (R—NH_2) group of another, forming peptides, polypeptides and proteins; basic to the synthesis of living matter:

$$R{-}\overset{O}{\overset{\|}{C}}{-}\boxed{OH + H}{-}\overset{H}{\overset{|}{N}}{-}R \underset{}{\overset{H_2O}{\rightleftharpoons}} R{-}\overset{O}{\overset{\|}{C}}{-}\overset{H}{\overset{|}{N}}{-}R$$

Peptide bond

Glycoside Bonds

Between hydroxyl or related groups in simple sugar molecules (*e.g.*, glucose) to form various disaccharides, and polysaccharides (polymers) like starch and cellulose. The glycoside bond is basic to the synthesis of many structural materials and energy-storing compounds in the cell:

$$R_2C(H){-}\boxed{OH \quad H}{-}O{-}C(H)(O{-}R)(R) \overset{H_2O}{\rightleftharpoons} R_2C(H){-}O{-}C(H)(O{-}R)(R)$$

Glycoside bond (intersaccharidic)

Note that all five reactions are indicated as being reversible. This is because most enzymic reactions are reversible unless too much energy is released by the reaction in one direction.

Hydrolysis

When the anhydrosynthetic process is reversed, the elements of water are reinstated and the constituent residues or subunits are separated and released in their original form. This process, called *hydrolysis,* is exceedingly important in biology because it is the basic reaction of virtually all processes of digestion of proteins, fats, polysaccharides (*e.g.,* starches) and many other compounds with anhydro bonds.

Hydrogen Bonds

As previously noted, hydrogen is unique in having only one electron, one proton and no neutron. The single proton exerts a strong, electrostatic, attractive force that extends beyond its single electron. Hydrogen therefore tends to exert an unbalanced electropositive effect in certain molecules or groups containing it. Such molecules or groups tend to be polarized. Strongly polarized groups or molecules (dipoles), like magnets, attract other polarized groups or molecules. When sufficiently strong the attraction results in the formation of a hydrogen bond through the sharing of a hydrogen atom (not an electron) between adjacent polarized groups or molecules; especially those containing oxygen or nitrogen in polarized groupings. For example, water molecules as well as ammonia molecules, all polarized groupings, become *associated* through hydrogen bonding:

$$\overset{+}{\underset{+}{}}H_2O^- \cdots \quad \overset{+}{\underset{+}{}}H_2O^- \cdots \quad \overset{+}{\underset{+}{}}H_2O^- \cdots$$

$$+{-}\overset{+}{\underset{+}{}}H_3N^- \cdots \quad +{-}\overset{+}{\underset{+}{}}H_3N^- \cdots \quad +{-}\overset{+}{\underset{+}{}}H_3N^-$$

Similarly, ammonia and hydrogen chloride form ammonium chloride through hydrogen bonding:

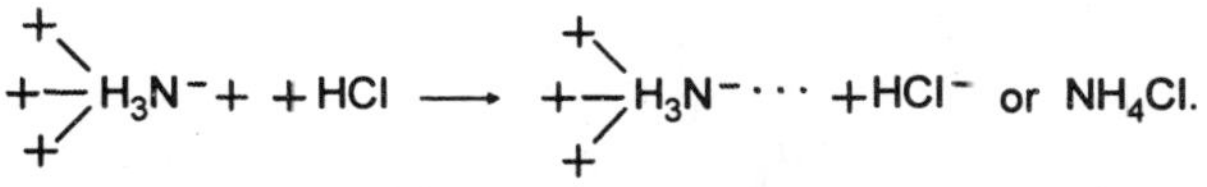

$$\overset{+}{\underset{+}{+}}\!\!>H_3N^- + \; +HCl \longrightarrow \overset{+}{\underset{+}{+}}\!\!>H_3N^- \cdots +HCl^- \text{ or } NH_4Cl.$$

(In the above diagrams hydrogen bonds are indicated by ...) In many combinations the proton of hydrogen is actually transferred to the other atom and the bonds are called proton bonds. As will appear, hydrogen bonds, although relatively weak, are essential in the phenomenon of inheritance and the reproduction of living cells. This is because they help to support the structure of genes (DNA).

STRUCTURE OF ORGANIC MOLECULES

Most of the organic molecules or compounds found in living cells are more or less complex derivatives of one large family of compounds, the alkanes or saturated hydrocarbons (hydrogen + carbon). These are the basis of the great series of *aliphatic* (Gr. *aleiphatos* = from fats or oils, *e.g.,* petroleum or hydrocarbon) compounds. The simplest saturated hydrocarbon is methane (CH_4), principal constituent of marsh and sewer gas. It is said to be saturated because all of its four valences are linked to hydrogen atoms. Hydrogen-carbon and carbon-carbon bonds are characteristic of organic compounds and are rare in inorganic compounds. Except methane, saturated hydrocarbons are straight chains of from two to scores of carbon atoms linked together by covalent bonds, each with its remaining valences saturated by a hydrogen atom:

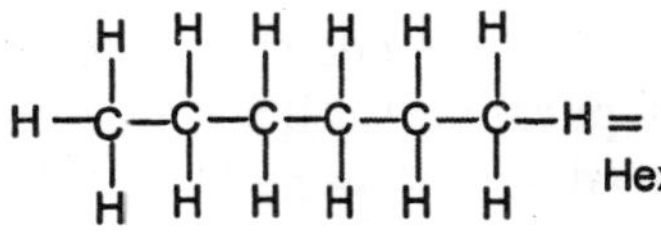

$$H-CH_2-CH_2-CH_2-CH_2-CH_2-CH_2-H = \text{Hexane} = C_6H_{14}$$

Compounds with this type of structure are termed *normal* hydrocarbons. They have the general formula C_nH_{2n+2}.

Simple modifications of this form are seen in branched chains or *isomers* (Gr. *isos* = like; *meros* = part):

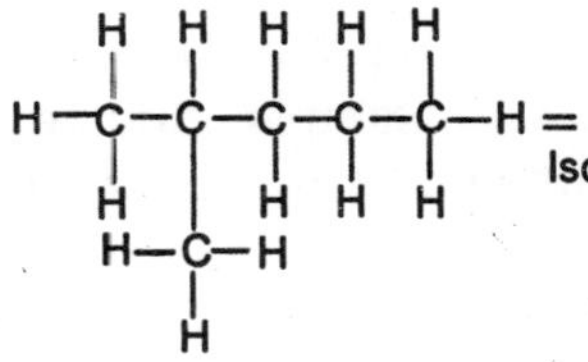

$$H-CH_2-CH(CH_3)-CH_2-CH_2-CH_2-H = \text{Isohexane} = C_6H_{14}$$

Isomers are compounds having the same empirical formula but different chemical or physical properties. Saturated hydrocarbon isomers also have the general formula C_nH_{2n+2}.

A second modification of the straight chain is the cyclic form in which the two ends of a chain are joined covalently:

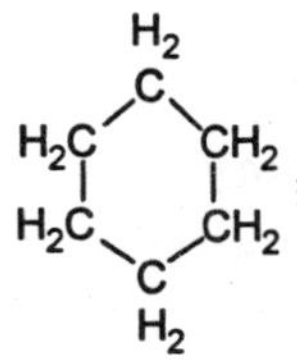

$$(CH_2)_6 \text{ (ring)} = \text{Cyclohexane} = C_6H_{12}$$

Two of the hydrogens are not needed in such molecules, and thus the general formula of the cyclic form is C_nH_{2n}. This molecule is suggestive of the six-sided ring compound, benzene:

$$\text{HC}\!=\!\text{CH}\ \text{ring}:\quad \begin{array}{c} \text{H} \\ \text{C} \\ \text{HC} \quad\quad \text{CH} \\ \text{HC} \quad\quad \text{CH} \\ \text{C} \\ \text{H}_2 \end{array} = \text{Benzene} = C_6H_{12}$$

Benzene, however, is the basis of the enormous family of *aromatic* compounds, *e.g.*, phenols. For convenience, benzene may be thought of as resembling cyclohexane in form but unsaturated, with three double bonds alternating with single bonds between the carbon atoms. Double and triple bonds between carbon atoms are also found in many hydrocarbons. Such hydrocarbons are said to be *unsaturated.* The double and triple bonds are increasingly unstable and easily broken. Hydrogen atoms of hydrocarbons may be replaced with hydrogen, hydroxyl, or with halogens: *i.e., substitution compounds.*

Functional Groups

The organic compounds that occur in protoplasm generally contain certain groups of atoms that are always linked together and are attached as a group, by covalent bonds, to the molecule of which they are a part. Such groups function as distinct entities. They are spoken of as radicals or functional groups, and they have an electric charge, positive or negative. Their charge is derived from the fact that when the covalent bonds that hold such a group to its molecule are broken, paired, shared electrons are separated. Both the molecular residue and the separated group then have unshared and unpaired electrons and are "free" radicals or ions. These, like electrons, under normal conditions exist free only theoretically or under special conditions. They tend to combine immediately with something else. The combining valence of a group, radical or ion, depends on the net charge of the group. Examples from inorganic chemistry are the ammonium "radical" ($^+NH_4$), valence 1; the sulfate radical ($^=SO_4$), valence 2; the phosphate radical ($\equiv PO_4$), valence 3. A free organic radical would be the methyl group: $^-CH_3$.

In organic chemistry different substances often 'have similar chemical potentialities because they contain the same functional groups. For example, glucose, a sugar, has at one end of the molecule a *hydroxyl* (^-OH) group, an arrangement that is characteristic of alcohols or organic *bases:*

$$\begin{array}{c} H_2 \\ \| \\ CH_3\text{—}C\text{—}OH \end{array}$$

(*Ethyl alcohol*)

At the other end is a carbonyl (=C=O) group, an arrangement that is distinctive of *aldehydes:*

$$\begin{array}{c} H \\ \| \\ CH_3\text{—}C\text{—}O \end{array}$$

(*Acetaldehyde*)

Glucose therefore, with its curious structure, has the combining properties both of an alcohol and of an aldehyde because its molecule contains those functional groups. Glucose is sometimes called an *aldose* (aldehyde-sugar). In a glucose isomer, fructose (fruit sugar), the carbonyl oxygen is attached to a carbon atom *within* the carbon chain rather than at the end of it:

	D-Glucose	D-Fructose
1	H—C=O	H_2COH
2	HCOH	C=O
3	HCOH	HCOH
4	HCOH	HCOH
5	HCOH	HCOH
6	H_2COH	H_2COH

When arranged in this way the group is called a *keto group* $(\mathrm{R{-}\overset{\overset{O}{\|}}{C}{-}OH})$, and sugars containing it are called ketoses. Note that the carbon atoms in the glucose chain are numbered, for reference, from 1 through 6.

If we oxidize the aldehyde group of glucose it changes to the carboxyl group $(\mathrm{R\overset{\overset{O}{/\!/}}{C}{-}OH})$. This is characteristic of all organic acids. These ionize to yield a hydrogen ion. By thus replacing the aldehyde group in glucose with the carboxyl group we form the *monocarboxylic sugar acid,* gluconic acid. If we also oxidize the alcoholic group at the other end of the glucose molecule we form a *dicarboxylic sugar acid* (a saccharic or glycaric acid).

Any basic alcoholic hydroxyl group usually reacts readily with any acidic carboxyl group to form an ester or organic salt, plus water. The formation of an ester is analogous to the reaction between the hydroxyl ion of an alkali (*e.g.,* Na^+OH^-) and the hydrogen ion of an acid (*e.g.,* H^+Cl^-) to form the salt, NaCl, and H_2O. Although alcoholic ^-OH generally acts as a group and acidic $(\mathrm{R{-}\overset{\overset{O}{\|}}{C}{-}OH})$ generally ionizes to give H^+ as above, in forming an ester it is the hydroxyl (not the H ion) of the $\mathrm{{-}\overset{\overset{O}{\|}}{C}{-}OH})$ group that combines with the alcoholic hydrogen:

$$\mathrm{CH_3\cdot \overset{\overset{O}{/\!/}}{C}{-}\boxed{OH + H}O{-}CH_2\cdot CH_3 \rightleftharpoons CH_3\cdot \overset{\overset{O}{/\!/}}{C}{-}O{-}CH_2\cdot CH_3 + H_2O}$$

This illustrates a difference between the organic hydroxyl group and the inorganic hydroxyl ion. It also illustrates the principle that the way an atom or group acts depends to a great extent on what other atoms it is combined with.

Using the glucose molecule again as our model, we may replace the hydroxyl group of carbon atom 2 with an amino group ($^-NH_2$), producing *glucosamine:*

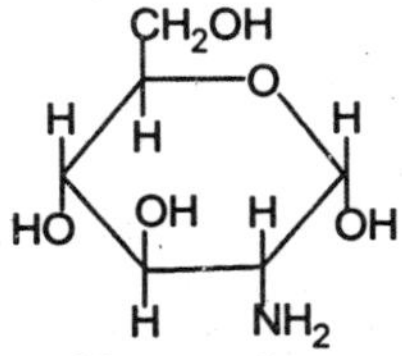

Glucosamine

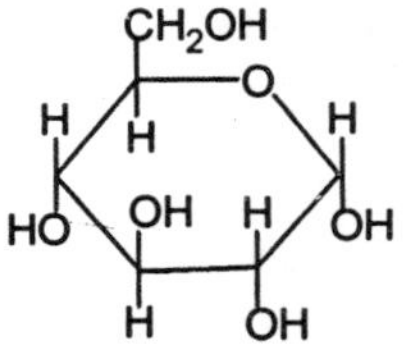

D-Glucose (Haworth cyclic formula)*

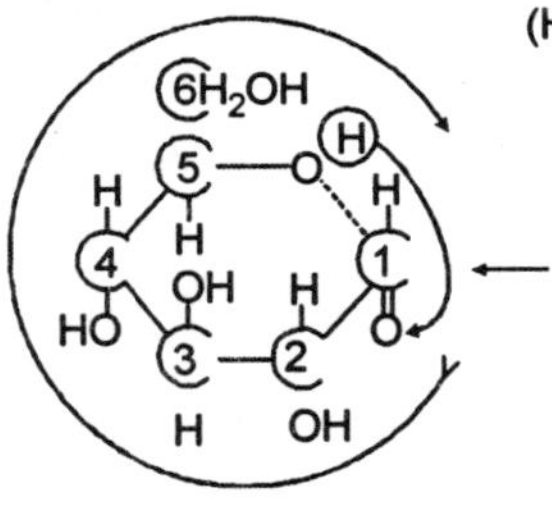

D- Glucose (cuclic from; see above)

This is an *amino sugar* and is found in the polysaccharide *chitin,* the rigid, structural substance in the skeletons of insects and Crustacea and in the cell walls of fungi, protozoa and some other protists. The large chemical family of amines, of which glucosamine is only one example, may be thought of as ammonia molecules (NH_3) with one or more of the hydrogen atoms replaced by organic groups (*e.g.,* methyl amine: $H_3C—NH_2$). The *amino group* ($^-NH_2$) is the ammonia residue. If the amino group is attached to the alpha carbon atom* of a carboxylic acid, such as propionic acid, we have an *alpha amino acid, alanine:* $CH_3\overset{NH_2}{\overset{|}{C}}HCOOH$. In alanine and, indeed, in all of the amino acids that occur in protoplasm, the amino group is attached to the *alpha carbon atom, i.e.,* next to the carboxyl carbon. Thus the general formula for amino acids that occur in nature is $R—\overset{NH_2}{\overset{|}{C}}H \cdot COOH$. The R may be very simple or very complex and may contain additional amino and carboxylic groups and cyclic and other groups. The amino group is basic and, with the acidic carboxyl group, makes amino acids *amphoteric* (able to react with acids or bases).

Since all amino acids have at least one basic amino group and at least one acidic carboxylic group, they can form bonds between the acidic and basic groups of adjacent amino acid molecules, producing long chains or polymers called *polypeptides,* the basis of protein structure. The bonds between the amino groups and the carboxylic groups are called peptide bonds.

Other molecules with both acidic and basic groups may react similarly. For example, the molecules of a fatty acid like beta-hydroxybutyric acid (β-hba), having acidic carboxyl groups and basic hydroxyl groups, can join, forming ester bonds between the carboxylic and hydroxyl groups of adjacent molecules. Long chains of the polymer, poly-β-hba, are formed:

---

* The straight-chain formula of the aldose is bent, as shown by long, outer, circular arrow, into a closed hexagon by bringing the $^-$OH group of carbon atom 5 into combination with the aldehyde group of carbon number 1 (shorter arrow):

$$\underset{\beta\text{-hydroxybutyric acid}}{\mathrm{HO\overset{CH_3}{\overset{|}{C}}H\cdot CH_2\cdot \overset{O}{\overset{\|}{C}}-\boxed{OH \pm H}\overset{H_2O}{}O\overset{CH_3}{\overset{|}{C}}\cdot CH_2\cdot \overset{O}{\overset{\|}{C}}-OH}} \longrightarrow \underset{\text{poly-b-hydroxybutyric acid}}{\mathrm{HO\overset{CH_3}{\overset{|}{C}}H\cdot CH_2\cdot \overset{O}{\overset{\|}{C}}\cdot O\cdot \overset{CH_3}{\overset{|}{C}}\cdot CH_2\cdot \overset{O}{\overset{\|}{C}}-OH}}$$

Another important functional group in protoplasmic substances is the sulfhdryl or mercapto groups (^-SH). It may be thought of as analogous to a hydroxyl group in which the oxygen has been replaced by sulfur. The ^-SH group is found in the amino acids cystine, cysteine and methionine, all of which are extremely important in the formation of enzymes and other essential structures in the cell. Other important functional groups will be mentioned in the appropriate place.

MACROMOLECULES

As mentioned elsewhere, most of the essential substances of living cells are built up of macromolecules. These are made of smaller molecules that are combined into macromolecules mainly by the process of anhydrosynthesis or *polymerization* (Gr. *polys* = many; *meros*—parts). By polymerization is meant the formation of compounds of high molecular weight by the combination of many small molecules that have identical or similar structures and functional groups that permit interlinkages. Each of the small molecules entering into any polymerization is called a *monomer* (Gr. *monos* = single). When two such molecules are joined chemically the result is a dimmer; when three are joined they constitute a *trimer.* A combination of more than three molecules in this way is called a *polymer*—a macromolecule. Polymerization is one of the commonest synthetic reactions occurring in living cells, and living cells consist largely of various kinds of polymers. If all of the units in a polymer are identical the polymer is called a *homopolymer* (Gr. *homos* = the same). Examples of homopolymers are starch, made of α-glucose units, and cellulose, made of β-glucose units—two very different substances, both made up of glucose isomers that differ slightly. If the units in a polymer are related but not identical substances, for example the various amino acids in proteins, the substance is said to be a *heteropolymer* (Gr. *hetero* = different). This illustrates the important point that in any given polymer all the *linkages* between the units are identical, though the *units* may be very different (as in a heteropolymer) provided they have the appropriate functional groups to form the linkage characteristic of that kind of polymer, *e.g.,* polysaccharide or polypeptide.

Polymerization

There are two general types of polymerization: (a) *addition polymerization* or *polymerization by association* and (b) *anhydropolyimrizaton* or *polymerization by condensation.*

Addition Polymerization

Addition polymerization commonly occurs spontaneously or under the activating (catalytic) action of oxygen or a few ions of hydrogen or hydroxyl or some metal. Add ion polymerization is one of the important processes in chemical evolution. Nothing is added to or removed from any one of the combining monomers. Thus, the empirical formula of an addition polymer (if a homopolymer) is theoretically a multiple of the empirical formula of the monomer; actually, the terminal groups may differ slightly.

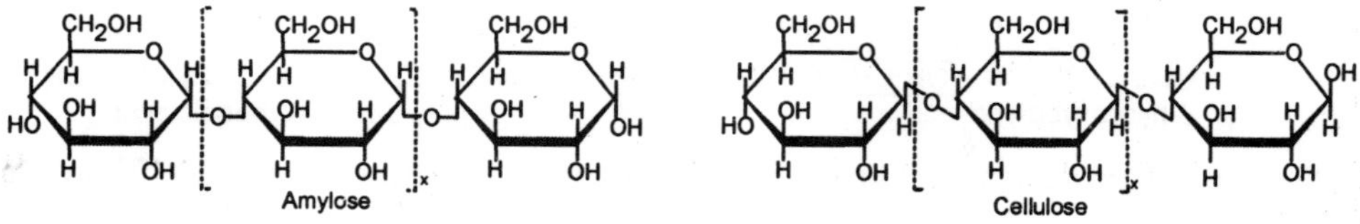

Fig. 1.3. Starch and cellulose formulas, showing marked similarities.

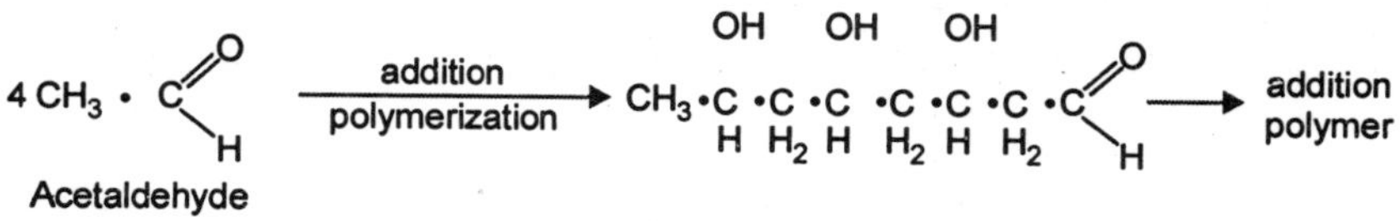

Fig. 1.4. In addition polymerization, molecules of acetaldehyde link together to form long polymers.

A simple example of addition polymerization is *aldol addition* or aldol condensation. (Do not confuse aldol condensation with condensation polymerization, which is here called *anhydro* polymerization to avoid confusion.) Molecules of aldehydes like formaldehyde and acetaldehyde (with alpha hydrogens), in the presence of H^+ or ^-OH, can link together to form dimers, trimers and long polymers with high molecular weights. By addition polymerization, formaldehyde forms solid *paraformaldehyde;* molecules of acetaldehyde may polymerize to form cyclic *metaldehye:*

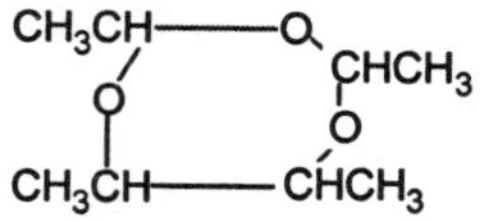

Ketones, alkenes and alkynes can also undergo addition polymerizations. It is evident that many complex and varied substances can be formed by simple addition reactions between relatively simple substances and under *rebiotic* conditions (spontaneously). Most addition polymers are easily depolymerized by heat.

Anhydro Polymerization (Condensation Polymerization)

In anhydro polymerization each monomer is combined with the next by removing H^+ from one and ^-OH from the other, forming water and causing the two residues to combine (polymerize). Because water is formed the process is sometimes called "anhydrosynthesis". Unlike addition polymers, the empirical formula of the anhydro polymer is not a simple multiple of that of the monomer since x number of molecules of water have been withdrawn in the process of polymerization.

Anhydro polymerization generally involves at least two kinds of enzymes: one or more to activate the units and one to withdraw the H^+ and ^-OH and cause the residues to combine. The process also requires energy. This is derived from elsewhere in the cell via the ADP-ATP and similar energy-transfer systems and is used in forming the bonds between the units of the polymer.

Energy Mechanisms in Polymerization

Condensation polymerization (anhydrosynthesis), whether of glucose, amino acids or other units, requires an input of energy; *i.e.*, polymerization is an endothermic process. The energy is derived ultimately from biooxidations of foodstuffs.

Energy derived from foodstuffs is transferred first into the phosphate bonds of ATP and similar compounds and stored there. These phosphate bonds are said to be energy-rich. When hydrolysis occurs during the process of polymerization, the energy is released.

In polymerization the energy released by breaking the energy-rich phosphate bonds is not lost. For example, in the polymerization of glucose molecules, each hexose molecule (or other monomer) is first activated or energized by combination with one energy-rich phosphate group. The transfer of the phosphate to the monomer is achieved via several intermediate steps involving transphosphorylases and phosphokinases (Gr. *kinema* = activity) or *synthetases.* In the case of starch formation glucose- 1-phosphate is formed by uridine diphosphate (UDP) - via uridine triphosphate (UTP). UDP and UTP constitute a nucleotide energy-transferring system working via, and analogous to, ADP and ATP. As each glucose-1-phosphate unit is added to the polymer chain, UDP is liberated and the energy of the phosphate bond goes into the making of the glycoside linkage between the glucosyl groups. UDP then receives energy-rich phosphate from ATP, becoming UTP; ATP becomes ADP. The ADP then receives new phosphate from substrate-level and oxidative-chain phosphorylations (and from photophosphorylations in photosynthetic cells) and is re-energized as ATP.

The role of ATP-ADP and UTP-UDP in the formation of starch furnishes an illustration of one common way in which energy from foodstuffs is transferred, by means of energy-rich bonds, into polymers. With some modifications to be discussed, the starch model holds also for the synthesis of proteins, RNA, DNA and lipids as described in the following paragraphs.

MONO-, DI-, TRI- AND POLYSACCHARIDES

A sugar consisting of single molecules (*e.g.*, glucose) is called a *monosaccharide.* Some sugars consist of two sugar molecules joined by glycosidic linkages. They are called *disaccharides.* Examples of familiar disaccharides are *lactose* (a molecule of glucose linked with a molecule of galactose), *sucrose* or ordinary cane sugar (one molecule of glucose and one of fructose) and *maltose,* the sugar in (malted barley (two molecules of glucose). Three-unit sugars are called *trisaccharides* or *trioses.* Trioses are of vital importance in glucose metabolism and in photosynthesis.

In plants "higher" than fungi the carbohydrate polymers starch and cellulose are among the most important structural and nutritional substances. Starch consists of masses of long chains of alpha glucosyl residues. By releasing glucose on hydrolysis starch functions as a reserve food, furnishing both energy and carbon. In the formation of cellulose, beta glucosyl residues are used instead of alpha residues. The strength of wood depends greatly on the tensile strength of the covalent bonds in the chains of glucosyl units. These cellulose fibrils are bundled together lengthwise with a tough "glue," a very complex substance called *lignin.*

In all of the "higher" plants the cell walls are cellulose. Masses of these, with lignin, constitute wood. In many bacteria the cell walls are heteropolymers made up of several amino acids and amino sugars. They are more fully discussed in the section on cell walls. Glycogen (animal starch), found mainly in animal livers and only rarely in plants, is also a homo-polymer of glucose, but the polymer chains are branched.

Several polysaccharides of biological importance are heteropolymers. Some of these, conjugated with lipids (lipopolysaccharides) and other substances, are major constituents of the cell walls of many species of bacteria. Most important of all, perhaps, are the polymers made up of phosphated molecules of the *pentoses* (sugars with five linked carbon atoms) *ribose* and *deoxyribose*. Ribose and deoxyribose phosphate polymers form the very backbone of the chains of nucleotides that make up the nucleic acids RNA and DNA described farther on in this chapter. These control and direct all heredity and all of the enzymic activities and resulting properties of every living cell.

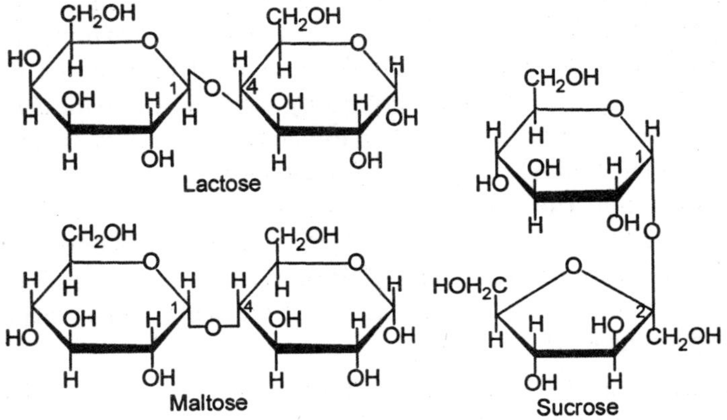

Lactose

Maltose

Sucrose

Fig. 1.5. Three disaccharides. These are common and important components in and sources of energy for living cells.

LIPIDS

This term includes fats, oils, waxes (not petroleum waxes), some fatty acids and polymerized fatty acids. Nearly all are complex esters or similar compounds of the higher fatty acids. Lipids are characteristically insoluble in water but soluble in organic solvents like ether, acetone, alcohol and chloroform. They are important constituents of cells, especially of cell walls and membranes.

Most natural fats are mixtures of triglycerides. A glyceride is a molecule of the tribasic (three replaceable ^{-}OH groups) alcohol, glycerol, combined by ester linkages with three molecules of fatty acid: commonly palmitic, stearic (both saturated), and oleic or linoleic (both unsaturated). Usually there are two or three different fatty acids per molecule. Melting points of fats are determined by the fatty acids. The longer chain fatty acids, *e.g.*, carnaubic acid (24 carbon atoms), produce tallow-like and harder fats and waxes. Fats that are fluid at about 22° C are commonly called "oils," though the term oil has no exact chemical meaning.

Esters, including lipids, are produced by reversible anhydrosynthesis or acid-base reactions; *i.e.*, the withdrawing (or restoring) of the elements of water (hydroxyl from the acid, hydrogen from the alcohol) and combination of the acid and alcohol residues as an ester. When fats are hydrolyzed in the presence of alkali (*e.g.*, potassium hydroxide) *saponification* occurs. The glyceryl receives the hydroxyl group; the fatty acid forms a salt with the potassium and becomes a soluble soap (*e.g.*, toilet soap). Soaps of calcium (*e.g.*, formed in hard water) and of most heavy metals form insoluble curds. In living cells fats serve mainly as reserve fuel globules. In higher animals fats in large, deposits also serve as thermal insulation.

Some waxes that are constituents of living cells are fatty acid esters of complex, monohydric (one OH group) alcohols other than glycerol. Some of these waxes, such as *cholesterol esters,* are important in the physiology of higher animals; *e.g.,* humans sometimes worry about cholesterol in the blood and arterial walls. In microorganisms, however, waxes are of lesser (known!) importance. In one group, *Mycobacterium,* which includes the tubercle bacillus, waxes are present in considerable amounts, though their exact physiological function has not yet been fully clarified. Waxes are important in methods of medical diagnosis (see acid-fast stain) and may confer some pathogenicity on acid-fast organisms like the tubercle bacillus.

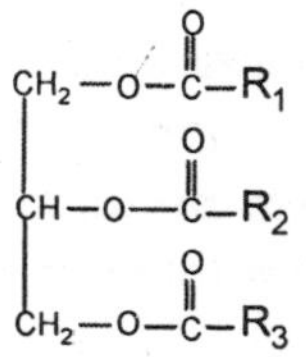

$$\begin{array}{l} CH_2-O-\overset{O}{\overset{\|}{C}}-R_1 \\ | \\ CH-O-\overset{O}{\overset{\|}{C}}-R_2 \\ | \\ CH_2-O-\overset{O}{\overset{\|}{C}}-R_3 \end{array}$$

Fig. 1.6. Formula of a triglyceride.

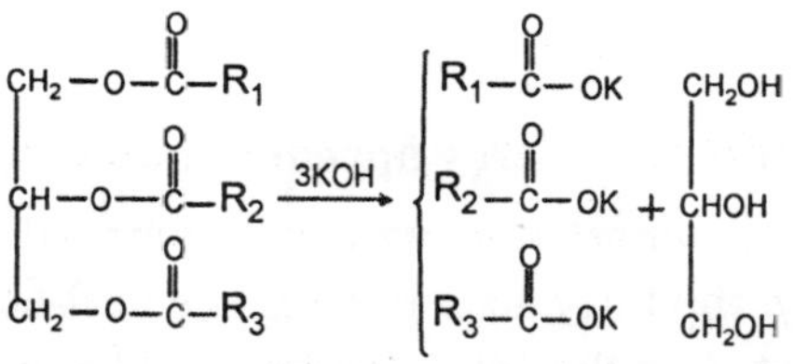

$$\begin{array}{l} CH_2-O-\overset{O}{\overset{\|}{C}}-R_1 \\ | \\ CH-O-\overset{O}{\overset{\|}{C}}-R_2 \\ | \\ CH_2-O-\overset{O}{\overset{\|}{C}}-R_3 \end{array} \xrightarrow{3KOH} \left\{ \begin{array}{l} R_1-\overset{O}{\overset{\|}{C}}-OK \\ R_2-\overset{O}{\overset{\|}{C}}-OK \\ R_3-\overset{O}{\overset{\|}{C}}-OK \end{array} \right. + \begin{array}{l} CH_2OH \\ | \\ CHOH \\ | \\ CH_2OH \end{array}$$

Fig. 1.7. Saponification of a triglyceride. Note that glycerol is liberated by saponification of any triglyceride.

Lecithins

Many triglycerides are not true fats but *lecithins.* Unlike fats, lecithins have a phosphate group attached to glycerol, with two fatty acid groups. Thus, lecithins belong to the group of *phospholipids.* Various additional groups may be attached to the phosphate group: inositol (the *phosphoinasitides),* ethanolamine (the *cephalins)* and so on. Many bacteria produce enzymes *(lecithinases)* that destroy lecithins. Several important diagnostic tests are based on this property.

Lipoproteins

A large number of compounds called *lipoproteins,* and others called *proteolipids,* that occur in living cells are lipids combined with proteins by linkages that are very little understood. Lipids are also often combined with carbohydrate complexes or polysaccharides *(lipopolysaccharides),* Lipoproteins are important constituents of the cell membranes of many types of cells. including important species of bacteria. They also occur in the outer coverings of some viruses that infect animals. In viruses the lipoproteins confer special and distinctive properties that will be mentioned in the section on viruses. In certain bacteria such as typhoid bacilli the lipopolysaccharides of the cell walls, complexed with polypeptides, have important pathogenic properties to be mentioned later.

One of the most important fatty-acid compounds is poly-beta-hydroxybutyric acid, a polymer of β-hydroxybutyric acid:

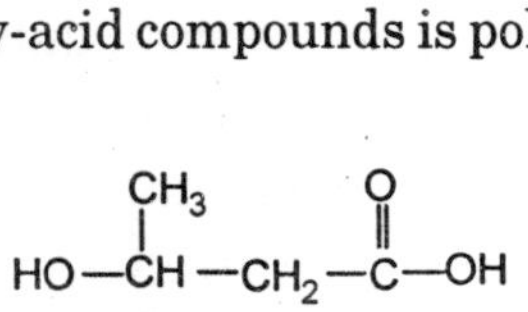

$$HO-\underset{}{\overset{CH_3}{\overset{|}{C}}}H-CH_2-\overset{O}{\overset{\|}{C}}-OH \quad \text{(phba).}$$

The units are joined by ester linkages between the alcoholic hydroxyl group of one molecule and the carboxyl group of the next, a molecule of water being eliminated (anhydrosynthesis) as in all typical ester groupings. Enzymes involved in such polymerizations are sometimes called synthases or synthetases. Poly-beta-hydroxybutyric acid is found as granules and globules in the cells of a number of important and common species of bacteria. It serves a purpose like that of fat or starch, or inorganic metaphosphates *(volutin), i.e.,* as a reserve food. Like starch, fat and volutin, it is insoluble and compactly stored as a polymer but readily hydrolyzed to release soluble, readily metabolizable, individual units: beta-hydroxybutyric acid, from its polymer. Similarly, soluble glucose is released from insoluble starch, soluble phosphorous compounds from insoluble volutin, fatty acids and glycerol from water-insoluble fats and oils.

In the synthesis of lipid polymers, energy for the linkages is derived from ATP as in polysaccharide synthesis, though not by direct formation of energy-rich phosphate bonds with the monomers. In lipid polymers the energy appears in an energy-rich acyl ($R-\overset{O}{\overset{\|}{C}}-R$) bond introduced into the fatty acid by combination with coenzyme A (CoA) (see Krebs cycle). The acyl group is then transferred from the fatty acid, leaving the acyl-bond energy. The energy is used in forming the ester linkage between the activated fatty acid and the alcohol (or alcoholic group) with which it combines.

NUCLEIC ACIDS

Unlike carbohydrates, lipids and proteins, nucleic acids (NA's) contribute little as energy sources or structural members of cells. Nucleic acids are like the pilothouse of a ship, contributing nothing as fuel or cargo, hull or superstructure, yet they guide the entire vessel, its contents and crew. As the name implies, NA's were formerly thought to occur only in the nucleus of cells. As will be explained, some NA's are strictly intranuclear, some are not. In the nucleus (or the bacterial nucleoid) NA's are loosely combined (conjugated) with proteins as *nucleoproteins.* The NA's are easily separated from their protein conjugates by mild hydrolytic processes.

Structurally, NA's are long-chain heteropolymers. Unlike polymers made up of simple molecules like glucose or amino acids, NA's are made up of complex units called *nucleotides: i.e.,* they are *polynucleotides.* Each nucleotide in NA is made up of three components: one molecule of phosphoric acid, one of the pentose ribose (or deoxyribose) and either a purine base *(adenine* or *guanine)* or a pyrimidine base *(cytosine, thymine* or *uracil).* Alternate forms of cytosine *(methyl cytosine* or *hydroxymethyl cytosine)* may occur in some forms of nucleotides (see farther on). The nucleotide units are joined by phosphate diester bonds between alternating phosphate and sugar groups: sugar-phosphate-sugar-phosphate-sugar and so on. To each sugar group a purine

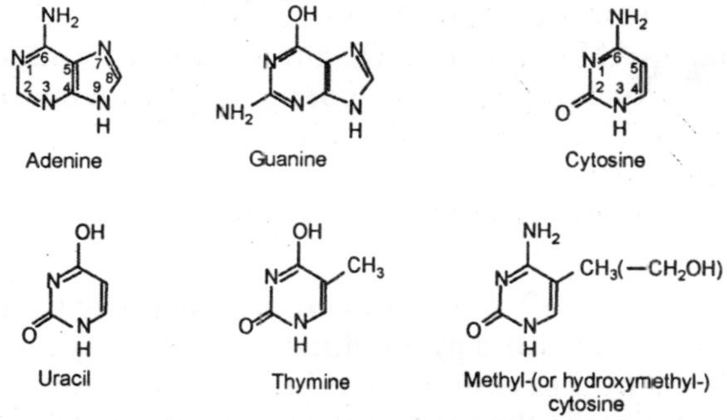

Fig. 1.8. Formulas of nucleotide bases.

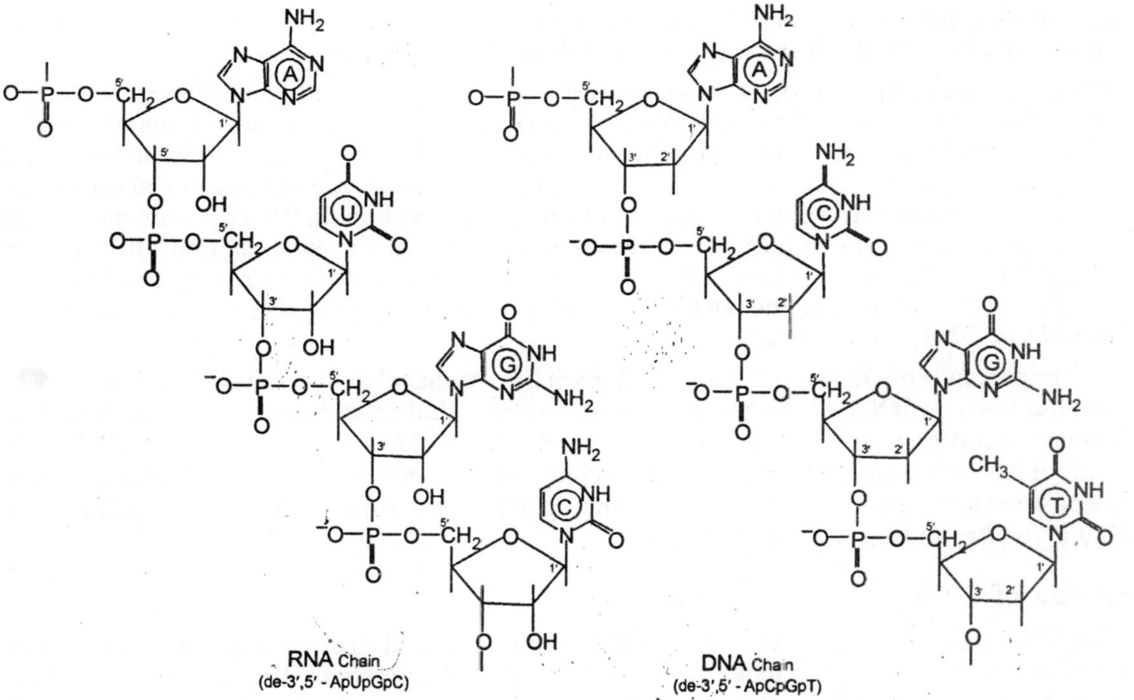

RNA Chain (de-3′,5′ - ApUpGpC)

DNA Chain (de-3′,5′ - ApCpGpT)

Fig. 1.9. Structure of sections of RNA and DNA chains. (A) = adenine; (T) = thymine; (G) = guanine; (C) = cytosine; (U) = uracil.

or pyrimidine base is attached by covalent C to N bonds. The chains of nucleotides in NA's are immensely long. In mammalian cells the number of nucleotides per chain is thought to be of the order of 10^9 = 1 billion.

DNA and RNA

There are two kinds of NA. One, ribonucleic acid (RNA), is so called because its sugar moiety is the pentose *ribose.* In the other, the sugar moiety is very similar but has one less oxygen atom than ribose and is called deoxyribose. This NA is called DNA. In RNA the purine bases may be adenine or guanine and the pyrimidine bases, cytosine or uracil. The bases in DNA are the same except that thymine occurs in place of uracil.

The nucleotides (base + pentose + phosphate) in RNA are adenylic acid (or adenosine-5′-monophosphate, familiar as AMP and in NAD), guanylic acid, cytidylic acid and uridylic acid; in DNA: deoxyadenylic, deoxycytidylic, deoxyguanylic and deoxythymidylic acids. All are *nucleoside phosphates.* While discussing nucleotides it is of interest to note that nucleotides other than adenylic acid also serve as coenzymes. Among these are uridylic acid or uridine-5′-monophosphate or UDP; guanylic acid or guanosine-5′-monophosphate or GDP. Adenylic acid is also found in FAD and CoA.

The full story of synthesis of the NA macromolecules logically begins with synthesis of the nucleotides and of the ribose, purines and pyrimidines of which they are composed. These details are not essential to the present discussion; we stipulate the ready formed nucleotides. The formation of the diester bonds linking the pentosephosphate groups of the various nucleotides into a long chain, the backbone of NA, occurs by the familiar anhydrosynthesis. The free nucleotides occur as triphosphates. In the presence of a NA polymerase enzyme system they join, each giving off pyrophosphate, leaving the energy of the phosphate bond in the polymer chain.

DNA, as is generally known, is the substance of genes, chromosomes and inheritance. In all "higher" cells DNA is confined to the chromosomal structures inside the nuclear membrane. In all bacterial cells DNA is confined to the chromosomal *equivalent,* the fibrillar nucleoid material which is *not* enclosed in a nuclear membrane. In both "higher" as well as in bacterial cells RNA appears in both nuclear and cytoplasmic regions. As will be explained, RNA does not determine heredity, but it acts as an intermediary between the DNA of the nucleus and all synthetic functions in the cell, which are carried on mainly in the cytoplasm. With regard to viruses, in any true virus (*i.e.,* excluding the bacterium-like "large viruses") there is only one kind of NA, either RNA or DNA, never both.

The importance of DNA or RNA in all forms of life can hardly be overstated. In all forms of life it is the DNA (or RNA) of the genes and chromosomes that undergoes the alterations called *mutations.* Mutations, with natural selection, are the basis of the great pageant of evolution of all forms of life on this earth, from primeval specks of subvital marine slime to modern astronauts and, who knows, to "moon men." An understanding of the structure, replication and alterations in NA's is the key to the *real* "facts of life".

Structure of DNA

The DNA macromolecules that constitute genetic material are intertwined *pairs* of very long chains (strands) of nucleotides. The two nucleotide polymer chains are connected together by weak hydrogen bonds between opposite or complementary purine and pyrimidine bases, much as the two uprights of a ladder are connected by rungs. In the famous Watson-Crick model of DNA, the ladder-like structure is twisted into a double helix.

A most important feature of DNA structure is the *pairing* of the purine and pyrimidine bases to each other. Adenine (A) of one chain is always connected by two hydrogen bonds with thymine (T) of the complementary strand; guanine (G) always is connected by three bonds with cytosine (C). The pairings occur in this way because this diminishes the energy associated with the structure to the greatest extent; hence, they are thermodynamically directed. The two chains of nucleotides in the double helix of DNA "run" in opposite directions: *i.e.,* they have opposite polarity; the deoxyribose residues are oppositely oriented. The pairs of bases, while always A to T and G to C, are highly irregular as to numbers and sequences. Any number or sequence is possible. Since the chains of DNA may contain many thousands of nucleotide units, it is evident that the number of possible combinations and arrangements of nucleotide pairs is virtually unlimited; we may have a sequence of pairs such as G—C, A—T, A—T, C—G, G—C, T—A, with many other sequences, almost ad infinitum. The helical form is maintained by electrostatic attractions between adjacent nucleotides.

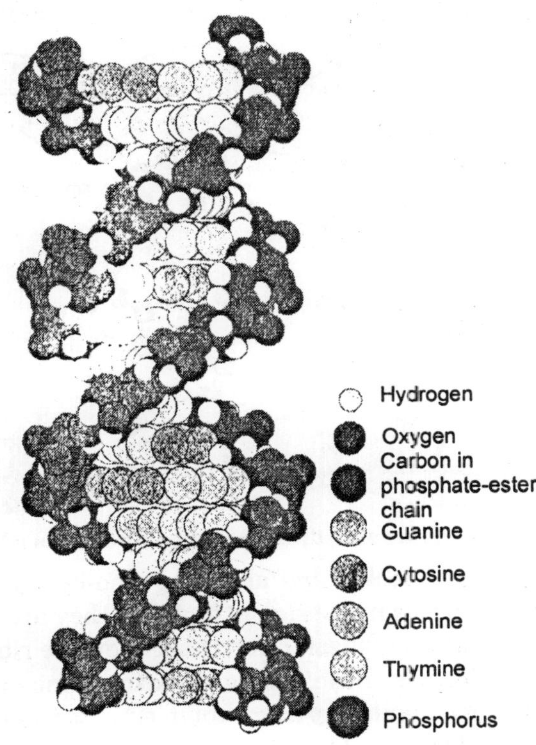

Fig. 1.10. A drawing of the DNA molecule, using solid circles to illustrate atoms. It can be seen that there are two helical grooves of unequal size on the outside of the DNA molecule.

Genetic Significance of DNA

The DNA of every species of cell has its own distinctive, invariable (except for rare *mutations,* as will be explained) and specific arrangement of nucleotides. The importance of this lies in the fact that each nucleotide, or any *genetically functional group* of from two to thousands of nucleotides, constitutes a *gene.* Thus your entire admirable self, fair reader, is merely an expression of the arrangement of nucleotides in the DNA of the sex cells (gametocytes) that have carried forward the genetic DNA from one generation of your family to the next from time immemorial. In brief, the helix of DNA is the basis of heredity; indeed, there is no (cellular) life without DNA. It has been called "The Thread of Life." Not the least remarkable feature of this unique vital machine is the fact that it is self-replicating both inside the cell and also outside the cell under laboratory conditions. So far as we are aware, without the replication of DNA life soon ceases.

Replication of DNA

Very briefly, when the process of duplication of the DNA double helix begins, the H-bonds between the A—T and G—C base pairs at a particular point in the double helix break, and the two individual helices separate at that point. This leaves unattached the H-bonding sites of the purine or pyrimidine bases in each separate chain, A, T, G or C. Therefore, each nucleotide residue in each separate strand of the unraveled helix immediately attracts, and combines with, a ready-formed nucleotide from the "pool" of nucleotides in solution in the fluid matrix of the cell; each A combines with a new T nucleotide; each G combines with a new C nucleotide and so

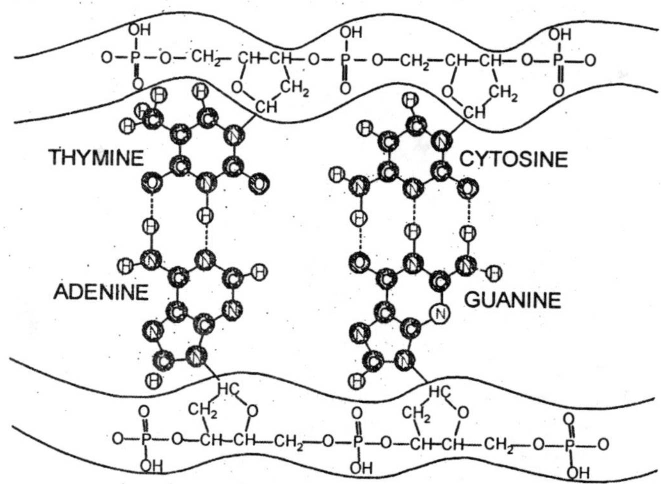

Fig. 1.11. Molecular drawing of components of DNA.

on. The new nucleotides, thus selected and arranged side-by-side according to the existing arrangement and sequence of bases in the strand to which they are attached, then join to form a nucleotide polymer by formation of diester bonds between the ribophosphate residues. This process continues along the length of the double helix from one end to the other. Thus each strand of the original double helix builds up a new complementary strand attached to itself by H-bonds as before. The nucleotide pairs of the new strand are necessarily in the same order and arrangement as before, since the order and arrangement of nucleotides in the old half of the new double helix was conserved and A could combine only with T, and G only with C. Thus the original double helix has now produced two complete, double helices, each an exact replica of the original. Each of the two new double helices consists of one chain of the old double helix and a new complementary chain. This mode of replication of DNA is spoken of as semiconservative.

Chromosome Replication

The mechanism of replication of DNA as a substance may be thus simply and superficially described, but the manner of replication of a bacterial chromosome, made up of DNA, presents many problems. The bacterial chromosome is a long thread of DNA formed into a closed circle. It consists of thousands of nucleotides. The opened circle of thread is estimated to be about 1 mm. (1000 μ) long. It is packed into a cell only about 2 by 8 μ in dimensions and it constitutes a single, doubly helical macromolecule with a molecular weight of the order of two billion. When it replicates, one or more still obscure mechanisms, called *replicators,* proceed along the

---

*tRNA has been said to "adapt" amino acids to the rRNA and is therefore sometimes also called *adaptor RNA* (aRNA).

† There are some exceptions that need not be considered at this point.

chromosome from a certain fixed point, separating the two strands of the helix as it (they?) go(es) and permitting the formation of complementary strands as previously described. A number of problems still exist concerning the mechanism. For example, what is the mechanism that triggers replication? How does the replicator unwind the helix through some 300,000 turns during the fission period of a cell like *Escherichia coli* that can mature and divide within 20 minutes? This involves some 15,000 turns per minute—a dizzy spin indeed! Further, the chromosome at one point is attached to the cell membrane and, if it unwinds, it involves some sort of swivel to permit the turning. Bear in mind also that the two helices run in opposite directions (have opposite polarities).

RNA commonly differs structurally from DNA in being single stranded. The exact mode of synthesis of RNA is still under investigation. However, double-stranded RNA and single-stranded DNA occur in some viruses. The single-stranded RNA in cellular organisms is of vital importance in protein synthesis and therefore in inherited characteristics, as explained in the following paragraphs.

Since about 1964 it has been observed that in some bacterial viruses the replication of RNA involves the formation, at least transitorily, of two double-stranded forms of RNA in addition to the single-stranded form. One is spoken of as a replicative form (RF), the other as replicative intermediate (RI). The significance of these findings is clearly of fundamental importance though complete elucidation awaits further investigation.

PROTEINS

Amino Acids; Peptides

Amino acids are organic acids that have an amino group ($^-NH_2$) attached to the alpha carbon atom, *i.e.*, the carbon atom next to the carboxyl (^-COOH) group (*e.g.*, alanine):

$$\begin{array}{l} CH_3 \\ | \\ C\text{—}NH_2 \\ | \\ COOH \end{array}$$

(D - alanine)

The alpha carbon atom is asymmetrical. Thus, as among carbohydrates, D and L steric forms of amino acids exist. They are referred to D- and L-serine:

$$\begin{array}{r} COOH \\ | \\ NH_2CH \\ | \\ CH_2OH \end{array}$$

(L-serine)

instead of to D- and L-glyceraldehyde, as are glucose and other carbohydrates.

It is of interest that the amino acids in natural proteins are nearly all of the L form. Some D amino acids, once regarded as "unnatural" because they had never been found in natural substances, occur only in bacteria and molds, notably in antibiotics produced by those organisms and in cell walls of bacteria. Not all of the amino acids that occur in cells are as simple as alanine:

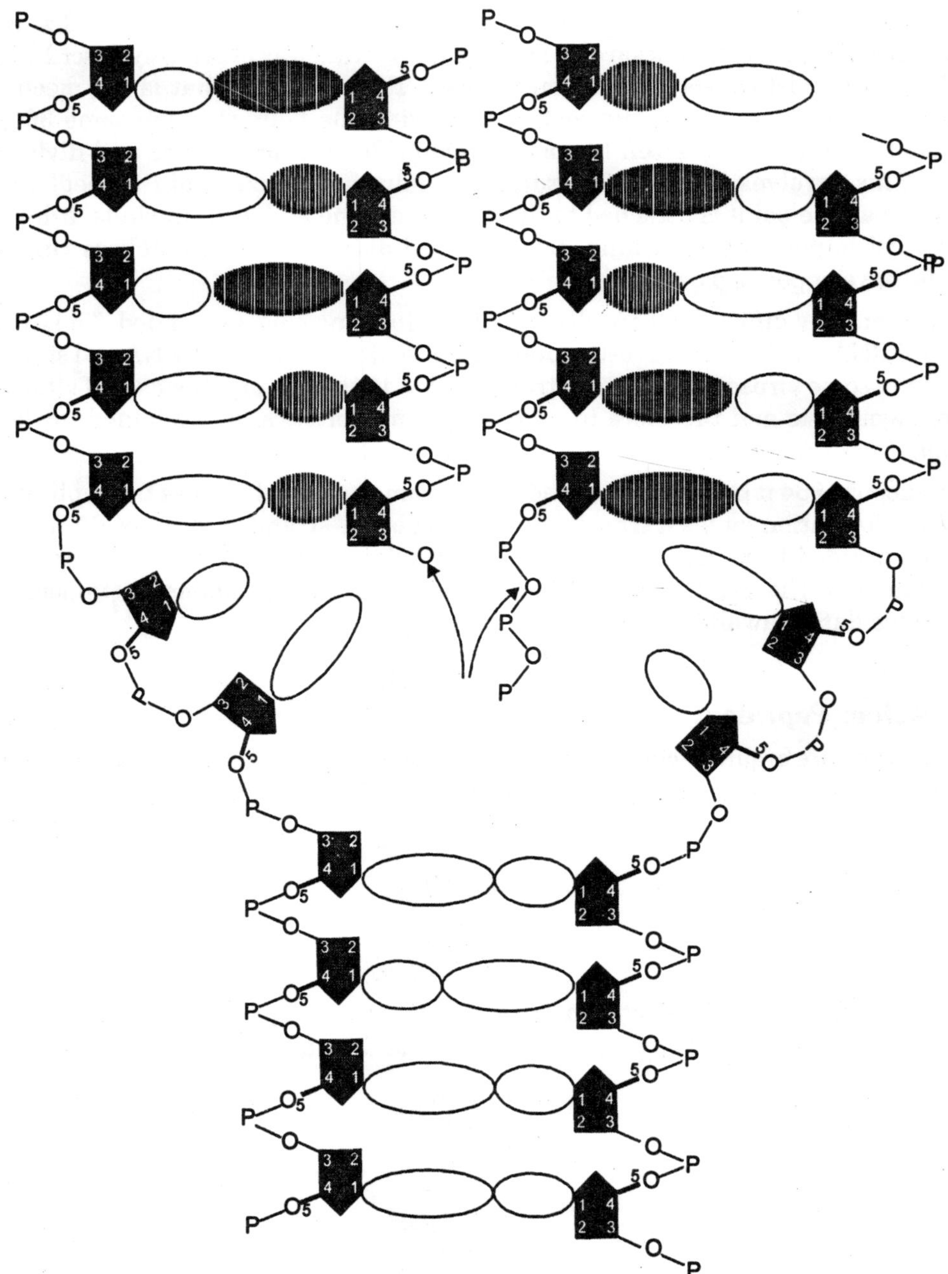

Fig. 1.12. Diagram of the form and mode of replication of the DNA double helix. The still-connected chains of nucleotides of the old double helix are at the lower part of the diagram; their separation is at the center. New nucleotides (bases shaded) are joining the nucleotides of the separated right and left parental (outermost) old chains as fast as the old chains separate. Note that pyrophosphate is split from each new nucleotide triphosphate as it joins the nucleotide chain. Note also the opposite polarities of the nucleotide arrangements in the complementary chains.

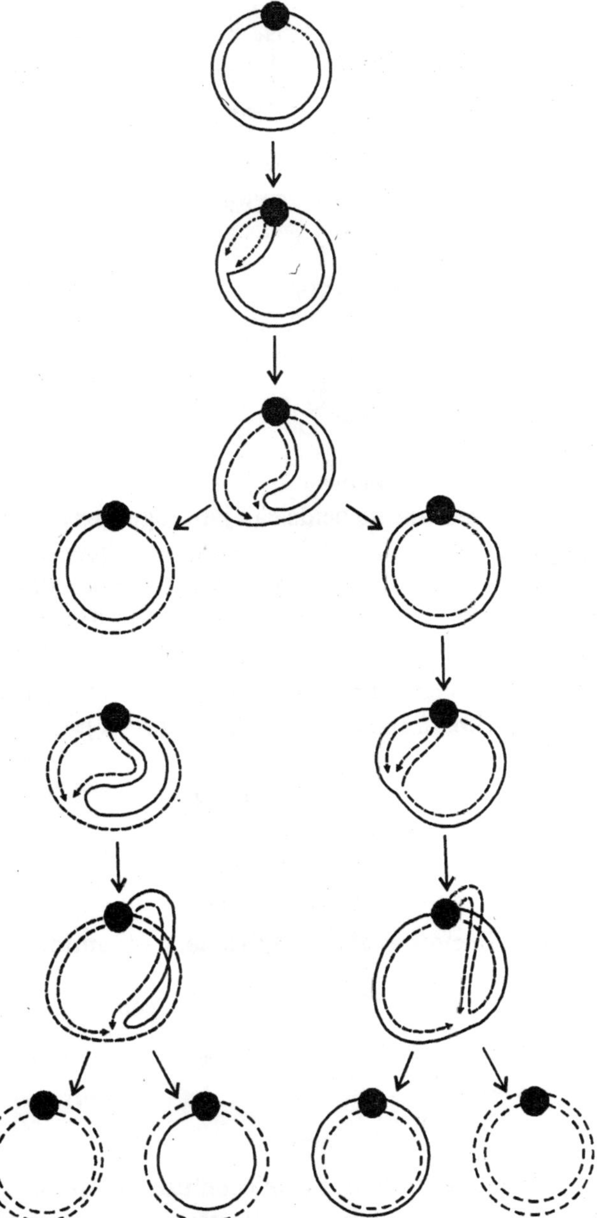

Fig. 1.13. Replication of bacterial DNA molecule. The two chains of the molecule, represented by concentric circles, are joined at a "swivel" (spot). DNA labeled with radioactive (isotope-containing) thymine is shown by dotted lines; part of one chain of the parent molecule is labeled by having taken up the isotopic thymine, as are two generations of newly synthesized DNA. Duplication starts at the swivel and proceeds counterclockwise. The arrowheads mark the points at which the two parental strands are being separated and new DNA is being synthesized in each chromosome. Two complete replications are represented.

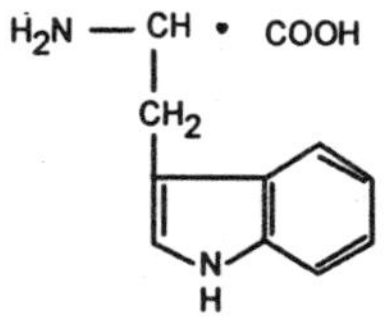

$H_2N—CH \cdot COOH$ with CH_2 side chain attached to indole ring (N–H)

tryptophan

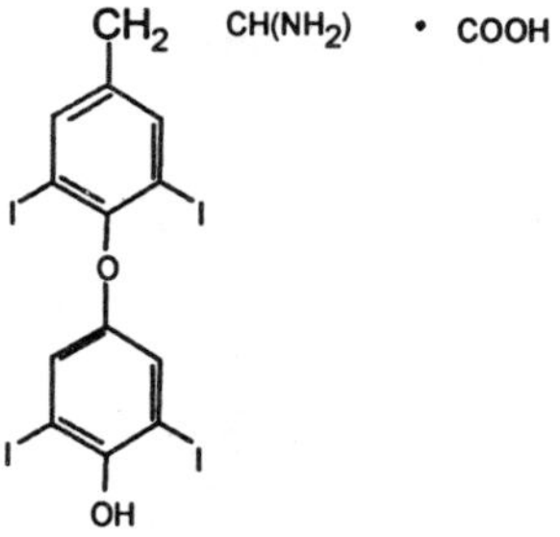

$CH_2 \quad CH(NH_2) \cdot COOH$ attached to diiodophenyl–O–diiodophenyl–OH

thyroxine

Like various monosaccharides, amino acids can join together by anhydro polymerization to form *peptides*. A combination of two amino acids constitutes a *dipeptide;* three make a *tripeptide;* many make a *polypeptide*. Proteins are enormously long-chain polypeptides. When protein macromolecules are broken, as by hydrolysis, into large groups of amino acids these are called *peptones* and *proteoses*. Complete hydrolysis of protein yields only amino acids.

The linkage between the amino acid residues in peptides including all proteins is between the carboxyl group of one and the amino group of the other—a peptide bond or linkage:

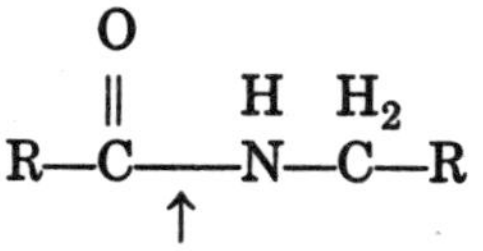

$$R—\overset{\overset{\displaystyle O}{\|}}{C}—\overset{H}{N}—\overset{H_2}{C}—R$$

↑
peptide bond

This is exemplified in the formation of glycylglycine, a dipeptide:

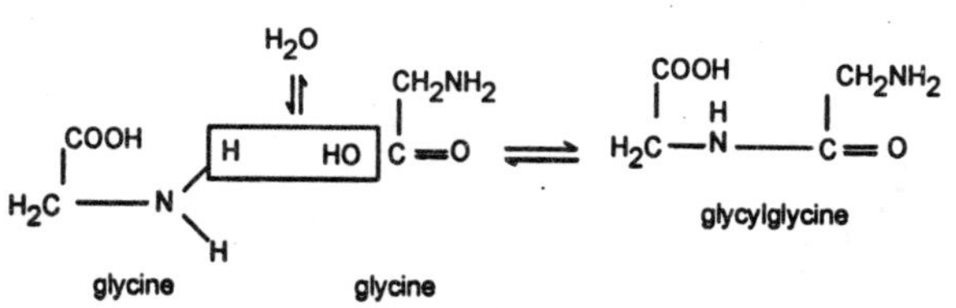

$$\underset{\text{glycine}}{H_2C(COOH)—N(H)—}\boxed{H \quad HO}\underset{\text{glycine}}{C(CH_2NH_2)=O} \overset{H_2O}{\rightleftharpoons} \underset{\text{glycylglycine}}{H_2C(COOH)—\overset{H}{N}—C(CH_2NH_2)=O}$$

Formation of the very long peptide chains that constitute proteins is a more complex process than the formation of a glucose homopolymer like starch, because all proteins are complex heteropolymers. Each of the score or more of the amino acids that go to form proteins is structurally different.

Protein Structure

The different amino acids that are joined to form proteins occur in innumerable sequences. The amino-acid sequence of a protein is said to constitute its *primary structure*. Secondly, the peptide chains may be held parallel to each other by H-bonds between adjacent R—NH and

O=C—R groups (R—NH • • • O=C—R) and by disulfide (R—S—S—R) bonds between adjacent sulfur-containing amino acids. Many parallel fibers thus joined can form protein sheets that may be flat or pleated. In another type of arrangement the polypeptide chains are twisted into helices such as the alpha helix. The helices are held in their coiled form chiefly by H-bonds between adjacent coils. A protein is said to be *denatured* when these bonds are broken, as by heating: the coil "unfolds" or straightens out. Many biologically extremely important substances (like keratins of hair, horn and fingernails) consist of protein complexes with such helical forms. These parallelings in sheets, twistings into helices and similar such arrangements of polypeptide chains constitute their *secondary structure*. A *tertiary structure* results when the sheets, helices or bundles are folded and twisted on themselves like the twisted yarn in a skein of wool or a bundled-up towel. The forms of the skeins or sheets are maintained largely by hydrogen and sulfide bonds. Because of tertiary structures we find proteins in the forms of globules, ropes, coiled fibers and many other arrangements. A quaternary structure results from aggregation of units that are in tertiary form. The subunits in quaternary structures are held together by disulfide bonds.

Specificity

The primary, secondary, tertiary and quaternary structural arrangements of protein molecules are not hit-or-miss. They depend on certain amino-acid sequences. Therefore they are genetically determined and relatively constant. Each protein has its own peculiar, distinctive and specific inherited structure; *i.e.*, it possesses the property of *specificity* (each protein differs from all others).

The number of different proteins possible from different arrangements of some 24 amino acids in the primary, secondary, tertiary and quaternary structures is obviously astronomical.

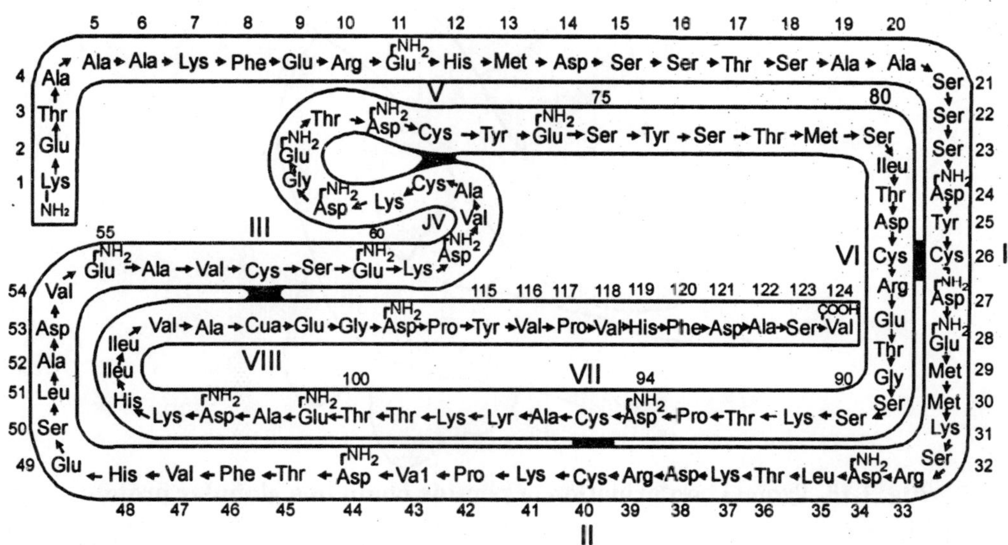

Fig. 1.14. The complete amino acid sequence of the enzyme ribonuclease. Standard three-letter abbreviations are used to indicate individual amino acid residues. Four disulfide bonds are seen, symbolizing a secondary structure.

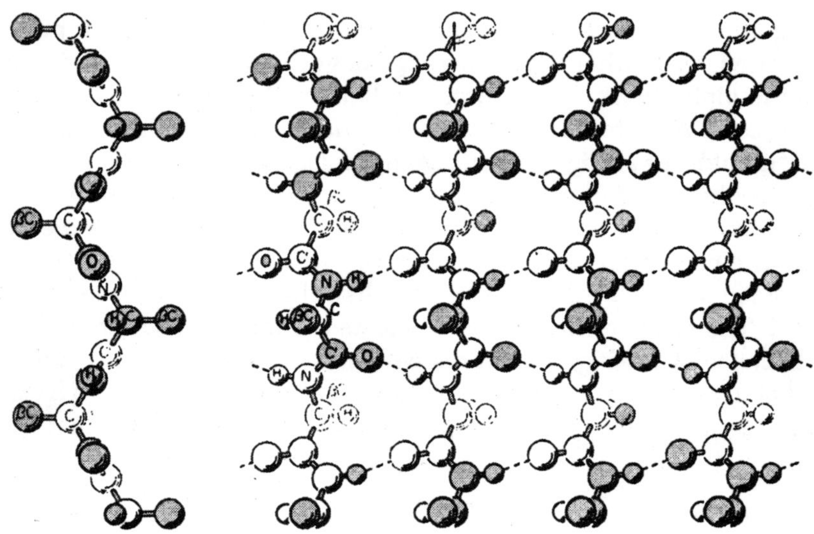

Fig. 1.15. The parallel-chain pleated sheet structure.

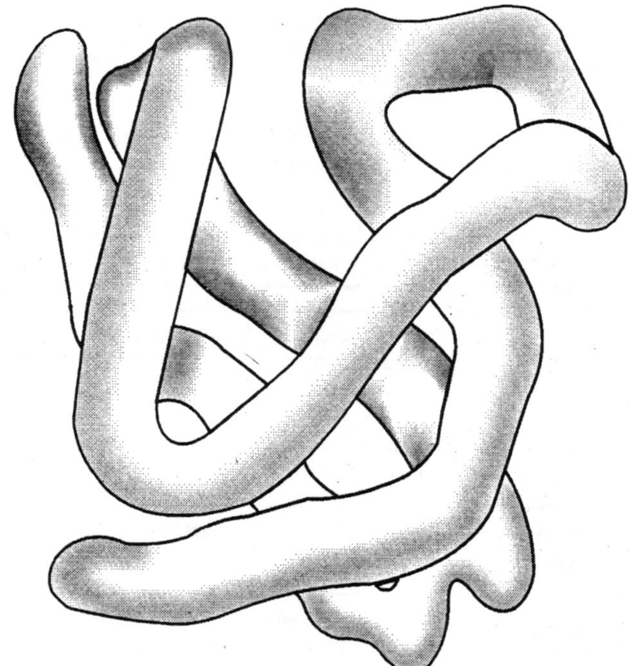

Fig. 1.16. Typical conformation of peptide chain in a globular protein.

When proteins are conjugated with other substances, e.g., metals, lipids and polysaccharides, the number of possible permutations and variations is virtually infinite. The property of specificity is of the greatest importance in all enzymic functions, in resistance to disease, in an individual's identity or in inheritance. Specificity will be discussed again in later sections. Denaturation breaks H-bonds and this results in loss of specificity and other properties of proteins.

Protein Synthesis

As previously explained, all essential biochemical reactions in the cell are catalyzed by *specific* enzymes, the specificity of which resides in the primary, secondary, tertiary and quaternary structures of their protein moieties (apoenzymes). In the preceding discussion it was stated that the structure of all proteins in the living cell, including enzymes, is determined by the nucleotide units in the DNA of each cell. The obvious question is, how? The answer, drawn from Nature only after years of the most exquisitely incisive, inductively imaginative and toilsome researches by hundreds of highly ingenious biochemists, molecular biologists, microbial geneticists and associated scientists, lies in: (a) the sequence of nucleotides (called the *genetic code)* that make up the DNA chains and (b) the mechanism by which this *code* or *genetic information* or *message* is transferred from the DNA polynucleotide in the nucleus to the protein-synthesizing mechanisms in the cytoplasm. These protein-synthesizing mechanisms are ultramicroscopic granules called *ribosomes* that consist of ribonucleoprotein (40 per cent protein, 60 per cent RNA). Ribosomes make up a great part of the finely granular, apparently diffuse substance of cytoplasm that is seen in specially prepared electron micrographs of cross-sections of cells. In "higher" cells, ribosomes are attached to membranous structures, the endoplasmic reticulum; in bacteria they seem to be diffusely scattered or attached to the cytoplasmic membrane and mesosomes.

rRNA and tRNA

The RNA of the ribosomes (rRNA) plays a basic role in the synthesis of proteins by providing a place for the work of specific *selection, arrangement* and *joining* of the amino acids to be done. Another form of RNA, called *transfer RNA* (tRNA) consists of relatively small molecules (70 to 100 nucleo-tides). It exists in solution in the fluid matrix of the cell and is therefore sometimes called *soluble RNA* (sRNA).* Each molecule of tRNA has a terminal nucleotide sequence that enables it to accept a single molecule of an *activated* amino acid and transfer it to a ribosome.

mRNA and the Genetic Code

Each ribosome has attached to it a third form of RNA that appears to be formed only by an enzyme, RNA-polymerase, under the direct influence of DNA. The exact details of its formation are not yet fully clarified. It might possibly be a derivative of DNA. However this may be, it exists as a single strand. This carries the identical sequence of nucleotides as a single strand of the DNA; *i.e.,* it is a one-sided transcription of the genetic code. This RNA, formed like a template of the DNA code, acts like a messenger to carry the DNA code or genetic message to the ribosomes. This RNA is called *messenger RNA* (mRNA).

Transcription of the Genetic Code

By the use of these three forms of RNA (tRNA, mRNA and rRNA), the construction of a protein polypeptide chain proceeds (in much simplified form) as follows:

Since each amino acid (NH_2R) that goes into a polypeptide (protein) chain has a different structure, a different specific enzyme is required to activate each. The energy for activation is derived from ATP bonds by the formation of the complex: Enz + ATP + NH_2R. Pyrophosphate is liberated from the ATP (yielding AMP) and the bond energy is left in the complex.

---

*First carbon atom in the chain after the carboxyl carbon atom.

The enzyme then catalyzes combination between the activated amino acid and a molecule of tRNA that has a sequence of three nucleotides (a *nucleotide triplet)* that is specific for that amino acid. Each molecule of tRNA has a distinctive triplet of nucleotides that enables it to combine with one, single, specific amino acid.

The activated amino acid is then transferred by its specific tRNA molecule, with energy derived from the high-energy compound, guanidine triphosphate (GTP; analogous in function to ATP), to a ribosome where the actual synthesis of the protein occurs. Probably many ribosomes act together in groups called *polysomes* or *polyribosomes.* These bodies are visible only with electron microscopes.

The rRNA itself appears to be nonspecific but it acts as a working place for the tRNA and the mRNA. As mentioned before, mRNA is a long, single chain of nucleotides whose sequence is the same as that of the nucleotide sequence (genetic code) in the DNA. Each activated amino acid brought by tRNA to the mRNA becomes attached to the strand of mRNA at a place in which the sequence of the three nucleotides in the tRNA corresponds exactly with that of a triplet of nucleotides in the mRNA. The tRNA triplet is thus oriented opposite the mRNA nucleotides as though the tRNA triplet were part of a complementary strand of DNA. For convenience in visualizing the relationships we may imagine that the tRNA is a three-nucleotide fragment of a single DNA strand while the mRNA is the complementary strand. This complementary relationship between the tRNA and the mRNA necessitates the pairing of the nucleotide triplets in the tRNA and mRNA according to the DNA code of which mRNA is a copy.

Each molecule of tRNA, with its attached amino acid, recognizes its place on the mRNA strand (which carries the DNA code). Thus the sequence of amino acids that are attached to the strand of mRNA by the tRNA is determined by the DNA code.

Final Synthesis

Once the activated amino acids are properly aligned in the coded order along the strand of mRNA, the polypeptide bonds are enzymically formed to join the amino acids. The now complete polypeptide chain is released from the ribosome or polysome as a fully fledged protein of highly specific structure. The protein probably then acts with some coenzyme in the daily work of the cell. The tRNA and the activating enzyme are liberated and repeat the process. The mRNA appears to remain attached to the ribosomes to act repeatedly as a template for the ordering of more polypeptide chains when the tRNA's shall bring the activated amino acids to it.

The Codon

Because of the almost infinite number of possible arrangements of the four kinds of nucleotides (letters or digits: A—T, C—G) in the DNA, and considering that one DNA macromolecule contains perhaps 100,000 nucleotides, an almost infinite number of three-letter "words" or nucleotide triplets, each representing an amino acid, can be arranged. Each word thus constructed of three nucleotides in DNA is called a *codon.*

Each protein thus bears an inherited structure and, because of its enzymic or other activities, confers an inherited characteristic on the cell of which it is a part.

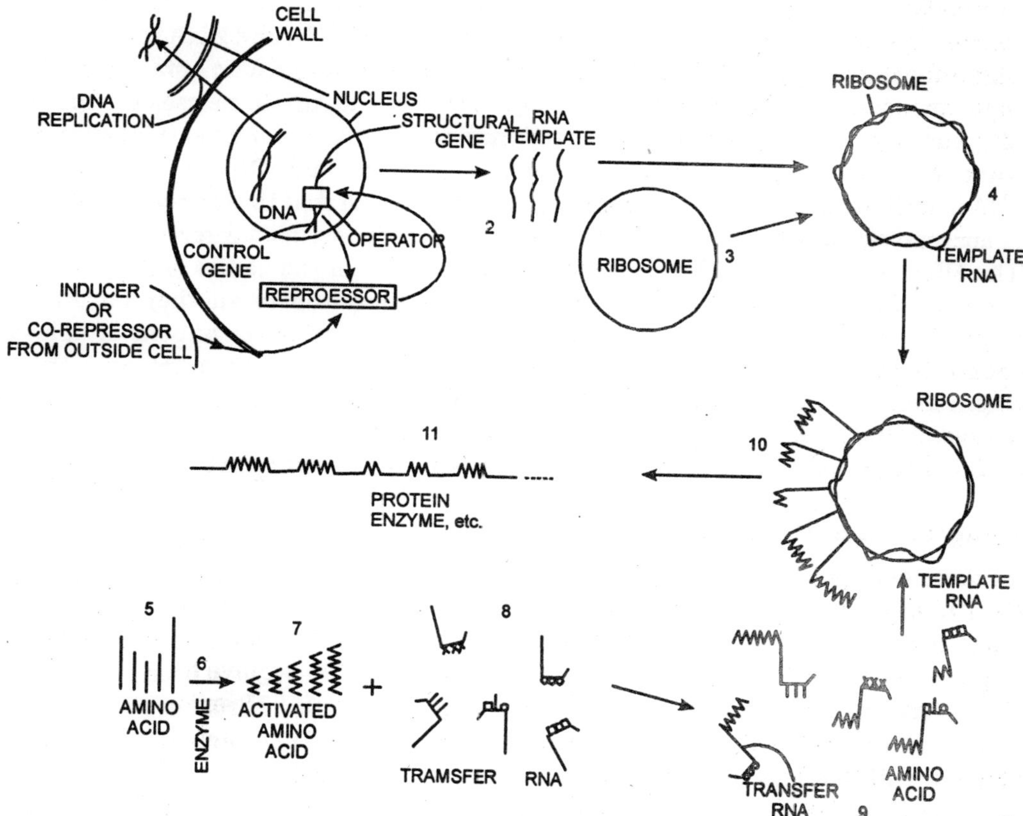

Fig. 1.17. Diagram of protein synthesis. At the upper left, DNA in the nucleus of the cell is seen forming and liberating strands of template RNA (mRNA; 2) under genetic control involving inducer, represser, structural gene, etc. The mRNA (2) becomes associated with ribosomes (3 and 4). Different amino acids (5, lower left), each activated by its specific enzyme (6, 7), become associated with specific nucleotide triplets that constitute transfer RNA (tRNA; 8). Each specific nucleotide triplet, with its specific amino acid (9) finds its corresponding nucleotide triplet on the strand of mRNA at the ribosome and attaches there (10). The amino acids thus ordered along the mRNA at the ribosome (10) are enzymically (polymerase) linked through peptide bonds (anhydrosynthesis) to form the polypeptide of protein (11) which may be a functional enzyme or a structural member of the cell.

Universality of the Genetic Code

A thought-provoking aspect of the genetic code as a biological mechanism is that DNA is constructed on the same general form throughout the living universe. Genes, codons, triplets, amino acids and nucleotide "words" are all involved in the same way and in the same process, though the nucleotide sequences occur in endless variations. The code, *i.e.,* the language of heredity, is said to be universal. The tRNA and mRNA from a bacterial cell will serve the same function and direct the synthesis of the same protein from the same amino acids in an artificial enzyme mixture derived from a mammalian cell.

Does this universality of the genetic language and mechanism point toward DNA as the common, primeval starting point of all life? There is evidence indicating that various amino acids, the purine and pyrimidine bases and other constituents of RNA and DNA could have been formed in lightning flashes under prebiotic conditions in vapors of water, ammonia and methane.* Thus a single, primeval DNA helix, spontaneously formed, could, in the course of thousands of millions of years, by mutations, genetic recombinations and natural selection, account for ourselves and all of the living things that now surround us. It is interesting to realize that each helix of DNA in each person was derived from ancestral DNA and that these ancestral strands of DNA have successively produced torch other by helix replication without interruption over eons of time. The DNA in each person has been "immortal," although the bodies of the individuals carrying the ceaselessly replicating strands from one generation to another have in time disintegrated. We may draw an analogy with a lighted torch carried for billions of miles and years since life began by successive relay runners who, having handed the torch on to the next runner, themselves drop panting or lifeless by the wayside. It is also conceivable that different forms of primeval DNA helices could have originated at different times and in different places on the earth, thus accounting for various types of animals and different human "races."

Deciphering the Genetic Code

One of the most magnificent pieces of decoding in human history, including Champollion and the Rosetta stone, was not done by international spies or supersleuths in Washington but by Benzer, Watson and Crick and others when they discovered in 1961 that each DNA code word consists of a sequence of three nucleotides (nucleotide triplets). These have been called codons. Each codon occurring in tRNA specifies one amino acid. Also in 1961 Nirenberg and Matthaei mixed, *in vitro,* ribosomes (separated from sonically disrupted cells by high speed differential centrifugations) with a suspension of cytoplasmic materials from the same sort of cells. The mixture contained, among other materials, enzymes, amino acids, tRNA, ribosomes and energy-transfer coenzymes such as ATP and GTP. Here were found all, except one, of the many elements necessary for the synthesis of polypeptides. The one missing ingredient was mRNA.

Now Nirenberg and Matthaei had synthesized a simple form of mRNA in which the entire chain was made up of only one kind of nucleotide, uridylic acid (U). This historic polynucleotide is now often called "poly-U." Be it noted that, although in RNA uracil replaces thymine, uracil and thymine are equivalent in respect to pairing. When U was added to the cellular "puree" just described, the cellular elements in the puree cooperated as they do in a living cell to form a polypeptide chain. But since there was only one kind of triplet or codon, UUU in the synthetic mRNA, the polypeptide formed consisted of only one kind of amino acid, phenylalanine.

$$C_6H_5-CH_2CH(NH_2)COOH$$

phenylalanine

It was thus revealed that the codon UUU meant phenylalanine; the first genetic code word ever to be deciphered! Of course, many eager researchers were at work deciphering other code words. The huge mass of data resulting from these studies fills many long shelves in libraries and carries the names of scores of brilliant scientists. Suffice it to say here that, in the same or similar ways, codons have been found since then for all the amino acids that occur in natural proteins. AAA means lysine, AGA means arginine, UUU means phenylalanine, UUA means leucine, and so on. It was found also that several amino acids have more than one codon, for example, leucine is specified by CUC, CUU, UUA and UUG. The code is therefore said to be *degenerate* in that each amino acid is not restricted to only one codon. This is true *in vitro,* but the code appears not to act degenerately *in vivo.*

Genetic Punctuation

In reading the entire sequence of triplets in a long chain of DNA (about 10^5 units) it became difficult to determine where one triplet ended and another began, *i.e.,* to know how the genetic message was punctuated. One might have a series of nucleotides like ACTGATGAGCAT. If we mark off triplets from the left, we find that they are ACT, GAT, GAG and CAT. They make "sense". But if we overlook the first A we find CTG, ATG, AGC and so on, *i.e.,* nonsense words. English words are used as an example but of course the triplet letters could be arranged in any sequence in a polypeptide.* In the above example using English words, a single error at the beginning would clearly result in making nonsense words such as CTG of many or all of the succeeding triplets and, in any case, in producing a faulty message (polypeptide) or none at all. After many hypotheses and experiments had been made it was found that a genetic message may be decoded by beginning at a point where the first triplet is not a nonsense word (*i.e.,* is an established codon [genetic word] for a known amino acid) and reading in triplets from there in both directions.

CHAPTER

2

Fungal Metabolites

A comprehensive list of fungal metabolites, with discussion on their biosynthetic origins and interrelationships, was published by Turner in 1971. Here we discuss the chemistry of all microbial metabolites which incorporate polyolefinic features within their structures. The Reporter has found it more convenient to discuss these compounds according to structural type rather than biosynthetic origin.

Polyolefinic Macrolides—Chainin : The structure of the antifungal pentaene antibiotic chainin, isolated from a *Chainia* species in 1968, has been deduced as structure 1(a) followed largely from a thorough mass spectral analysis of the octa-acetates of 1(a) and the corresponding decahydro-derivative, and also of the hydrocarbon and methyl ester reduction products of 1(a). Some structural studies on two homologous antibiotics, designated nor- and homo-chainin, present as minor constituents in *Chainia,* were also presented in the same publication, but no definite conclusions regarding their structures were made. A striking feature of the structure assigned to chainin is that it is almost identical with that assigned to filipin 1(b). The only difference between the two antibiotics is that chainin has a 2-n-butyl substituent where filipin has a 2-α-hydroxy-n-hexyl substituent.

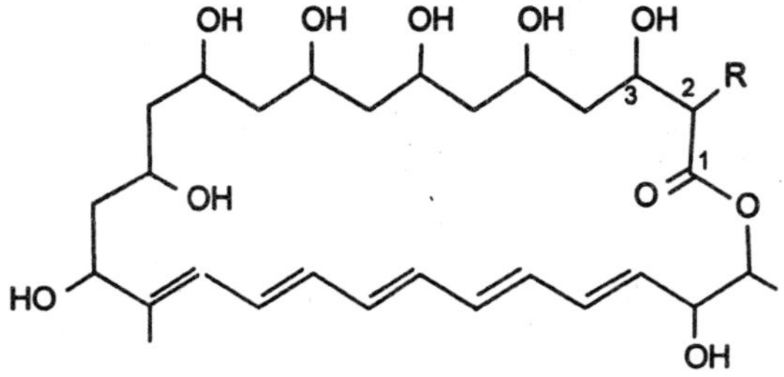

(1) a; R = Bu^n

b; R = — $CH^1(OH)C_5H_{11}$-n

Filipin : Filipin was recently shown to be a mixture of at least four components, which were designated filipins I, II, III, and IV; according to mass spectral studies the four compounds were present in the approximate proportion 4:25:53:18. Rinehart and collaborators have re-examined the structure of filipin by methods similar to those employed in deducing structure 1(a) chainin. Their data are in agreement with the proposed structure 1(b) for filipin III, and indicate that filipin IV is probably stereoisomeric at C-3 or C-1′ with filipin III, and that filipin II has a structure similar to 1(b) but lacking either a C-l′ or C-3 hydroxy-group.

Flavofungin : Full details have been published on the structure of flavofungin from *Streptomyces flavofungini.* The antibiotic was shown to be composed of a mixture of (2a) and (2b) the approximate ratio 10:1. Comparison of flavofungin with 'mycotocin' from *S. rubber* led to the conclusion that the two constituents of flavofunginhaye the same gross structures as those of mycotocin [where (2a) and (2b) are present to equal amounts], but no conclusions could be made concerning their absolute configurations. It is possible that the two antibiotics differ in absolute configuration at one or more asymmetric centres. The biosynthesis of flavofungin from acetate-propionate has been studied earlier.

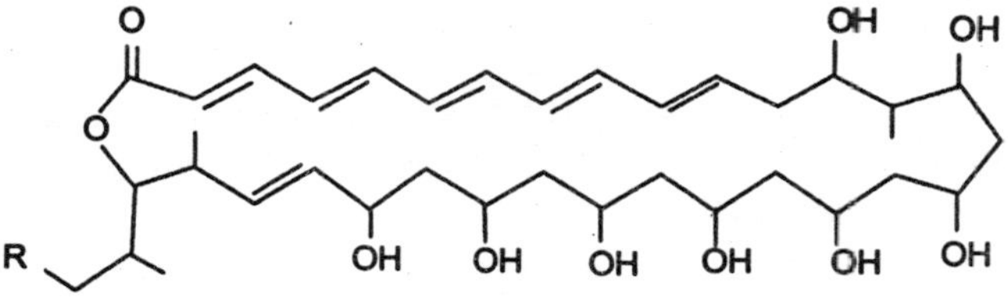

(2) a; R = H
b ; R = Mc

Polifungin : Polifungin is a new tetraene antibiotic complex produced by *Streptomyces noursei* var. *polifungini* ATCC 21581. Four biologically active components are present in the complex, and three of these were shown to be identical with the three tetraene components found in pharmaceutical grade nystatin, which has structure (3).

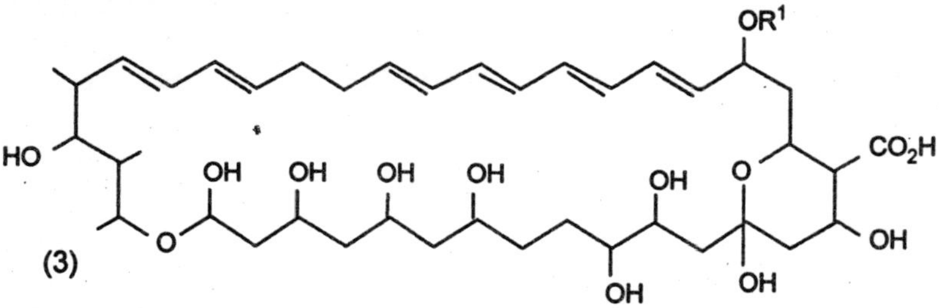

(3)

Amphotericin B and Related Macrolides : Structure (4) was recently established for amphotericin on the basis of *X*-ray crystallographic analysis.

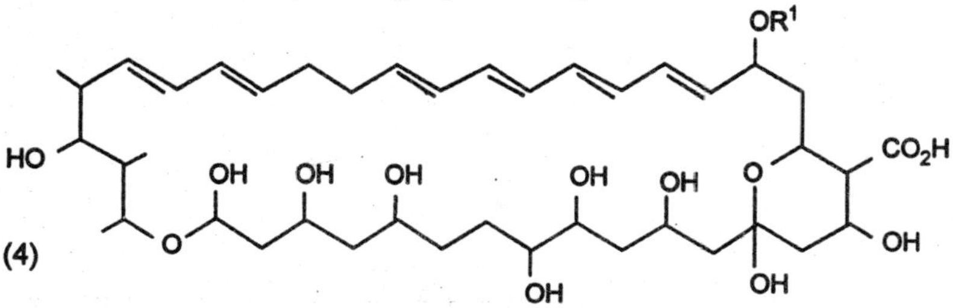

(4)

N-Acylation and esterification reactions have been reported on the antibiotic as part of a general survey to investigate the effect of chemical modifications on the retention of antifungal activity of the parent molecule.

Spectroscopic studies have indicated that the closely related macrolides nystatin (3), amphotericin B (4) pimaricin (5a) and lucensomycin (5b) exist in solution as cyclic hemi-acetals. The presence of the hemi-acetal structure clearly imposes restraints on the sizes and conformations of the macrolide rings which could be important in influencing the ease of formation and stabilities of the sterol-polyene complexes formed from the macrolides and sterols.

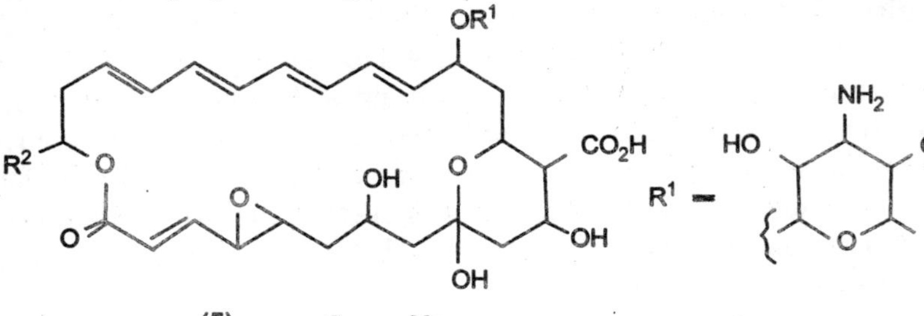

OR[1] CO_2H OH O OH O R^2 O O OH OH R^1 = HO NH_2 OH O

(5) a ; R^2 = Mc

b ; R^2 = Bu"

The comparative chemotherapeutic activities of several heptaene macrolide antibiotics, including amphotericin B, have been examined, and some biological properties of polyene macrolide ester salts have been reported.

Fungimycin and Candicidin : Fungimycin, an antifungal heptaene produced by *Streptomyces coelicolor* var. *aminophilus,* was shown earlier to contain an *N*-methyl-*p*-aminophenyl moiety. Biosynthetic studies have now that the aromatic ring is formed from shikimic acid, with *p*-aminobenzoic acid as the direct terminal intermediate; methionine provides the *N*-methyl group. The biosynthesis of the *p*-aminophenyl moiety in candicidin, another heptaene antibiotic produced by *S. griseus* IMRU 3570, follows the same pathway. The full chemical structures of these two macrolides are not known with certainty.

Venturicidines : Brufani *et al.* have published full details of their chemical and spectroscopic examinations of the macrolide antibiotics venturicidin A and B from *Streptomyces aureofaciens.* In combination with an *X*-ray analysis, structures (6a) and (6b) have been determined for the two macrolides.

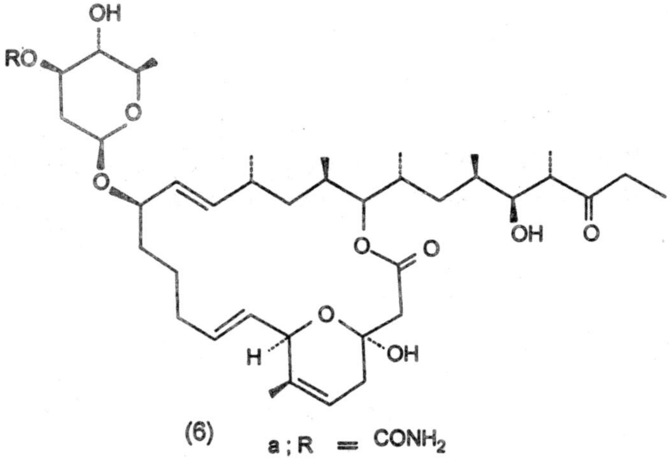

OH RO O O O O OH O O H OH

(6) a ; R = $CONH_2$

b ; R = H

Chlorothricins : The structure of chlorothricolide methyl ester, derived from the aglycone of the antibiotic chlorothricins (from *Streptomyces antibioticus),* has been deduced (7).

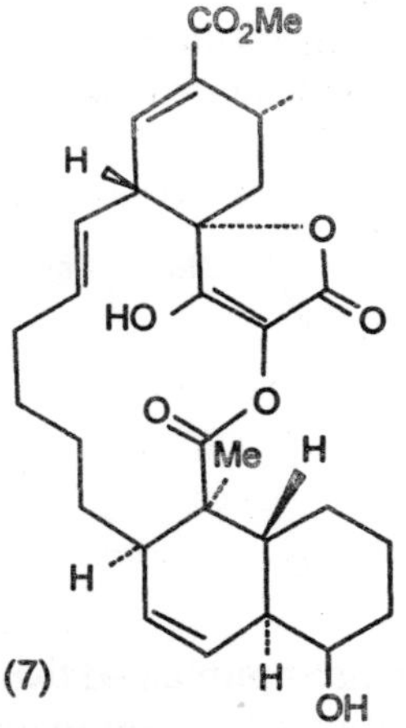

CO$_2$Me
H
O
HO
O
O
O
Me
H
H
(7)
H
OH

Oligomycins. The three-dimensional structure of oligomycin B has been determined, and the molecule is composed of a 26-membered ring to which is condensed a substituted tetrahydropyran ring which in turn is linked to another tetrahydropyran ring by a spiro-junction.

Leucomycins. A review of the chemistry and biological activity of leucomycins has been published, and Omura *et al.* have described some conformational studies on the 16-membered-ring (9) lactone which forms the aglycone of the antibiotic.

Leucomycins belonging to the Ac-group (9; R = Ac) have been found to show more antimycoplasma activity than those of the Fr-group (9; R = H).

Erythromycins : A shunt metabolite of erythromycin biosynthesis has been isolated from the fermentation broth of a block mutant of *Streptomyces erythraeus,* and structure (10) was determined for the compound.

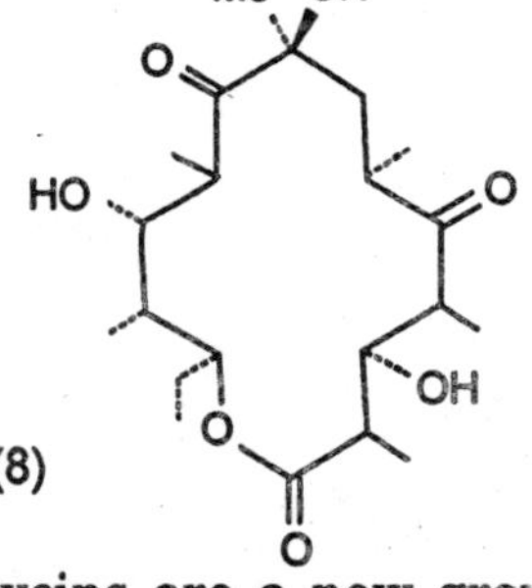

Me OH
O
HO
O
OH
O
(8)
O

Maridomycins : The maridomycins are a new group of macrolide antibiotics from *Streptomyces hygroscopicus* and are designated maridomycins I—VI. The aglycones of the antibiotics have the general structure (11).

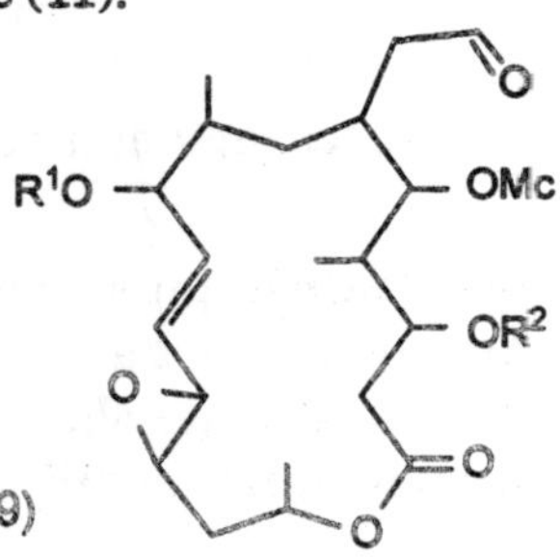

O
R^{1}O
OMc
OR^2
O
O
(9)
O

Angolamycin (Shincomycin A) : The aglycone of angolamycin from *Streptomyces erythraeus* has been shown by mass spectrometry to possess structure (12). This structure is closely similar to that determined earlier for the aglycone of tylosin, *viz.* (13).

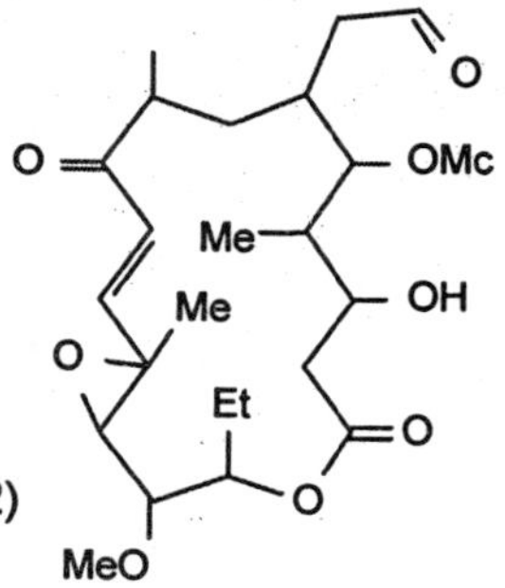

(12) 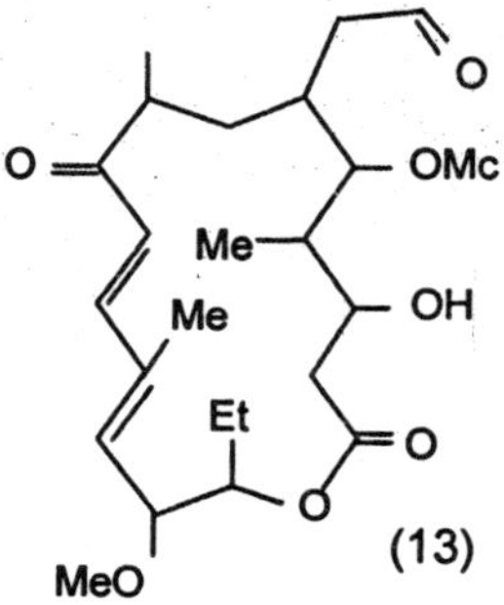 (13)

Zearalenone : The Merck group have published two papers in a programme directed towards chemical modification of zearalenone (14), a metabolite of *Gibberella zeae* which has shown hormonal and growth-promoting activities. Bolliger and Tamm have reported the isolation of the minor metabolites (18 – 17) from cultures of *G. zeae,* and found that the production of (14) and its congeners was very dependent on the conditions of the culture. Photolysis of (14) and of zearalenol (18) produces the corresponding

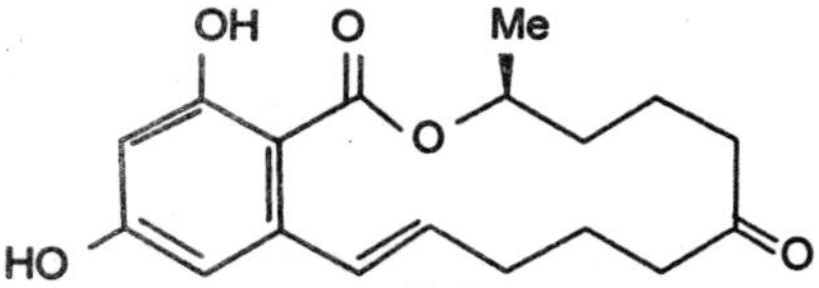

(14) 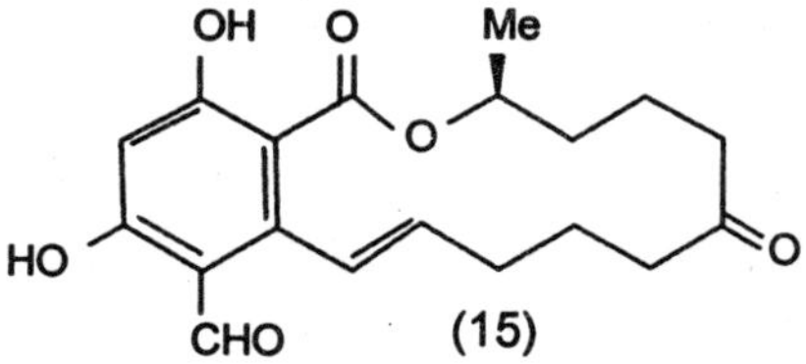 (15) 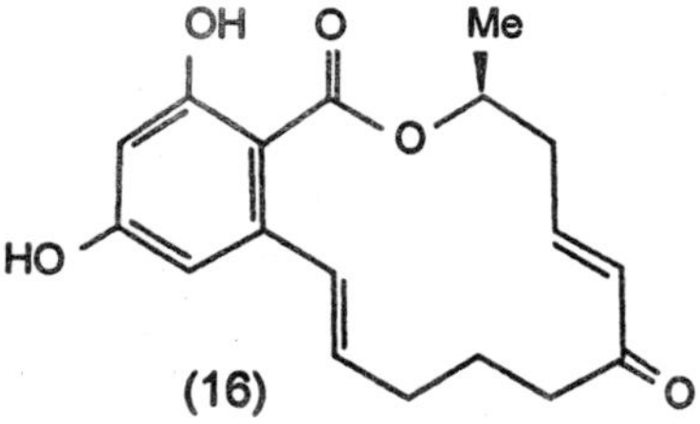 (16) 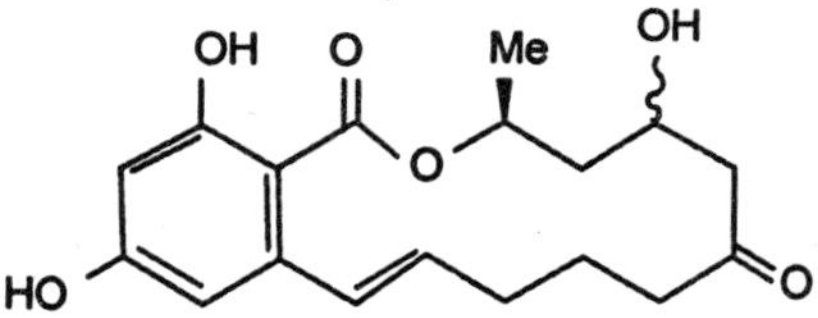 (17) 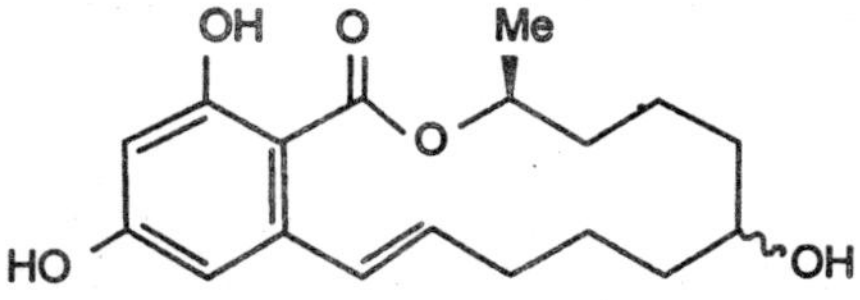 (18) 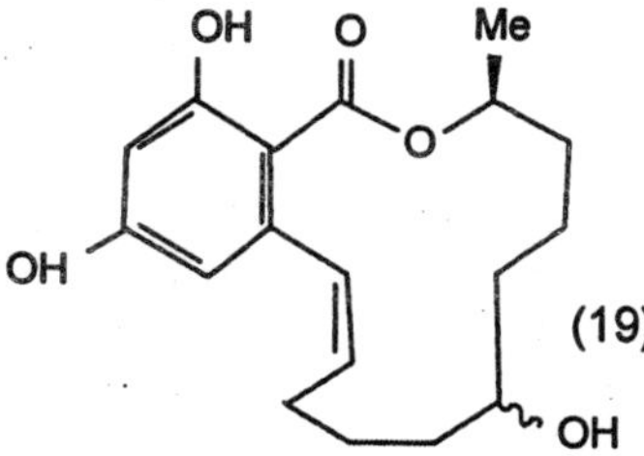 (19)

Z-isomers; *Z*-zearalenol (19) showed much greater uterotropic activity than E(18), and E and 2(14) showed comparable activity.

Pyrenophorin : A total synthesis of racemic pyrenophorin (25) an antifungal and cytostatic metabolite of the plant pathogenic fungi *Pyrenophora avenae* and *Stemphylium radicinum,* has been outlined by Colvin *et al.* Reduction of γ-valerolactone produced lactol (20), which gave dithian (21) on treatment with propane-1,3-dithiol and BF_3-Et_2O. Reaction of the protected alcohol of (21) with base and ethyl formate then gave aldehyde (22), which was converted by a Wittig reaction into *E*-alkene (23). After removal of the alcohol-protecting group from (23) the free alcohol was converted into the P-ylide (24), which produced diene (25) upon Wittig reaction with (24). Removal of the protecting groups in (25) followed by conversion of the resulting hydroxy-acid into the corresponding imidazolide and cyclization in the presence of DBN then gave smoothly (26) converted into pyrenophorin (27).

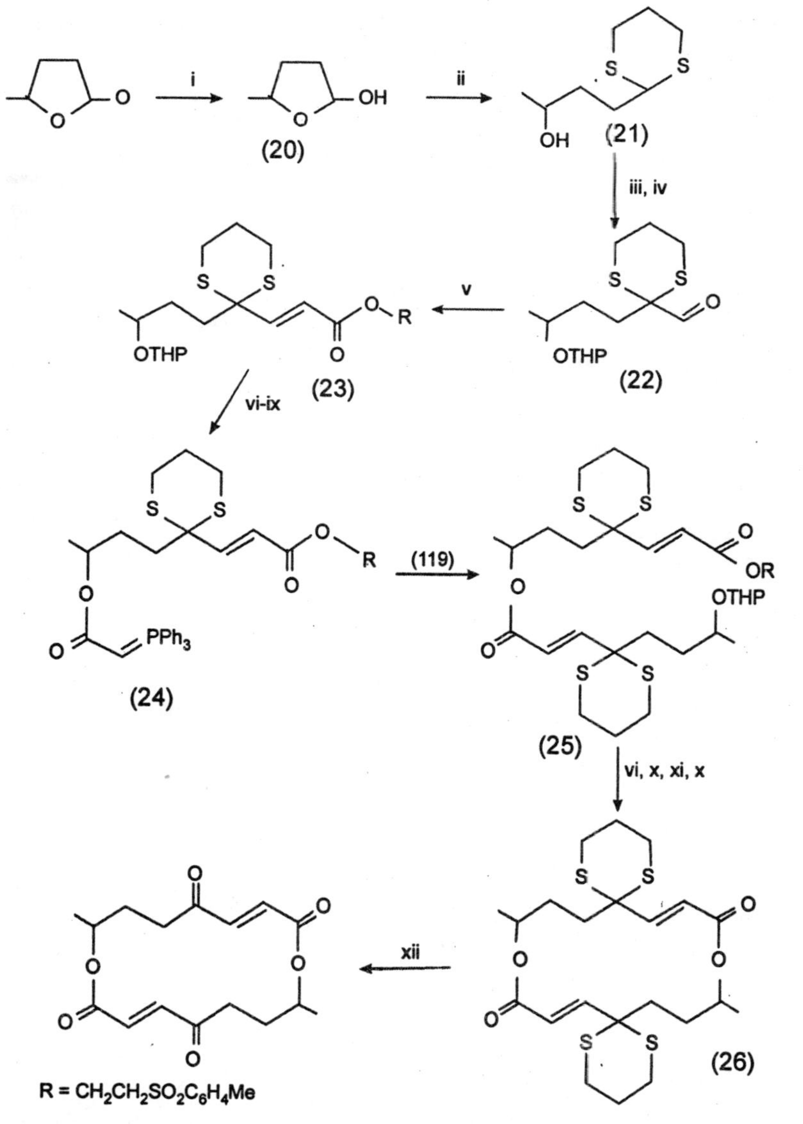

(20) (21) (22) (23) (24) (25) (26) (27)

R = $CH_2CH_2SO_2C_6H_4Me$

Reagents : i, $NaAlH_4$; ii, $(HSCH_2)_2CH_2$-BF_3; iii, dihydropyran-H^+; iv, Bu^nHLi-HCO_2Et; v, $MeC_6H_4SO_2CH_2CH_2CO_2CH{=}PPh_3$; vi, H^+; vii, $BrCH_2CO_2H$; viii, PPh_3; ix, NaOH; x, 1,5-diazabicyclo-[4,3,0]non-5-ene; xi, imidazolide formation; xii, *N*-chlorosuccinnimide—$AgNO_3$

Polyolefinic Macrocycles (Non-macrolide)—The Cytochalasins : The cytochalasins are fungal metabolites which produce unusual effects on mammalian cells in tissue-culture.

Cytochalasins A (28) and B (29) are metabolites of *Helminthosporium dematioideum* and of a *Phoma* sp.,* and Cytochalasins C (30) and D (31) are metabolites of *Metarrhizium anisopliae. cytochalasin* D has also been isolated from *Zygosporium masonii.* In the period under review, Turner and collaborators have reported two new members of this class of compounds, designated Cytochalasins E and F. Cytochalasin F is a minor metabolite of *H. dematioideum,* which was shown to have structure (33), isomeric with cytochalasin B. Cytochalasin E is produced by *Rosellinia necatrix,* and chemical degradation combined with spectral data have established structure (32) for this compound, incorporating an unusual vinyl carbonate residue.

The Cytochalasins are clearly related biosynthetically, and it has been suggested that the macrolide rings of A and B are derived from a carbocyclic precursor. It is possible that the carbonate residue in E is formed by insertion of two oxygen atoms into a carbon chain, the similarity of the large ring of E and that of C and D adding credulity to this possibility. It has also been suggested that A, B, C, and D are derived from a precursor which has a double bond in the cyclohexane moiety, as arranged in E and F.

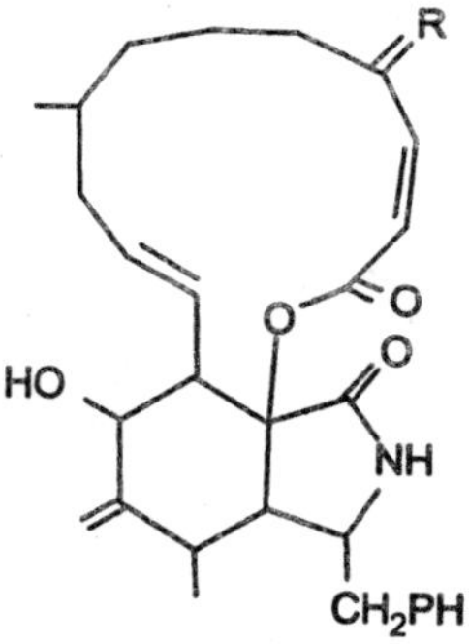

(28) R = O

(29) R = H, OH

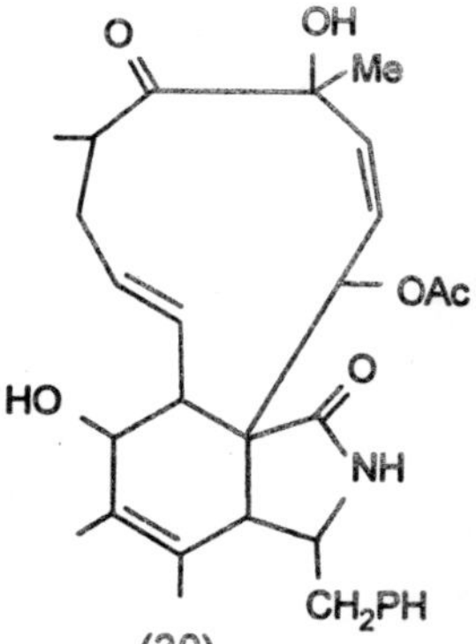

(30)

Binder and Tamm have examined the protection of the hydroxy-groups in cytochalasin B (29), and the corresponding dihydro- and dodecahydro-derivatives, by studying the reactions of these molecules with various methylating reagents. Cytochalasin B produced the corresponding imino-methyl ether exclusively, whereas the dihydro- and dodecahydro-derivatives were shown to produce mixtures of *N*-methyl and *O*-methyl derivatives.

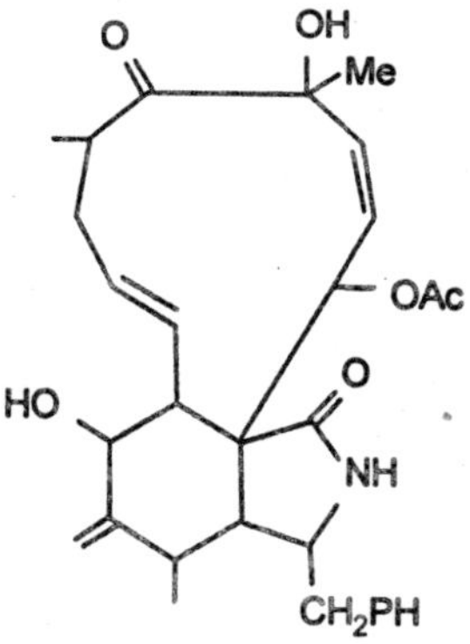

(31)

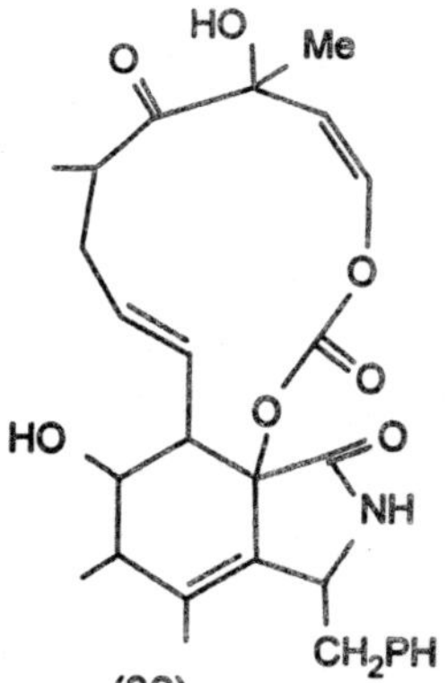

(32)

The Nonadrides and Related Compounds Biosynthesis : The elegant work which led to the constitution and geometry of the three fungal metabolites (the 'nonadrides') glaucanic (34a), glauconic (34b), and byssochlamic (35) acids was described in a series of papers by Barton and Sutherland and their collaborators in 1965. In one of these papers it was proposed that the carbon skeletons (34) and (35) might be derived in Nature by coupling of two identical C_9 units *i.e.* (37), *i.e.* formed by condensation of

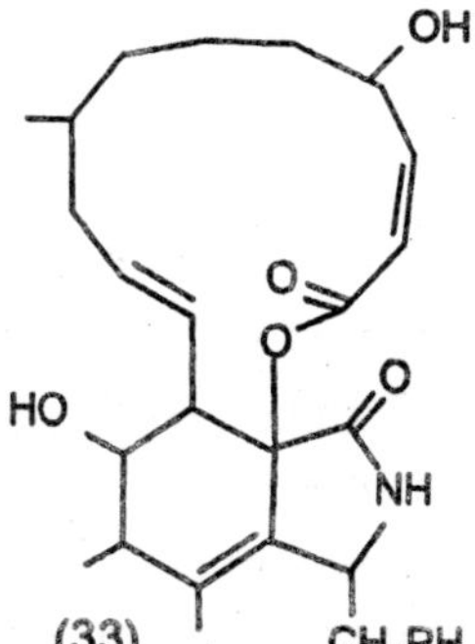

(33)

hexanoate (from acetate) with oxaloacetate to yield an intermediate substituted citric acid, which was

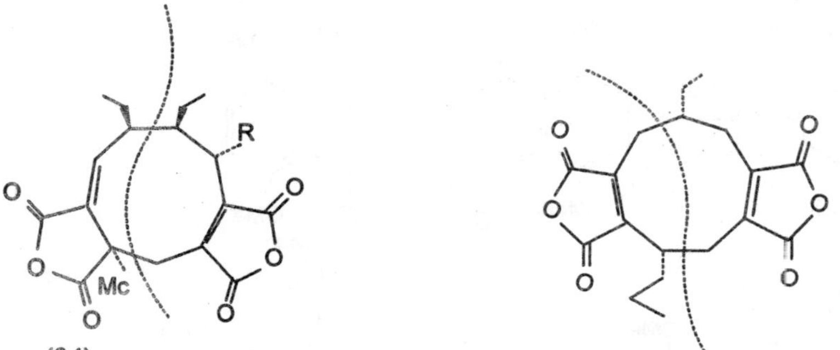

(34) a ; R = OH
b ; R = H

(35)

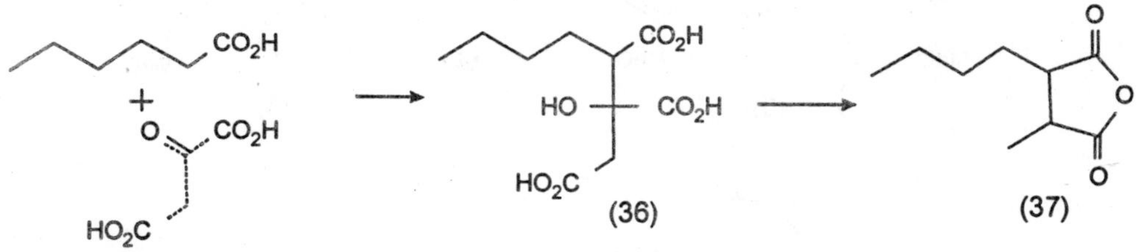

(36)

(37)

decarboxylated. Biosynthetic data, obtained from feeding experiments with labelled acetate, succinate, glucose, and *N*-butyl(methyl)-maleic anhydride, published in 1968 supported this hypothesis for the biosynthesis of (34). In 1972, full details of the synthesis of the anhydride (38) which appeared to be the most likely C_9 unit which dimerizes, of its incorporation (55%) into *Penicillium purpurogenum* were published by Sutherland *et al.* These data then leave little

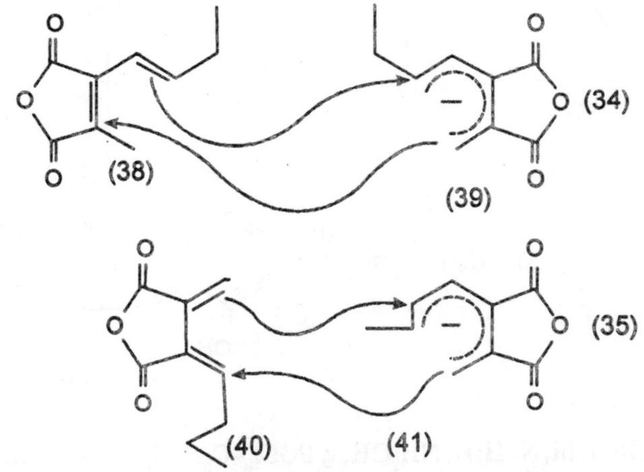

(34)

(38)

(39)

(35)

(40)

(41)

doubt that (34) is formed in Nature from two molecules of (38) and a mechanism for this combination involving a [6 + 4] cycloaddition of the anion (39) with the *cisoid*-anhydride has been suggested. Reaction *via* an *endo*-transition state would produce the correct stereochemistry in (34). Sutherland *et al.* have also described the base-catalysed *in vitro* dimerization of (38) to

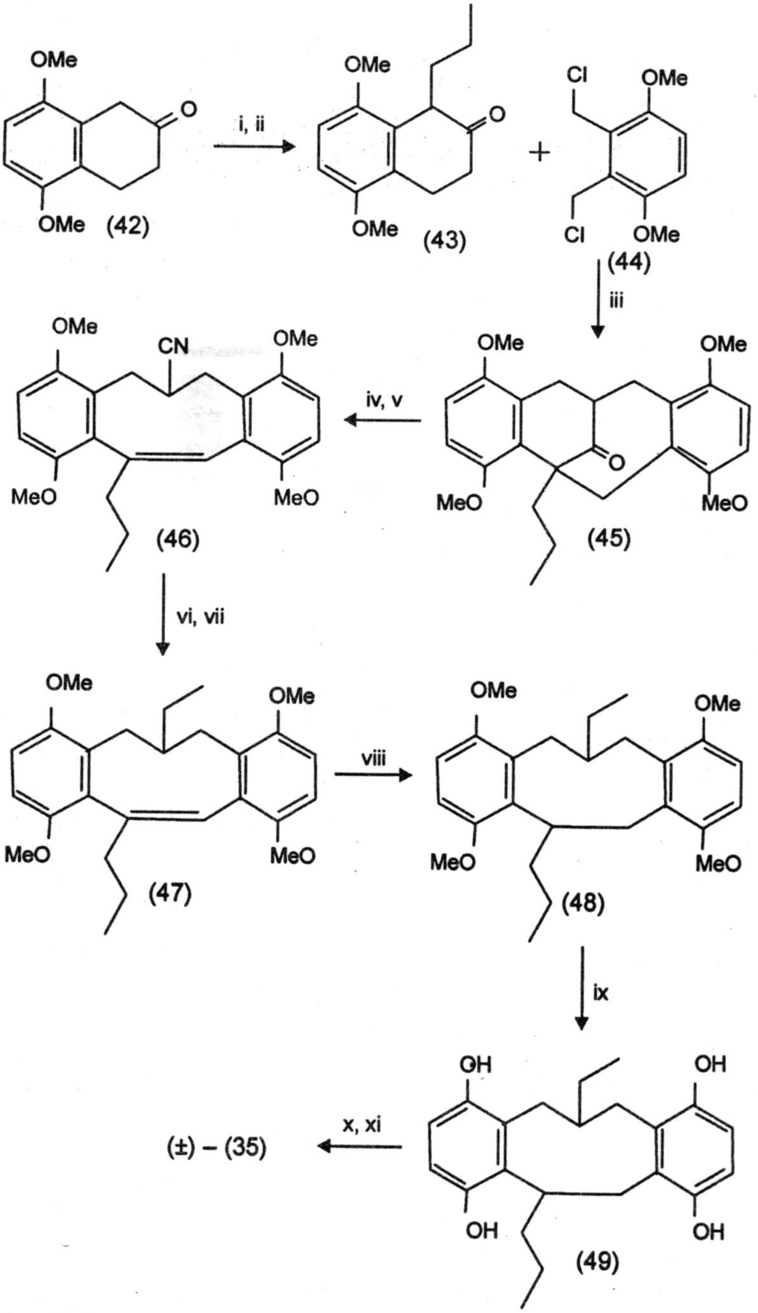

Regamts :i, Pyrrolidine; ii, Pr^nI; iii, NaH; iv, NH_2OH; v, $POCl_3$–C_5H_5N; vi, McL; vii, Wolff-Kischner; viii, Li–NH_3; ix, BBr_3; x, $KMnO_4$; xi, $Pb(OAc)_4$

produce an epimer in 5 % yield. It has been proposed that the byssochlamic acid skeleton could arise in Nature in a similar fashion from the diene (40) and the stereoisomeric anion (41).

Synthesis : The first total synthesis of a nonadride, byssochlamic acid (35) has been outlined by Stork *et al.* In this elegantly conceived route to (±) (35) the pyrrolidine enamine from 2-tetralone (42) is first propylated to (43) which is then elaborated to (45) by cycloalkylation with (44). Conversion of (45) into the oxime and fragmentation then produced the nine-membered ring system (46). Transformation of the nitrile group into an ethyl substituent then produced (47), which readily formed the corresponding dihydro-compound (48). The synthesis of (35) was then completed following demethylation to oxidation of the latter with permanganate and lead tetra-acetate.

Scytalidin : This is the name proposed for a new fungitoxic metabolite isolated from a culture medium which has supported growth of *Scytalidium* sp., an imperfect fungus. Structure, (51) which is similar to those determined for the nonadride metabolites (34) and (35), was established for this compound on the basis of spectroscopic and chemical investigations. Its biosynthesis, by analogy with the nonadrides, is considered to involve coupling of two C_u units (50), probably formed by decarboxylation of the condensation product from an octanoic acid derivative and oxaloacetate.

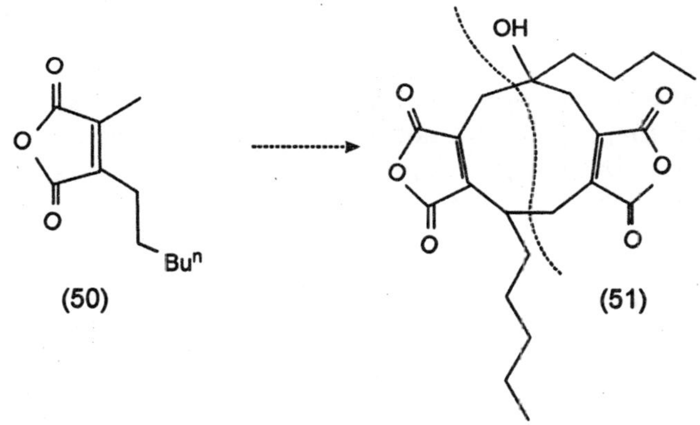

(50) (51)

Ansamycin Antibiotics—The ansa macrolide antibiotics, formerly composed of the rifamycins, streptovaricins, tolypomycins, and geldanamycin, but also now including the maytansines, have aroused considerable interest as antiviral and antimicrobial agents, and as inhibitors of RNA tumour virus reverse transciptases. The chemistry *etc.* of this interesting class of antibiotics has been the subject of a recent review by Rinehart.

Streptovaricins : Rinehart and collaborators have re-examined the structures earlier proposed for the streptovaricins. On the basis of chemical degradation data combined with ^{13}C n.m.r. and, ultimately, *X*-ray crystallographic analysis, the new structure (52c) has been determined for Streptovaricin C, the major component of the Streptovaricin complex. Structures (52a), (52b), (52c), (52d), (52e), (52f) and 53 then followed for Streptovancins A, B, D, E, F, and G, respectively.

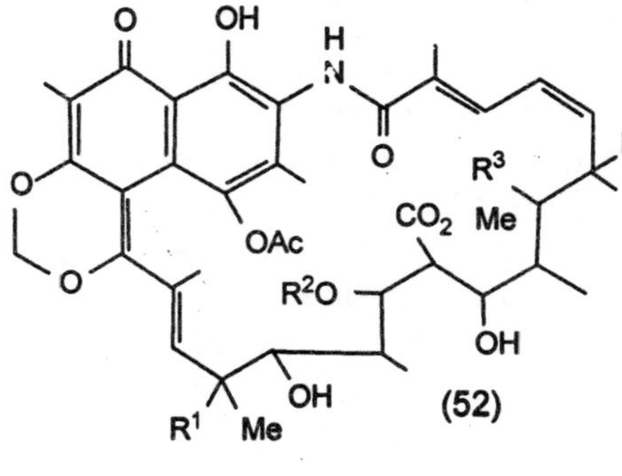

(52)

(52) a; R^1 = OH, R^2 = Ac, R^3 - OH, R^4 = OH
b; R^1 - OH, R^2 = Ac, R^3 = OH, R^4 = H
c; R^1 = OH, R^2 = H, R^3 = OH, R^4 = H
d; R^1 = H, R^2 = H, R^3 = OH, R^4 = H
e; R^1 = OH, R^2 = H, R^3 = =O, R^4 = H
f; R^1 = OH, R^2 = H, R^3 = OH, R^4 = OH

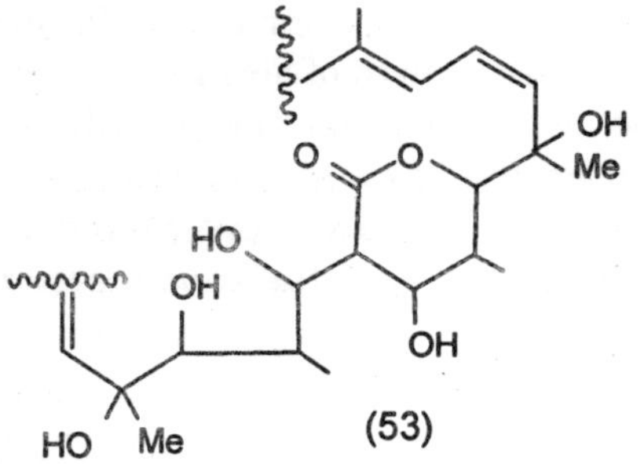

OH Me O O HO OH OH HO Me

(53)

Maytansine and Homologues : Three potent anti-leukaemic ansa macrolides, designated maytansine, maytanprine, and maytanbutine, have been isolated from *Maytenus ovatus* Loes. Spectral and *X*-ray crystallographic analysis showed that maytansine had structure the homologous structures (54b) and (54c) for maytanprine and maytanbutine followed from comparative spectral data.

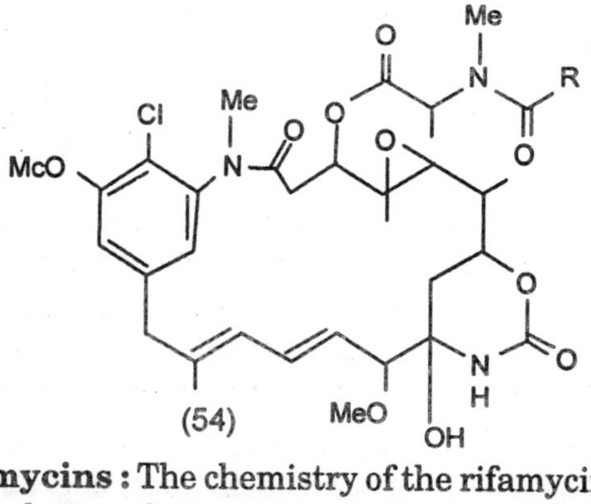

Me O N R Cl Me O O McO N O O O O N O H MeO OH

(54)

Rifamycins and Tolypomycins : The chemistry of the rifamycins has been reviewed, and Kishi *et al.* have reported the isolation of a new ansa antibiotic from *Streptomyces tolypophorus,* molecular formula $C_{43}H_{54}N_2O_{14}$ and designated Tolypomycin Y.

Pyran-pyranoid Compounds—Aureotine (Aureothin) : Two research groups have independently examined separate features of the biosynthesis of Aureotine (≡Aureothin) (58), an antibiotic produced by *Streptomyces thioluteus.* Feeding experiments with tritiated *p*-aminophenylalanine *p*-aminocinnamic acid *p*-aminobenzoic acid (55), and *p*-aminophenylserine (57) showed that [but not (55) or *p*-$NH_2C_6H_4CO_2H$] are involved in the biosynthesis of the nitroaromatic portion of (58), which suggested the partial route. Studies by Yamazaki *et al.* of the incorporation of L-methionine into (58) the four 'extra' methyl groups in the side-chain are not derived from methionine, which thus ruled out an acetate-malonate route (a) to the side-chain in aureotine. In fact, ^{13}C n.m.r. studies showed that the 'extra' methyls in (58) were all derived from the methyl of propionate, but since both acetate and propionate are incorporated, the Japanese workers consider that an acetate-propionate route (*b*) to the side chain is most probable.

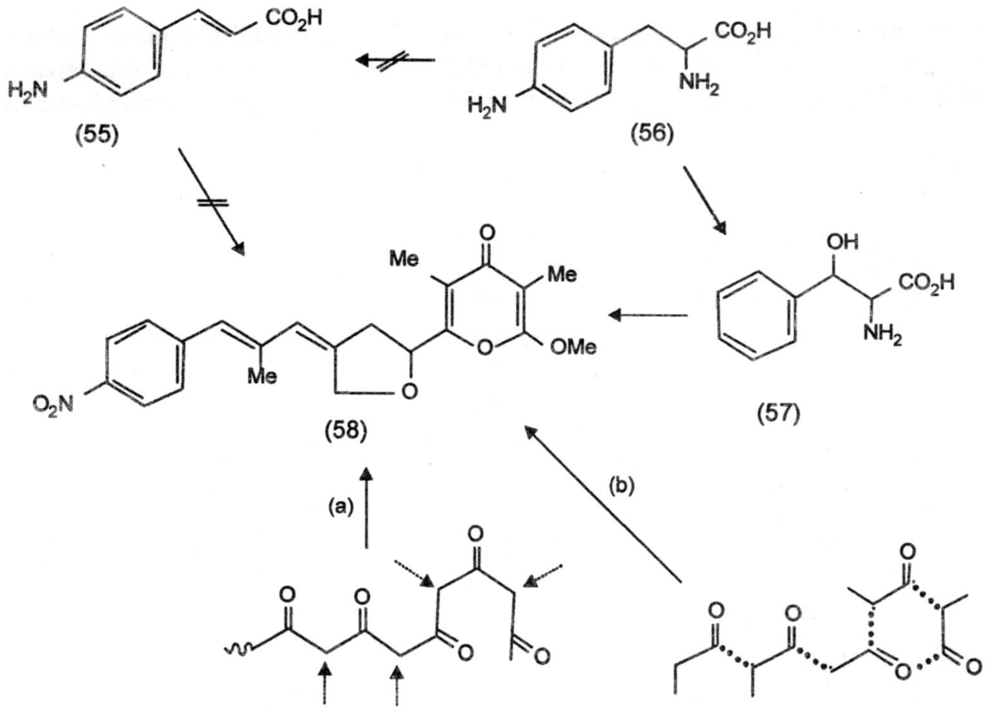

(55) (56) (57) (58) (a) (b)

Citreoviridin : Nagel *et al.* have shown that *Penicillium pulvillorum* produces citreoviridin (59) a yellow fluorescent, neurotoxic polyene, previously isolated from *P. citreoviride*. In another publication, Nagel et al. presented evidence supporting an acetate pathway to (59) the five methyl groups (marked*) being derived from methionine.

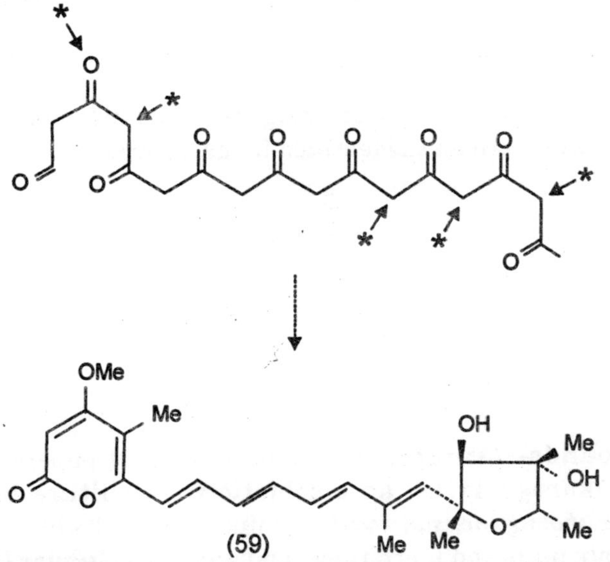

(59)

Patulin : Scott and his collaborators have published full details of their work on the biosynthesis of patulin, (60) an antibiotic isolated from cultures of *Penicillium patulum*. Much of this work was published in preliminary form some time ago, and the main pathway to (60) 6-methylsalicyclic acid. The isolation of deoxypatulinic acid (61) P. *patulum* was reported during the period under review, and clearly the role of (61) in patulin biosynthesis merits investigation.

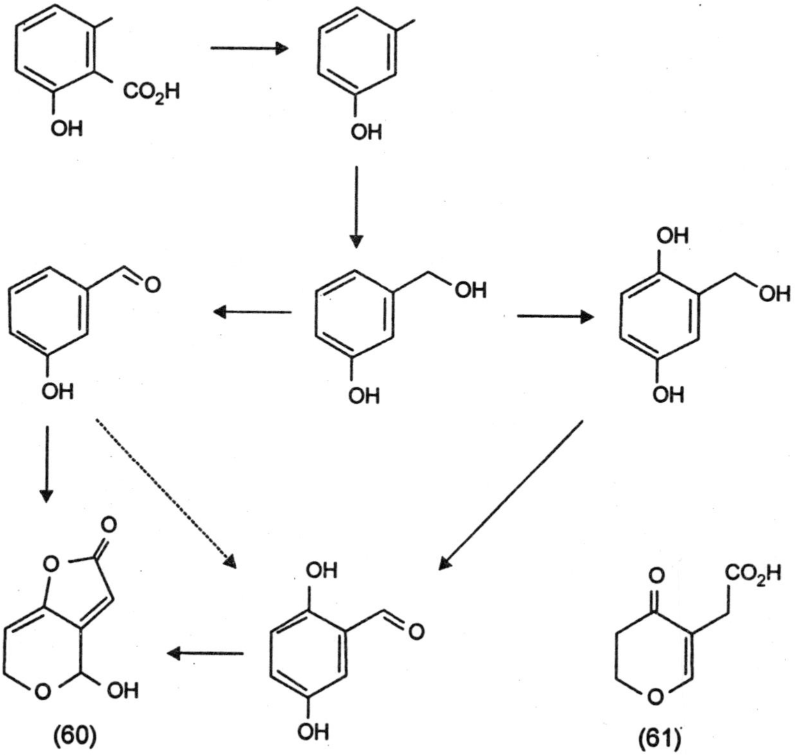

(60) (61)

LL-Z1220. The novel cyclohexene diepoxide ring structure (62) has been deduced for a new antibiotic, called LL-Z1220, from an unidentified fungal species.

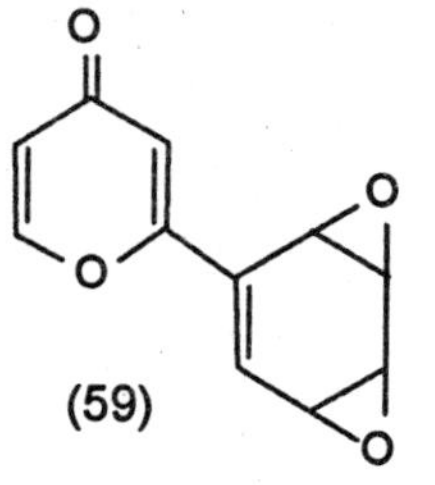

(59)

Alicyclic Compounds—Ikarugamycin : In a series of papers, the full structure and stereochemistry of Ikarugamycin, an antibiotic from culture broth of *Streptomyces phaeochromogenes* var. *ikaruganensis*, was determined as (64). Its biosynthesis is considered to involve two hexa-acetate units and L-ornithine, and an intramolecular Diels-Alder reaction of a hypothetical intermediate (63) has been put forward as a suggested route to the *trans-anti-cis*-decahydra-*asym*-indacene skeleton [*i.e.* the A, B, and C rings of 64].

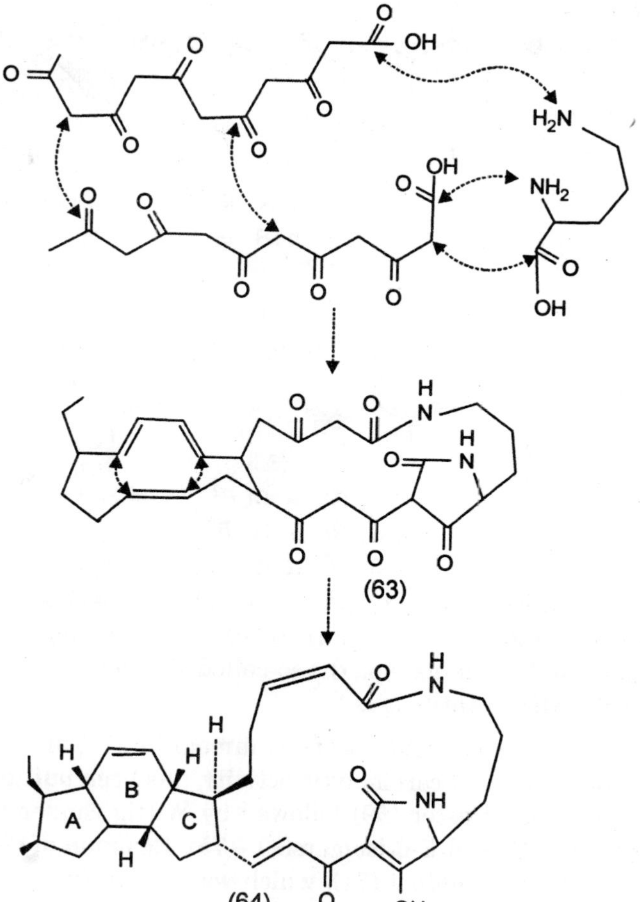

(63)

(64)

Ascofuranone and Related Compounds : Ascofuranone is a new antibiotic which has been isolated from *Ascochyta viciae*. The structure (65) determined for this compound is closely similar to that of ascochlorine (66a) the same source.

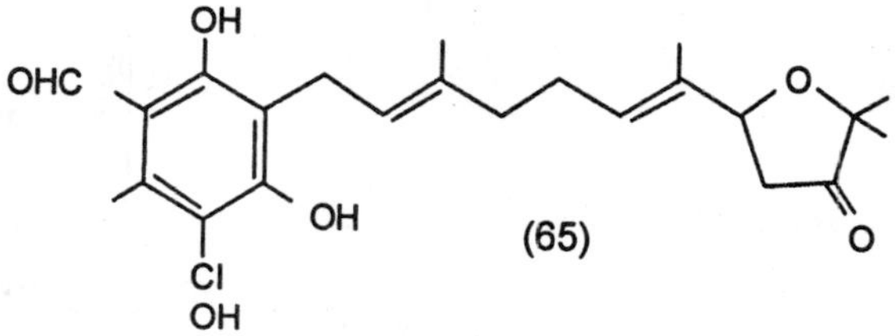

(65)

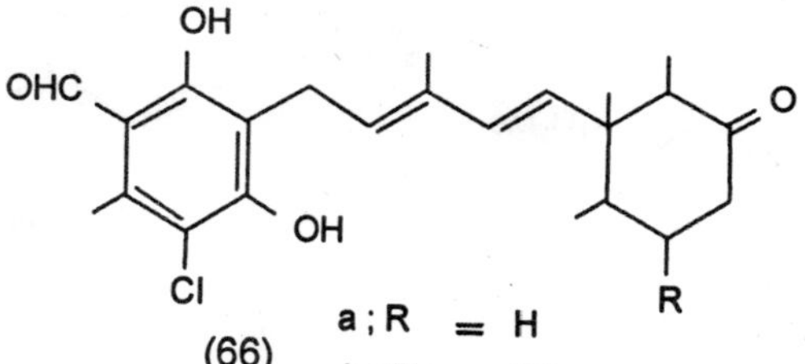

(66) a ; R = H
b ; R = OAc

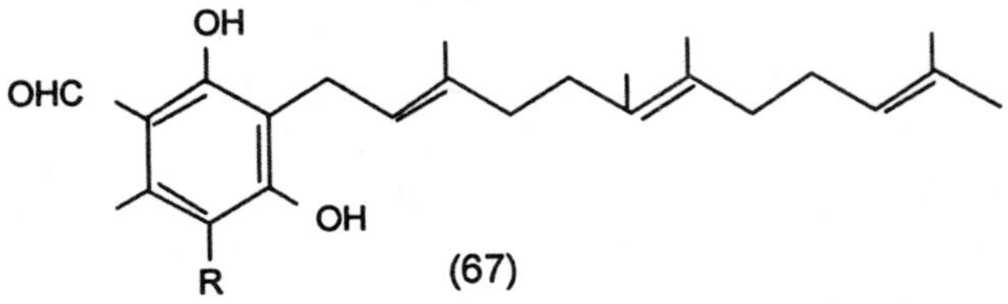

OH
OHC
OH
R
(67)

a ; R = Cl
b ; R = H

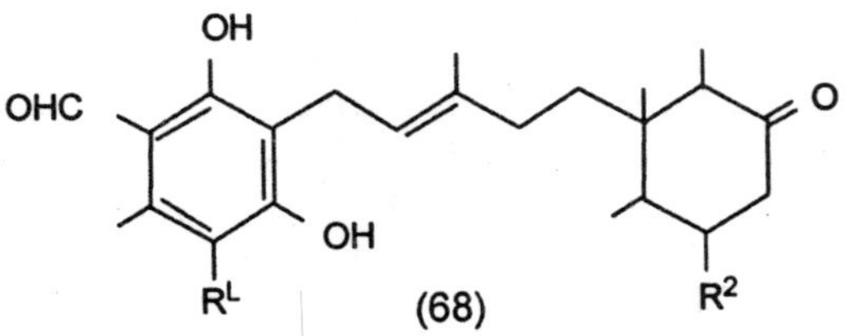

OH
OHC
OH
R^L
(68)
R^2

a ; R^1 = Cl, R^2 = H
b ; R^L = H, R^2 = H
C ; R^1 = Cl R^2 = OAc

Five metabolites of *Nectria coccinea* have been to have the closely related structures (66a) ascochlorine. Compounds (66 a), (66 b), (67 a), (67 b) and (68 a) and (68 b) have previously been reported in an unidentified *Fusarium* sp., the so-called 'ilicicolins A—H', from *Cylindrocladium ilicicola*, are identical with the antibiotics

Fumagillin : The first total synthesis of (±)-fumagillin (75) an antibiotic from *Aspergillus fumigatus* with antiparasitic and carcinolytic activity, has been outlined by Corey and Snider. Selenium dioxide oxidation of ester (69) followed by Wittig condensation with allyl P-ylide, produced a 1:1 mixture of *Z*- and E-6-isomers(±) –(75). Reaction of (70) with α-bromoacrolein then produced the Diels-Alder adduct (71), which was smoothly converted into (72) following reduction, silylation, and epoxidation. Hydrolysis gave di-epoxide, which produced upon hydroxylation with OsO_4. Elaboration of (74) then produced of (70).

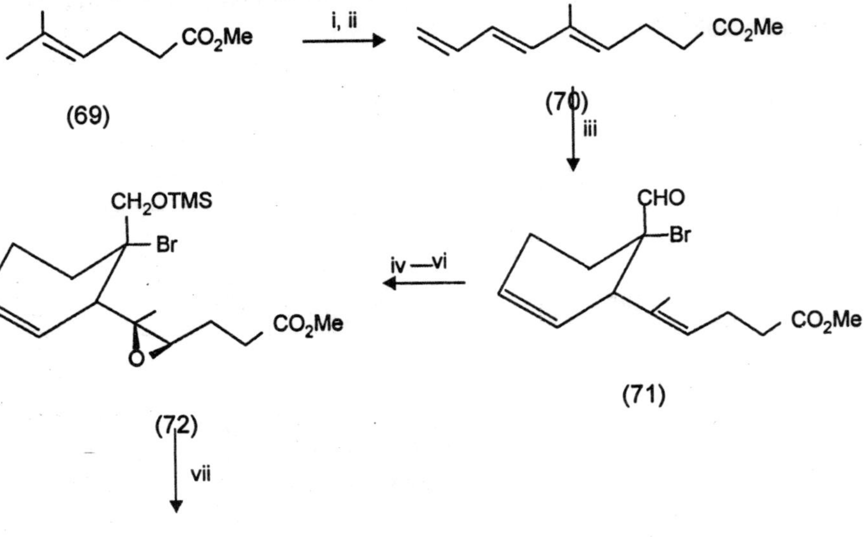

CO_2Me —i, ii→ CO_2Me

(69) (70)

iii

CH_2OTMS
Br
CO_2Me
(72)

iv—vi

CHO
Br
CO_2Me
(71)

vii

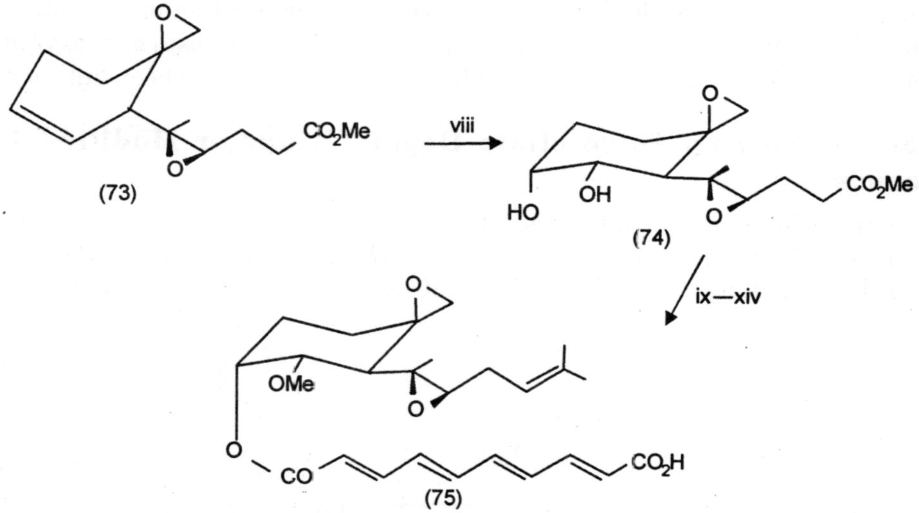

(73) CO$_2$Me viii (74) CO$_2$Me HO OH ix—xiv (75) OMe CO$_2$H

Reagents: i, SeO_2; ii, CH_2=CH-CHPPh$_3$; iii, CH_2=C(Br)-CHO; iv, $NaBH_4$; v, Me_3SiCl; vi, m-$ClC_6H_4CO_3H$; vii, $Bu_4N^+F^-$; viii, OsO_4; ix, MeI; x, MeLi; xi, Ac_aO-C_5H_5N; xii, dehydration; xiii, saponification; xiv, ClCO(CH =CH)$_4$COCl

SCHEME 13

6 Cyclopentenone Polyolefinic Compounds

Jasmone—Several new routes to *cis*-jasmone, (78) an important constituent of the essential oil of jasmine, have been described. Ellison and Woessner have outlined a convenient route to the

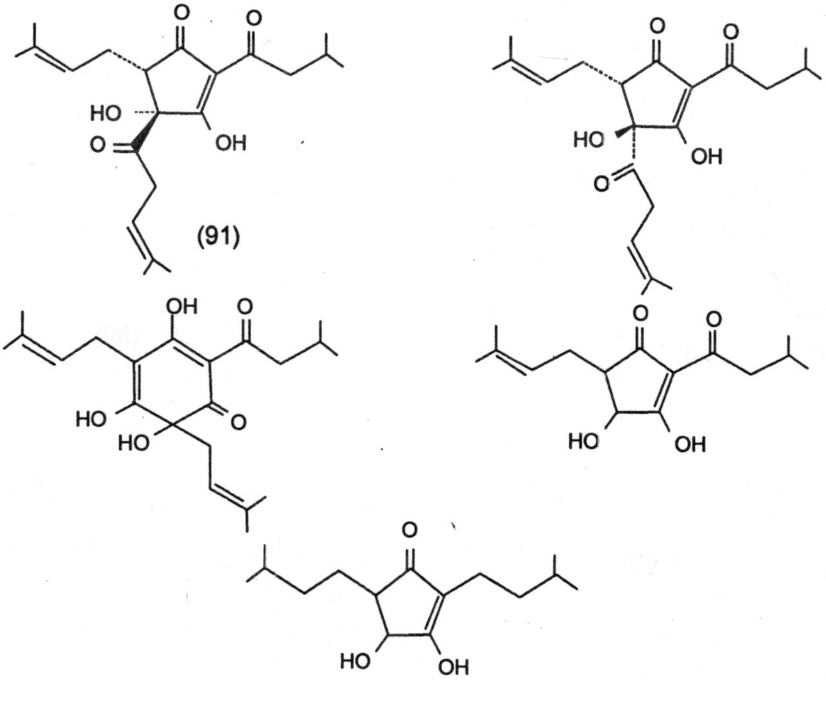

(91)

1,4-diketone (77). Naya and Kotake have described the application of G.C.-M.S. to the rapid analysis of the volatile components in hops, and alternative syntheses of deoxyhumulones, the biogenetic congeners of the humulones and lupulones in hops, have been outlined.

Naturally Occurring Polyolefinic Degraded and/or Modified Isoprenoid Compounds

Gyrinal—Schildknecht *et al* have isolated a C_{14} dional from the pygidial glands of *Gyrinidae* (Coleoptera; Carabidae), for which they have assigned structure (100) proposed the name 'gyrinal'; 210 mg of the compound was obtained from 2500 Gyrinidae, and structure (100) was deduced largely on the basis of proton n.m.r. spectra.

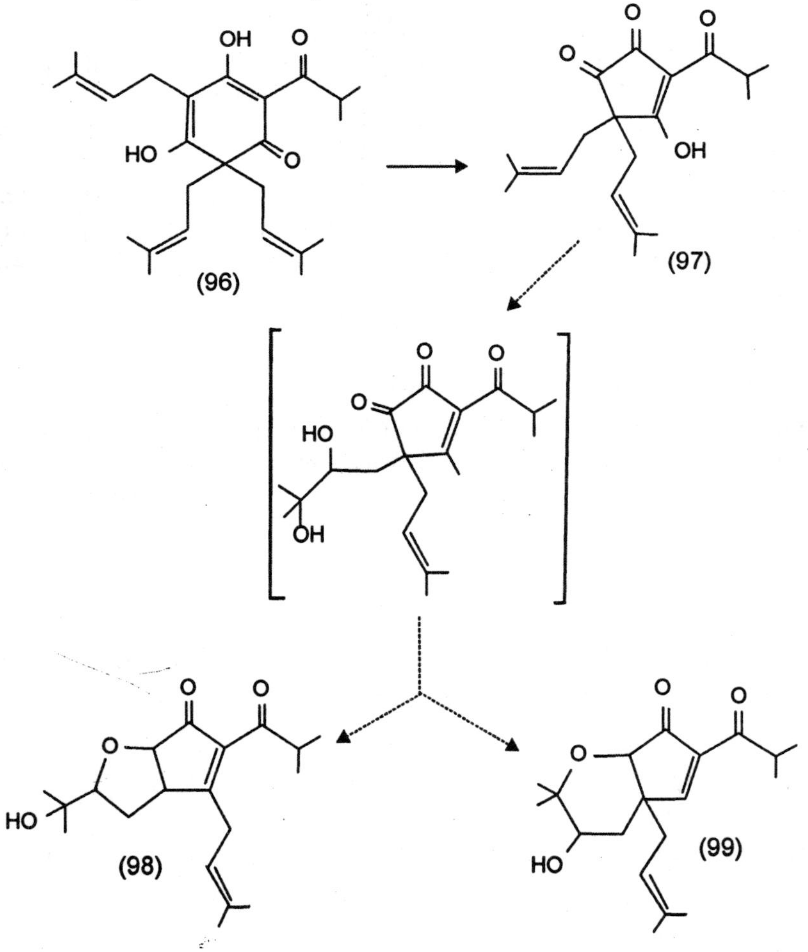

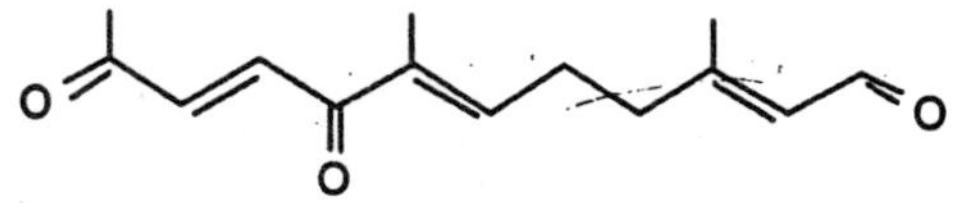

(96)

(97)

(98)

(99)

(100)

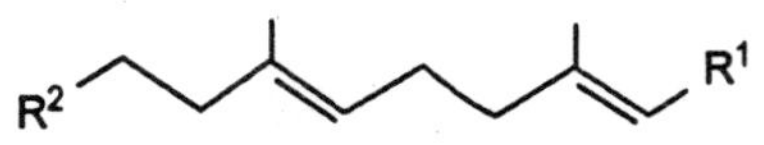

(101)

a; R^1 = R^2 = CO_2H

b; R^1 = CO_2H, R = CH_2OH

c; R^1 = R^2 = CH_2OH

Pheromonal Secretion of Monarch and Queen Butterfly—New synthetic routes haye been outlined to the degraded terpenoid compounds isolated (101 a) and (101 b) from the pheromonal secretion of the Monarch butterfly (*Danaus plexippus*) and of the related diol (101 c) from the male Queen butterfly (*Danaus gilippus berenice*).

Quiesone—Quiesone is the name proposed for a potent germination inhibitor of *Peronospora tabacina,* recently isolated from tobacco leaves infected with this pathogen. Structure (102) 5-isobutyroxy-β-ionone, was assigned to the new compound.

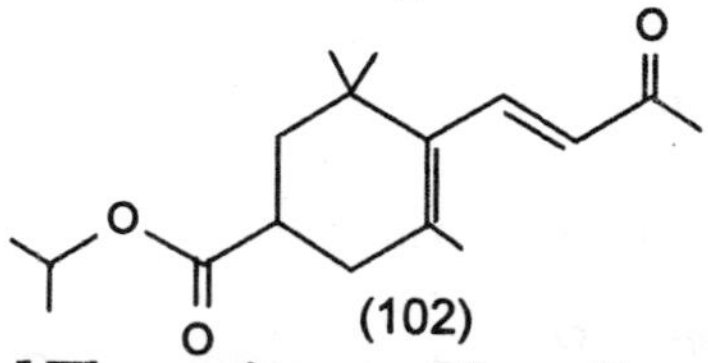

(102)

Blumenols, Abscisic Acid, and Theaspirone—Three C_{13} compounds, designated blumenol A, B, and C, have been isolated from the bark of *Podocarpus blumei* Endl., and are shown to have structures (103), (104)a, and (104b). Interestingly, each of the blumenols is reported to have the opposite stereochemistry to that of abscisic acid at the ring chiral centre. However, the absolute configuration of natural (+)-abscisic acid has only recently been revised and established as *S*, and it is probable that natural abscisic acid and the blumenols have the same stereochemistry at the ring chiral centre after all. The absolute configuration of the related (–)-theaspirone has been recently determined as (106).

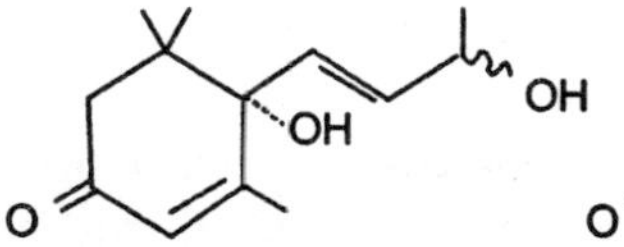

(103) 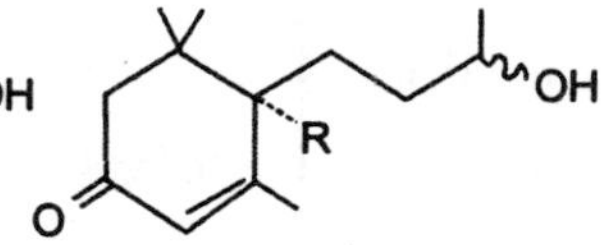 (104) a ; R = OH
b ; R = H

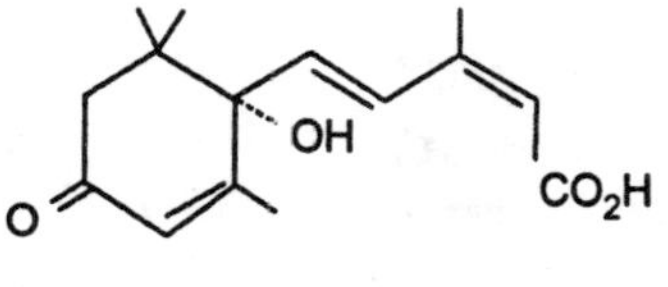

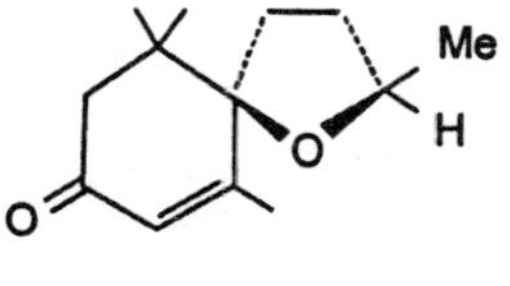

Geijerone—Analysis of the essential oil of *Juniperus communis* L. has shown that it contains the C_{12} ketone (107) which has been called geijerone.

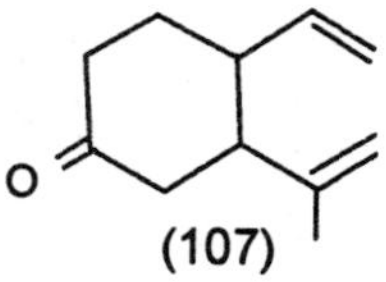

(107)

Constituents of Tobacco—Extensive studies of the volatile constituents of Greek tobacco (*Nicotiana tabacum* L.) have been made by the Swedish Tobacco Company. In addition to methyl esters of the C_8—C_{20} straight-chain fatty acids, which represent the largest group of oxygenated constituents found in Greek tobacco, several isomeric megastigmatrienones (109 and 113) dehydration products of 3-oxo-α-ionol (205), have recently bee nreported.

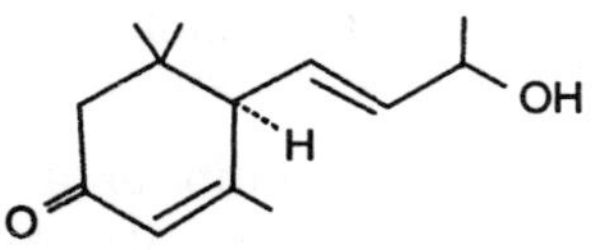

(108) 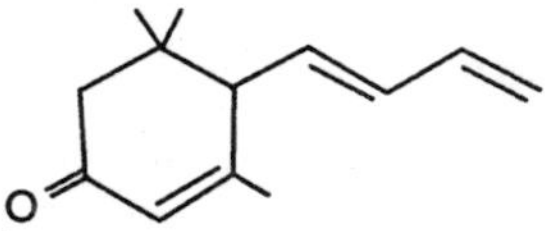 (109) 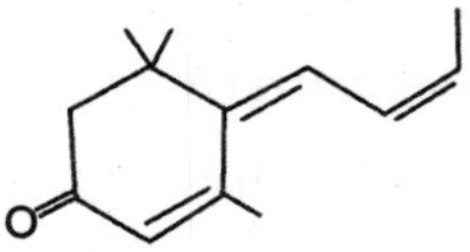 (110)

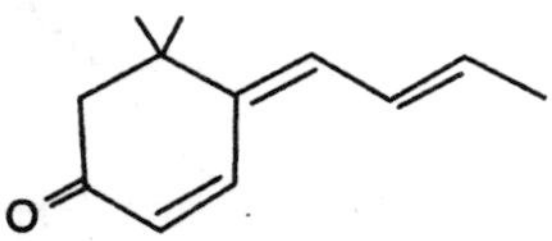

(111) 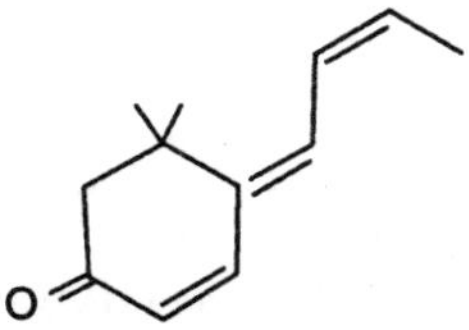 (112) 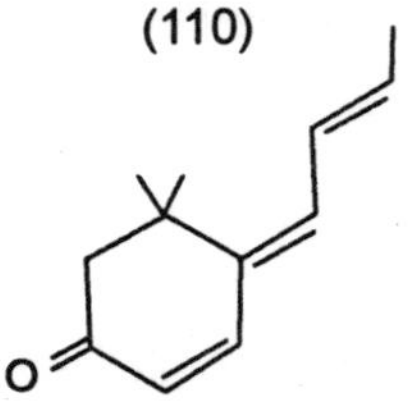 (113)

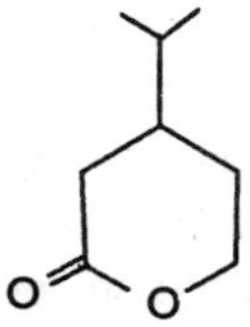

(114) 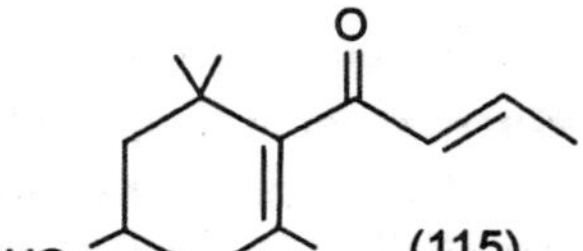 (115)

Kaneko and Harada have reported the presence of lactone (114) and of 4-hydroxy-*E*-*β*-damascone (115) in cigar leaf tobacco.

Miscellaneous Natural Polyolefins

Lactonic Compounds from *Litsea japonica*—Six new compounds, designated 'litsenolides', have been isolated from the roots of *Litsea japonica.* Spectral and chemical data established structures (116 a), (116 b) (116 c), (117 a) (117 b), and (117 c) and for these compounds, and their biosynthesis is considered to proceed from a C_{14} fatty acid (to 'a' and 'b' series) or a C_{16} one (to 'c' series) and pyruvyl-CoA.

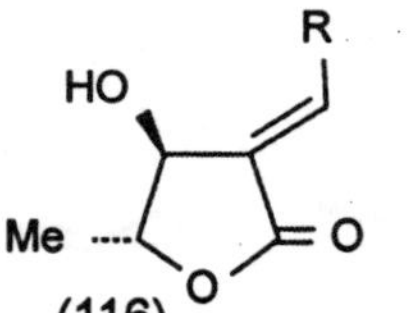

(116) 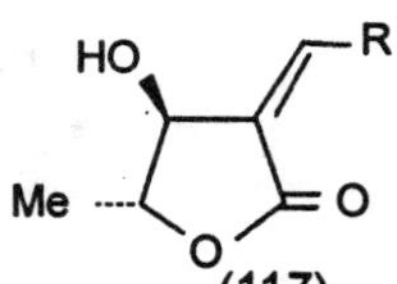 (117)

a ; R = $(CH_2)_9CH = CH_2$

b : R = $(CH_2)_9C \equiv CH$

c ; R = $(CH_2)_{12}Me$

Constituents of *Alnus pendula*—The structure (118) proposed for a new arylpolyene ketone named 'alnustone', isolated from flowers of male catkins of *Alnus pendula,* has been confirmed by synthesis.

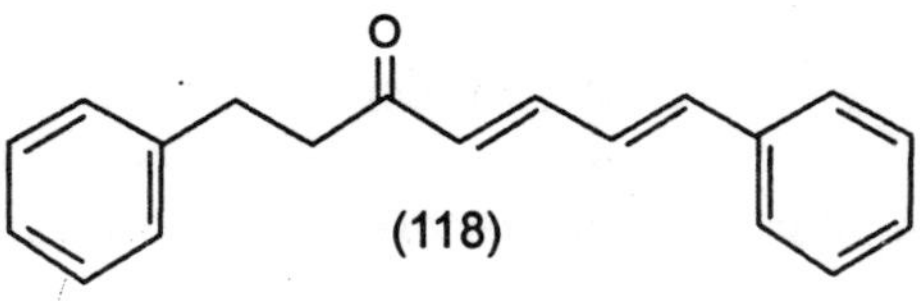

(118)

Constituents of Bartlett Pears—The synthesis of ethyl and methyl *E*-2,*Z*-6-dodecadienoates, two constituents of Bartlett pears, has been described, these esters have proved essential to the reconstruction of good pear flavour.

Piperinic Acid and its Isomers—The four isomers of piperinic acid have been formed by irradiation of piperinic acid (119) obtained from hydrolysis

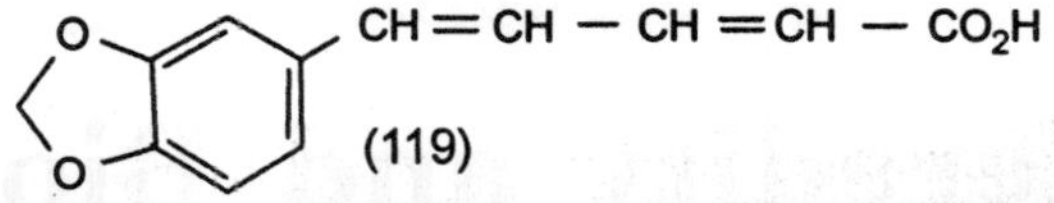

CH=CH — CH=CH — CO_2H

(119)

of pepper extract, the isomers were separated by counter-current distribution and characterized. The four isomers (119) were synthesized two years ago by Grewe *et al.*, but there seems to be some disagreement between this work and the results of the present study.

Constituents of *Cryptocarya bourdilloni*—A novel 5′,6′-dihydrochalcone (cryptocaryone) has been isolated from *C. bourdilloni* Gamb, and has been given structure. The compound co-occurs with the lactone cryptocaryalactone, it is believed to be the first of its kind to be found in Nature.

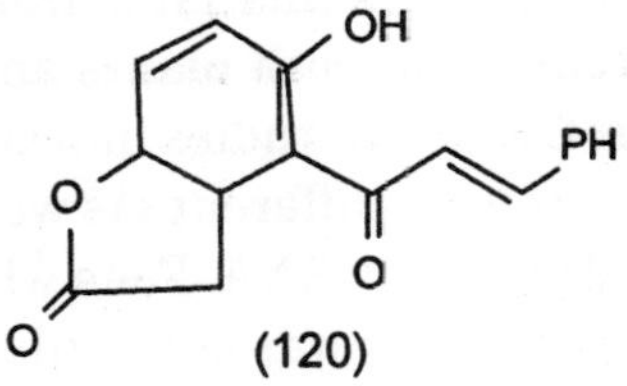

(120)

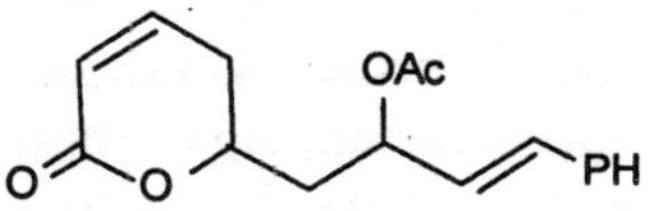

(121)

Constituents of *Thalictrum aquilegifolium*—The major cyanogenic glycoside of *T.jjquilegifolium* has been isolated and its structure has been determined as (122).

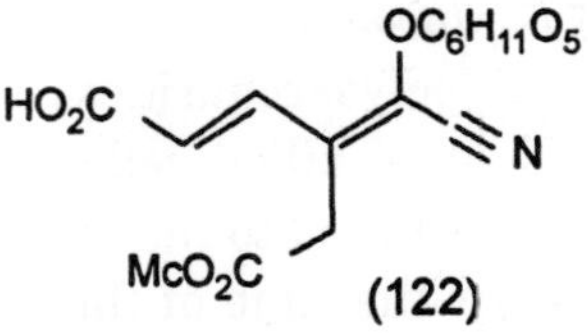

(122)

Sex Pheromone of *Laspeyresia pomonella* L—Some controversy exists over the structure of the sex pheromone produced by the virgin female codling moth *Laspeyresia pomonella* L. The E-8,*E*-10-dodecadien-1-ol structure (123) was proposed for the pheromone in 1971, largely on the basis of electroantennograms of model compounds, and Descoins and Henrick have recently reported a synthesis of (123) which has been found to be very attractive to male moths in the field. Independent spectroscopic and chemical degradation data, obtained by McDonough *et al.*, have, however, led to formulation (124) the attractant, and this structure was also confirmed by independent synthesis.

(123)

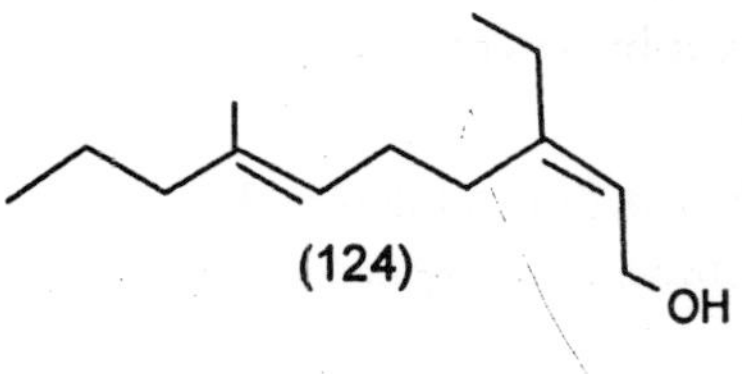

(124)

CHAPTER 3 Heredity and Biochemistry

HISTORICAL BACKGROUND

The discovery of deoxyribonucleic acid dates to 1869, when Miescher isolated a new chemical substance from white blood cells that he obtained from pus and later from sperm cells. The material, which became known as nucleic acid, occurred in both plants and animals, thymus glands and yeast cells being among the best sources. Chemical studies indicated that the nucleic acids isolated from thymus glands and from yeast cells were different. As we now know, thymus nucleic acid was primarily DNA and yeast nucleic acid primarily RNA. For a while it was suspected that animals contained only DNA and plants only RNA, and it was not until the early 1940s that it was established that both substances were present in all organisms.

DNA as the Genetic Material

In 1928, Griffith, using cells of *Diplococcus pneumoniae,* showed that genetic information that controls properties of the capsular polysaccharides could be transferred from one strain of bacteria to another. A material present in killed cells and in cell-free extracts permanently altered the capsular properties of cells exposed to the material. This *transformation* of bacteria remained a mystery for many years. At the time of the experiments there was no hint of the genetic role of nucleic acids, which were generally regarded as strange materials. Furthermore, the covalent structure of nucleic acids was uncertain.

A popular idea was that a tetranucleotide served as a repeating unit for some kind of regular polymer. Genes were most often thought to be protein in nature. However, in 1944, Avery and associates showed that purified DNA extracted from pneumococci could carry out transformation. The transforming principal appeared to contain little protein. It was not inactivated by proteolytic enzymes but was inactivated by deoxyribonuclease. Avery was 67 at the time of this discovery, refuting the popular contention that all important scientific discoveries are made by young people. Other experiments also pointed to the conclusion that DNA was the genetic material.

DNA was found localized in the nuclei of eukaryotic cells. The absolute amount per cell was constant for a given species. Studies of bacteriophage replication pointed strongly to DNA as the genetic material. In 1952 Hershey and Chase showed that when a phage particle infects a cell the viral DNA enters the bacterium, but the protein "coat" remains outside. This was demonstrated by preparing two types of labeled bacteriophage T2. One contained ^{3}-P, which had been incorporated into the DNA, and the other ^{35}S incorporated into the proteins. Cells of *E. coli* were infected with the labeled phage preparations and were then agitated violently in a blender to shear the phage particles off of the bacteria. Over 80% of the ^{35}S was removed from the bacteria by this treatment, but most of the ^{32}P entered the bacteria and could be recovered in the next generation of progeny bacteriophage.

The Double Helix

Development of newer methods of investigation of the chemical composition of nucleic acids led Char-gaff to an important discovery. Although the base composition of DNA was extremely variable from organism to organism, the molar ratio of adenine to thymine was nearly always 1:1, as was that of cytosine to guanine.[15] This observation provided the basis for the concept of base pairing in the structure of DNA. The final important information was supplied by X-ray diffraction studies of stretched fibers of DNA, which showed that molecules of DNA were almost certainly helical structures containing more than one chain.

The crucial experiments were done by Rosalind Franklin and M.H.F. Wilkins, whose data were used by Watson and Crick in 1952 in constructing their model of the double helical structure. Once established, the structure of DNA itself suggested both the nature of the genetic code and a replication mechanism. *The genetic code had to lie in the sequence of nucleotides. Base pairing provided a mechanism by which the two mutually complementary strands could be separated. Biosynthesis of a new complementary strand alongside each one would result in a precise replication of each gene and of the entire genome.* In a similar fashion, RNA could be synthesized alongside a DNA "template" and could then be used to direct protein synthesis. The presence of RNA in the cytoplasm had been linked to protein synthesis by experiments done in the early 1940s.

After the discovery of the double helix, the concept followed quickly that DNA was the master "blueprint" from which secondary blueprints or *transcripts* of RNA could be copied. The RNA copies, later identified as *messenger RNA* (mRNA), provided the genetic information for specifying protein sequence. The flow of information from DNA to RNA to proteins could be symbolized.

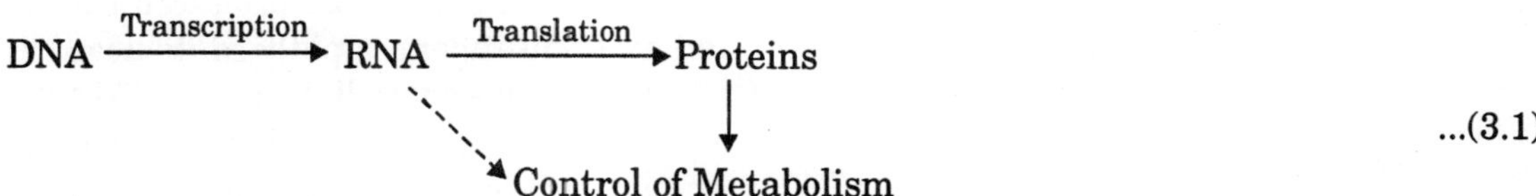

Proteins, in one way or another, control nearly all of metabolism. This includes the reactions that form the nucleotide precursors of the nucleic acids and that lead to polymerization of the amino acids and nucleotides, reactions catalyzed by *protein enzymes* and *ribozymes*. Thus, the flow of information from DNA to proteins is only part of a larger loop of metabolic processes. Genetic information flows from DNA out into the cell, and copies of the master blueprint are passed nearly unchanged from generation to generation. The simple concepts implied by Eq. 3.1

quickly caught the imagination of the entire community of scientists and led to a rapid blossoming of the field of biochemical genetics.

Ribonucleic Acids and Proteins

By 1942 it was clear from ultraviolet cytophotometry developed by Caspersson and from cytochemical work of Brachet that RNA had something to do with protein synthesis. Use of radioautography with ^{3}H-containing uridine showed that RNA was synthesized in the nucleus of eukaryotic cells and was transported out into the cytoplasm. Ribosomes were discovered by electron microscopists examining the structure of the endoplasmic reticulum using ultrathin sectioning techniques. Their presence in cells was established by 1956, and the name ribosome was proposed in 1957. At first it was difficult to study protein synthesis *in vitro* using isolated ribosomes.

No *net* synthesis could be detected until Hoagland and associates measured the rate of incorporation of ^{14}C-labeled amino acids into protein. This sensitive method permitted measurement of very small amounts of protein synthesis in cell-free preparations from rat liver and paved the way toward studies. Immediately after the Watson-Crick proposals were made in 1953, it was generally thought that ribosomal RNA (rRNA), which constitutes up to 90% of the total RNA of some cells, carried the genetic message from the nucleus to the cytoplasm. By 1960 it seemed unlikely. For one thing the size and composition of rRNA was similar for different bacteria, despite differences in base composition of the DNA. It had been concluded that a relatively unstable, short-lived form of RNA must carry the message. Ribosomal RNA, however, was quite stable.

Messenger RNA

In 1956, Volkin and Astrachan detected a rapidly labeled and labile RNA in phage-infected bacterial cells. Studies of enzyme induction also suggested the existence of mRNA. Many bacteria, including *E. coli,* when grown on glucose as the sole source of energy and then suddenly switched to lactose, are unable to utilize the new sugar immediately. However, transfer to lactose induces, within a period of two minutes, the synthesis of new proteins needed for the metabolism of this sugar. Among the new proteins is a *permease* for lactose and a *β-galactosidase* that cleaves the disaccharide to glucose and galactose. When the lactose is exhausted, the level of the induced enzymes drops almost as quickly. These results suggested that the RNA that carries the genetic message for synthesis of the new enzymes must be unstable. It must be produced rapidly in response to the presence of the inducing sugar and must disappear rapidly in its absence. In 1961, Jacob and Monod postulated messenger RNA (mRNA) as a short-lived polynucleotide. An abundance of additional evidence supported the proposal. For example, RNA molecules produced after infection of *E. coli by* bacteriophage T4 underwent hybridization with denatured DNA of the bacteriophage. Furthermore, this virus-specific mRNA became associated with preexisting bacterial ribosomes and provided the template for synthesis of phage proteins. The experiment provided direct evidence for transcription of mRNA from genes of the viral DNA.

Transfer RNA

Crick suggested, in 1957, that special "adapter" molecules might be needed to align amino acids with their codons in the RNA transcript. He thought that the adapters might be polynucleotides. At the same time chemical studies of the RNA of cells revealed that a low-molecular-weight RNA made up 15% of the total RNA of *E. coli*. This RNA was recognized in the same year as constituting the needed adapters, when Hoagland demonstrated enzymatic "activation" of amino acids and their subsequent incorporation into protein. The name transfer RNA (tRNA) was proposed.

Deciphering the Genetic Code

The genetic code consists of triplets of DNA base pairs (codons), each corresponding to a single amino acid. The triplets are consecutive, do not overlap, and are not separated by any "punctuation." Although this represented one of the simplest possibilities for a code, it required much effort more than a period of more than 10 years to prove it. Even after the triplet nature of the genetic code became evident, many questions remained. Were all of the 64 possible codons used by the living cell? If so, were they all used to code for amino acids or were some set aside for other purposes? How many codons were used for a single amino acid? Was the code universal or did different organisms use different codes? How could one decipher the code?

Despite the complexity of these questions, they seem to have been almost completely answered. In an important experiment Nirenberg and Matthaei, in 1961, isolated ribosomes from *E. coli* and mixed them with crude extracts of soluble materials, also from *E. coli* cells. The extracts included tRNA molecules and aminocyl-tRNA synthases. The 20 amino acids, ATP, and an ATP-generating system (PEP + pyruvate kinase) were added. Nirenberg showed that under such conditions protein was synthesized ribosomes in response to the presence of added RNA.

For example, RNA from tobacco mosaic virus was very effective in stimulating protein synthesis. The crucial experiment, which was done originally simply as a "control," was one in which a synthetic polynucleotide consisting solely of uridylic acid units was substituted for mRNA. In effect, this was a synthetic mRNA containing only the codon UUU repeated over and over. The ribosomes read this code and synthesized a peptide containing only phenylalanine. Thus, poly(U) gave polyphenylalanine, and *UUU was identified as a codon specifying phenylalanine.* The first nucleotide triplet had been identified! In the same manner CCC was identified as a proline codon and AAA as a lysine codon. Study of mixed copolymers containing two different nucleotides in a random sequence suggested other codon assignments. A few years later, after Khorana had supplied the methods for synthesis of oligonucleotides and of regular alternating polymers of known sequence, the remaining codons were identified. Another important technique was based on the observation that synthetic trinucleotides induced the binding to ribosomes of tRNA molecules that were "charged" with their specific amino acids.

For example, the trinucleotides UpUpU and ApApA stimulated the binding to ribosomes of ^{14}C-labeled phenylalanyl-tRNA and lysyl-tRNA, respectively. The corresponding dinucleotides had no effect, an observation that not only verified the two codons but also provided direct evidence for the triplet nature of the genetic code. Another powerful approach was the use of artificial RNA polymers, synthesized by combined chemical and enzymatic approaches. For example, the polynucleotide CUCUCUCUCU led to the synthesis by ribosomes of a regular alternating polypeptide of leucine and serine. The codon assignments, as we now know them, for each of the 20 amino acids remain as usual.

In addition to those codons assigned to specific amino acids, three are designated as *chain termination codons:* UAA, UAG, and UGA. These are frequently referred to as "nonsense" codons.

The termination codons UAA and UAG are also known as *ochre* and *amber*, respectively, although these names have no scientific significance. The codons AUG (methionine) and much less often GUG (valine) serve as the *initiation codons* in protein synthesis. Consequently, the *N*-terminal amino acid in most newly synthesized eukaryotic proteins is methionine, and in bacterial proteins it is *N*-formylmethionine. As explained. *N*-formylmethionyl-tRNA is specifically bound to initiation sites containing the AUG codon in bacterial mRNA-ribosome complexes. A number of studies suggested that the genetic code as worked out for *E. coli* might be universal. For example, in the laboratories of Wittman and of Fraenkel-Conrat, RNA extracted from tobacco mosaic virus was treated with nitrous acid, a procedure known to deaminate many cytosine residues to uracil. Such treatment could change the codon UCU (serine) to UUU (phenylalanine) and the codon CCC (proline) to CUC (leucine).

When the nitrous acid-treated RNA was used to infect tobacco plants and virus particles were prepared in quantity from the resultant mutant strains, it was found that the amino acid sequence of the virus coat protein had been altered, and that many of the alterations were exactly those that would be predicted. Likewise, the amino acid substitutions in known defects of hemoglobin could be accounted for, in most cases, by single base alterations. Thus, hemoglobin S arose as a result of the following change in the sixth codon of the globin β chain gene: GAG (Glu)→GTG (Val).

Another argument favoring a universal code was based on the observation that mRNA coding for a globin chain could be translated by ribosomes and tRNA molecules from *E. coli*. The resulting protein was authentic mammalian globin. As often happens, a well-established conclusion may have to be modified. There are exceptions to the universal genetic code in mitochondrial DNA and in some protozoa.

GENETIC METHODS

Our present knowledge of molecular biology has depended greatly on the methods of genetics. The following introduction begins with a consideration of mutations.

Mutations

Changes in the structure of DNA occur only rarely. The average gene may be duplicated 10^6 times before some mistake results in a single detectable mutation. Nevertheless, by using bacteria or bacterial viruses, it is possible to screen enormous numbers of individuals for the occurrence of mutations. If one million virus particles are spread on an agar plate under conditions where mutation in a certain gene can be recognized, on the average one mutant is found. The most common mutations are *base-pair switches* or *point mutations* that result from incorporation of the wrong base during replication or repair. In these mutations one base of a triplet codon is replaced by another to form a different codon, causing the substitution of one amino acid by another in the corresponding protein as was seen for hemoglobin S. Changes involving replacement of one pyrimidine by another (C→T or T→C) or of one purine by another are sometimes called *transition mutations,* whereas if a pyrimidine is replaced by a purine, or vice versa, the mutation is known as a *transversion.* An example is the previously mentioned A → T in hemoglobin S. Transition mutations are by far the most common, one possible cause being pairing with a minor tautomer of one of the bases.

For example, A could pair with a minor tautomer of C, causing a mutation from T to C. Note that substitution of an incorrect base in one strand will lead, in the next round of replication, to correct pairing again but with an AT pair replaced by GC, or vice versa, in one of the daughter DNA duplex strands. A base substitution does not always cause an amino acid replacement because of the "degeneracy" of the code, *i.e.,* the fact that more than one codon specifies a given amino acid. From the observed rate of appearance of point mutations (one mutation per 10^6 gene duplications), we can estimate that one mutation occurs per 10^9 replications at a single nucleotide site. Point mutants tend to "back mutate," often at almost the same rate as is observed for the forward mutation. That is, one in 10^9 times a mutation of the same nucleotide will take place to return the code to its original form. The phenomenon is easy to understand. For example, if T should be replaced by C because the latter formed a minor tautomer and paired with A, the mutation would appear in progeny duplexes as a GC pair. When this pair was replicated, there would be a finite probability that the C of the parental DNA strand would again assume the minor tautomeric structure and pair with A instead of G, leading to a back mutation. Although the rates of spontaneous mutation are low, they can be greatly increased by mutagenic chemicals or by irradiation.

It is perfectly practical to measure the rates of both forward and back mutation. When this was done, it was found that certain chemicals, *e.g.,* acridine dyes, induce mutations that undergo reverse mutation at a very much lower frequency than normal. It was eventually shown that these mutations resulted either from *deletions* of one or more nucleotides from the chain or from *insertions* of extra nucleotides. Deletion and insertion mutations often result from errors during genetic recombination and repair at times when the DNA chain is broken. Mutations involving deletion or insertion of one or a few nucleotides are called *frame-shift mutations.* Messenger RNA is read by the protein synthesizing machinery from some starting point. The codons are read three bases at a time, and the proper amino acid corresponding to each codon is inserted. *When a deletion or insertion in the mRNA is met, all subsequent codons may be misread because the reading frame has shifted forward or backward by one or two nucleotides.* The protein synthesized bears little resemblance to that formed by the nonmutant organism and is usually completely nonfunctional.

Mapping the Chromosome of a Bacteriophage

Intensive work on the "T-even" phage T2, T4, and T6 begun in 1938 by Delbruck and associates. The genetic information for these viruses is carried in a single linear DNA molecule, which in the case of T4 contains ~ 1.7×10^5 base pairs (170 kb), enough for about 200 genes. Before the sequences of the viral DNA were known, the positions of more than 60 of these genes were mapped in the following way. When a bacteriophage infects a cell of *E. coil,* it injects its DNA through the cell wall and into the cytoplasm. About 20 minutes later the cell bursts, and ~ 100 fully formed replicas of the original virus particle are released. This rapid rate of production of progeny is so fast that it is possible to carry out in a test tube in 20 min a genetic experiment that would require the entire population of the earth if done with humans. The approach is explained nicely by Seymour Benzer, the man who first mapped the fine structure of a gene.

Bacteriophage particles, like bacteria, can be "plated out" on agar plates, which must contain a uniform suspension of bacteria susceptible to the virus. Wherever a virus particle lies, a bacterium is infected. Soon the infection spreads to neighboring bacteria with production of a transparent "plaque". The number of active virus particles present in a suspension can be determined easily by plating and counting of the plaques. Mutant bacteriophages can be identified in various ways. Some biochemical traits affect the appearance of the plaque. Other easily detected traits include alteration in the specificity toward strains of the host bacterium.

A key discovery that made genetic mapping possible for bacteriophage was that *genetic recombination between two phage particles can take place within a host bacterium.* When large numbers of bacteriophages of two different mutant strains were grown and were mixed together in excess with many bacteria, a few of the progeny phage were found to contain both mutant traits in the same virus and an equal number were "wild type." Although recombinations between mutations that are located close together in the DNA are rare, their frequency still greatly exceeds that of new mutations. While this type of experiment gave no hint about the nature of the events involved, it showed that recombination had occurred. Study of recombination frequencies between different strains of phage soon revealed that some sites of mutation are *closely linked.* Recombination between these sites occurs only rarely.

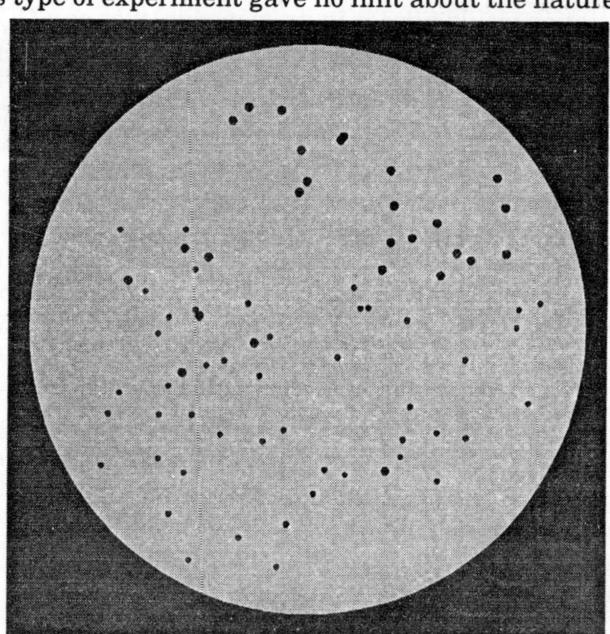

Fig. 3.1. Plaques formed by bacteriophage ϕ11 growing on *Staphylococcus aureus.* Each transparent (dark) plaque is the result of lysis of bacteria by the progeny of a single bacteriophage particle.

Other sites are weakly linked, and recombination occurs often. This behavior was reminiscent of that established many years earlier for genes of the fruit fly *Drosophila,* maize, and other higher organisms. Recombination by "crossing over" in the chromosomes of *Drosophila* was established by Morgan and associates in 1911. The basic idea behind chromosome mapping in any organism is the assumption that *recombination frequencies between two mutations are directly proportional to the distance between them on the genetic map.* For the T4 phage a recombination frequency of 1% is taken as one unit. The total T4 map is 700 units long. The fact that this is greater than 100% means that if genes are located at opposite ends of the chromosome multiple recombination events can occur between them. However, a maximum of 50% crossing over is observed for distant gene pairs, and the approximate linearity of map distance and recombination frequency holds only for distances of 10 units or less.

How can recombinant bacteriophage be identified rapidly? Benzer used two strains of *E. coli,* the B strain and the K strain, as hosts. Mutants in gene *rII* form characteristic plaques on strain B but do not grow on strain K. To determine the recombination frequency between two different *rII* mutants, the viruses were added to liquid cultures of B cells (in which they replicate), and recombination was allowed to occur. Recombination permitted the emergence not only of a phage containing *both* mutations but also of a wild-type phage in which both mutations had been eliminated by the recombination process. Since only recombinants of the latter type grow in strain K, it was possible to detect a single recombinant among one billion progeny. Since the total DNA length in phage T4 is 166 kb, there are 237 base pairs for each of the 700 units of map length. Thus, a recombination frequency at 0.01% between two mutations meant that the two mutations were no more than three base pairs apart in the DNA.

Benzer concluded that he had observed the expected recombination frequencies for mutations even of immediately adjacent bases in the DNA. To make fine genetic mapping practical, a series of bacteriophage containing deletion mutations involving large segments of the *rII* gene were isolated. Using these, it was possible to establish in which segments of the gene a particular mutation lay. Then, recombination experiments with previously identified mutations in that same general region allowed the mutations to be pinpointed. In this way Benzer identified over 300 sites of mutations within the *rII* region. He concluded that the minimum distance between two mutable sites was compatible with the Watson-Crick structure of the gene.

The Cistron

How can one tell whether two mutations are in the same gene or in nearby or adjacent genes? The answer can be supplied by a test of *complementation*. If two mutant bacteriophage are altered in different genes, they can often reproduce within a host if the bacterium is infected with both of the mutant phage. Since each one has a good gene for one of the two proteins involved, recombinant phage, in which all of the gene functions are fulfilled, will be formed. On the other hand, if both mutant phages are defective in the same gene (although at different locations), they usually cannot complement each other at high frequencies in a coinfection. The experiment is referred to as a *cis-trans comparison*. The coinfection with the two different mutants is the trans test.

A control, the *cis* test, uses a recombinant containing both of the mutations in the same DNA and coinfection with a standard phage. Normal replication is expected in this instance. When the complementation (or *cis-trans*) test was applied to various mutants in the *rII* region, it became clear that there are two genes, *rIIA* and *rIIB*. The name **cistron** was proposed by Benzer to represent that length of DNA identifiable in this fashion as a genetic unit. For most *purposes the terms gene and cistron are nearly synonymous.* When mapping of the *rII* region was done, there was no information about the functions of the proteins specified by these two cistrons. However, both the *rIIA* and *rIIB* proteins have since been shown to become incorporated in the membranes of phage-infected bacterial cells. There they affect the ease of lysis of the infected cells and, in that manner, cause *rII* plaques to be larger and to have sharper edges than standard plaques.

Nutritional Auxotrophs

The beginning of biochemical genetics is often attributed to Beadle and Tatum, who, in 1940, discovered mutants of *Neurospora* with specific growth requirements. They X-rayed one parent strain to form mutants, then tested individual spores for their possible need of a specific nutrient for growth. The 299th spore tested required pyridoxine! Many other mutants requiring vitamins, amino acids, and nucleic acid bases were then discovered. A few years later similar *nutritional auxotrophs,* as the mutants are called, were discovered for *E. coli.* Ordinary wild-type cells of *E. coli* can grow on a minimal medium containing a carbon compound as a source for energy together with inorganic nutrients.

Irradiation with ultraviolet light or treatment with mutagenic chemicals produces many mutant cells that fail to grow on such a minimal medium. However, addition of one or more specific compounds, such as an amino acid or vitamin, usually permits growth. Selection of such nutritional auxotrophs can be accomplished by plating out large numbers of the irradiated or chemically treated cells on a solid, rich nutrient medium. Colonies *(clones)* are allowed to develop by multiplication of the individual bacteria. The auxotrophs are selected by *replica plating.* In this procedure a sterile velveteen pad is pressed against a nutrient agar plate containing small colonies of bacteria and is used to "print" replica plates containing a minimal medium.

The colonies on the initial and replica plates are compared and the colonies of auxotrophs (which do not grow on the minimal medium) are selected. In a second stage, the auxotrophs may be replica plated to minimal medium supplemented with various nutrients (amino acids, purines, pyrimidines, vitamins, etc.). Selection is made easier by pretreatment of the irradiated cells suspended in minimal medium with penicillin. Penicillin kills the growing cells, but the auxotrophs, which do not grow on the minimal medium, survive. The penicillin is then destroyed by adding penicillinase (a β-lactamase) leaving a suspension much enriched in the percentage of auxotrophic mutants. A nutritional auxotroph of a bacterium often has a defect in a gene specifying a protein needed for the biosynthesis of the required nutrient.

Individual genes recognized in this way are named with a genetic symbol. For example, gene *trpA* specifies one of the two protein subunits of tryptophan synthase. Other kinds of mutations, *e.g.,* those affecting motility or other properties of the cells, can also be detected and are given appropriate symbols. A few of these genetic symbols are indicated on the *E. coli* chromosome map and many others are used throughout this book. On the basis of such nutritional experiments Beadle, by 1945, had proposed his famous one-gene-one-enzyme hypothesis.

Establishing the Correspondence of a Genetic Map and an Amino Acid Sequence

Although the studies of the *rII* region of the T4 chromosome established that genetic mapping could be carried to the level of individual nucleotides in the DNA, it was still necessary to prove a linear correspondence between the nucleotide sequence in the DNA and the amino acid sequence in proteins. This was accomplished by Yanofsky and associates through study of the enzyme tryptophan synthase of *E. coli.* Tryptophan synthase consists of two subxunits, α and β, the former containing only 268 amino acids and encoded by the *trpA* gene. To obtain a fine structure map of the A gene, a large series of tryptophan auxotrophs unable to grow in the absence of added tryptophan were isolated. Genetic crosses were carried out with the aid of a *transducing bacteriophage* Plke. Transducing bacteriophage, while multiplying in susceptible bacteria, sometimes incorporate a portion of the bacterial chromosome into their own DNA. Then, when the virus infects other bacteria, some of the genetic information can be transferred through recombination into the chromosome of bacteria that survive infections.

Use of a series of deletion mutants, as in the *rII* mapping, permitted division of the A gene into a series of segments, and observation of recombination frequencies permitted fine structure mapping. The second part of the proof of colinearity of DNA and protein sequences was the determination of the complete amino acid sequence of tryptophan synthase and peptide mapping of fragments of the mutant enzymes. From the peptide maps it was possible to identify altered peptides and to establish the exact nature of the amino acid substitutions present in a variety of different tryptophan auxotrophs. When this was done, it was found that those mutations that mapped very close together had amino acid substitutions at adjacent or nearly adjacent sites in the peptide chain. The same problem was approached by Sarabhai and associates through the nonsense mutations, which lead to premature chain termination during protein synthesis.

During late stages of the infection of *E. coli* by phage T4, most of the protein synthesis is that of a single protein of the virus head. Synthesis of protein by infected cells was allowed to proceed in the presence of specific ^{14}C-labeled amino acids. Then cell extracts were digested with trypsin or chymotrypsin, the head-protein peptides were separated by electrophoresis, and autoradiograms were prepared. A series of T4 nonsense mutants that mapped within the head-protein gene were shown to give rise to incomplete head-protein chains. The peptide fragments were of varying lengths. By examining the radioautograms prepared from the enzymatically fragmented peptides, it was possible to arrange the mutants in a sequence based on the length of peptide formed and to show that this was the same as that deduced by genetic mapping.

More recently the colinearity of codon and amino acid sequences has been verified repeatedly by comparison of experimentally determined nucleotide sequences in RNA and DNA molecules with the corresponding amino acid sequences for thousands of proteins. Before the triplet nature of codons had been established, Crick and associates used frame-shift mutations in a clever way to demonstrate that the genetic code did consist of triplets of nucleotides. Consider what will happen if two strains of bacteria, each containing a frame-shift mutation (*e.g.,* a -1 deletion), are mated. Genetic recombination can occur to yield mutants containing *both* the frame-shift mutations. It would be difficult to recognize such recombinants because, according to almost any theory of coding, they would still produce completely defective proteins. However, Crick *et al.* introduced a third frame-shift mutation of the same type into the same gene and observed that the recombinants containing all three deletions (or insertions) were able to synthesize at least partially active proteins. Thus, while introduction of one or two single nucleotide deletions completely inactivates a gene, deletion of three nucleotides close together within a gene shortens the total message by just three nucleotides. The gene will contain only a short region in which the codons are scrambled. The reading frame for the remainder of the protein will not be changed.

The protein specified will often be functional because it has a normal sequence except for a small region where some amino acid substitutions will be found and where one amino acid will be completely missing.

Conditionally Lethal Mutations

Studies of plaque morphology and of nutritional auxotrophs are directed narrowly at one gene or group of genes. It is desirable to have a general means of detecting mutations in the many other genes present within cells. However, most mutations are *lethal,* and this effect cannot be overcome by adding any nutrient. Lethal mutations are very common in higher organisms but since eukaryotic cells have pairs of homologous chromosomes, they can be carried in one chromosome and the individual survives. With bacteria and viruses there is only one chromosome, and lethal mutants cannot survive. Nutritional auxotrophs can be described as *conditionally lethal mutants;* they survive only if the medium is supplemented with the nutrient, whose synthesis depends upon the missing enzyme.

Other kinds of conditional lethal mutations permit study of almost every gene in an organism. For example, *temperature-sensitive* (*ts*) mutants grow perfectly well at a low temperature, *e.g.*, 25°C, but do not grow at a higher temperature, *e.g.*, 42°C. Many temperature-sensitive mutations involve an amino acid replacement that causes the affected protein to be less stable to heat than is the wild type protein. Others involve a loss in protein-synthesizing ability for reasons that may be obscure. Temperature-sensitive mutations occur spontaneously in nature, an example being the gene that controls hair pigment in Siamese cats. The gene (or gene product) is inactivated at body temperatures but is active in the cooler parts of the body, such as the paws, tail, and nose, with the result that the cat's hair is highly pigmented only in those regions. Screening for conditionally lethal temperature-sensitive mutants of bacteriophage T4 permitted isolation of hundreds of mutants involving sites at random over the entire viral chromosome.

Complementation studies permitted assignment of these to individual genes, which at first were identified only by number. Later specific functions were associated with the genes. For example, the product of gene 42 was identified as an enzyme required in the synthesis of hydroxymethyl-dCMP. Genes 20-24, among others, must code for head proteins because mutants produce normal tails but no heads. Gene 23 codes for the major head subunits, while gene 20 has something to do with "capping" the end of the head. These mutants produce cylindrical "polyheads" in place of the normal heads. Mutants of genes 25-29 have defective base plates and do not form tails, while mutants 34-38 lack tail fibers.

The specific ways in which some of these gene products are assembled to form the base plates and tails of the phage are indicated in the figure. The positions of the *rIIA* and *rIIB* genes are also shown in Fig. 3.2. A second type of conditionally lethal mutation leads to alteration of an amino acid codon to one of the three *chain termination codons* UAG, UAA, and UGA. These are often called *nonsense mutations* in contrast to *missense mutations* in which one amino acid is replaced by another. A chain termination mutant synthesizes only part of the product of the defective gene because of the presence of the termination codon.

A remarkable aspect of chain termination mutations is that they can be *suppressed* by other mutations in distant parts of the virus or bacterial chromosome. Many otherwise lethal mutations of bacteriophage T4 were discovered by their ability to grow in mutant strains of *E. coli,* which contained *suppressor genes,* and their inability to grow in the normal B strain. Three different suppressor genes *supD, supE,* and *supF* were found to suppress mutations that formed UAG. These are commonly known as *amber* suppressor genes. A second group of *ochre* suppressor genes, including *supB* and *supC,* suppressed mutations that formed codon UAA. Suppressors for mutations that form codon UGA have also been found.

Like the temperature-sensitive mutants, *amber* and *ochre* mutants can be obtained in almost any genes of a bacterial virus. Chain termination mutants of unessential genes in bacteria can be recognized by transferring the genes by conjugation or by viral transduction into a strain (sup^+) that contains a desired suppressor gene. Conditionally lethal mutants have been of great value in developing our understanding of the genetics of bacterial viruses. They have also provided a powerful technique for approaching complex problems of bacterial physiology. For example, we may ask how many genes are required for a bacterium to sense the presence of a food and to swim toward it.

Even though few clues as to the basic chemistry underlying these phenomena can be obtained in this way, the use of temperature-sensitive mutants and complementation tests permits us to establish the total number of genes involved in these complex processes and to map their positions on a bacterial or viral chromosome. This is often an important step toward a more complete understanding of a biological phenomenon.

The Nature of Suppressor Genes

How can one mutation be suppressed by a second mutation at a different point in the chromosome? Rarely, a mutation is suppressed by a second mutation within the *same* gene. Such *intragenic complementation* sometimes occurs when a mutation leads to an amino acid replacement that disrupts the structural stability or function of a protein. Sometimes a mutation at another site involving a residue, which interacts with the first amino acid replaced, will alter the interaction between the two residues in a way that restores function to the protein.

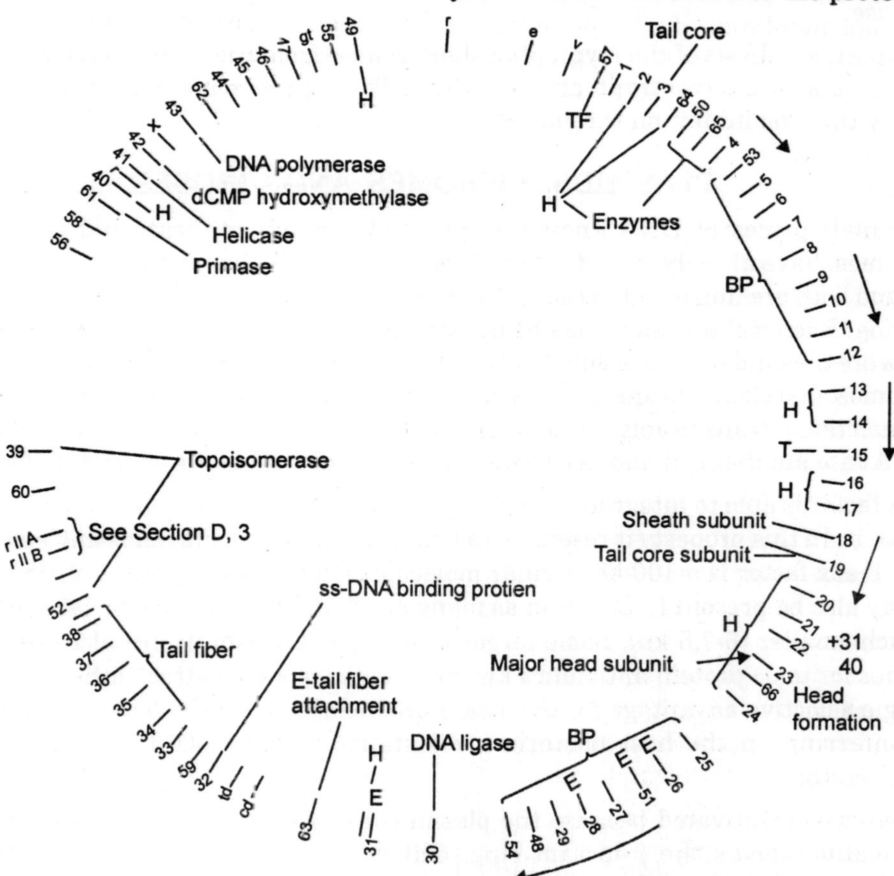

Fig. 3.2. Linkage map for bacteriophage T4. Abbreviations : H, head; T, tail; TF, tail fiber; BP, base plate; and E, proteins with enzymatic rather than structural functions. The bold arrows indicate some genes that are transcribed together (operons).

For example, if the first amino acid side chain is small and is replaced by mutation with a larger side chain, a second mutation leading to a decrease in the size of another side chain may permit the protein to fold and function properly. An example was found among mutants of tryptophan synthase. Mutants in which Gly 211 of the α chian was replaced by Glu or Tyr 175 of the same chain by Cys both produced inactive enzymes. However, the double mutant with both replacements synthesized active tryptophan synthase. It is known now that these residues are adjacent to one another and form part of the binding site for the substrate indole-3-glycerol phosphate. Only the double mutant permits the substituted side chains to pack properly.

In other cases, intragenic suppression involves changes in the subunit interactions in oligomeric proteins. These changes may affect the formation of correct quaternary structures of the proteins. As discussed in the preceding section, the best known suppressor genes are those that suppress chain termination mutations. These genes often encode mutant forms of tRNA molecules, which allow incorporation of an amino acid rather than chain termination to occur. Suppressor genes are not limited to bacteria. For example, the vermilion eye color mutation of *Drosophila* leads to a loss of brown eye pigments because of the inactivity of tryptophan 2,3-dioxygenase.

However, synthesis of the tryptophan dioxygenase from the *vermilion* mutant is inhibited by $tRNA_2^{Trp}$, one of the two tryptophanyl tRNAs. The suppressor mutation alters the tRNA in such a way that the inhibition is relieved.

PLASMIDS, EPISOMES AND VIRUSES

The small pieces of DNA known as plasmids, which replicate independently of the chromosomes, have already been discussed briefly Plasmids share a number of properties with viruses, and both are important to the techniques of contemporary molecular biology and genetic engineering. Bacterial plasmids may be present as one or several copies for each chromosome. Episomes are plasmids that are able to become integrated into the bacterial chromosome. Some extrachromosomal elements are episomes in one host and plasmids in another. Bacterial plasmids may be infectious (transferable) or noninfectious. In the former case, they arguable to transfer their DNA into another cell and are Known as sex factors (F agents or fertility factors).

A sex factor is able to integrate into *a* chromosome and later to come out and transfer other genes with it. In this property it resembles a transducing phage. Plasmids and episomes vary in size. The F sex factor is a 100-kb circular molecule of supercoiled DNA. Colicinogenic factors, which may also be present in *E. coli* in as many as 10-15 copies per bacterial chromosome, are often much smaller (6-7.5 kb). Some larger colicinogenic plasmids are also sex factors. They carry genes for toxic protein antibiotics known as *colicins* which attack other strains of *E. coli*, providing a selective advantage for the strain producing the colicin. They also carry a gene or genes conferring on the host bacterium resistance to antibiotics such as penicillin and chloramphenicol.

Penicillin is inactivated because the plasmids carry a gene encoding a penicillinase that hydrolytically cleaves the β-lactam ring. Chloramphenicol inactivated by the action of chloramphenicol *O*-acyltransferase.

Bacterial Sex Factors

Bacteria usually reproduce by a simple cell division. The DNA in the chromosome is doubled in quantity and the cell divides, each daughter cell receiving an identical chromosome. However, in 1946 Lederberg and Tatum showed that sexual reproduction is also possible. They studied nutritional auxotrophs of *E. coli* strain K-12, which lacked the ability to synthesize amino acids or vitamins. When cells of two different mutants were mixed together and allowed to grow for a few generations, a few individual bacteria regained the ability to grow on a minimal medium. Since each of the two strains had one defective gene, the creation of an individual with neither of the two defects required combining of genetic traits from both strains.

The existence of bacterial conjugation was recognized. Later it was established that true *genetic recombination* had occurred, *i.e.*, genes from the two mating cells had been integrated into a single molecule of bacterial DNA. This transfer of DNA between bacterial cells requires the presence of a plasmid sex factor (F agent), whose presence confers a male character to the individual cell. The F agent is large enough to contain about 90 genes and has a length of ~ 30 μm, ~ 2.5% that of the *E. coli* chromosome.

Among other things, the *E. coli* F agent contains the genes needed to direct the synthesis of the *F pili* (sex pili). These tiny appendages, 8.5 nm in diameter, grow out quickly during a period of 4-5 min to a length of about 1.1 μm. The end of an F pilus becomes attached to a female cell (a bacterium lacking the F agent) and may induce the transfer of DNA into the female cell. The mechanism of transfer has been uncertain. It may involve formation of a cytoplasmic bridge between cells in close contact. The pilus may be retracted into the membrane of the male cell, pulling the two cells close together. The DNA probably flows through the pilus into the female. On rare occasions an F agent becomes integrated into the chromosome of a bacterium. Both the F agent and the chromosome have been shown by electron microscopy to be circular.

The integration process requires the enzymatic cleavage of the DNA of both the chromosome and the F agent and the rejoining of the ends in such a way that a continuous circle is formed. Different F agents can be incorporated into the chromosome at different points around the circle. A strain of bacteria containing an integrated F agent is known as an *Hfr* (high frequency of recombination) strain. When an *Hfr* strain conjugates with an F^- (female), replication of the entire male chromosome commences at some point near the end of the integrated F agent, and genes of the bacterial chromosome followed by those of the F factor are transferred into the female.

Only a single strand of DNA (customarily referred to as the *plus strand)* is transferred from the donor cell and into the recipient cell. There the complementary *minus strand* is synthesized to form a complete double-stranded DNA molecule bearing the genes from the *Hfr* cell. Only rarely does a copy of the entire chromosome of the donor cell enter the female cell. More often the DNA strand, or perhaps the pilus itself, breaks and only part of the chromosome is transferred. Partial chromosome transfer from a male cell transforms the F^- cell into a partial diploid *(merozygote)* containing double the usual number of some of the genes.

Within this partial diploid genetic recombination between the two chromosomes takes place. The end result of the recombination process is that the daughter cells formed by subsequent division contain only single chromosomes with the usual number of genes. However, some genes come from each of the two parental strains. Thus, an F^- mutant unable to grow on a medium deficient in a certain nutrient may receive a gene from the male and now be able to grow on a minimal medium. Even though the number of such recombinants is small, they are easily selected from the very large number of mutant bacteria that are mixed together initially. One result of DNA transfer from Hfr into F^- bacteria is sometimes the introduction of a complete copy of the F agent into the female bacterium.

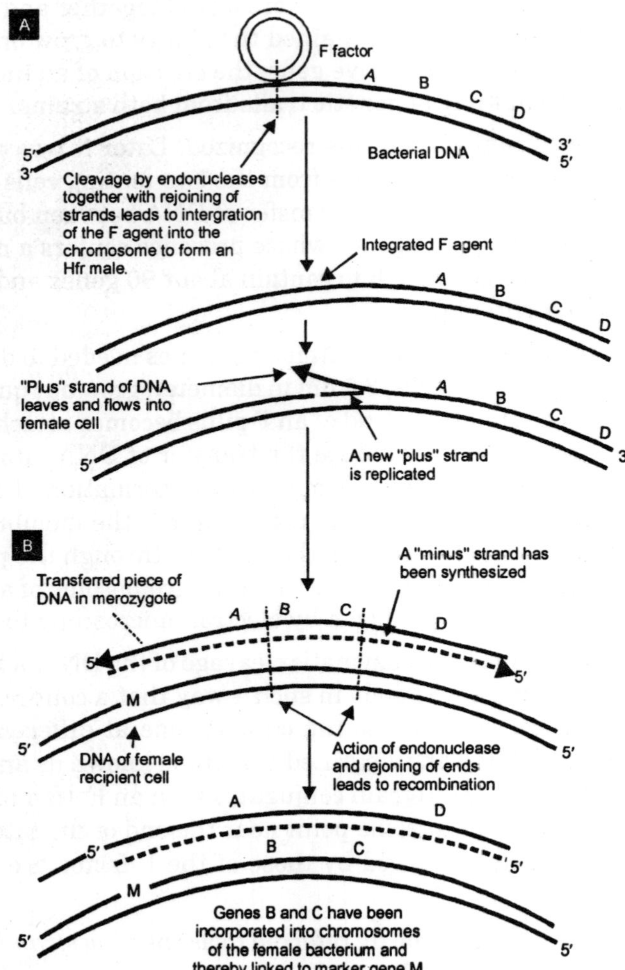

Fig. 3.3. Integration of an F agent into a bacterial chromosome and transfer of some bacterial genes into another cell. (A) Incorporation of the F agent into *E. coli* genome and transfer of the "plus" strand of DNA out to a female recipient cell. (B) Genetic recombination between a piece of transferred DNA and the genome of the recipient cell.

Since this converts the recipient into a male, Brinton referred to "bacterial sex as a virus disease." In fact, infectious plasmids and viruses display many similarities. For example, filamentous bacteriophages adsorb to the F pili of male bacteria and the DNA, flowing in a direction opposite to that in bacterial conjugation, enters the cell.

The bacteriophage carry genes for the protein subunits of their coats, while F factors carry genes for synthesis of pilins. Pilins accumulate within the cell membrane and are extruded to generate F pili, just as viral sub-units are extruded to form the virus coats. There is also a close similarity between episomes that can be integrated into bacterial chromosomes and the *temperate bacteriophages* considered in the next section.

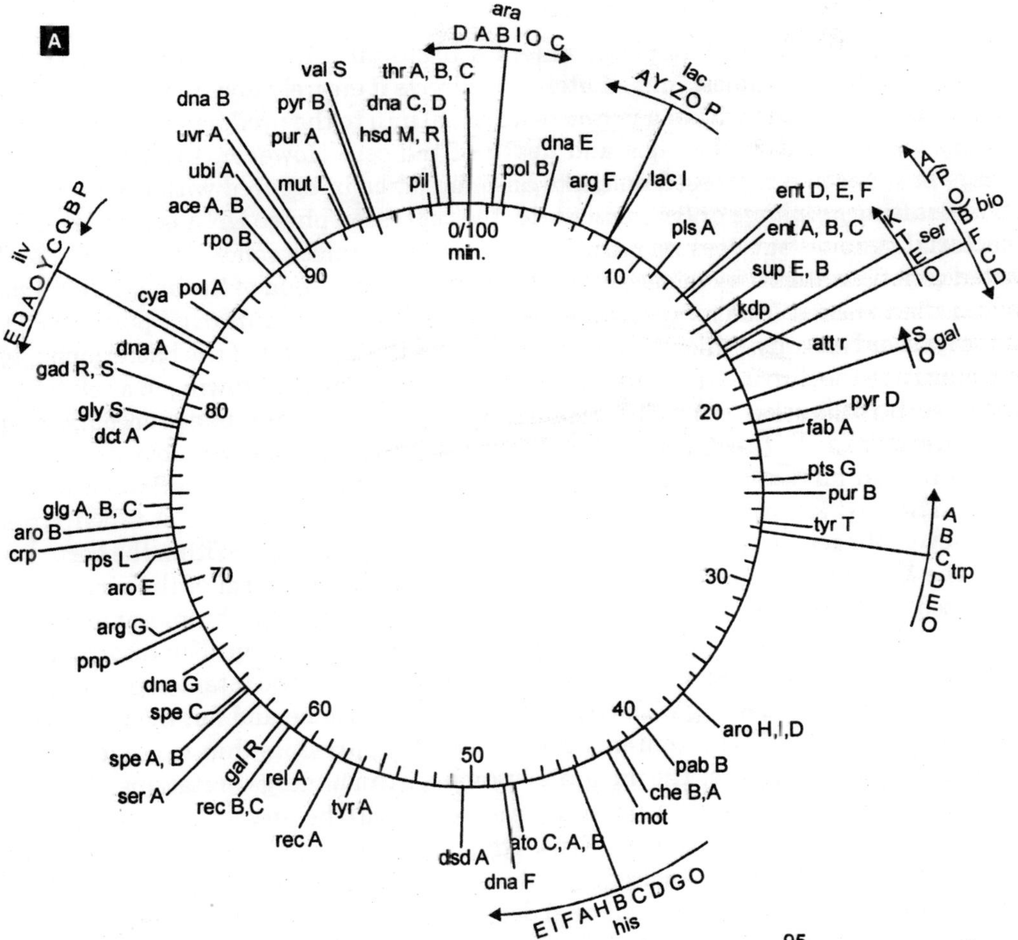

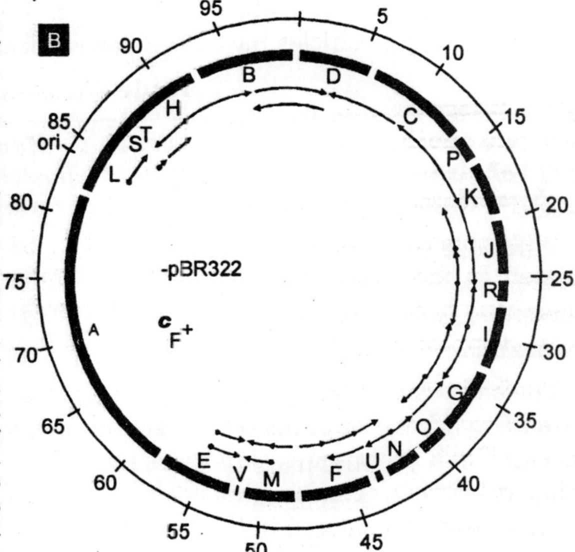

Fig. 3.4. (A) Genetic map of *E. coli*, strain K-12 published in 1976. The scale in minutes is based on the results of interrupted conjugation experiments. The *thr* locus is arbitrarily taken as zero. The genetic symbols are defined. Only a single strand of the DNA molecule is normally transcribed into RNA over any given region. Because of the antiparallel nature of the DNA helix, transcription from one strand proceeds clockwise and transcription from the other counterclockwise (the *r* strand is conventionally taken as the one that is transcribed in the clockwise direction when the chromosome is oriented as in the drawing). The directions of transcription of certain operons are indicated by the arrows. The 1983 version of this map contained 1027 loci and we now know the location of all of the more than 4300 genes. (B) Physical map of the *E. coli* chromosome obtained by study of large 20-100 kb fragments obtained by digestion with restriction endonuclease *Not*I and separated by pulsed field gel electrophoresis. The fragments A-V were ordered using genetic information and overlapping fragments from partial *Not*I digests (arrows). On the outside the genetic map has been superimposed after distortion of the scale in minutes to make the two maps coincident. The 100-kb F⁺ plasmid and the 4.5-kb plasmid pBR322 are also drawn to show relative sizes.

Temperate Bacteriophage; Phage Lambda

When DNA from a typical bacteriophage enters a bacterial cell, it seizes control of the metabolic machinery of the cell almost immediately and directs it entirely toward the production of new virus particles. This leads within a period of about 20 min to the production of one or two hundred progeny viruses and to the lysis and death of the cell. However, the DNA from a temperate phage may become repressed and, like an F factor, be integrated with the bacterial genome. In the resulting *prophage* or *lysogenic* state, the repressed phage DNA is replicated as part of the bacterial genome but does no harm to the host cells unless some factor "activates" the incorporated genetic material by release of the repression. Replication of the phage and lysis of the bacterium then ensues. Temperate phage may also exist as plasmids (*e.g.*, plasmid P1). The decision for lysis or lysogeny, which is very important for the survival of the bacteriophage, is governed primarily by the nutritional status of the host. For a bacterium growing in a relatively rich environment such as the colon, lysis will increase the chances of daughter phage encountering host bacteria. However, in soils, *E. coli* grows very slowly, and a bacteriophage capable of entering a lysogenic state has an increased chance of survival until the host bacterium finds a richer growth medium. The best known temperate phage is *phage lambda* of *E. coli.* A tailed virus resembling the T-even phages, phage λ has a smaller (~48.5 kb) DNA genome. Within the bacterial cell the ends of the λ, DNA may be joined to form a circular replicative form of the virus. In ~ 30% of the infected cells the λ DNA becomes integrated into the *E. coli* chromosome at the special site, *att* λ, which is located at 17 min on the *E. coli* chromosome map. The incorporated phage DNA now occupies a linear segment amounting to about 1.2% of the total length of the *E. coli* chromosome. It is replicated along with the rest of the chromosome and for, the most part, goes unnoticed. The host, *E. coli* K12, contains useful *amber* suppressors that make it easy to detect mutations in the bacteriophage. The integrated prophage can undergo mutations of almost any type, including large deletion mutations, and can still be investigated through complementation studies with other strains of virus. Thus, a family of modified *defective λ phage* was developed. When the *K* prophage is excised from the bacterial chromosome, adjacent bacterial genes are occasionally carried with it. This allowed development of *λ transducing phage*, which can carry genes and transfer them into bacteria lacking these genes. More recently an important series of cloning vehicles have been derived from phage λ.

Table 3.1. Some Genes of *E. coli*.

Gene Symbol	*Mnemonic*	*Map position (min)*[a]	*Phenotypic trait affected*
*ace*A	Acetate	89	Isocitrate lyase
*ace*B	Acetate	89	Malate synthetase A
*ara*A	Arabinose	1	L-Arabinose isomerase
*ara*B	Arabinose	1	L-Ribulokinase
*ara*C	Arabinose	1	Regulatory gene
*ara*D	Arabionose	1	L-Ribulose-5-phosphate-4-epimerase
aral	Arabinose	1	Initiator locus
*ara*O	Arabinose	1	Operator locus
argF	Arginine	6	Ornithine carbamoyltransferase

*arg*G	Arginine	68	Argininosuccinic acid synthetase
*aro*B	Aromatic	73	Dehydroquinate synthetase
*aro*D	Aromatic	37	Dehydroquinate dehydratase
aroE	Aromatic	71	Dehydroshikimate reductase
*aro*H	Aromatic	37	DAHP synthetase (tryptophan-repressible isoenzyme)
*aro*J	Aromatic	37	Probable operator locus for *aro*H
*ato*A	Acetoacetate	48	Coenzyme A transferase
*ato*B	Acetoacetate	48	Thiolase II
*ato*C	Acetoacetate	48	Regulatory gene
*att*λ	Attachment	17	Integration site for prophage λ
*bio*A	Biotin	17	Group II; 7-oxo-8-aminopelargonic acid (7 KAP) → 7,8-diaminopelargonic acid (DAPA)
*bio*B	Biotin	17	Conversion of dethiobiotin to biotin
*bio*C	Biotin	17	Block prior to pimeloyl-CoA
*bio*D	Biotin	17	Dethiobiotin synthetase
*bio*F	Biotin	17	Pimeloyl-CoA → 7 KAP
*bio*O	Biotin	17	Operator for genes *bio*B through *bio*D
*bio*P	Biotin	17	Promoter site for genes *bio*B through *bio*D
*che*A	Chemotaxis	42	Chemotactic motility
*che*B	Chemotaxis	42	Chemotactic motility
crp		73	Cyclic adenosine monophosphate receptor protein
cya		83	Adenylate cyclase
*dct*A		79	Uptake of C_4-dicarboxylic acids
*dna*A	DNA	82	DNA synthesis; initiation defective
*dna*B	DNA	91	DNA synthesis
*dna*C	DNA	99	*dna*D; DNA synthesis; initiation defective
*dna*E	DNA	4	*pol*C, DNA polymerase III and mutator activity
*dna*F	DNA	48	*urd*A; ribonucleoside diphosphate reductase
*dna*G	DNA	66	DNA synthesis
*dsd*A	D-Serine	50	D-Serine deaminase
*ent*A	Enterochelin	13	2,3-Dihydro-2,3-dihydroxybenzoate dehydrogenase
*ent*B	Enterochelin	13	2,3-Dihydro-2,3-dihydroxybenzoate synthetase
*ent*C	Enterochelin	13	Isochorismate synthetase
*ent*D,E,F	Enterochelin	13	Unknown steps in conversion of 2,3-dihydroxybenzoate to enterochelin

*fab*A		22	b-Hydroxydecanoylthioester dehydratase
*gad*R		81	Regulatory gene for *gadS*
*gad*S		81	Glutamic acid decarboxylase
*gal*E	Galactose	17	Uridine diphosphogalactose 4-epimerase
*gal*K	Galactose	17	Galactokinase
*gal*O	Galactose	17	Operator locus
*gal*T	Galactose	17	Galactose 1-phosphate uridyltransferase
*gal*R	Galactose	61	Regulatory gene
*glg*A	Glycogen	74	Glycogen synthetase
*glg*B	Glycogen	74	α-1,4-Glucan: α-,1,4-glucan 6-glucosyltransferase
*glg*C	Glycogen	74	Adenosine diphosphate glucose pyrophosphorylase
*gly*S	Glycine	79	Glycyl-transfer RNA synthetase
*his*A	Histidine	44	Isomerase
*his*B	Histidine	44	Imidazole glycerol phosphate dehydrase: histidinol phosphatase
*his*C	Glycogen	74	Imidazole acetol phosphate aminotransferase
*his*D	Histidine	44	Histidinol dehydrogenase
*his*E	Histidine	44	Phosphoribosyl-adenosine triphosphate-pyrophospho-hydrolase
*his*F	Histidine	44	Cyclase
*his*G	Histidine	44	Phosphoribosyl-adenosine triphosphatepyro-phosphorylase
*his*H	Histidine	44	Amidotransferase
*his*I	Histidine	44	Phosphoribosyl-adenosine monophosphate-hydrolase
*his*O	Histidine	44	Operator locus
*dsd*M	Host specificity	98	Host modification activity: DNA methylase M
*hsd*R	Host specificity	98	Host restriction activity: endonuclease R
*ilv*A	Isoleucine–valine	83	Threonine deaminase (dehydratase)
*ilv*B	Isoleucine–valine	83	Acetohydroxy acid synthetase I
*liv*C	Isoleucine–valine	83	α-Hydroxy-β-axo acid reductoisomerase
*ilv*D	Isoleucine–valine	83	Dehydrase
*ilv*E	Isoleucine–valine	83	Aminotransferase B
*ilv*O	Isoleucine–valine	83	Operator locus for gene *ilv*A,D,E
*ilv*P	Isoleucine–valine	83	Operator locus for gene *ilv*B
*ilv*Q	Isoleucine–valine	83	Induction recognition site for *ilv*C
*ilv*Y	Isoleucine–valine	83	Positive control element for *ilv*C induction
kdp	K accumulation	16	Defect in potassium ion uptake

*lac*A	Lactose	8	Thiogalactoside transacetylase
*lac*I	Lactose	8	Regulator gene
*lac*O	Lactose	8	Operator locus
*lac*P	Lactose	8	Promoter locus
*lac*Y	Lactose	8	Galactoside permease (M protein)
*lac*Z	Lactose	8	b-Galactosidase
mot	Motility	42	Flagellar paralysis
*mut*L	Mutator	93	Generalized high mutability (AT → GC)
*pab*B	p-Aminobenzoate	40	Requirement
pil	Pili	98	Presence or absence of pili (fimbriase)
*pls*A	Phospholipid	11	Glycerol-3-phosphate acyltransferase
pnp		68	Polynucleotide phosphorylase
*pol*A	Polymerase	85	DNA polymerase I
*pol*B	Polymerase	2	DNA polymerase II
*pts*G	Phosphotranserase	24	Catabolite repression system
*pur*A	Purine	93	Adenylosuccinic acid synthetase
*pur*B	Purine	25	Adenylosuccinase
*pyr*B	Pyrimidine	95	Aspartate carbamoyltransferase
*pyr*D	Pyrimidine	21	Dihydroorotic acid dehydrogenase
*rec*A	Recombination	58	Ultraviolet sensitivity and competence for genetic recombination
*rec*B	Recombination	60	Ultraviolet sensitivity, genetic recombination; exonuclease V subunit
*rel*A	Relaxed	59	Regulation of RNA synthesis
*rpo*B	RNA polymerase	89	RNA polymerase: β subunit (*rif* gene)
*rps*L	Ribosomal protein, small	72	Ribosomal protein S12 (*str*A gene, streptomycin resistance)
*ser*A	Serine	62	3-Phosphoglyceric acid dehydrogenase
*ser*O	Serine	20	Operator locus
*ser*S	Serine	20	Seryl transfer RNA synthetase
*spe*A	Spermidine	63	Arginine decarboxylase
*spe*B	Spermidine	63	Arginine ureohydrolase
*spe*C	Spermidine	63	Ornithine decarboxylase
*sup*B	Suppressor	15	Suppressor of echre mutations
*sup*E	Suppressor	15	Suppressor of amber mutations (*su*-2)
*thr*A	Threonine	0	Aspartokinase I-homoserine dehydrogenase I complex
*thr*B	Threonine	0	Homoserine kinase
*thr*C	Threonine	0	Threonine synthetase
*trp*A	Tryptophan	27	Tryptophan synthetase, A protein
*trp*B	Tryptophan	27	Tryptophan synthetase, B protein

*trp*C	Tryptophan	27	*N*-(5-Phosphoribosyl) anthranilate
*trp*D	Tryptophan	27	Phosphoribosyl anthranilatertransferase
*trp*E	Tryptophan	27	Anthranilate synthetase
*trp*O	Tryptophan	27	Operator locus
*tyr*A	Tyrosine	56	Chorismate mutase T-prephenate dehydrogenase
*tyr*T	Tyrosine	27	Tyrosine transfer RNA_1 (*su*-3 gene; amber suppressor)
*ubi*A	Ubiquinone	90	4-Hydroxybenzoate → 3-octaprenyl 4-hydroxybenzoate
*uvr*A	Ultraviolet	91	Repair of ultraviolet radiation damage to DNA, UV endo-nucelase
*val*S	Valine	95	Valyl-transfer RNA synthetase

MAPPING OF CHROMOSOMES

Let us now consider how knowledge of bacterial sex factors and of phage A, permitted the mapping of bacterial chromosomes. Together with the use of restriction endonucleases these techniques gave us the first precise physical maps of bacterial chromosomes and pointed the way toward the determination of complete genome sequences.

The Chromosome Map of E. coli

There are about 4,639,221 nucleotide pairs in the circular DNA molecule that is the chromosome of *E. coli* strain K-12. We now know the complete sequence, which includes all of the individual genes that are present. However, our first knowledge of the location of these genes in the chromosome depended upon construction of a *linkage map*. Construction of this map, with 126 genes, began with the study of nutritional auxotrophs whose defective genes are located at many points on the chromosome. By 1983, 1027 genes had been mapped. In 1997, when the complete nucleotide sequence became known, 4288 protein coding genes could be recognized. The map in Fig. 3.4 established 30 years earlier, largely by use of interrupted bacterial mating. In this procedure *Hfr* cells carrying specific mutations are mixed with wild-type F cells, and conjugation is allowed to proceed for a certain length of time. Then the cells are agitated violently, *e.g.*, in a Waring blender. This breaks all of the conjugation bridges and interrupts the mating process.

Mating is interrupted at different times, and the recipient bacteria are tested for the presence of genes transferred from the donor strain. Using this technique it was found that complete transfer of the chromosome takes ~ 100 min at 37°C, and that the approximate location of any gene on the chromosome can be determined by the length of time required for transfer of that gene into the recipient cell. It is a little more complex than this. Because complete chromosome transfer is rare, substrains of *E. coli* K-12 with an F agent integrated at different points were used. With certain F factors those genes lying clockwise around the circle immediately beyond the point of integration are transferred quickly and with high frequency. The *time-of-entry map* in Fig. 3.4A based not only on interrupted matings but also on the use of *transduction* by bacteriophage P1.

Transduction by phage permits the tranfer of a short fragment of DNA, about 2 min in length, on the *E. coli* map. Joint transduction, *i.e.,* joint incorporation of two genes into the chromosome of the receptor, occurs with a frequency related to the map distance between these two genes. Thus, finer mapping was done within many segments of the *E. coli* chromosome. Meanings of the gene symbols used in the figure are given in Table 3.1. Similar maps were prepared for *Salmonella typhimurium* and *Bacillus subtilis.*

Restriction Endonucleases

Many of the procedures for cloning genes, synthesizing more copies of a DNA ("amplifying" the DNA), making genetic maps, and generating mutants, are dependent upon *restriction endonucleases*. The name comes from a property of bacteria, which often can digest and destroy DNA of invading viruses or DNA that has been injected during mating with a bacterium of an incompatible strain. Investigation of this phenomenon, known as *restriction*, revealed that the DNA of viruses that are able to replicate within a particular host is *marked* in some fashion at specific sites in the molecule. The marking often consists of the presence of methyl groups. Properly methylated DNA is not degraded, but unmethyiated DNA is cleaved by a highly specific endonuclease at the same sites that are normally methylated. Each species of bacteria (and often an individual strain within a species) has its own restriction enzymes.

Restriction enzymes are very specific and cut DNA chains at unique base sequences. Three types are recognized. Type I restriction enzymes, such as those encoded in the chromosome of *E. coli,* are large 300- to 400-kDa proteins composed of at least three kinds of polypeptide chain. They bind at specific sites of a foreign DNA and apparently cleave the chain randomly nearby. They require ATP, Mg^{2+}, and *S*-adenosylmethionine and have the unusual property of promoting the hydrolysis of large amounts of ATP. The significance of these properties is still unknown

The type II restriction endonucleases, which are the ones most widely used in molecular biology, are relatively small 50- to 100-kDa monomeric or dimeric proteins.

About 2400 different enzymes with 188 different specificities had been isolated by 1995. The sites of attack, in most instances, are nucleotide sequences with a twofold axis of local symmetry. For example, the following sites of cleavage have been identified for two restriction endonucleases encoded by the DNA of R-factor plasmids of *E. coli* and for a restriction enzyme from *Hemophilus influenzae.* In the diagrams ↓ are sites of cleavage, * are sites of methylation, and • are local twofold axes (centers of palindromes); N can be any nucleotide with a proper base paring partner. Restriction enzymes often create breaks in each of the two strands in positions symmetrically arranged around the local twofold axis. This is what we might expect of a dimeric enzyme that binds in the major or minor groove of the double helix, each active site attacking one of the polynucleotide chains.

E. coli R factor (*Eco*RI)

```
         ↓ *
5′–N N N G A A T T C N N N
            •
3′–N N N C T T A A G N N N
                 * ↑
```

E. coli R factor (*Eco*RII)

```
            ↓ *
5′–N N N N C C A G G N N N N N–3′
              •
3′–N N N N G G T C C N N N N N–5′
                 * ↑
```

H.influenzae (*Hind*III)

```
          *↓
5′–N N N A A G C T T N N N–3′
               •
3′–N N N T T C G A A N N N–5′
                   ↑ *
```

In fact, the two 277-residue subunits of the *Eco*RI enzyme bind primarily in the major grooves of the DNA, one active site on each strand. Each recognition unit makes 12 hydrogen bonds to the DNA. Each base pair forms two of these hydrogen bonds. Four arginines and two glutamates participate. This provides a net charge of +2, which may be important for electrostatic interaction of the protein with backbone phosphate groups of the DNA. The binding of the protein affects the conformation of the DNA, widening the major groove from that in B DNA, and causing a torsional kink with some unwinding of the double helix. It appears that the specificity for the GAATTC hexanucleotide is in part a result of direct complementary interactions between functional groups in the major groove, bound water molecules, and amino acid side chains from the enzyme.

Table 3.2. Some Commonly used Restriction Endonucleases, Their Sources, and Cleavage Sites

Enzyme	*Source*	*Cleavage Site*
AluI	*Arthrobacter luteus*	5′- -AG↓CT - - 3′
*Bam*HI	*Bacillus amyloliquefaciens* H	G↓GATCC
*Bcl*I	*Bacillus caldolyticus*	T↓GATCA
*Bgl*II	*Bacillus globigii*	A↓GATCT
*Cfr*10I	*Citrobacter freundii*	Pu↓CCGGPy
*Eco*RI	*Escherichia coli*	G↓AATTC
*Eco*RV	*Escherichia coli*	GAT↓ATC
*Hae*III	*Haemophilus aegypticus*	GG↓CC
*Hind*III	*Haemophilus influenzae*	A↓AGCTT
*Kpn*I	*Klebsiella pneumoniae*	GGTAC↓C
*Mbo*I	*Moraxella bovis*	↓GATC
*Pst*I	*Providenica stuartii*	CTGCA↓G
*Sal*I	*Streptomyces albus*	G↓TCGAC
*Sau*3AI	*Streptococcus aureus*	↓GATC
Sau 3AI	*Streptococcus aureus*	GGCCNNNN↓NGGC

Methylation of the 6-amino groups of the adenines in the center of the recognition sequence prevents cleavage by the *Eco*RI endonuclease, but appears to alter the interaction with the protein only slightly. Although they often share little sequence similarity and have quite different specificisties, many restriction enzymes have similar three-dimensional structures as well as mechanisms of action. This is true for the *Eco*RI, BamHl, *Eco*RV, and *Cfr*10I enzymes, and presumably many others.

The specifically shaped and tightly packed active sites in the enzyme-substrate complexes ensure specificity. For example, the *Eco*RV endonuclease cleaves DNA at its recognition site at least a million times faster than at any other DNA sequence. As mentioned in Chapter 12, restriction endonucleases require a metal ion, preferably Mg^{2+}, and probably act via a hydroxyl ion is generated from Mg^{2+}-OH_2 at the active site. Three conserved active site residues, Asp 91, Glu 111, and Lys 113, in the *Eco*RI endonuclease interact with the DNA near the cleavage site. Lys 113 is replaced by Glu 113 in the *Bam*Hl enzyme. The corresponding conserved residues in the smaller *Eco*RV enzyme are Asp 74, Asp 90, and Lys 92. They are shown in the following diagram that represents one of several possible metal-ion dependent mechanisms.

The metal-coordinated hydroxyl ion is generated by proton transfer to the -NH_2 group of Lys 92 and carries out an in-line attack on the backbone phospho group of thymidine at the cleavage point. At least two metal ions are needed, and three may be present, as shown in this diagram from Sam and Perona.

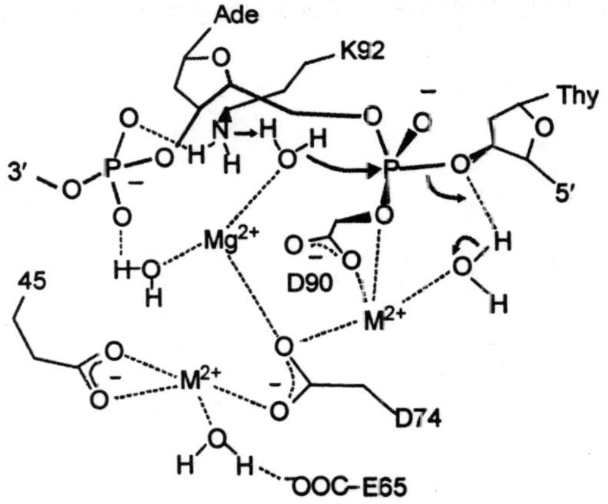

Restriction enzymes that cleave DNA at a large number of specific sequences are available commercially. Another group of restriction enzymes have similar recognition sequences but cut the dsDNA that they recognize at a specific neighboring site rather than within the recognition sequence. An example is *Fok*I, which recognizes the nonpalindromic $\begin{matrix}\text{GGATG}\\\text{CCTAC}\end{matrix}$, but cuts the chains 9 and 13 base pairs to the right. This enzyme has been used by Szybalski and associates to devise a system for cutting ssDNA precisely at a desired point and converting it to ds fragments.

Restriction Mapping

The calibration of the *E. coli* genetic map in minutes was a temporary expedient. It was followed by *physical maps* expressed directly as micrometers of DNA length (total length ~1.6 mm) or thousands of nucleotide units (kb). A physical map obtained by *restriction enzyme mapping* is shown in Fig. 3.4B. To obtain this map DNA fragments were prepared using specific restriction endonucleases. The fragments were aligned using genetic markers, and their lengths were estimated by their electrophoretic mobilities.

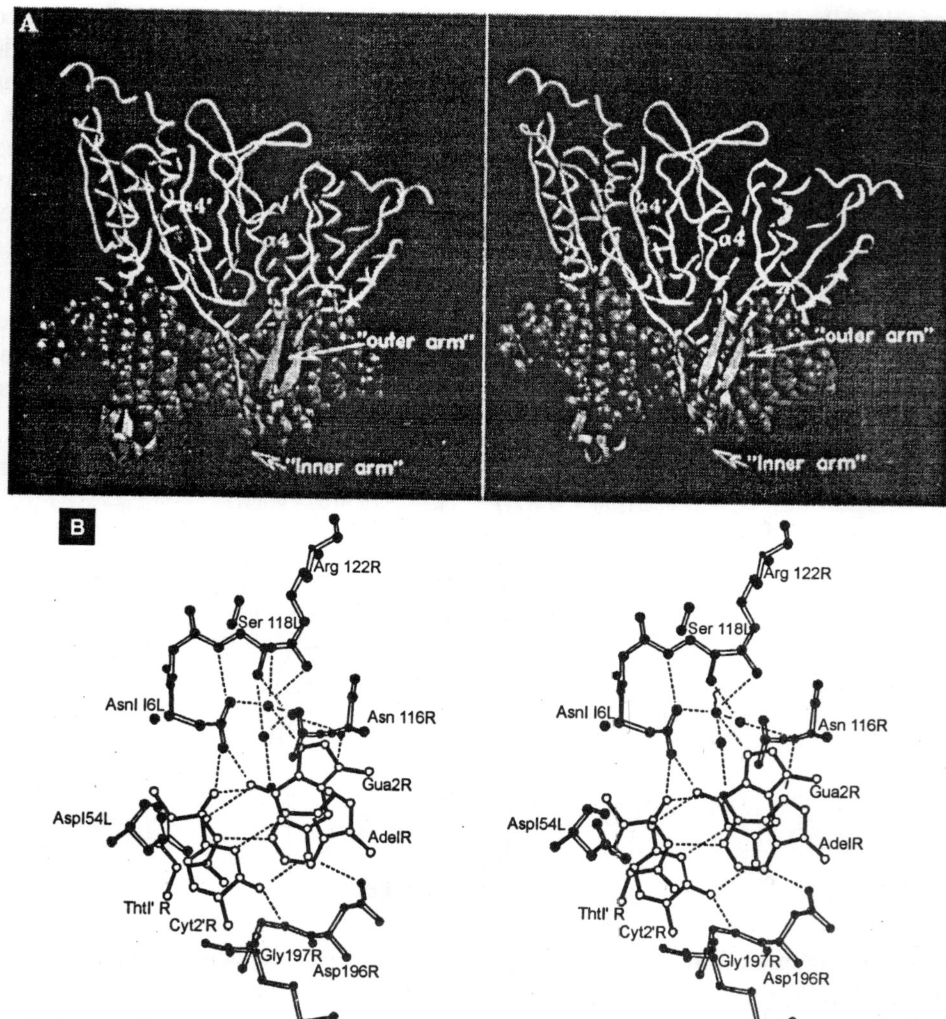

Fig. 3.5. (A) Stereoscopic ribbon drawing of the dimeric *Eco*RI restriction endonuclease in a complex with DNA. The equivalent helices marked α4 and α4′ point into the major groove of the DNA double helix while the inner and outer "arms" wrap around the DNA. (B) Stereoscopic view of the two base pairs T•A and C•G of the right end of the recognition motif 5′-CGATCC-3′ bound to the *Bam*HI restriction endonuclease. The third base pair C•G lies below the two that are shown. Notice the numerous hydrogen bonds, some of which bind atoms of the DNA directly to atoms of the protein and also hydrogen bonds to water molecules (filled circles). The tight packing of complementary charged and dipolar groups of protein, nucleic acid, water, and Mg^{2+} ions (not seen in this drawing) thoughout the complex accounts for the high specificity of these enzymes.

The time-of-entry map has been added on a distorted scale. There are, on the average, 46.4 kb of DNA per minute, but this amount varies around the chromosome between 38 and 61 kb/min. Mapping with restriction endonucleases was for many years an essential step in determination of the complete sequence of apiece of DNA. To make a restriction map the DNA, which may have been cut from a chromosome by a restriction endonuclease, is cloned. This permits isolation of a large amount of the DNA, which is then cut by other restriction endonucleases with differing specificities.

Overlapping fragments resulting from the cleavages by single restriction enzymes are ordered to provide a map such as that of yeast mitochondrial DNA. Any piece from the mapped DNA can now be cloned, and the exact sequence determined. The development of pulsed-field electrophoresis, with its ability to separate DNA fragments 2000 kb or greater in length, allows restriction mapping with enzymes that cut at rare intervals to give very large fragments. For example, the *Not*I restriction endonuclease cuts the 4.7 Mb *E. coli* K12 genome into 22 fragments that were used to construct the complete restriction map of Fig. 3.4B.

Sequences of many viruses and plasmids, mitochondrial and plastid DNAs, and several bacterial genomes have been determined by use of restriction mapping and sequencing of the restriction fragments. Restriction fragment patterns have also been important to determine eukaryotic genome maps including the first genetic linkage map, the first physical map of the human genome, and the complete human genome sequences.

Electron Microscopy

Physical mapping by electron microscopy has been applied to bacteriophage, which can be obtained with large deletions in various parts of the genome. The method can also be applied to cloned pieces of DNA. DNA might be isolated from two different phage strains, for example, from wild-type λ and from a mutant phage with a particular gene or genes deleted. The λ DNA can be denatured readily and separated into *r* strands and *l* strands by isopycnic centrifugation. If the isolated *l* strand of one strain is mixed with the *r* strand of another strain and annealed, a double-stranded DNA will be formed and, if there is a deletion in one strain, the homologous region in the normal λ DNA will form a single-stranded loop that can be visualized in the electron microscope.

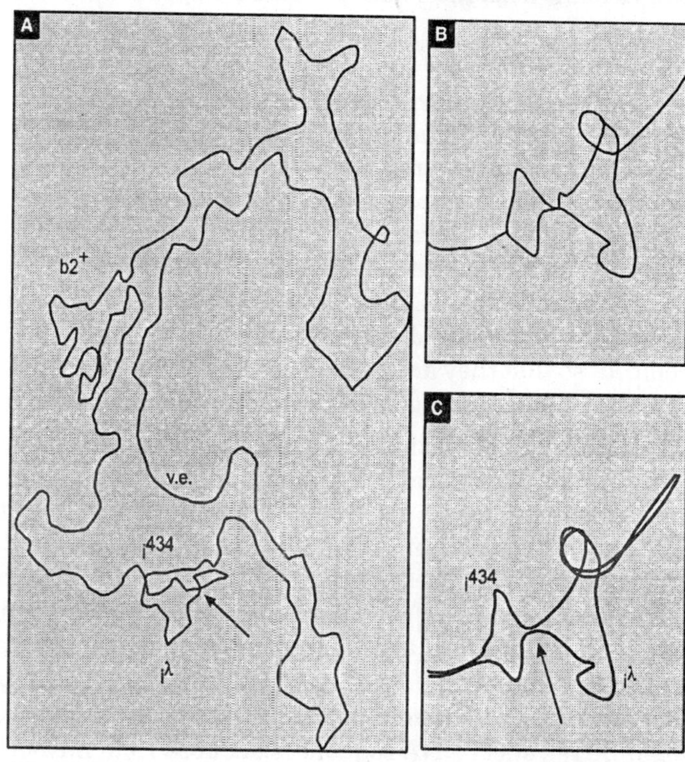

Fig. 3.6. (A) Electron micrograph of a heteroduplex NDA molecule constructed from complementary strands of phages λb2 and λ*imm*434. In λb2 a segment of λ DNA has been deleted producing a deletion loop (labeled b2) and in λ*imm*434 a piece of DNA from phage 434 has been substituted for λ DNA resulting in a "nonhomology bubble" (labeled *i*/*i*λ). The vegetative (cohesive) ends of the DNA are labeled v.e. (B) Enlargement of the nonhomology bubble. (C) Interpretative drawing of view in (B). Arrow marks a short (20-150 nucleotide) region of apparent homology.

Figure 3.6 shows an example of a micrograph of such a *heteroduplex* molecule with a deletion loop and also a "bubble," where a segment of nonhomologous DNA has been substituted in one strand. Since distances can be measured accurately on the electron micrographs, rather precise (±50-100 bp) physical maps can be obtained. The chromosome map of phage λ was mapped in this way initially; now its complete nucleotide sequence is known. Another electron microscopic method is useful for location of AT-rich regions that denature readily. In a suitable concentration of formamide these regions melt to form visible single-stranded *denaturation loops* similar to the bubbles in Fig. 3.6. An important technique is to hybridize pieces of mRNA with DNA. If this is done with denatured (single-stranded) DNA and processed mRNA, which has been transcribed from genes with intervening sequences, the intervening sequences will form single-stranded loops in the DNA-RNA hybrid. A related technique depends upon the increased stability of DNA-RNA hybrids in high concentrations of formamide. Under these conditions an RNA segment will hybridize with its complementary strand of the DNA duplex displacing the other strand of DNA, which then appears as a visible *R-loop*. Intervening sequences appear as undisturbed DNA duplexes.

Optical Mapping

The long DNA molecules of bacterial or eukaryotic chromosomes are easily broken by vigorous stirring. However, if handled carefully very large fluorescently stained DNA molecules of 0.4-1.4 Mb lengths can be stretched out on a glass surface, and their lengths measured by optical microscopy. The technique depends upon binding of one of the ends of the DNA to the glass surface. If biotin is covalently attached to one end of the DNA, it will bind to a steptavidincoated plate. However, unaltered DNA also binds to a vinyl-silane or a trichlorosilane coating under suitable conditions. If DNA is incubated with a submerged silanized coverslip, which is then pulled out of the liquid mechanically at a constant speed, the DNA molecules are "dyamically combed" so that they are aligned for easy observation and measurement. DNA can be transferred directly from agarose gels used for electrophoretic separation to the plates.

Used in combination with restriction enzymes these procedures allow rapid automated construction of physical maps. An example of whole-genome optical mapping is provided by the restriction map of the genome of the radiation-resistant bacterium *Deinococcus radiodurans*. The genome consists of two circular DNA molecules of 2.6 and 0.415 megabases and a smaller 176-kilobase DNA. These were mapped without the laborious subcloning required by conventional restriction mapping. For example, the *E. coli* restriction map required analysis of 3400 phage clones. Mapping of eukaryotic chromosomes has involved additional methods which are discussed in Sections E and G,1. These incluse *radiation hybrid mapping*, use of *meiotic recombination*, identification of *restriction fragment length polymorphisms* (RFLPs, and use of *expressed sequence tags* (ESTs), short DNA sequences deduced from mRNA molecules transcribed from the DNA.

CLONING, MODIFYING, AND TRANSFERRING GENES

A true revolution in biology and in medicine is in progress as a result of our ability to clone, sequence, mutate, and manipulate genes at will. A diploid cell contains only two copies of many genes, and these two copies are often not identical. In a gram of any tissue, which may contain 10^9 cells, we will have only ~ 1 ng of a 1 kb (1 kilobase) gene. To isolate this gene we would have to fish it out from a huge excess of other genes. Present-day cloning techniques offer a way to locate the gene, increase its quantity by many orders of magnitude, learn its sequence, induce any desired mutations at any points, and transfer the gene into other organisms in such a form that it can be expressed, *i.e.,* be transcribed and induce synthesis of proteins. The methods have taken years to develop and continue to be improved. Only some basic procedures and concepts are described here. Numerous manuals as well as commercial "kits" are available.

Joining DNA Fragments

The cloning and manipulation of genes usually depends upon the precise cutting of DNA into discrete fragments by restriction endonucleases. Many restriction enzymes generate *cohesive ends* (sticky ends). Thus, EcoRI produces DNA fragments with the single-stranded "tails" shown here at the 5′-ends of the cut duplexes:

5′- - - - - G 3′ 5′ pAATTC - - - - - 3′

3′- - - - - ĊTTAAp5′ 3′ Ġ - - - - - 5′

These cohesive ends can be used to join together different restriction fragments. It is easy to see that the complementary single-stranded tails can form base pairs to regenerate the original hexanucleotide sequence cleaved by *Eco*RI. There will still be nicks between G and A at the specific cleavage points, but these nicks can be closed enzymatically using DNA ligase. Thus, the original DNA cleaved by *Eco*RI can be reformed, or another piece of DNA that also has tails generated by *Eco*RI can be grafted onto an end. Many of the other enzymes cohesive ends. Five of them (*Bam*HI, *Bel*I, *Bgl*II, *Mbo*I, and *Sau*3A) have at the center of their recognition sites the same tetranucleotide: GATC. Enzymes *Sau*3A and *Mho* I are called *isoschizomers* because they have just the same 4-base recognition sequence and also yield the same restriction patterns.

Notice that they will both cut all of the *Bam*HI, *Bel*I, or *Bel*II sites, but *Bam*HI and *Bgl*II will not cut all *Mbo*I or *Sau*3A sites. However, cohesive ends made by any of these enzymes can be joined. The gaps left during the joining of certain of the fragments can be ligated enzymatically. *Sau*3A will cut at either methylated or unmethylated sites but not cut at methylated sites. Two enzymes listed in Table 3.2 form 3′-cohesive ends rather than 5′-cohesive ends.

In addition, there are three *(Alu*I, *Eco*RV, and *Hae*III) that cut at the local twofold axis; they form no cohesive ends but leave *blunt ends* (flush ends). Blunt end fragments are also much used in genetic engineering. "Linkers" that provide cohesive ends can be added. The *Sfi*I endonuclease cuts between two 4-bp palindromes in a 13-bp recognition sequence.

Some Useful Enzymes

Several enzymes of use in cloning are listed. Among these are the *ligases* that allow DNA fragments to be joined. They act on DNA strands with adjacent 3′-OH and 5′-phosphate termini. The *E. coli* ligase seals single stranded nicks using NAD^+ as an energy source. It is therefore able to ligate DNA fragments with cohesive ends. The T4 DNA ligase, which is obtained from *E. coli* infected with phage T4, not only can seal nicks but can ligate pieces of DNA with blunt ends. Its activity is linked to cleavage of ATP. If two DNA strands are joined, but with gaps in one or both strands, the gaps can be filled efficiently by the 109-kDa *DNA polymerase* I from *E. coli*.

Most often the 76-kDa *Klenow fragment*, which is lacking the 5′ → 3′ exonuclease activity, is used. *T4 DNA polymerase* has similar properties. A problem with DNA fragments with cdohesive ends is that they spontaneously form closed circles, a process that may compete with a desired joining to another piece of DNA. One solution to this problem is to hydrolyze off the 5′-phosphate groups with an *alkaline phosphatasc.* This prevents formation of covalently closed circles. However, pieces of DNA that retain their 5′-phosphate groups can be ligated to these dephosphorylated pieces. *T4 polynucleotide kinase* can be used to put a phospho group back onto the 5′ end of a chain. A ^{32}P end label can be added to such a polynucleotide using ^{32}P-labeled ATP.

Forming Homopolymeric tails

Chromosomal DNA may be cleaved with restriction enzymes that leave blunt ends or it may be cleaved randomly by shearing. In either case the blunt ends can be treated first with *λ-exonuclease*, then with *terminal deoxynucleotidyl transferase* isolated from calf thymus. The exonuclease treatment cuts off a few nucleotides from the 5′ termini leaving short single-stranded 3′-OH termini. The terminal transferase, a polynucleotide polymerase that acts on ssDNA, is nonspecific and requires no template. Using an appropriate nucleotide triphosphate, it will add a single-stranded tail of either deoxyribonucleotides or ribonucleotides to the exposed 3′ termini of a polynucleotide of three or more residues. If deoxyATP is used, a 3′ poly(dA) tail will be added to each 5′ terminus. Such a poly(dA)-tailed DNA fragment can be annealed and ligated to DNA carrying poly(dT) tails. This approach has been used widely to insert a piece of DNA into a cloning vehicle. For example, a circular plasmid can be opened by a single cleavage with *Eco*RI or other suitable restriction enzyme.

The opened plasmid is treated with exonuclease, and poly(dA) tails are added. The piece of DNA to be cloned is tailed with poly(dT). After annealing and ligation recombinant plasmids carrying the *passenger DNA* will be formed. If DNA is cleaved with *Pst*I *or Kpn*I, the resulting 3′ cohesive ends can be extended with a poly(dC) tail. If the cloning vehicle also has a site for *Pst*I or *Kpn*I, it can be opened and poly(dG) tails can be applied. A useful feature is that after annealing, filling in the gaps, and ligation the original *Pst*I or *Kpn*I sites are restored. This provides for easy recovery of the cloned fragments.

Preparing Material for Cloning

DNA may be prepared for cloning by (1) random cleavage by shearing or by enzymatic attack, (2) cleavage by one or more restriction endonucleases, (3) preparation of cDNA from mRNA, or (4) nonenzymatic chemical synthesis of DNA segments. The use of random cleavage has largely been replaced by cleavage with restriction enzymes. A major problem is the separation of the very large number of different restriction fragments formed from a large piece of DNA or from an entire genome. Considerable simplification comes from separation of individual eukaryotic chromosomes before the library is prepared.

Careful purification of DNA to be used in cloning is helpful. This may be done by electrophoresis in agarose or polyacrylamide gels or using HPLC. One technique is to embed cells directly in a gel, to diffuse in proteases and restriction enzymes that lyse the cells and release the DNA, and cleave it, and then to conduct the electrophoresis. DNA fragments of very large size can be separated. In addition to isolation of a fragment to be cloned the cloning vehicle must be prepared. This often involves release of a plasmid by lysis of bacteria that carry it and isolation using a suitable column.

Likewise, after the DNA has been cloned and the content of DNA has been increased by growing a large bacterial culture, the plasmids must be released and purified, and the cloned DNA excised with a restriction enzyme. Alternatively, the cloning vehicle may be a virus, which must be isolated and disrupted to release the DNA. Use of the polymerase chain reaction (PCR) allows cloning and amplification of DNA fragments with a minimum of purification.

Cloning Vehicles (Vectors)

Many cloning vehicles, more commonly referred to as *vectors*, originated with naturally occurring, independently replicating plasmids or viruses (replicons). More recently artificial chromosomes have been developed as cloning vehicles. Plasmids and viruses have been extensively engineered to provide convenience and safety. A large number of specialized vehicles have been devised. Only a few will be described briefly here. Suitable books and manuals must be consulted for details.

Table 3.3. Some Enzymes Used in Molecular Cloning

Name	*Source*	*Reaction*
T4 DNA polymerase	*E. coli* infected with bacteriophage T4	5′ → 3′ chain growth 3′ → 5′ exonuclease
E. coli DNA polymerase I and Klenow fragment	*E. coli*	5′ → 3′ chain growth 3′ → 5′ exonuclease 5′ → 3′ exonuclease (lacking in Klenow fragment)
Reverse transcriptase	RNA tumor viruses, *e.g.*, avian myoblastosis virus	5′ → 3′ DNA chain growth
Ribonuclease H	*E. coli*	Cuts RNA in DNA-RNA hybrid
Lambda and T7 exonucleases	Bacteriophages	
Bal 31 nuclease	*Alteromonas espejiano*, a marine bacterium	Degrades both 3′ and 5′ termini of dsDNA
T7 RNA polymerase	Bacteriophage T7	DNA-dependent RNA polymerase
Terminal deoxyribo-nucleotide transferase	Thymus gland, plants	Limited 5′ → 3′ chain growth; template independent addition of tails to DNA fragments
T4 DNA ligase	*E. coli* carrying an engineered λ phage	Ligation of DNA, either blunt or cohesive ends; used ATP
E. coli DNA ligase	*E. coli*	Ligation of DNA with cohesive ends; uses NAD^+ as energy source
RNA ligase	Bacteriophage T4	Ligation of RNA and DNA
T4 polynucleotide kinase	Bacteriophage T4	Phosphorylation of 5′-OH terminus of a polynucleotide (DNA or RNA)
*Eco*RI methylase	*E. coli*	Transfer CH_3 from *S*-adenosylmethionine to adenines in *Eco*RI sites

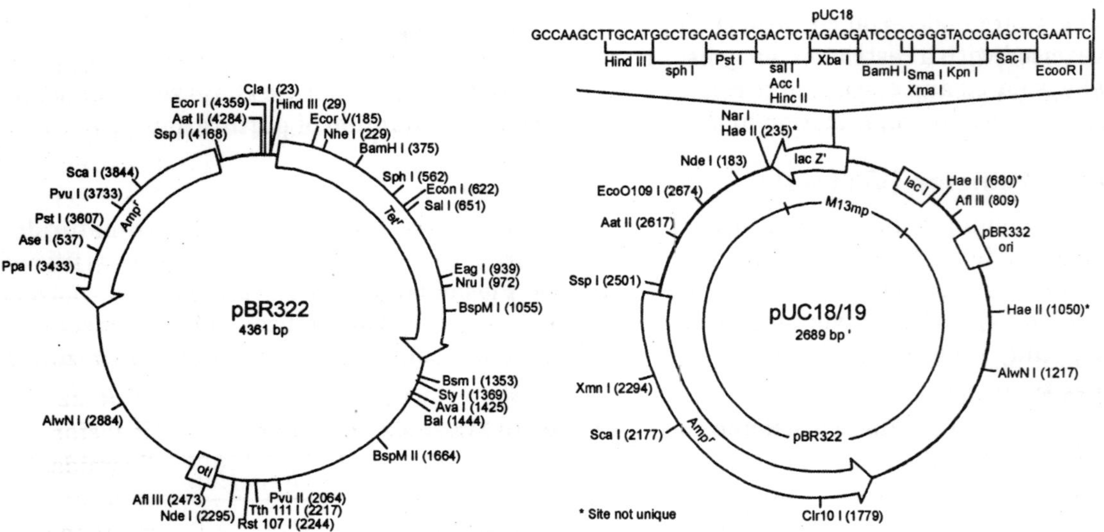

Fig. 3.7. Genetic map of cloning plasmids pBR322 and pUC18. Abbreviations: ori, origin of replication; Ampr, ampicillin resistance gene; Tet^r, tetracycline resistance gene. Other abbreviations are for sites cleavted by specific restriction endonucleases. The nucleotide sequence numbers and directions of transcription are also indicated.

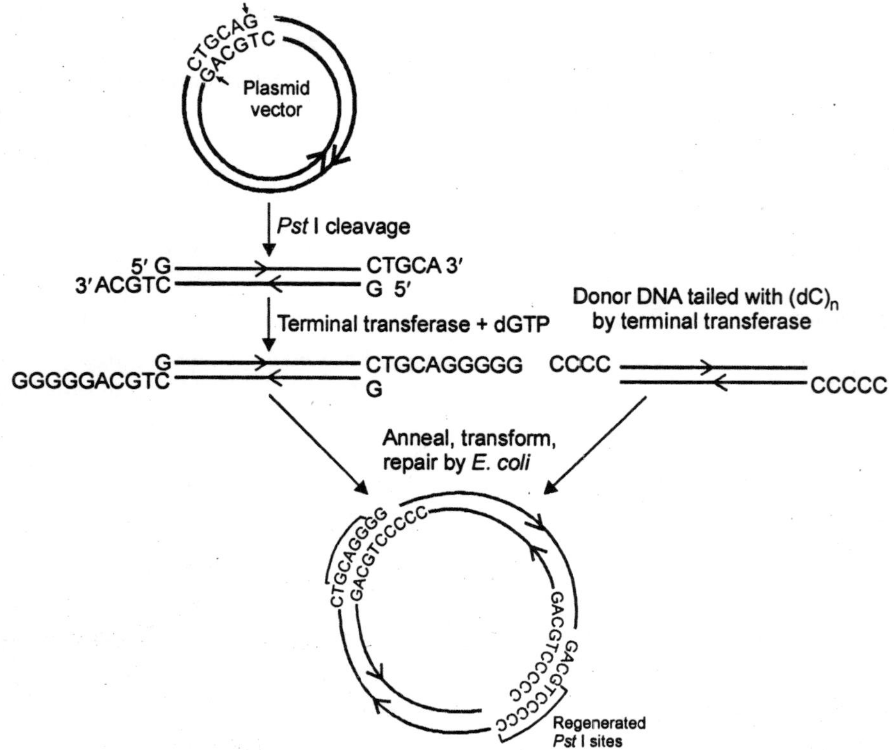

Fig. 3.8. Regeneration of a *Pst* I site by dG:dC tailing with terminal transferase.

Plasmids Related to ColE1

The small colicinogenic plasmid, ColE1 of *E. coli*, is attractive for cloning because there are 20 copies per bacterial cell. If chloramphenicol is added during the logarithmic phase of growth, the *E. coli* cells will make 1000-3000 copies. Thus, a relatively small culture of the bacteria containing the cloned DNA in a ColE1 plasmid will yield a large amount of the desired DNA. In the past one of the most widely used cloning vehicles has been plasmid pBR322, which was derived from a close relative of ColE1. The genome size of the original plasmid was reduced by deletion of genes unnecessary for its successful replication. Transposons that might permit accidental transfer of DNA to other organisms were inactivated. Unneeded or undesirable restriction enzyme sites were eliminated, and useful restriction sites were introduced by point mutations. The resulting pBR322 has only one site of cleavage each for *Bam*HI, *Sal*I, *Pst*I, *Pvu*I, and *Eco*RI. These are at known positions in the 4363-nucleotide plasmid. Plasmid pBR322 contains two different antibiotic-resistance genes that were brought in from bacterial R-factors. These are used in selecting bacterial colonies that carry the desired recombinant plasmids. One of these is the β-lactamase gene, which confers resistance to ampicillin (Amp^r); the other provides resistance to tetracycline (Tet^r). Their positions are indicated in the pBR322 gene map as is the essential origin of replication (*ori*).

The drug resistance genes are used as follows. If a unique restriction site such as that for BamHI or *SalI* that lies within the Tet^r gene is used to introduce the passenger DNA, the resistance to tetracycline is lost but that to ampicillin is retained. Thus, after incubation with the recombinant plasmids under conditions that favor their uptake by the host bacteria, the bacteria are plated onto an ampicillin-containing medium. Only those harboring the pBR322 plasmid with its Amp^r gene can grow. After these have produced small colonies, a replica plate is made on a tetracycline-containing medium. On this medium the desired recombinants do *not* grow because the Tet^r gene has been inactivated. This allows selection of colonies containing passenger DNA.

A further selection procedure is required to establish that the piece of DNA inserted into the recombinant plasmid is one that is desired.

Typical Cloning Procedure with pBR322

In much simplifed form the procedure might go as follows: (1) Purchase or isolate plasmid. (2) Cleave plasmid with *Bam*HI; heat at 70°C to inactivate the enzyme. (3) Treat with alkaline phosphatase to remove the 5′-phospho groups. (4) Mix with passenger DNA with cohesive ends generated by *Bam*HI, anneal, and join with DNA ligase. Although the resulting circular recombinant DNA contains a nick as a result of the missing 5′-phospho group, it will be taken up by bacteria and repaired. (5) Incubate joined DNA with cells of host *E. coli* that have been made permeable to DNA by treatment with Ca^{2+} ions. This type of transformation is called *transfection* and is widely used in cloning. (6) Plate transfected cells onto agar containing the first antibiotic, in this case ampicillin. (7) Make replica plate on medium containing second antibiotic, in this case tetracycline. (8) Screen selected colonies for desired DNA fragment. In one procedure a small sample from each of selected colonies is placed onto spots on a nitrocellulose filter.

Several colonies can be placed on one filter and the bacteria lysed, hybridized with a radioactive probe, and then viewed by autoradiography.

Selecting Clones Using β-galactosidase in pUC Cloning Vehicles

The newer **pUC** vehicles contain the origin of replication and the ampicillin-resistance gene from pBR322. In addition, a segment of DNA from the *E. coli lac* operon has been grafted into an intergenic region. It contains the lac control region as well as the coding sequence for the first 145 residues of β-galactosidase. Within bacteria containing the pUC DNA an N-terminal portion of β-galactosidase is synthesized. The specially designed host cell contains (in an episome) the gene for another defective p-galactosidase, one lacking the N-terminal portion. This defective enzyme, together with the N-terminal portions encoded in the pUC DNA, forms an active galactosidase. When the chromogenic substrate 5-chloro-4-bromo-3-indolyl-β-D-galactoside ("X-gal") and a suitable inducer, such as isopropylthio-β-galactoside (IPTG), are present, the unoccupied plasmid vehicles generate blue colonies. However, if passenger DNA is inserted within the galactosidase gene segment, formation of the enzyme will be disrupted and white colonies will appear. The restriction sites for cloning are placed in a *polylinker* near the 5′ terminus of the galactosidase gene. The polylinker has been carefully engineered to maintain the correct reading frame and to avoid disruption of the galactosidase activity. It contains several different restriction sites for insertion of passenger DNA. Such insertion does destroy the galactosidase activity allowing the user to detect the recombinant DNA from the white plaques.

Filamentous Bacteriophages

An important series of cloning vehicles have been derived from the circular replicating forms of the filamentous bacteriophage M13. Although the genome contains only a short intergenic region that can be deleted, up to 50 kb of passenger DNA can be inserted into these vehicles. Since long inserted sequences may be deleted spontaneously, M13 is most useful for cloning about 300- to 400-nucleotide chains. Many of the M13 vehicles also use the β-galactosidase blue-white screening technique. These modified viruses are highly infective, but the infected *E. coli* cells are not killed.

Rather they produce large numbers of virus particles with single-stranded DNA representing one of the two DNA chains of the parental phage. These are widely used for sequencing by the chain-termination procedure of Sanger *et al.* The procedure requires a primer sequence. If M13 recombinants are sequenced, the primer consists of a synthetic oligonucleotide that can be annealed to the galactosidase gene fragment at its 3′ end just in front of the DNA segment that is to be sequenced.

Lambda Cloning Vehicles

Many cloning vehicles have been derived from the 48,502-bp DNA from the temperate *E. coli* bacteriophage λ. The DNA from phage particles are taken up efficiently by *E. coli* cells, much more so than by transfection. The virus has a complex life cycle. Within the phage head the λ DNA exists in a folded linear form with 12-base 5′ cohesive ends. After entrance into the bacterial cell the DNA cyclizes through its cohesive ends and is ligated by the *E. coli* ligase. Replication of the circular forms ensues. Later in the cycle, rolling circle replication produces long concatamers with several phage genomes joined as a single chain. This DNA is "packaged" into new phage heads. As this is done, a nuclease cuts the concatamers at *cos* sites forming monomeric genomes with cohesive ends. This is the *lytic cycle* of the phage. In the alternative *lysogenic cycle* the DNA becomes integrated into the *E. coli* genome.

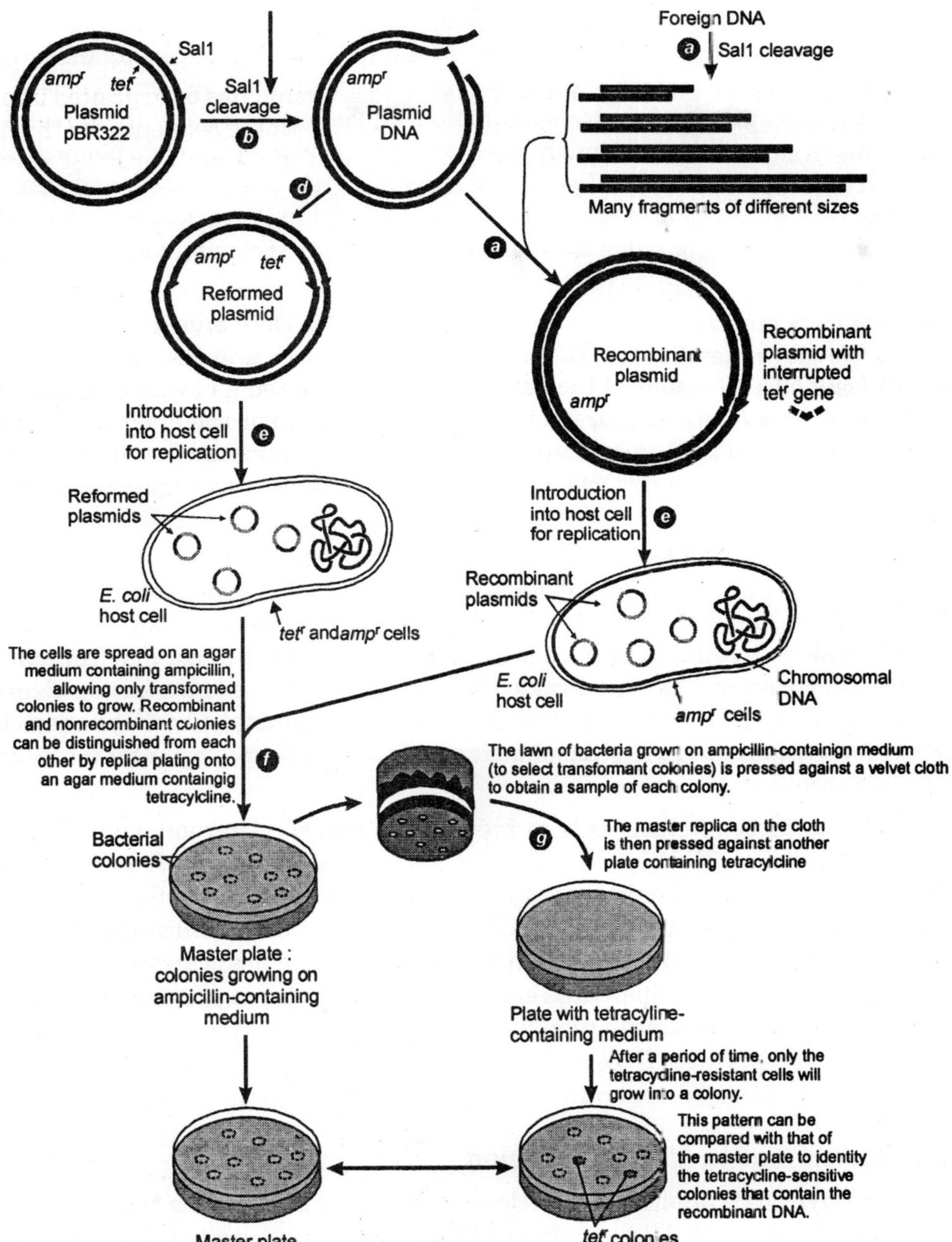

Fig. 3.9. A classical scheme for cloning DNA in a pBR322 plasmid vehicle. A DNA sample is digested (step *a)* with one of the restriction endonucleases (*e.g., Sal*I), that cuts the cloning vehicle within the *Tet*r gene. The plasmid is also cut with the *SalI* restriction enzyme (step *b*). After mixing with the digested DNA sample and annealing, recombinant molecules are formed (step *c*). Some plasmids are reformed (step *d*). Both recombinant and reformed plasmids transform *E. coli* cells (step e). The transformed cells are plated on an agar medium containing ampicillin (step *f*). Only cells containing the ampicillin-resistance gene grow (step *g*). A replica plate is made and is pressed onto another plate containing tetracycline. The recombinant colonies do not grow on the medium because the tetracycline-resistance gene has been interrupted. By comparing the two plates recombinant colonies can be selected. These can be tested, using a suitable probe, to determine whether they carry a desired gene.

Maintenance of the lysogenic state depends in part on gene *c*I, which encodes a repressor that prevents expression of the genes required in the lytic pathway. Since only the lytic cycle is needed for cloning, it is convenient to place a cloning site within the *c*I gene. The screening of recombinant phage particles is done by examination of plaques. A phage without inserted DNA will be able to undergo both the lytic and lysogenic cycles and will form turbid plaques. However, if passenger DNA is inserted in the *c*I gene, the lysogenic cycle is prevented, and clear plaques are formed. Of the ~ 50 genes present in native λ, only about half are necessary for replication in the lytic cycle. Thus, it is possible to delete about 1/3 of the genome to make room for more passenger DNA.

However, to form mature phage particles the length of the DNA must be at least 75% of the native length. No more than 110% of the native amount may be present. The total DNA must fall between 38 and 53 kb in length. To accommodate these packaging requirements *replacement vectors* containing unnecessary "stuffer DNA" between two lambda "arms" are used. The unneeded stuffer piece has the same kind of restriction site or sites at each end so that it can easily be cut out and replaced by the passenger DNA. This permits cloning of DNA segments up to about 22 kb in length.

Lambda vehicles have all been engineered to eliminate undesired restriction sites and to reduce the number of sites for *Eco*RI and other restriction enzymes commonly used for cloning. The widely used *Charon series* have been further engineered so that they will grow only in strains of bacteria that cannot survive in the human intestinal tract. For example, amber mutations are incorporated into genes needed for phage assembly, and the bacterial hosts must contain an amber suppressor gene. The bacteria are also nutritional auxotrophs with absolute requirements for thymidine and diaminopimelic acid in the medium. The latter compound is not found in the intestinal tract.

The purpose of these alterations is to prevent the spread of recombinant DNA into the environment. To sequence DNA carried in a lambda vehicle or to study it in other ways, it is often necessary to cut it with restriction enzymes, to prepare a restriction fragment map, and to subclone the fragments into a plasmid vehicle. Lambda vehicles, which will automatically transfer the passenger DNA into an M13 vehicle when propagated in a host carrying a special helper virus, have been devised. The helper virus encodes proteins that recognize and cleave sequences that mark the initiation and termination of M13 DNA synthesis. These are used to mark the ends of the passenger DNA. As DNA synthesis occurs the displaced passenger DNA and M13 genes are excised, circularized, and converted into a replicating form of an M13 cloning vehicle.

Cosmid Vehicles and in Vitro Packaging

Cosmids are hybrids of a plasmid vehicle and phage λ. They contain the *cos* sites that are cleaved during packaging of λ DNA. A cosmid cleaved at a restriction site will form upon ligation a range of different sized DNA molecules that contain *cos* sites on both sides of a piece of passenger DNA, which may be up to 45 kb in length. This can be cut at the *cos* sites and packed into heads using an *in vitro* packaging system. In this system the unassembled subunits of the phage particle are produced in special strains of bacteria and are allowed to assemble and package the cosmid DNA. Since cosmids contain relatively large pieces of DNA, they are useful for preparing sequentially over-lapping clones that allow the investigator to "walk" along the DNA hunting for a target gene. Cosmid vehicles have been designed to allow both efficient genomic walking and restriction mapping.

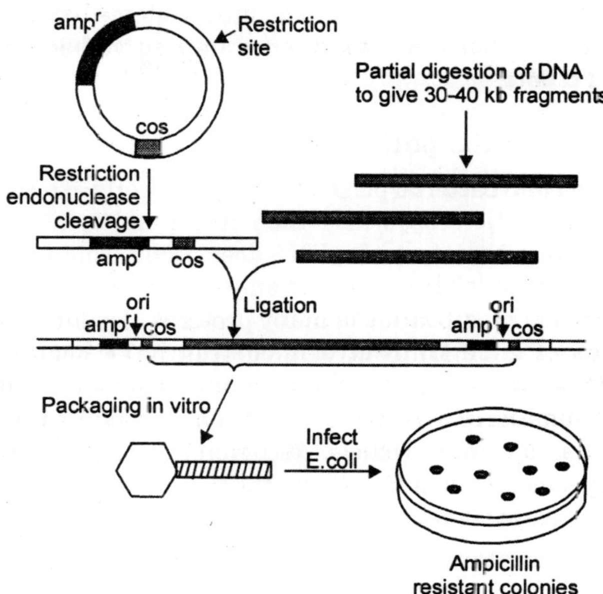

Fig. 3.10. Cloning DNA in cosmids.

Expression of Cloned Genes in Bacteria

A major goal in recombinant DNA technology is the production of useful foreign proteins by bacteria, yeast, or other cultured cells. Protein synthesis depends upon both transcription and translation of the cloned genes and may also involve secretion of proteins from the host cells. The first step, transcription, is controlled to a major extent by the structures of promoters and other control elements in the DNA. Since eukaryotic promoters often function poorly in bacteria, it is customary to put the cloned gene under the control of a strong bacterial or viral λ promoter. The latter include the λ promoter P_L and the *lac* and *trp* promoters of *E. coli*. These are all availble in cloning vechicles. It is often useful to create hybrid proteins fused to the *E. coli* β-galactosidase gene.

If another gene is spliced in at either the N terminus or the C terminus of the galactosidase (*lac*Z) gene but is kept under control of the *lac* promoter, the resulting hybrid protein will have galactosidase activity, which can be used for screening. In addition, the hybrid protein will often react with antibodies directed against the protein whose gene is being cloned. Another kind of hybrid fuses the cloned gene to that of β-lactamase, for example at the *Pst I* site of plasmid pBR322. The β-lactamse activity will be gone, but the hybrid protein will be secreted because β-lactamase normally is secreted into the periplasmic space and its *N*-terminal signal sequence is now fused to the cloned protein.

Engineering of a suitable site for cleavage by a protease can release the foreign protein in an active form. A variety of other *expression systems*, often using *reporter gene* products, have been developed. Bacteria often degrade foreign proteins by hydrolytic attack. One way in which such damage has been minimized is to clone multiple fused copies of the gene for the desired product. The resulting polyprotein may be resistant to degradation, and if the gene has been correctly engineered, may be cut apart by cyanogens bromide or by a specific protease. For many years most cloning was done in *E. coli*, but cloning systems have now been developed for many other bacteria including *Bacillus* and other grampositive bacteria, and also for yeast, insect cells, animals, and plants.

Cloning and Transferring Eukaryotic Genes

Eukaryotic genes cloned into bacterial plasmids are often poorly expressed. It is advantageous to clone such genes in eukaryotic cells, where the cutting and splicing of hnRNA to remove intervening sequences during formation of mRNA does occur. This permits expression of the cloned gene, something that is possible in bacteria only if cDNA that lacks the introns is cloned. The need for posttranslational modification of many proteins also interferes with expression in bacteria. Many methods of gene transfer and cloning have been developed. The yeast *Saccharomyces cerevisiae* is often an ideal host for cloning. It grows rapidly in either its haploid or diploid stage. Some strains carry a 2-μm 6.3-kb circular plasmid with 50-100 copies per cell. This has been developed as a cloning vehicle. Recombinant plasmids can be used as *shuttle vehicles* for transferring genes cloned in *E. coli* into the yeast plasmid. Genes may also be cloned as minichromosomes, such as *yeast artificial chromosomes* (YACs).

Artificial chromosomes contain origins of replication from yeast, human, or bacterial chromosomes as well as telomeres and centromeres. YACs have been widely used and became popular because they can accomodate 600 kbp or more of DNA. However, their use in the human genome project resulted in serious problems of instability. Bacterial artificial chromosomes (BACs), which accommodate only 200-300 kbp, *P1 artificial chromosomes* (PACs), and human minochromosomes are more stable. Another problem, which affects the use of yeast for production of eukaryotic proteins, is the tendency for poor removal of introns.

Plant Genes

Much of the cloning in higher plant cells has made use of the *Ti plasmid* of *Agrobacterium tumefaciens,* a soil bacterium that enters wounds in dicotyledenous plants causing tumors known as *crown galls*. A related species *A. rhizogenes* harbors a similar plasmid that causes "hairy root" disease. The infecting bacteria respond to the synthesis of certain phenolic compounds such as *acetosyringone*, which are produced in plant wounds, by entering the plant cells. Only bacteria carrying the Ti (tumor-inducing) plasmid cause tumors. The plasmid carries a 13-kb region (transferred region) known as T-DNA that encodes enzymes for production of the auxin indoleacetic acid, the cytokinin isopentenyl-AMP, and the compounds known as *opines*. These are reduction products of Schiff bases of amino acids and 2-oxoacids or sugars.

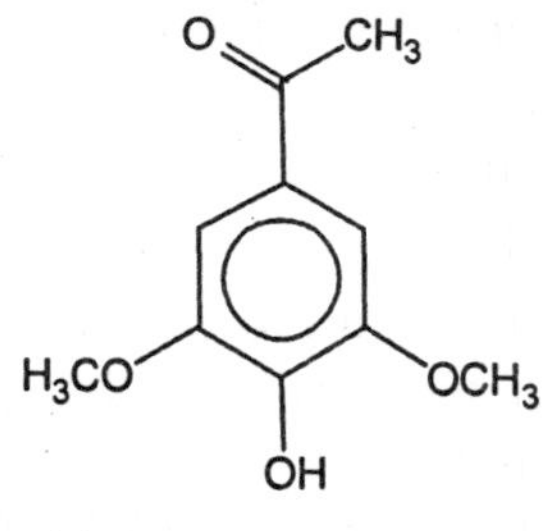

O CH$_3$

H$_3$CO OCH$_3$

OH

Acetosyringone, a compound that induces infection by Agrobacterium

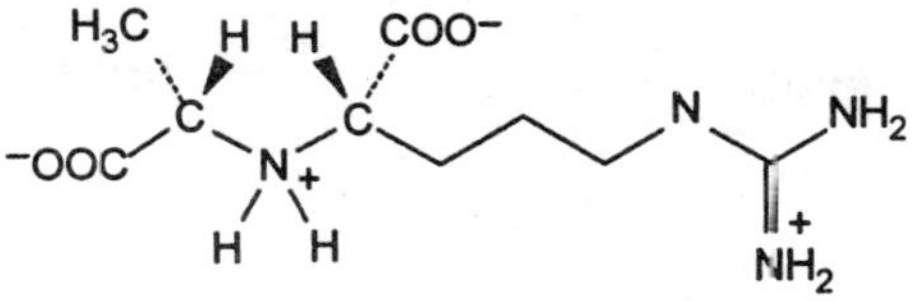

Octopine: derived from pyruvate and arginine

Histopine: derived from pyruvate and histidine

Lysopine: derived from pyruvate and lysine

Nopaline: derived from 2-oxoglutarate and arginine

Leucopine: derived from 2-oxoglutarate and leucine

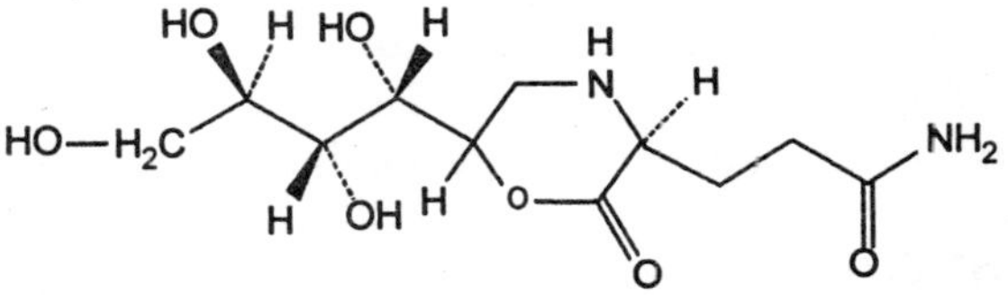

Agropine: derived from D-mannose and L-glutamine

The auxin and cytokinin, whose production is normally controlled, are now formed in large amounts and cause uncontrolled tumor growth.

The opines are used by *Agrobacterhiu as* a unique source of energy and of metabolites for biosynthesis. The host plant cells, however, cannot catabolize the opines. Upon entrance into a plant cell the T-region is excised from the Ti plasmid and can become integrated into the DNA of the host plant much as occurs during bacterial conjugation. By deleting the genes for synthesis of auxin, cytokinin, and opines, the Ti plasmid loses its tumorigenic property but is still able to transfer genes into the plant genome. Fusion of an *E. coli* plasmid vehicle into the modified Ti plasmid creates a useful plant shuttle vehicle. Also see below under "physical methods."

Transferring Genes with Engineered Animal Viruses

There is special interest in transferring cloned DNA into human cells to correct genetic defects. Transfer of human genes into other animals is also important for a variety of reasons. For example, human proteins of therapeutic value could be produced in animal milk. Animals can be engineered to have defects that mimic those in humans and which can then be studied in animals. A number of different viruses have been used to transfer foreign genes into eukaryotic cells and also to create stable plasmids for cloning.

One of the first was the 5380 bp SV40. The relatively small size of SV40 limits the amount of DNA that can be incorporated. However, many of the functions of this virus can be performed by simultaneous infection with a *helper virus*, often an adenovirus which is itself defective. With the helper virus making proteins essential for SV40 replication, all but 85 base pairs at the origin of replication of SV40 can be deleted and replaced by other DNA. More recently engineered *adenoviruses* and *retroviruses* have been widely used. In both cases genes required replication of the viruses have been deleted.

Adenovirus vehicles can carry 7 kbp or more of passenger DNA. They are efficiently taken up into endosomes and into the nucleus by both replicating and nonreplicating cells. The DNA is not incorporated into the genome of the host cell. Replication-deficient avian *retroviruses*, or *retrotransposons*, can carry up to 9 kbp of passenger DNA and are incorporated into the host genome by the process of homologous recombination. Both adenovirus and retroviral vehicles suffer from serious problems. The adenovirus proteins can induce fatal inflammatory reactions, and DNA transfer from the retroviral vehicles is inefficient.

A nonpathogenic human parvovirus, the *adenoassociated virus* (AAV), is also being developed as a gene therapy vector. Another possibility is to engineer a *lentivirus*, even HIV, to provide efficient integration. Safety concerns have delayed human testing of this possibility. A group of insect viruses, the *baculoviruses*, are being used in cultured insect cells for large-scale production of proteins. The cloned genes are placed under the control of the promoter region of the gene that encodes the major viral coat protein. The *baculovirus* vehicles can carry over 15 kbp of passenger DNA and may also be useful in human gene therapy. Filamentous bacteriophages have been reengineered for the same purpose. Nonviral methods for DNA incorporation into the human genome may utilize *transposons* or mobile *group II introns*.

Physical Methods of Gene Transfer

Genes can often be transferred without the use of a cloning vehicle. This is especially important for certain plant cells, such as those of cereal grains, for which transfer of genes via the Ti plasmid has been difficult. If DNA, which may be in a plasmid, is coprecipitated with calcium phosphate, it can often be taken up directly either by animal cells or by plant protoplasts. Polycations also facilitate DNA uptake; cationic *liposomes* seem to be especially effective. In the widely used *electroporation* technique a short electrical pulse of a few hundred volts/cm is applied to create transient pores in the plasma membrane through which the DNA can enter a cell.

Chromosomes can be transferred by cell fusion and either entire chromosomes, isolated DNA, or cloned DNA can be transferred into egg cells by microinjection. Following transfer of DNA or of intact chromosomes, recombination sometimes leads to stable incorporation of some of the transferred genes into the host cell's genome. A very important technique is the use of high-velocity microprojectiles shot from a particle gun.

Spherical tungsten particles of ~0.5-1 μm diameter are coated with recombinant plasmids containing the genes to be transferred. The particles are then shot, using a gunpowder cartridge, into intact recipient cells. The particles penetrate cell membranes, mitochondria, and chloroplasts without serious damage if the number of particles is not too great. The technique is very useful for genetic engineering of plants.

Genomic Libraries

The human genome contains about 3×10^6 kb of DNA, about the average amount for a eukaryotic organism. If the entire genome is digested to completion with a restriction endonuclease, whose cleavage sites are distributed more or less randomly, the resulting restriction fragments constitute a "library" for that genome. If the average length of a fragment is 17 kb, about 1.8×10^5 unique fragments will be produced. To make a practical library these must be cloned into a suitable vehicle. Derivatives of phage λ or cosmids are most often used.

The cloned fragments can then be packaged using the *in vitro* packaging system to form infectious phage particles, which can be propagated as plasmids in *E. coli* cells. What is the probability that a given fragment among the total produced will be found in one of the recombinant phage in the library? From simple probability theory the number of clones that must be isolated and screened is given by Eq. 3.2. Here N is the number of clones needed, p is the probability of having the desired fragment in the library, and f is the fractional proportion of the genome represented by the fragment sought. For our example of a 17-kb fragment of the human genome this is 17/3000. From this equation one can calculate that to have a 99% chance of finding our fragment (p = 0.99) we need 8×10^5 clones.

$$N = \text{In}(1-p)\ \text{In}\ (1-f) \qquad ...(3.2)$$

What the equation does not show is that there will probably be some long fragments that cannot be cloned in the selected vehicle. These will be missing from the library. Large numbers of clones obtained can be screened rapidly by colony hybridization using a labeled DNA probe. Thus, if it is desired to isolate a gene for a particular protein and some part of that protein has been sequenced, a synthetic DNA probe can be made. The phage containing the recombinant fragments can be plated, and after plaques form a nitrocellulose filter can be laid on the plate to form a replica. After release of DNA from the phage, denaturation by NaOH, and neutralization the single-stranded DNA fragments can be hybridized with the probe.

Another screening method uses a probe carried in a plasmid that promotes homologous recombination between the probe sequence and restriction fragments with a similar sequence.

A problem that arises in preparing genomic libraries is that certain sequences, *e.g.,* those involving highly polymorphic regions and inverted repetions, often cannot be propagated in most lambda cloning vehicles. In addition togenomic libraries, *cDNA libraries* can be prepared from mixed mRNAs. The total RNA of cells is isolated and passed through an affinity column containing oligo(dT) chains. These bind to the 3′-poly(A) tails of the mRNAs, allowing them to be isolated. The mixed mRNAs can then be cloned using a poly(AT)-tailed plasmid vehicle and a reverse transcriptase.

Probes

The first step in screening the recombinant DNA in a library is use of some probe for detecting the desired DNA fragment. The most direct way is to synthesize a radioactive or fluorescent labeled oligonucleotide complementary to a short known sequence in the protein. The number of codons for a single amino acid varies from 1 to 6. It is therefore desirable to prepare a probe complementary to segments of DNA containing a high proportion of codons for Trp and Met (1 codon each) and Asn, Asp, Cys, Glu, Gin, His, Lys, Phe, and Tyr (2 codons each). A popular procedure is to synthesize a mixture of probes containing all of the possible nucleotide sequences coding for the selected sequence of amino acids.

The probe may be a mixture of more than 1024 different nucleotide. Antibodies are another popular type of probe. Antibodies to a specific protein may be utilized in isolating mRNA from ribosomes that are making that protein. Thus one or more strongly binding antibodies may have already been obtained before the library clones are to be screened. To use this technique for screening recombinant DNA, the cloning must be done in a vehicle that causes expression of the gene, *e.g.,* as proteins fused to *E. coli* galactosidase. One type of expression library is created by insertion of the cDNAs into copies of a bacteriophage gene that permits the expressed proteins to be displayed on surfaces of the phages.

Studies of Restriction Fragments

By cleavage with the correct restriction enzyme, cloned DNA fragments can be released from the vehicle in which they are carried in the library. What can be done with these fragments? The first obvious use is to sequence a gene that has been located with a probe. In many instances the gene will be longer than the cloned piece. However, the isolated restriction fragment, if labeled and denatured, becomes a highly specific probe for locating other restriction fragments that overlap it. For example, a fragment from an *Eco*RI library may bind to two or more fragments from a *Hind*III library. This permits "walking" along the chromosome to locate adjacent fragments. Cosmid vehicles that facilitate ^{32}P-labeling at the ends of the passenger DNA are useful.

A related approach called "jumping" depends upon converting very large DNA restriction fragments into circular molecules, digesting with restriction enzymes, and cloning the junction fragments of the circles. These fragments contain segments that may have been separated by as much as 100 kb in the genomic DNA, and enable the investigator to walk or jump again from a new location.

Locating Mutations

The study of restriction fragments provides a way of locating many mutations. For point mutations in genes for known proteins, sequencing reveals the exact defect. Some mutations, especially deletions, lead to changes in the lengths of restriction fragments. If the mutation causes loss of a restriction site, a longer piece of DNA will be present than in a digest of normal DNA. Such differences in length of restriction fragments are usually referred to as *restriction fragment length polymorphism* (RFLP, usually used in the plural as RFLPs). These polymorphisms are readily detectable by differences in mobility on gel electrophoresis. They can often be mapped to particular chromosomes by hybridization *in situ* study of naturally occurring translocations of chromosome fragments, or other techniques.

An example is provided by the human hemoglobin abnormalities known as *thalassemias*. Here, deletions remove certain restriction sites leading to observation of RFLP. Occasionally a point mutation is linked to RFLP at a nearby site. For example, in the United States most carriers of the sickle cell trait have a 13-kb *Hpa*I fragment that carries the globin S gene. However, noncarriers have their globin gene on a 7.6-kb *Hpa*I fragment. Although the association is fortuitous, the linkage between the hemoglobin S gene and the mutated restriction site is broken only rarely by crossing-over during meiosis. RFLPs have been linked to many other human genetic defects and have also provided the basis for the first linkage maps of human chromosomes.

Positional Cloning

By using enzymes that cut at relatively rare sites, genomic DNA can be cut into very large restriction fragments. From studies of inheritance within families carrying specific genetic traits it is sometimes possible to find linkages between those traits and polymorphic restriction fragments. This has been accomplished for a number of defective human genes including those responsible for sickle cell anemia, cystic fibrosis, Duchenne muscular dystrophy, Huntington's disease, X-linked chronic granulomatous disease, neurofibromatosis (elephant man's disease), the hereditary cancer retinoblastoma, and others. By 1997 nearly 100 hereditary disease loci had been located by positional cloning. These astonishing successes provided a major impetus for what became the Human Genome Project. Serious problems were met in actually locating these disease genes.

Crossing-over is infrequent, occurring only at about 50 locations during each meiosis. Therefore, linkage analysis does not tell us with any precision how close the linked gene is to a known DNA probe within a restriction fragment that may be up to 2000 kb in length. Finer restriction mapping or chromosome walking can be used to locate the precise piece of DNA that is defective. This can still be a formidable problem. However, if the defective protein can be identified it can be sequenced. A specific oligonucleotide probe can be made for its gene and can be used to establish the exact chromosome location.

Directed Mutation

In addition to the developments of cloning and sequencing of DNA, a third technique is essential to the present revolution in molecular genetics. That is the ability to mutate any gene at any point in a specific way. Because of its precise nature the technique is called *directed mutation* in this book. However, the term *site-directed mutagenesis* is often used. Mutations can be introduced randomly in DNA in many ways including treatment with nitrous acid, bisulfite, formic acid, or hydrazine or by incorporation of nucleotide analogs. Efficient procedures have been devised for isolating the mutants. For many purposes *oligonucleotide-directed mutation* is the preferred technique. An oligonucleotide of ~16-20 nucleotide length is synthesized with a sequence complementary to the coding strand containing the desired site of mutation. At that site the codon for the new amino acid is present.

Despite this mismatch the oligonucleotide can be successfully hybrudized with a single-stranded DNA such as that cloned in an M13 vehicle. Now the Klenow fragment of DNA polymerase I or a viral DNA polymerase is used to convert the single-stranded circular DNA into a double-stranded replicating form. Many of the single-stranded progeny will contain the mutated DNA. They can be screened with a lableled probe made from the oligonucleotide used to induce the mutation. It will hybridize most tightly to the correctly mutated gene. Use of the PCR reaction simplifies the procedure.

In another screening procedure the template strand is synthesized in an M13 phage vehicle using uracil rather that thymine. The circular heteroduplex obtained after synthesis with T4 DNA polymerase and T4 DNA ligase is taken up by *E. coli* cells, which select against the uracil-containing strand and, therefore, in favor of the mutated strand. A third approach is to completely synthesize a gene for the protein under study. The sequence does with restriction enzyme cleavage sites that permit easy excision and readdition of particular fragments.

Synthetic fragments containing various mutations can then be grafted in at will. Genes of this type have been made for rhodopsin and related proteins and for numerous other proteins. What are the uses of directed mutation? As we have seen in previous chapters, the technique is being used in every area of biochemistry to bring new understanding of protein functions and of the chemical basis of disease. Together with complete synthesis of genes it provides the basis for genetic engineering of specific proteins of plants, animals, and microorganism. Many protein products can probably be improved. For example, enzymes can be made more stable. Specificities can be changed, but this is difficult.

Targeting and Replacing Genes

One goal of human genetic therapy is to replace a defective gene in body cells with a good gene. It this really possible? It is essential to engineer the DNA that is to be transferred, so that it contains all of the components needed for efficient expression in the host following its incorporation as a *transgene*. A transcription-initiation region with suitable promoter, both 5′ and 3′ untranslated regions, start and stop codons, and polyadenylation site must all be present. It is hoped that a correctly constructed promoter region will allow the transgene to be picked up by the machinery of homologous recombination and be incorporated into the host's DNA and expressed in the appropriate tissues. Only some cells will take up the new gene and discard the old. However, it may happen enough to benefit a patient. Targeted gene replacement has been very successful in mice.

Knockout Mice

If targeted DNA is injected into a fertilized mouse egg, there is a chance that the mouse will have the targeted gene replaced in one chromosome, and that it will be stably transmitted to some of its progeny. If the transgene is totally nonfunctional, the mouse will be a "knockout mouse," suffering from a hereditary defect that can be transmitted through carriers such as its mother. The standard knockout technique is to inactivate the gene of choice in cultured embryonic stem cells and to inject these into mouse embryos. Some progeny will carry the inactivated gene in their germ cells. A refinement of the technique utilized the *Cre recombinase* or related enzymes to selectively remove pieces of DNA from genes in specialized tissues of mice. By 1996 several hundred different knockout mice had been created. Nevertheless, interpretations of results of gene knockout are sometimes complex.

THE GENETICS OF EUKARYOTIC ORGANISMS

Whereas DNA synthesis takes place almost continuously in a rapidly growing bacterium, replication of DNA occupies a more limited part of the *cell cycle* of eukaryotes. In a mammalian cell mitosis proper may require about one hour. It is followed by the "gap" period, G_1, whose length is variable and depends upon the cell type, the nutritional state of the cell, and other factors. About 10 h is typical. During the S phase (~9 h) active DNA replication takes place. This is followed by a second gap (G_2) that occupies 4 h in the 24 h cell cycle. The length of the different segments of the cell cycle varies widely among different organisms. Indeed, the concept of a cell cycle can be criticized. It is only in a rapidly growing culture that all, or most, cells can follow the same cycle. In the adult body most cells are inhibited from division (or are not stimulated to divide) most of the time.

Mitosis

The distribution of chromosomes to daughter cells of somatic cells undergoing division is accomplished by mitosis whose successive phases are referred to as *prophase*, *metaphase*, *anaphase*, and *telophase*. As the chromosomes condense during prophase, it is seen that each one actually consists of two separate entities coiled together. These are the identical *chromatids*, which are formed from the two identical double-stranded DNA molecules formed by replication of the DNA of the chromosome during the S phase of the cell cycle. As the folding of the chromosomes occurs (during prophase), the nuclear envelope completely fragments or dissolves in many species. An important event that *precedes* the main stages of mitosis is the formation of *poles* in the cell.

In animal cells, the poles are formed by the *centrioles*, which move apart and take up positions at opposite sides of the cell. Each of the centrioles is accompanied by a smaller "daughter" centriole lying at right angles to the larger parent.

In plant cells, which lack centrioles, a more diffuse pole is formed. As the cell prepares for mitosis, fine microtubules (15 nm diameter) can be seen radiating from the poles. At the end of prophase the microtubules run from one to the other to form the *spindle*. Microtubules also become attached to the chromosomes at the *centromeres*. At metaphase the chromosomes are precisely lined up in the center of the cells to form the *metaphase plate*. Now each centromere divides, permitting the sister chromatids to be completely separated. A protein complex *cohesin*, which holds the sister chromatids together, undergoes proteolysis by a *separase* at this stage. During anaphase the separated chromatids, now referred to as *daughter chromosomes*, move to opposite poles as if pulled by contraction of the spindle fibers. Telophase is the final stage in which new nuclear envelopes are formed around each set of daughter chromosomes.

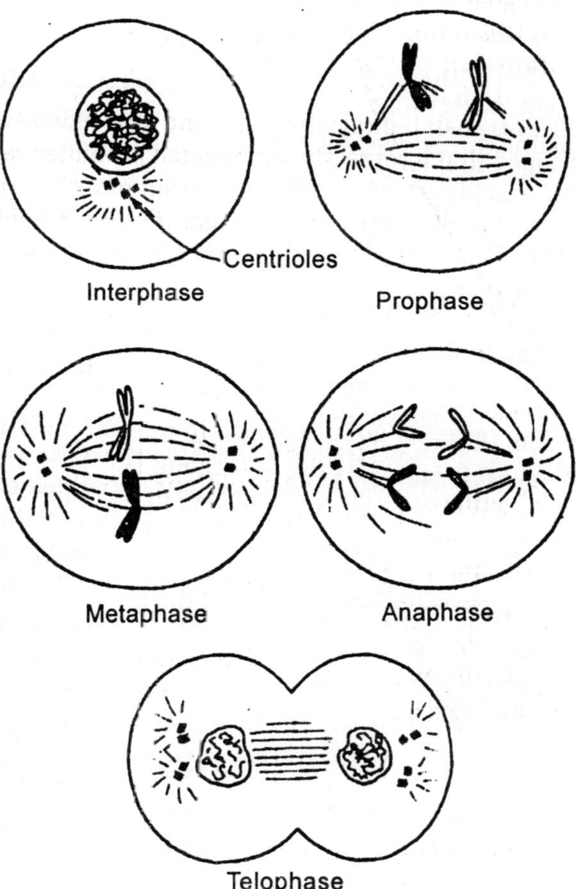

Fig. 3.11. Mitosis. Illustrated for a cell with one homologous pair of chromosomes.

In humans and many other species the cell pinches in two. In plants and fungi new plasma membranes and cell walls are constructed through the center of the cell. A partitioning of mitochondria and of Golgi compoenents between cells must also occur. The foregoing description overlooks the extreme complexity of mitosis, each stage of which must occur with precision and in the correct sequence. The replication of DNA, which takes place in the S phase of the cell cycle must be completed before mitosis begins. This is followed by condensation of the DNA into chromosomes, breakdown of the nuclear membrane, assembly of the *kinetochores* by which the chromosomes attach to the spindle, assembly of the spindle, attachment of chromosomes to the spindle, segregation of the chromosomes to opposite poles in anaphase, and finally the cleavage of the cell.

Cyclin-dependent Kinases

The cell cycle is controlled by a series of complexes of the 30 – to 45-kDa proteins called *cyclins* with *cyclin-dependent protein kinases* (CDKs). These kinases contain ~300-residue catalytic cores that resemble protein kinase A. like that kinase they transfer phospho groups from ATP to serine and threonine side chains of target proteins. The kinases are inactive until a complex with the appropriate cyclin is formed and is activated by phosphorylation. One or more additional kinases are required for this activation. Each stage in the cell cycle is controlled by one or more different cyclin-CDK complexes.

One of the best known CDKs is human CDK2, which functions in a complex with cyclin A during the S phase of the cycle. Binding of the cyclin greatly alters the conformation of CDK2, opening the catalytic cleft and exposing threonine 160. Its hydroxyl group can be phosphorylated by the action of the CDK-*activating kinase* (CAK) with a 100-fold increase in catalytic activity. Control of the CDKs also depends upon proteins that act as specific inhibitors and upon precise elimination of both cyclins and inhibitors via the ubiquitin system.

Checking for Completion of Replication and for DNA Damage

The first checkpoint in the cell cycle is the *start* or *G_1 DNA damage checkpoint*. Replication of DNA does not begin until the cell has had time to repair as much damage to DNA as is possible. As mentioned, the cancer suppressor protein p53 is an essential component of the checking process. A second checkpoint, the *G_2 checkpoint*, at the end of the G_2-phase requires verification that all DNA of all chromosomes has been replicated, checked for damage, and repaired if necessary. Control of these processes is accomplished by a mechanism first identified in fission yeasts, such as *Schizosaccharomyces pombe,* and which has been conserved in metazoa. The CDK known as *cdc2* (fission yeasts), CDC28 in budding yeast, or CDC2 or CDK1 in mammals, functions from G_2 through the DNA damage checkpoint to mitosis sometimes described as the master cell-cycle switch. The catalytic subunit of cdc2, a serine/threonine protein kinase, is often called p34 edc2. It is inactive unless complexed with a suitably phosphorylated cyclin and phosphorylated on Thr 161 by the action of kinase CAK.

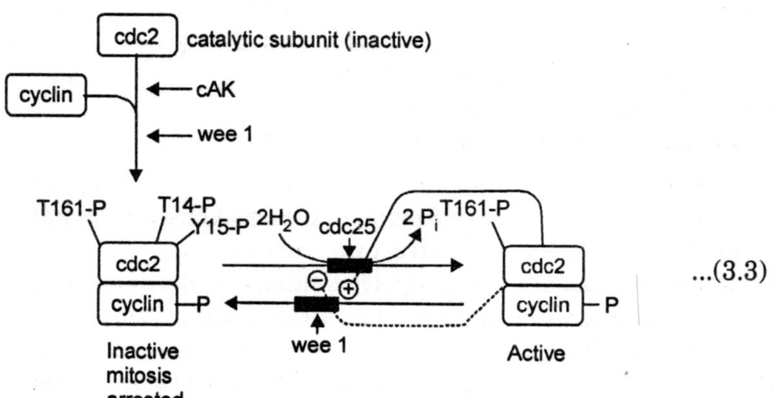

...(3.3)

However, when cdc2 becomes phosphorylated on Tyr 15 (in *Schizosaccharomyces pombe)* by the action of protein kinase wee 1 on Thr 14 by another kinase, cdc2 is inhibited and mitosis is arrested at the G_2 checkpoint. This allows time to verify completion of replication as well as for repair before the replicated DNA strands are separated in mitosis. Hydrolytic removal of the phospho groups from Tyr 15 and Thr 14 of cdc2 by the action of phosphatase cdc25 then allows mitosis to begin. This phosphatase is also controlled by a phosphorylation-dephosphorylation cycle. The checkpoint kinase *Chk1* phosphorylates and inhibits cdc25.

When damaged DNA is present, protein p53 accumulates, just as at the G_1 checkpoint, and activates transcription of the seven phosphoserine-binding proteins of the 14-3-3 family. These bind to the phosphorylated cdc25 phosphatase preventing activation of cdc2 by dephosphorylation. In fission yeast one of the 14-3-3 proteins, *Rad 24*, apparently binds to the phosphorylated cdc25 and induces its export from the nucleus preventing mitosis. Dephosphorylation of cdc25 and return to the nucleus allows mitosis to occur.

A somewhat different regulatory mechanism is used by the budding yeast *Saccharomyces cerevisiae*. In animal cells there are three cdc25 isoforms. In each case we probably have only a glimpse of a very complex process of checking for DNA damage and repair.

Mitotic Spindle Formation and the Spindle Assembly Checkpoint

Separation of the two copies of each replicated chromosome depends upon the spindle fibers. Their formation is preceded by replication of centrioles, if present, and formation of the two poles of the cell. Both γ-tubulin and acidic Ca^{2+}-binding proteins called *centrins* are involved. The microtubules appear to grow outward from the poles with their minus ends at the poles and their plus ends available for binding to the kinetochores, specialized protein complexes that assemble around the centromeric DNA. Each chromosome has two kinetochores, one for each daughter chromatid. These must be attached to spindle fibers coming from opposite poles. It has usually been assumed that random encounters of microtubule plus ends with kinetochores leads to correct linkage.

However, neither centrosomes or kinetochores are always essential to spindle assembly, and self-assembly occurs by motordriven sorting according to polarity of the microtubules. In addition to microtubules formation of spindles may require specialized matrix proteins. The spindle isn't formed until replication is complete. A small G protein called *Ran*, a relative of Ras, regulates spindle formation. Ran, in turn, depends upon a nucleotide exchange factor called RCC1 (*regulator of chromatin condensation*). RCC1 may signal that the replicated DNA is folding into chromosomes indicating that replication is complete. At the *spindle assembly checkpoint* the cell verifies that the metaphase spindle has been assembled correctly. All of the microtubules that pull the sister chromatids toward one pole or the other must be correctly attached to a kinetochore. In addition, there are interdigitated microtubules coming from both poles.

Specific motor molecules then push the two poles apart. During the assembly and complex movements of the spindle both cytosolic dynein and four different types of kinesin-like motor molecules are required. One of these is Kar3. After assembly of the spindle is complete, a signal must be sent to the mitotic apparatus to move into anaphase. A clue to the nature of the signal has come from the observation that a single kinetochore lacking a spindle fiber connection causes arrest of mitosis. Apparently unattached kinetochores send a "wait" signal, perhaps via the cytoskeleton.

Anaphase

After the spindle has been checked, a sudden loss of cohesion between the sister chromatid pairs allows them to move toward the opposite poles. This process is catalyzed by the *anaphase-promoting complex* (*APC*, or cyclosome) and its activator protein *Cdc20*, a large multiprotein complex. The APC also promotes proteolytic breakdown of cyclins and other proteins by a ubiquitin- and proteosome-dependent mechanism. Its E_3-ubiquitin ligase targets the mitotic cyclins and other proteins for destruction. A specific E_2 ubiquitin-conjugating enzyme required for degradation of cyclin B and exit from rmitosis.

The centrosome also plays an active role in cytokinesis, the final step in cell. The cell cycle encompasses so many different processes that it is clearly impossible to describe it by the single diagram or by the text written here. The cycle is influenced by a host of growth factors and external stimuli, many of which act on transcription of cyclins and other essential proteins. Transcription factors such as those of the Fos/Jun (AP-1) family in response to the MAP cascade are among those that control the transcription of cyclins.

However, during mitosis most transcription of any genes is repressed. Among other factors influencing the cell cycle is the size of the cell and the availability of nutrients including purine and pyrimidine nucleotides. Lack of cholesterol decreases the cdc2 kinase activity and causes apoptosis. A cell cycle regulator in *S. pombe* known as *suc1* is essential for cell cycle progression. Although its three-dimensional structure is known, its function (like that of its human homolog CksHs2) is uncertain.

Meiosis

The mechanism by which chromosomes are distributed during the formation of *gametes* (egg and sperm cells) is known as meiosis. Formation of gametes involves a halving of the chromosome content of a cell, each gamete receiving only one chromosome of each homologous pair. Genes found in the same chromosome are said to be *linked* because of their tendency to be passed together to the offspring.

Genes present in different chromosomes are not linked, and their inheritance follows the pattern of *random segregation* established in Mendel's famous studies. The simple fact that the genetic material is put up in several different packages (chromosomes) is sufficient to provide for considerable mixing of genetic information between different individuals in sexual reproduction. However, it doesn't provide a means for exchanging genes on the same chromosomes. Mixing of genetic information within chromosomes occurs by genetic recombination occurring during *crossing-over*, an aspect of meiosis with an essential biological role.

In the S phase preceding meiosis, DNA is duplicated just as it is prior to mitosis. This provides sufficient genetic material to produce *four haploid cells*. These are formed during meiosis by two consecutive cell divisions. Crosssing-over occurs prior to the first of these divisions, at the four-strand stage. The two homologous chromosomes of a pair come together to form what is called a *bivalent* or *tetrad* made up of four chromatids. For each chromosome, at least one chromatid is seen to come into intimate contact with a chromatid in the other homologous chromosome at points known as *chiasmata*.

During metaphase of the first meiotic cell division the homologous chromosomes, each still containing two chromatids, separate. Each chromatid now carries with it some genetic information that was previously found in the other member of the homologous pair and vice versa. Now, without further replication of DNA in the second meiotic cell division, the chromatids separate to form haploid cells. The process of crossing-over provides a means by which genes that are linked on the same chromosome can be separated, providing offspring with mixtures of genetic traits other than those predicted by simple Mendelian theory.

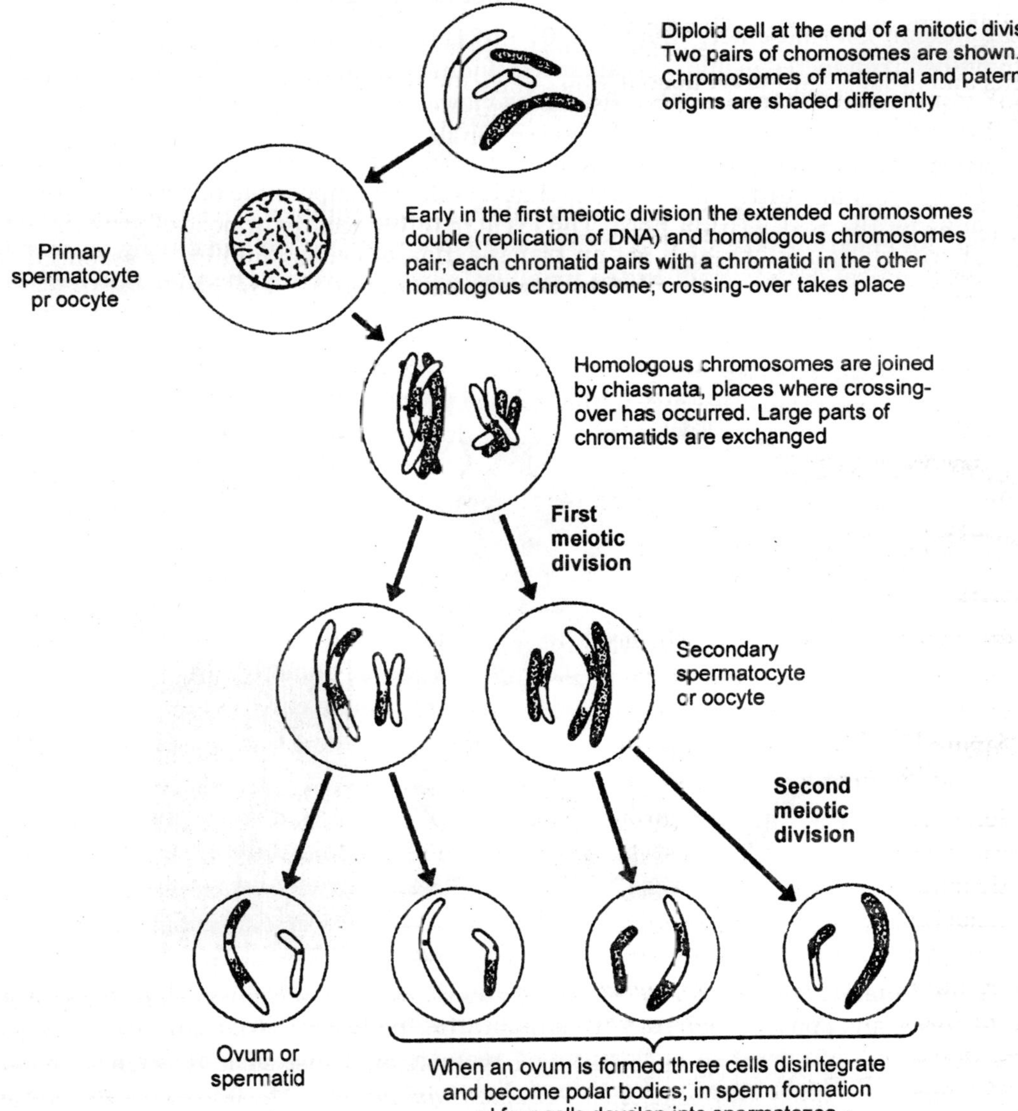

Fig. 3.12. Meiosis. Cell division leading to formation of haploid gametes.

The effects of crossing-over were first studied extensively by T. H. Morgan with the fruit fly *Drosophila melanogaster.* Morgan discovered his first mutant, a white-eyed fly, in 1910. The first genetic maps were made by assuming a direct relationship between the frequency of crossing-over and the linear distance between genes in a chromosome. Thus, the same approach to genetic mapping that was used later with *E. coli, i.e.,* the measurement of recombination frequencies, was applied much earlier to crossing-over in the chromosomes of *Drosophila*. Extensive genetic maps involving many mutations were obtained for the four chromosomes of this organism, and similar techniques have been applied to many other organisms. The unit of distance in these chromosome maps is the *morgan* (named for T. H. Morgan). One centimorgan is the distance that allows recovery of 1% of recombinant progeny.

In the human chromosomes, this is ~1000 kb (1Mb). Biochemical and genetic studies of meiosis have been conducted in many organisms including fission and budding yeasts, *Drosophila*, starfish, *Xenopus*, and the mouse. Meiosis can be viewed as a modification of mitosis but with the added initial step of crossing over and recombination. In addition, the S-phase of the cell cycle is absent in the second meiotic division.

Meiosis may occur at different stages of the life cycle of organisms. An important advantage in using fungi for genetic studies is that, like prokaryotes, they are haploid during much of their life cycle. Biochemical defects such as the inability to synthesize a particular nutrient can be recognized readily at this stage. At the same time genetic crosses can be made, and crossing-over frequencies can be measured and used for genetic mapping.

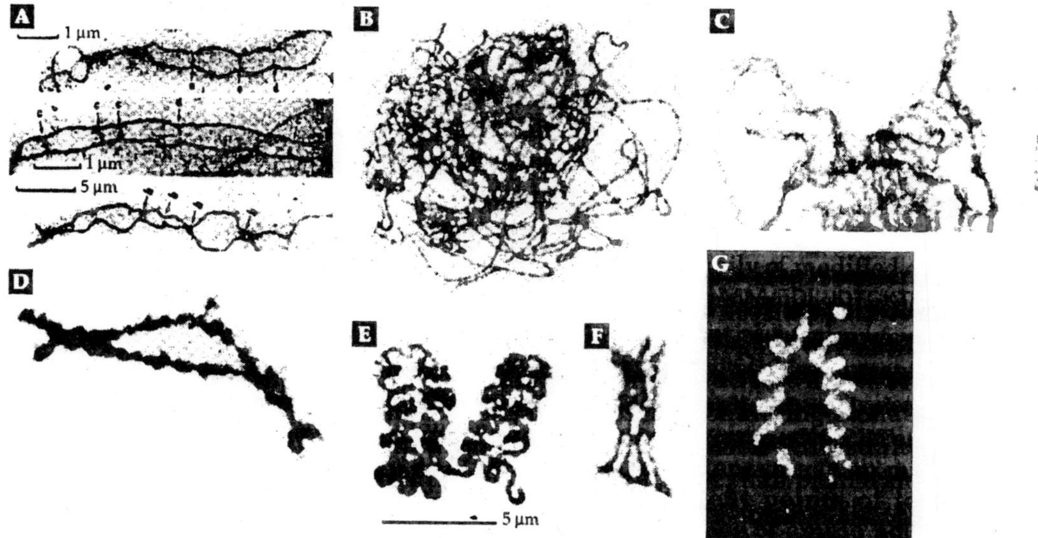

Fig. 3.13. Synaptonemal complexes. (A) Aligned pairs of homologous chromatids lying ~0.4 μm apart in *Allium cepa*. Arrows indicate "recombination nodules" which may be involved in initiating formation of crossovers. Portions of meiotic chromosomes of lily are shown at successive stages: (B) Pachytene. (C) Portion of diplotene nucleus. (D) A bivalent at diplotene. (E) Two bivalents at diakinesis. Pairs of sister chromatids are coiled with appropriate handedness. (F) Sister chromatid cores are far apart in preparation for separation. A chiasma is present between the two central strands. (B) through (F) courtesy of Stephen Stack. (G) Pair of sister chromatids coiled with opposite handedness at metaphase. These are immunostained with anti-topoisomerase II antibodies.

The onset of meiosis may vary in time with diploid organism. Meiosis in males may occur quickly. However, after chromatid pairing and recombination meiotic divisions of oocytes are arrested at the G_2 stage, in some species for many years. The arrest is ended by hormonal stimulation, *e.g.*, via progesterone acting through a cyclin B=cdc2 complex. *Meiosis-activating sterols*, intermediates in lanosterol metabolism accumulate and are thought to participate in control of meiosis. In haploid strains of fission yeasts sexual development is induced by starvation, especially of nitrogen. Cells of opposite mating types then fuse to form zygotes, which usually undergo meiosis immediately. Starvation is apparently signaled by the *cyclic AMP-protein kinase A cascade*. During the prophase of the first meiotic division (meiosis I) two homologous pairs of partially "condensed" chromosomes must find each other and pair with appropriate orientation. A protein in the telomeres of the chromosomes seems to be involved. The key structure in meiotic crossing-over is the ribbonlike *synaptonemal complex* formed by the pairs of homologous chromatids. This complex, in which a proteinaceous core or axial element separates the greatly extended chromatid pairs is fully formed in the *pachytene stage* of meiosis. Formation of the synaptonemal complex is preceded by development of a few double-stranded breaks in the DNA. The 3′ ends of the DNA chains then invade the homologous chromatids to initiate the exchange via *Holliday junctions*.

The first meiotic division is a long process, for example, lasting ten days in mouse spermatocytes. As cells pass to the metaphase stage of meiosis I, the chromosomes become much more compact, but the attachments between the homologous chromatids are still visible as chiasmata. The chromatids then separate, the two homologs appearing as coils of opposite handedness. Within these coils the pairs of sister chromatids continue to be held together at their centromeres through metaphase and anaphase of meiosis I. One or more specific proteins, which must release their hold in meiosis II, are required:

A leading cause of embryonic and fetal death as well as severe mental and physical problems after birth is incorrect segregation of chromosomes during meiosis. If a gamete contains two copies of any chromosome the embryo will have three. Down syndrome, trisomy of chromosome 21, is the most frequent example.

The spindle assembly checkpoint during meiosis I may sometimes be at fault.

Polytene Chromosomes

While most cells of higher organisms are normally diploid, the chromosome number may sometimes be doubled or increased even more. A cell with twice the diploid number of chromosomes is tetraploid, and with higher multiples of the haploid number it is polyploid. Plant breeders have succeeded in producing many tetraploid varieties of flowering plants often with increased size. One tetraploid mammal, the red viscacha rat, is known. While most of our body cells are diploid, we, too, have polyploid cells. For example, some are always found in the liver. The most spectacular example of an increase in the normal DNA content of cells is provided by the giant *polytene chromosomes* of dipteran (fly) larvae. The DNA of cells in the salivary glands and some other parts of these organisms doubles as many as 11-14 times without cell division to give a several thousandfold (*i.e.*, 2^{14}-fold) increase. The supercoiled, duplicated DNA molecules all line up side by side in a much more extended form than in ordinary chromosomes. They can be seen readily by light microscopy. The total length of the four giant chromosomes of *Drosophila* is ~2 μm, compared to 7.5 cm in diploid cells.

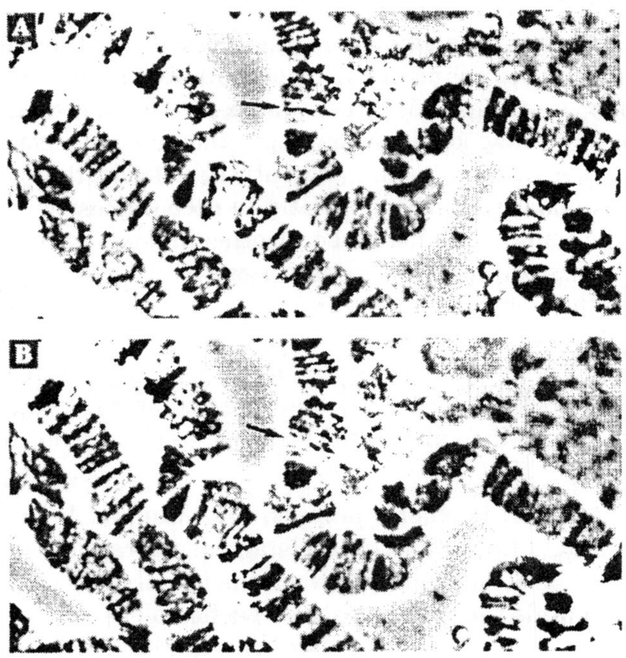

Fig. 3.14. Microdissection of a *Drosophila* salivary gland chromosome. (A) Before the cut. (B) After cutting a medium size band. The arrow indicates the site of the cut.

The giant chromosomes have a banded structure, ~5000 bands being visible along the length of the chromosome. An average band contains ~ 36 kb of DNA in each of the strands. Since it has been possible to correlate visible changes in the appearance of these bands with particular mutations in the DNA, study of polytene chromosomes provided a second important method of mapping genes of the fruit fly. The maps produced by the two methods agree well. Another use of polytene chromosomes is micro-dissection of DNA for cloning.

A piece of DNA containing 100-400 kb can be cut desired spot, cleaved with restriction enzymes, and cloned. Since it has been estimated there may contain only ~9000 -17,000 genes there may be 2-3 genes per band in these chromosome technique has been extended to human and other mammalian chromosomes.

Cytoplasmic Inheritance

Not all hereditary traits follow the Mendelian patterns expected for chromosomal genes. Some are inherited directly from the maternal cell because their genes are carried in the cytoplasm rather than the nucleus. There are three known locations for cytoplasmic genes: the mitochondria, the chloroplasts, and certain other membrane-associated sites. An example of the last is found in "killer" strains of yeast. Cells with the killer trait release a toxin that kills sensitive cells but are themselves immune. The genes are carried in *double-stranded RNA* rather than DNA, but are otherwise somewhat analogous to the colicin factors of enteric bacteria. Similar particles (κfactors) are found in *Paramecium.*

THE HUMAN GENOME PROJECT

The discovery of site-specific restriction endonucleases in 1970 and the development of efficient DNA sequencing methods in 1977 sparked a revolution in biology. On Oct 1, 1990, the 15-year project to sequence the complete human genome was officially begun. The project is far ahead of schedule. "First drafts" of the genome were published in 2001. A more complete version will include *annotation*, a listing of predicted exons, mRNA transcripts, functions of recognizable genes, etc. Such documentation was provided first for chromosome 22 and is now available for other human genes. It was a surprise to discover, at least initially, that only 30,000-40,000 protein-coding genes could be recognized. Earlier 50,000-150,000 genes had been predicted.

A nearly complete version of the human genome is anticipated for spring of 2003. Included in the genome project are completion of the sequences for *Caenorhabditis elegans, Drosophila melanogaster,* and the mouse (*Mus musculus*). However, substantial difficulties remain in filling hard-to-sequence gaps, and some new approaches may need to be developed. More recently the project has been extended to include many additional species. Among them are the rat, the zebrafish, the pufferfish *Fugu rubripes*, the human malaria parasite *Plasmodium faleparum,* the rodent malaria *Plasmodium yoelii, yoelli* and the mosquito *Anopheles gambia,* which carries the parasite. Plant genome sequences are also being determined.

The Mammalian Genome and Human Health

Human beings and other mammals all have about 3500 Mbp of DNA containing perhaps 40,000 protein-coding genes. However, ~97% of the genome is repetitive DNA and other DNA that, at present, has no recognizable function. The genome project will reveal all of these sequences and will doubtless provide us with many surprises. A primary goal of the genome project is to understand the relationships between gene sequences and human diseases and health.

Until recently little was known about the locations or structures of genes in human chromosomes. Human genetic diseases provided the first clues. Although systematic genetic experiments cannot be done with human beings, almost the entire population is under some kind of medical care. Genetic defects are being detected more and more often, and the inheritance within families is traced with increasing frequency. Consequently, a huge body of knowledge of mutations to the human genome is available. By 1995 ~4500 human genetic disorders had been discovered. Many more have been found since then. The first mapping of human genetic defects came with the recognition that sex-linked traits are encoded on the X-chromosome. Some linkage analysis was also possible from studies of inheritance within families.

For example, among individuals who have two X-linked traits, *e.g.,* color blindness plus one other, naturally occurring crossing-over occasionally breaks the linkage in some individuals within the family. Somatic cell fusion provided an additional approach. Human lymphocytes can be fused with rodent cells under the influence of inactivated Sendai virus, which causes the cells to adhere and then to fuse. From such human-mouse or human-hamster hybrid cells, strains in which the nuclei have also fused can be selected. Although such cell lines can be propagated for many generations, they tend to lose chromosomes, especially those of human origin.

By observing loss of particular human enzymes or other proteins (separable from the hamster enzymes by electrophoresis), it was possible to assign genes to specific chromosomes. This also required identification of the chromosomes lost at each stage in the experiment. New staining techniques made it possible to identify each of the 22 pairs of human autosomes as well as the X and Y chromosomes. Using a variety of techniques, 500 loci in human chromosomes had been mapped by 1983. Separation of individual human chromosomes with a fluorescence-activated cell sorter permitted the preparation of libraries of cloned DNA from individual chromosomes.

Fragments, of an average length of 4 kbp, from digestion by *EcoR* I or *Hin*dIII were packaged into the Charon 21A cloning vehicles and "amplified" by culturing infectious phage particles in *E. coli*. However, this approach was inadequate for sequencing the entire genome. By 1987 the use of yeast artificial chromosomes (YACs) and later bacterial artificial chromosomes (BACs) and radiation hybrid mapping provided the means for sequencing the 34- to 260-Mbp human chromosomes. Also essential was the development of high-speed automatic sequencing machines and of computers adequate to assist in compiling the sequences.

Sets of overlapping and redundant, clones define acontinuous sequenced segment that is called a *contig*. As the contigs are enlarged the gaps are filled. In 1987, the first global genetic linkage map of the human genome was published. It was based on 403 polymorphic loci, of which 393 were RFLPs studied in 21 three-generation families. By 1996 a linkage map containing 5264 loci, in the form of polymorphism in *short tandem* repeats $(AC/TG)_n$ represented by *microsatellite* DNA, was available. Also mapped by 1998 were 2227 *single nucleotide polymorphisms* (SNPs), where the two alternative bases occur with a frequency. This was increased to 1.4 million by 2001. As many as 5.3 million common SNPs may occur, each with a frequency of 10-50%. On the average there may be one SNP for every 600 base pairs. These SNPs account for a large fraction of the diversity in human DNA.

By 1995 a physical map with >15,000 *sequence tagged sites* (STSs) with an average spacing of 199 kb had been constructed. An STS is defined as a cloned fragment in a YAC (or BAC) library that has been amplified by PCR and tested to establish that it contains a known locus. DNA sequences known as *expressed sequence tags* have also been mapped. Using STSs, RFLPs, ESTs, and the growing contigs, more than 16,000 human genes had been mapped by 1996. By 1998 more than 1,060,000 ESTs had been reported. In 1999, the nucleotide sequence of the smallest human chromosome, the 33.4-Mbp chromosome 22, was completed. It contains at least 545 genes and 134 pseudogenes. In the entire genome >1000 seed contigs had been assembled, and completion of the first phase of the Human Genome Project was in sight. In 2001, a whole-genome clone-based physical map wās published.

Understanding Gene Sequences

The vast array of sequence data coming from the human genome and from genes of other species are deposited, as they are reported in the scientific literature, in the Human Genome Central. It can be found on the Web at the National Center for Biotechnology Information (NCBI) at http://www.ncbi.nlm.nih.gov/ genome/guide and European Bioinformatics Institute (EBI) at http://www.ensembl.org/genome/central/. The data have been doubling every 15 months. By 2000 almost 1.3 rnillion human ESTs were included as were sequence data on more than 25,000 other species. A problem is that there are many errors in the data with some seriously incorrect sequences. It is hard to manage the mass of new data, but in time most of the errors will probably be corrected. Books and computer programs are available to help understand genomes. The widely used programs BLAST and FASTA, available on the Web or in computational packages, routinely compare protein-coding genes with known genes in order to predict function. An inexpensive high-powered desktop computer and an internet connection will enable a person to do complex biocomputing. Genes must be transcribed, and most transcripts must be spliced, modified, and translated by the ribosomal machinery. Genes cannot be fully understood without considering these processes.

Human Variation

Between any two unrelated human beings there are on the average one base pair that is different out of every 500-1000 nucleotides. This amounts to ~4 × 10^6 differences in the whole genome. In addition to these single nucleotide polymorphisms (SNPs) there are many differences in the >100,000 nearly randomly dispersed short tandem repeats of microsatellite DNA. The latter form the basis of DNA fingerprinting, and together with the SNPs are helpful in tracking genetic diseases. One of the more difficult goals is to identify genes that confer increased susceptibility to cancer and to the complex syndromes of diabetes and mental problems.

Searching for correlations with SNPs has proved difficult. A new approach termed *haplotype mapping* may permit correlations of disease susceptibility with larger blocks of conserved DNA sequence known as haplotypes. Some controversy surrounds possible uses of these maps, as also is the case for proposals to conduct extensive genetic testing of populations including newborn infants. Another planned project is the sequencing of 1000 individual human genomes. Studies of DNA have also shed light on human evolution.

Mitochondrial DNA, as well as a variety of nuclear genes, is being studied in attempts to establish approximate dates of evolutionary divergence. If we assume a constant rate of incorporation of nucleotide substitutions during evolution, we can use sequence data as a *molecular clock* to construct phylogenetic trees such as that, which is based on the gene for 16S ribosomal RNA. A problem is that mutation rates of genes are not all the same, and there are not yet enough data to draw firm conclusions. Estimates based upon the maternally transmitted mitochondrial DNA suggest that we are all descended from Africans who lived 100,000 to 200,000 years ago.

Limited data from the Y chromosome, which is transmitted through males, agree with the mitochondrial DNA data. However, study of the pyruvate dehydrogenase gene has been interpreted to indicate a more ancient divergence of African and other peoples. Analysis of DNA from a fossil bone shows that the mtDNA of Neandertal people differed from modern mtDNAs in an average of 26 positions. From this figure the mitochondrial DNA molecular clock predicts that the extinct Neandertal line diverged from ours ~ 550,000 to 690,000 years ago. The most ancient mtDNA found is Australian. Unfortunately, DNA in fossils is very unstable and has not been recovered from older fossils, *e.g.,* from dinosaurs. However, the search for ancient DNA goes on. We diverged from our closest ape relative, the chimpanzee, about 4-6 million years ago.

Chimpanzee and bonobo DNA sequences are ~98.8% identical to those of humans. If differences in inserted and deleted segments of genes are included, however, the identity drops to 95%. One of the longest DNA sequences to be compared among humans and apes is a 7-kb length around the pseudogene in the β globin cluster. In this sequence chimpanzees are closest to humans with gorillas being the next closest. We may well ask in what way we differ from these apes? Some specific differences have been found. Notably, human beings do not hydroxylate the glycolyl groups of sialic acids to form *N*-glycolylneuraminic acid residues on glycoproteins. Could this really be the most important difference between us and the apes? More genomic analysis may tell. What does DNA analysis tell us about race?

Most investigators conclude that there is only one human race with no detectable boundaries between the group commonly referred to as races. As Paabo put it, "in terms of the variation at most loci, we all seem to be Africans, either living on that continent or in recent exile. The differences in skin color seem to reflect adaptation to the environment in which people live. Variations in the level of melanocyte-stimulating hormone receptor, one regulator of skin color, are especially high in Africans. DNA analysis has also been useful in tracing human migration. For example, a genetic marker in the Y chromosome is carried by 85% of native Americans, suggesting that they are all descended from a man who lived ~20,000 years ago, probably an immigrant from Siberia.

Contrary to usual assumptions women, more often than men, seem to have spread their DNA to new locations in the world. Studies of cattle and of the wild ox reveal information about domestification of these animals about 10,000 years ago in Europe, Asia, and Africa.

Other Evolutionary Relationships

Studies of chromosome banding, chromosome maps, restriction fragments, and detailed sequences provide many insights into relationships among species. For example, the chromosome banding pattern and also DNA sequences show close similarities between human beings and the mouse. The latter is often regarded as the premier organism for the study of mammalian genetics and development. Dense genetic maps are available for both the mouse and rat as are moderate-resolution maps for livestock, companion animals, and other mammals. Comparative gene maps are being constructed for more than 40 mammals and other species of animals, plants, and fungi. Comparisons of these genomes reveal much of interest.

For example, the pufferfish *Fugu ruripes* has a genome only 1/9 the size of the human genome. However, both species seem to contain about the same number of genes. Many of them can be directly correlated and some human-disease-causing mutations have been identified first in the Fugu. Evolutionary history is being rewritten in molecular terms. Comparison of sequences of individual proteins allows evolutionary relationships of their genes to be traced. Many families of *homologous* genes can be identified. These include both *orthologs*, genes in different species that have evolved from a common ancestor, and *paralogs*, genes related by gene duplication within a genome. Orthologs have the same function in different organisms, but paralogs have different functions within a single species. Sophisticated molecular clock studies suggest that the evolution of metazoan organisms began earlier than had been supposed.

Ancestral primates appeared at least 65 million years ago. Gene sequence data for many species suggest that a great variety of mammals lived 100 million years ago in the age of dinosaurs, a view also supported by new fossil evidence. Metazoans appeared earlier than the "Cambrian explosion" generally thought to have occurred ~550 million years ago. New geochemical data suggest that cyanobacteria diverged from other bacteria as early as 2.1×10^9 years ago. Gram-negative bacteria diverged from gram-positive microbes ~3.2×10^9 years ago, *Salmonella* from *Escherichia* only 0.1×10^9 years ago. However, DNA analysis shows that within this latter time period many genes from other microorganisms have been inserted into the *E. coli* chromosome and into other bacterial chromosomes. Some of these transfers have occurred with the help of bacteriophages.

A puzzle is the fact that among eukaryotic cells the enzymes catalyzing the genetic *information transfer* via transcription and protein synthesis resemble those of archaea. However, *operational enzymes* that catalyze other basic metabolic processes tend to be more similar to those of eubacteria such as *E. coli.* So much gene transfer between organisms has occurred that it is difficult to establish the earliest parts of a phylogenetic tree. Another factor that confuses our study of bacteria is that less than 1% of all living microorganisms have been grown in pure cultures. We are still dangerously ignorant of the complexity of the microbial world, which both threatens us with diseases and sustains our environment.

We do have complete genome sequences for more than 60 different bacteria with hundreds more expected within a few years. They contain from 0.58 Mbp (*Mycoplasma genitalium*) to 8.7 Mbp for the antibiotic-producing *Streptomyces coelicolor,* a size shared by the legume symbiont *Bradyzhizobium japonicum*? The latter is one of many bacteria that have genomes split into two or more parts, often a major chromosome plus one or more plasmids. For example, the 6.7-Mbp genome of *Sinorhizobium melitoti,* an alfalfa symbiont, comprises a 3.65-Mbp chromosome and 1.35- and 1.68-Mbp megaplasmids. The 5.67-Mbp genome of *Agrobacterium tumefaciens,* much used in genetic engineering, consists of a circular chromosome, a linear chromosome, and two plasmids. The genome sequence revealed a close similarity to those of the above-mentioned legume symbionts. Sequences of many pathogenic bacteria are known. These include the causative agents of cholera, typhoid fever, plague, brucellosis, leprosy, tuberculosis, and anthrax. Also included is a virulent strain of *Streptococcus pneumonia* (respiratory infections, ear aches, meningitis), *Pseudomonas aeruginosa* (a common "opportunistic" pathogen), *Listeria monocytogenes,* which causes a severe food-borne disease the pathogenic *E. coli* 0157:H7, and a tiny mucosal pathogen *Ureaplasma urealyticum*? Many surprises were found in the genome sequences. For example, *E. coli* 0157:H7 is related to the laboratory strain *E. coli* K-12, and the two strains share a ~4.1 Mbp common "backbone" sequence.

However, the pathogenic strain has hundreds of "islands" of additional DNA spread throughout the genome and amounting to ~1.54 Mbp. Many of these carry genes associated with virulence. Specialized features appear in virtually every genome. For example, the mycoplasma *U. urealyticum* contains genes for enzymes that allow the bacterium to obtain almost all of its ATP from hydrolysis of urea. Both *Caulobacter crescentus* and *E. coli* develop specialized structures involved in motility. *C. crescentus* and also spore-forming bacteria have alternative developmental plans. At the low end of the genome size range are species of *Buchnera,* which are endocellular symbionts of aphids. Their genome is only slightly larger than that of *M. genitalium.*

The extremely salt-tolerant *Halobacterium* and the heat-tolerant *Thermoplasma volcanium* are among the archaea for which complete genome sequences are known. Species of *Xanthomonas,* whose genomes have been sequenced, are economically important plant pathogens. *Xanthomonas campestris* is also grown commercially to produce xanthan gum. Determination of the genomes of mycobacteria has been challenging. Among these slowgrowing organisms is *Mycobacterium tuberculosis,* the causative agent of human tuberculosis. A large fraction of its ~4000 coding genes is devoted to metabolism of lipids and to the synthesis of unusual proteins and lipids of its cell wall. The larger genomes code for many proteins of unknown function, but over 80% of the "ORFs" (presumed genes) of *Haemophilus influenzae* have been identified, as have their presumed functions. The encoded proteins appear to catalyze 488 metabolic reactions on 343 different metabolites. Together these systems provide a *metabolic genotype*. Results of such analyses are accumulating in *metabolic data bases*.

What is the minimum number of metabolic reactions needed for support of life? Transposon insertions can inactivate all but from 265 to 350 of the protein-encoding genes of *M. genitalium* without killing the bacteria. From comparisons of a variety of bacterial chromosomes it seems likely that -256 of these genes are truly essential. This conclusion leads to an interesting question. Is it ethical to now try to generate such a minimal bacterium? Are there hazards, *e.g.,* that its escape might endanger our health or the environment? On the other hand, genetic engineering of bacteria, which is already practiced, can provide useful improvements in bacteria used in foods and in industry. Modeling with the aid of data available on the World Wide Web is leading to development of new mathematical descriptions of metabolic networks. An ambitious new project is to model the entire *E. coli* cell. Many experimental data will be required and it has been estimated that ten years will be needed. The effort involves investigators in many laboratories and will be at least ten times as complex as the determination of the human genome.

Metabolic Studies of Eukaryotic Cells

The yeast, *Saccharomyces cerevisiae,* contains ~ 6200 genes of which, until recently, only 40% had been assigned a function. Now a variety of methods are being employed to understand this little fungus. A useful approach is to systematically inactivate or "knock out" genes. Davis and associates used a PCR-based strategy to delete one gene at a time of 2026 yeast genes. Of these genes 1620 were found not essential for growth in a rich medium. Ross-McDonald and coworkers engineered a transposon, a 274-bp derivative of the *E. coli* Tn3, and allowed it to be inserted into the genes of yeast cells by homologous recombination. The transposon carried DNA for a short peptide tag in the form of a specific immunological epitope that could identify the transformed cells.

More than 11,000 strains with disruptions in nearly 2000 genes were obtained. These and other deletion mutants are now available for study. A second yeast, *Schizosaccharomyces pombe,* has assumed major scientific importance in studies of the cell cycle and of metabolism. Its 13.8-Mbp genome is only a little smaller than that of *S. cerevisiae,* but it has ~1400 fewer recognized genes, a total of 4824. There are smaller eukaryotic genomes. That of the tiny marine chordate *Oikopleura dioica* may be only 51 Mbp. However, there may be a total of ~15,000 genes. Transposon-induced mutations have also been created in nearly one-fourth of the ~12,000 genes of *Drosophila melanogaster.* Studies of the expression of genes in both *Drosophila* and in the nematode *Caenorhabditis elegans* are directed toward understanding of development and differentiation. Of the nematode's predicted 19,293 genes, only 7% have been studied at the biochemical level. To understand what happens during development we need methods for studying simultaneously the expression of all of these genes. One approach is to look at messenger RNAs that are formed at different times during development. More than 9000 mRNAs have been identified in cells of C. *elegans,* and their patterns of expression have been observed using DNA microarrays. A similar technique has been applied to yeast. DNA sequences of fragments of ~ 6400 yeast genes were amplified by PCR and printed onto a glass plate to form a "DNA chip." From mRNAs formed at different times during growth, fluorescent cDNA copies were made, and their amounts were checked by use of the DNA chip. Another approach is to look directly at the proteins formed. Walhout and coworkers have devised a large-scale automated system for cloning all of the genes of *C. elegans,* expressing the protein products, and testing them in the yeast two-hybrid system for protein-protein interactions.

Another project is to use large-scale sequence comparisons between proteins of *C. elegans* and of other organisms to identify nematode genes that encode extracellular matrix proteins involved in cell adhesion and to trace their evolution. The genomes and the metabolism of the two insects *Drosophila melanogaster* and *Anopheles gambia* can now be compared. Many differences can be seen but almost half of the genes are orthologs. Many of these can be related also to those of pufferfish, mice, humans, and other species. Among plant genomes that of *Arabidopsis thaliana* has been studied most. The sequences of its five chromosomes have been determined and analyzed. About 25,498 genes encoding proteins from 11 different families have been found.

More than 14,000 ESTs were established from cDNAs. Many of the genes represent new families, some of which may be peculiar to plants. However, many others are homologous to those of *C. elegans* and *H. sapiens*. For example, developmentally important homeotic genes, marked by a *homeo sequence* as in animal genes, are present as are thousands of cell surface receptors. However, only Ser/Thr and histidine kinases are present in *Arabidopsis*. No tyrosine kinases have been identified. The ~125-*Arabidopsis* genome is tiny compared to the 3000 Mbp of DNA present in the genomes of maize and of many other plants. However, the rice genome, only ~15% as large as that of maize, has been chosen for complete sequencing.

Draft sequences for the genomes of two subspecies of rice (*Oryza sativa L.*), ssp. *Indica* and ssp. *japonica*, were published in 2002. The much larger genomes of other cereal grains (2500, 4900, and 16,000 for maize, barely, and wheat, respectively) will probably all be sequenced within a few years. The genes of maize are being mapped and studied using transposon-tagging methods. The smaller, more compact genomes may have just as many genes as the longer ones but have less repetitive DNA. Some ferns have 307,000 Mbp of DNA, nearly 100 times that of a human; bony fishes have -307,000 Mbp and the ameba >200,000 Mbp. These organisms appear to lose unneeded repetitive DNA faster than those with smaller genomes.

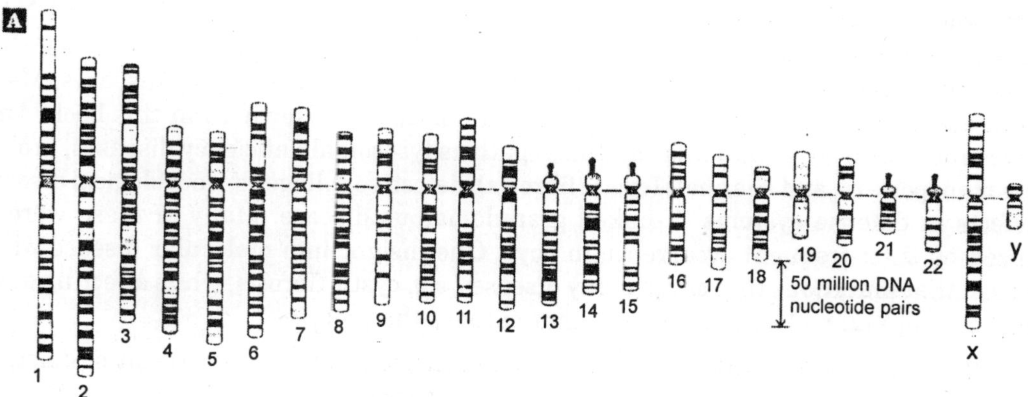

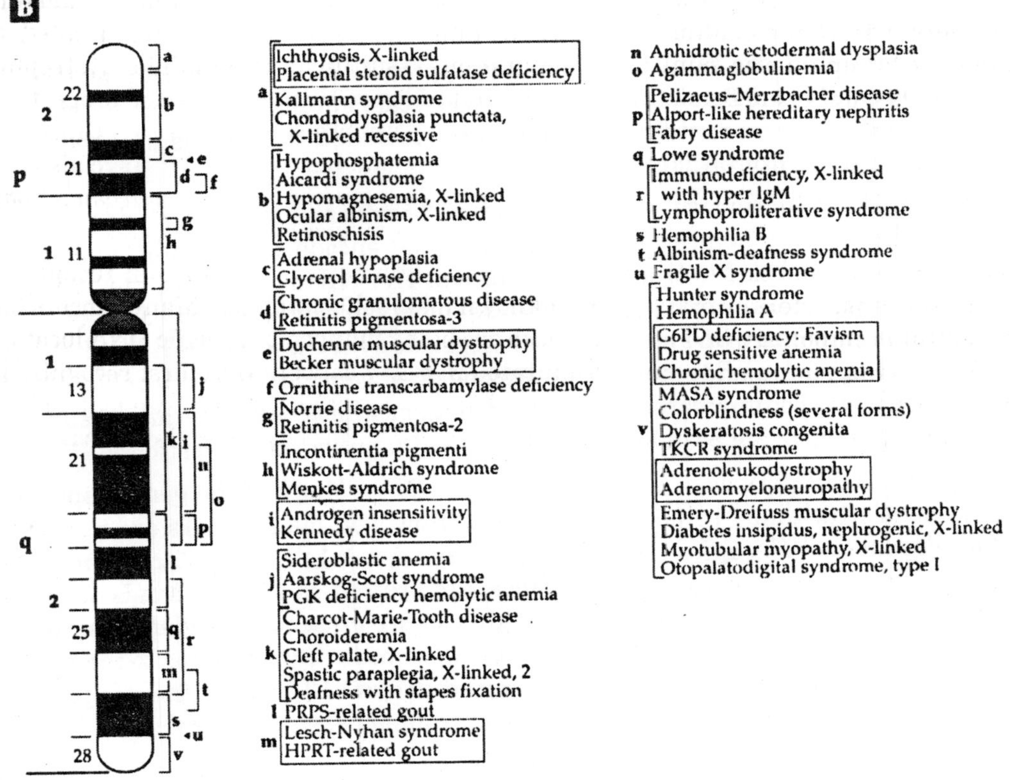

Fig. 3.15. (A) Numbering and staining patterns (G-banding) of human chromosomes. The horizontal line marks the centromers. Chromosome 13, 14, 15, 21, and 22 have nucleolar organizers located at the constrictions in their short arms. From Alberts *et al.* Adapted from U. Franke. (B) Map of genetic defects identified in X chromosomes by 1995.

Understanding Human Genetic Diseases

Genetic diseases have always been with us, but it was not until 1949 that the first disease, sickle cell anemia, was understood at the molecular level. A single base substitution in DNA and the resultant single amino acid substitution in hemoglobin causes this disastrous disease. It was soon recognized that defects in single proteins are causes of other inherited diseases. Many of the hundreds of other known genetic disorders are discussed elsewhere in this book. Among them are muscular dystrophies and cardiomyopathies lysosomal deficiency diseases, problems with ion transporters, and channels (cystic fibrosis), defective collagens, neurological disorders, and defects in defense systems (X-linked granulomatous disease. Many of these were first recognized by their frequent occurrence in boys. One insight into molecular disease was the recognition that mutations that cause many diseases, *e.g.* cystic fibrosis, while affecting a single protein, occur at many places in the gene that encodes the protein.

Not all mutations are caused by base substitutions; they are often a result of deletion or insertion of DNA. A whole group of diseases are caused by accumulation of repetitive DNA, often of nucleotide triplets, within genes or in control regions of genes. Another important insight is that understanding a newly discovered and very rare disease may help us to understand other related disorders. For example, after the Duchenne muscular dystrophy gene was located, the encoded protein dystrophin was found to have mutations resulting in other milder dystrophies. Mutations are only rarely beneficial, but we know that many mutations alter the properties of proteins very little. We can anticipate that most genes may undergo mutations that cause some loss of good health and vitality without being diagnosed as causing disease. We have also come to understand that many complex diseases such as diabetes, *polycystic ovary syndrome*, *Crohn's disease* (inflammatory bowel disease), and schizophrenia are in fact multiple diseases.

Diabetes is a syndrome that can arise from causes such as defective insulin receptors or defective glucose transporters or from as yet unknown metabolic problems. Many cancers have a strong hereditary component, and many of these are being mapped to specific DNA locations. Some specific proteins, such as *Ras* that are mutated in many cancers were first recognized as avian oncogenes. The tumor suppressors *Rb* (retino-blastoma protein, and *p53* are major sites of mutation in cancer. Since cancers contain multiple mutations, they are complex diseases.

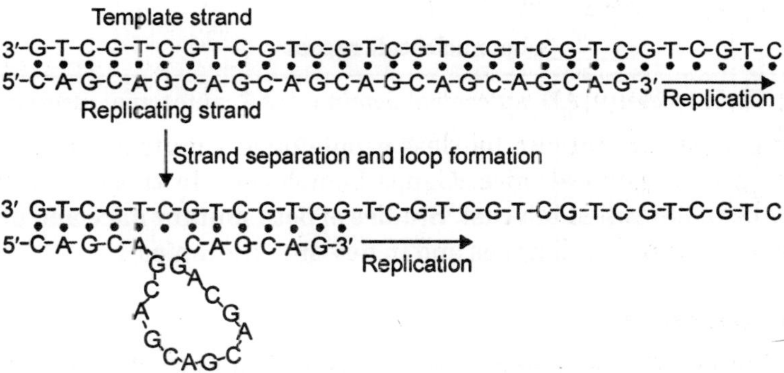

However, many specific susceptibility loci are being located, including some for breast cancer prostate cancer, and familial adenomatous polyposis, a hereditary disease leading to colon cancer. Cancer has long been known to be associated with chromosome instability including deletion and insertion mutations at simple repeat sequences, frame-shift mutations, DNA breakage, translocation, and losses or gains of whole chromosomes. From a practical viewpoint the understanding that we are gaining will help us to provide better treatments of genetic diseases. At present almost every human genetic disease can be mimicked in a knockout mouse.

Table 3.4. Some Human Triplet Repeat and Related Diseases

Name	*Repeat sequence*	*Transmission*[b]	*Location*
Fragile X syndrome (FRAXA)	$(CGG)_n$	XD	by 5′ side of gene *FMR*-1
			+ C_pG island methylation
Type E (FRAXE)	$(CGG)_n$	XR	5′ side of gene *FMR*-2
Synpolydactyly	$(GCG)_n$		HOXD13 gene (polyalanine)
Myotonic dystrophy (DM1)	$(CTG)_n$	AD	3′ untranslated region of gene

—			for CAMP-dependent protein kinase
DM2	$(CCTG)_n$	AD	Intron in a zinc finger protein gene
Friedreich ataxia	$(GAA)_n$	AR	Intro in frataxin, a mitochondrial protein
Huntington disease (HD)	$(CAG)_n$	AD	Huntingtin (polyglutamine)
Spinocerebellar ataxia, SCA-1	$(CAG)_n$	AD	Ataxin-1 (polyglutamine)
SCA-2	(CAG)n	AD	Ataxin-2 (polyglutamine)
Machado-Joseph disease (SCA-3)	$(CAG)_n$	AD	Ataxin-3
SCA-6	(CAG)n	AD	Calcium channel
Spinobulbar muscular atrophy (Kennedy disease; SBMA)	$(CAG)_n$	XR	Atrophin (androgen receptor)
Dentato-rubro-pallido-luysian atrophy (DRPLA)	$(CAG)_n$	AD	Cystatin B
Progeressive myoclonus epilepsy	(G+C)-rich oligonucleotide repeat		

[b] XD, dominant; XR, X recessive; AD, autosomal dominant; AR, autosomal recessive

A turnabout is that rare human hairless people have a mutation in a gene homologous to the well-known *hairless* gene of mice. Genes homologous to those of many human genetic defects have also been identified in yeast. Studies of both mutant mice and of mutant yeast cells can also help in understanding diseases and in devising therapies.

Triplet Repeat Diseases

With the exception of Down disease (extra chromosome 21, affecting 1 in 600 children) the most prevalent cause of mental retardation is the *fragile X syndrome*, which affects ~1 of 2000 newborn males. A fragile site; where the X chromosome breaks easily, marks the location of the defect. Identification of nearby RFLPs led to cloning of the gene in 1991. The defect was found to be a repeated trinucleotide sequence 5′-$(CGG)_n$ in the DNA. The value of n varies, in most healthy individuals averaging ~30. However, for some normal individuals n may be 200-300. These persons may transmit the fragile X disease to their offspring in whom n increases from one generation to the next with increasing severity of the disease. There may be 2000 or more CGG triplets. The defect lies at the 5′ end in an untranslated part of the gene for the fragile X mental retardation protein (FMRP).

A cytoplasmic RNA-binding protein, FMRP, may enter the nucleus and have an as yet uncertain function. In rare cases the fragile X syndrome arises from deletions or missense mutations in the *FMRP* gene. There are actually two similar genes *FMRP*1 and *FMRP*2. Mutations in the latter are associated with a milder form of fragile X disease. FMRP genes with expanded $(CGG)_n$ tracts are not expressed, evidently because the mutation induces methylation of an adjacent "CpG island" as well as of sites within the FMRP gene. Both FMRP proteins are apparently needed for normal brain function. The FMRP1 defect is genetically dominant, and female heterozygotes also suffer from the fragile X syndrome. However, FMRP2, also encoded on the X chromosome, is recessive. FMRP appears to function in neuronal dendrites where it binds to polysomal aggregates and participates in regulation of translation of mRNA. Like DNA, mRNA may contain *G quartets*. Twelve or more additional triplet repeat diseases, many with neurological symptoms, have been identified.

These involve other trinucleotide repeats 5′-$(GCG)_n$, 5′-$(CTG)_n$, 5′-$(GAA)_n$, and 5′-$(CAG)_n$. In *synpolydactyly*, an inherited developmental defect causing malformation of hands and feet, an expansion of a GCG trinucleotide occurs within the gene *HoxD*13. This results in incorporation of a polyalanine tract near the N terminus of the protein. *Myotonic dystrophy* DM1 results from expansion of CTG to 6 kbp or more within the untranslated 3′ region of a gene for cAMP-dependent protein kinase. The mRNA transcripts accumulate in the nucleus and may bind to a CUG-binding protein that is involved in splicing other mRNAs, thereby poisoning the cell. DM2 is caused by expansion of a CCTG quartet in an intron of a zinc-finger protein. Expansion of the GAA triplet is associated with the neurological disease *Friedreich ataxia*, which has a prevalence of ~1 in 50,000. The defect lies within an intron found in the gene for the 210-residue mitochondrial protein *frataxin*. The function of the protein is unknown.

However, the defect in Freidrich ataxia leads to a deficit in mitochondrial ATP synthesis. Studies of a corresponding protein in yeast suggests that frataxin is an iron storage protein. Apparently the DNA defect in the intron interferes with splicing of the mRNA transcript. The polyGAA strand in the triplet repeat region of the DNA is able to form various alternative structures including a parallel (GAA) • (TTC) duplex? Such structures in mRNA may interfere with proper splicing. A defect in an α-tocopherol-transfer protein causes a similar set of neurological symptoms. Oxidative damage may therefore be a component of this. Expansion of $(CAG)_n$ sequences causes a series of neurodegenerative diseases the commonest of which is *Huntington disease* (HD). In Huntington's disease the $(CAG)_n$ tract is found in the *huntington* gene. The protein is enormous, with over 3140 residues. It is essential for nerve development, but the function remains uncertain. The first exon at the HD gene encodes a polyglutamine tract of 6-35 residues, which when expanded to 36-100 or more causes the disease.

A similar situation holds for the *spinocerebellar ataxias* (SCA) and other $(CAG)_n$ diseases. The encoded proteins have a variety of functions. The SCA-1 protein ataxin-1 functions in the nuclear matrix. The X-linked *spinobulbar muscular atrophy* (SBMA) gene encodes an androgen receptor protein. The polyglutamine sequences inserted in those proteins seem to be toxic, but the mechanism of toxicity is uncertain. It may result from aggregation of the proteins within cells. It has also been suggested that cleavage of these chains by *caspases* (cysteine proteases) may produce truncated proteins, which induce apoptosis. A form of epilepsy appears to be a result of repeats of a (G + C)-rich sequence that may be a dodecamer.

Dinucleotide repeats and other "mini-satellite" DNA sequences are also associated with instability of DNA and may undergo expansion. A pentanucleotide repeat $(CCTTT)_n$ is associated with increased expression of the nitric oxide synthase gene NOS2A. Persons with $n = 14$ were found to have enhanced resistance to development of diabetic retinopathy. This seems to be a case of a beneficial "gain of function" mutation. How do repeat sequences expand from one generation to the next?

There is probably more than one mechanism. One is *strand slippage* during DNA replication. If single-strand loops are present at any time one strand could be displaced (slipped) relative to the other. Replication could then either expand the repetitive sequence or cause a deletion. Expansion could also occur by gene conversion during homologous recombination during DNA repair. Repeat sequences may also prevent proper formation of nucleosomes.

Cloned Genes and Diagnosis

The first areas in which cloned DNA has affected medicine are in the diagnosis of genetic diseases and in the production of medicinally useful proteins in bacteria, yeast, or cultured cells. One of the first applications was for the diagnosis of the sickle-cell trait by use of PCR on DNA isolated from blood. This was followed quickly by methods for recognition of other inherited diseases and by automated procedures. Intense efforts are now being made to develop DNA "chips" (usually small glass plates with an array of DNA fragments bound to the surface) that can recognize a great variety of defects. For example, DNA polymorphisms in the 16.6-kbp human mitochondrial genome can be recognized by a plate containing 135,000 oligonucleotide probes assembled in a regular grid by photolithography and solid-state synthesis. Cancers may be classified quickly.

Many systems are under development for binding and recognizing genomic DNA, cDNA, or mRNA using microchip arrays of up to 400,000 or more oligonucleotides on a 2 cm × 2 cm plate. One chip combines PCR with use of "zip-code primers" that direct the PCR products to specific zip-code addresses on the chip to give a universal array able to detect low abundance of mutations in any gene of interest. Mass spectrometry is also being harnessed to identify oligonucleotides bound at any address on a chip. Commercial chips are expensive, $100-2000 apiece, and good for only one use. The price will fall. To build a machine to make your own chips go to http//cmgm.stanford.edu/pbrown/mguide/. An alternative to DNA chips is to miniaturize DNA sequencing and analysis machinery.

Using nanoliter droplets of fluid passing through microchannels built by photolithographic techniques of computer chip construction, these nanolaboratories may be the size of a credit card but able to cleave DNA and conduct PCR reactions, gel electrophoresis, and sequence determinations.

Vaccines, Hormones, and other Medicines

A myriad of products of recomzbinant DNA technology are already in use in medical diagnostics. In the past problems have arisen because vaccines and hormones can cause allergic reactions and may harbor viruses. A small percentage of diabetics are allergic to animal insulins, but human insulin produced in bacteria is now available. A number of children receiving human growth hormone isolated from cadavers contracted the fatal Creutzfeld-Jakob disease, neurological disorder caused by a prion. One way in which AIDS has been spread is through contamination of the blood-clotting factors VIII or IX needed by hemophiliacs. These sources of contamination are being eliminated by the use of bacterially produced proteins. Recombinant DNA techniques can be used in two ways in the production of vaccines.

The first is to find a protein in the virus or other infective agent that is a good inducer of antibody formation, *i.e.,* a good antigen. This protein, or even a fragment of it, can then be produced from its cloned gene or can be made synthetically. Since the cloned protein can be purified highly, it may make a superior vaccine to those made from killed cells, inactivated virus particles, or mixtures of proteins. The first commercial vaccine of this type was against viral hepatitis B, a major cause of liver cancer. Particles consisting of viral envelope proteins can be produced in yeast and be used for vaccination. A DNA encoding these proteins has been transferred, using the Ti plasmid into lettuce. Human volunteers produced anti-hepatitis antibodies after eating the lettuce. This suggests that vaccination through ingestion of antigenic proteins in food crops may be feasible.

Injection of a small piece of DNA carrying the gene may also lead to antibody formation. Using either purified antigens, proteins, or DNA, it may be possible to develop effective vaccines against Rocky Mountain spotted fever, a rickettsial disease, and against malaria, for which there are no satisfactory vaccines. The vaccinia virus, formerly used to vaccinate against smallpox, has been engineered for use against other diseases. Much of its 187-kb DNA can be excised and replaced with passenger DNA.

The virus particles are stable and highly infectious. The vaccinia virus is unusual in carrying genes for both RNA and DNA polymerases and other proteins that permit it to undergo replication and transcription of its own genes in the cytoplasm of the infected cells. One application of a recombinant vaccinia virus is oral vaccination of wild foxes and racoons against rabies. A major advance based on cloned genes is the production of new medicines previously unavailable or available in only small amounts. Among these are the *interferons* and many hormones such as the *interleukins* produced by lymphocytes and the *atrial natriuretic hormones*. Another candidate is the α1-*protease inhibitor*.

Perhaps better inhibitors than the natural one can be devised and produced in bacteria.

Gene Therapy

A few years ago it seemed like fantasy, but there is little doubt that we will soon be able to routinely treat some genetic illnesses by introducing new copies of genes into the body. A current goal is to insert cloned genes into body cells (somatic cells) to correct specific hereditary defects. For example, juvenile diabetics would benefit from introduction of genes for insulin production into cells that could replace their atrophied pancreatic beta cells. At present we don't know how to do this. However, genes have been transferred into human beings lacking adenosine deaminase and showing severe *combined immunodeficiency* and those lacking hypoxanthine guanine phosphoribosyltransferase and displaying the *Lesch-Nyhan syndrome*. Corrections of *Gaucher disease* and other deficiencies of lysosomal enzymes is also an early goal. By 1986 more than 5000 children had received bone marrow transplants from close relatives to correct severe combined immunodeficiency caused by a defective adenosine deaminase gene.

However, the patients must receive chemotherapy or irradiation to suppress their immune system before the transplantation. Hospitalization may last for 30-60 days. Using genetic therapy some bone marrow cells can be removed from the patient with a needle. The cells can be treated to introduce the corrected genes using a suitable retrovirus. Clones of corrected bone marrow stem cells, which will give rise to lymphocytes in the body, can be selected, cutured and reintroduced into the patient. Retrovirus vehicles that locate and become integrated at appropriate sites in the genome have been developed. Use of homologous recombination introduces the cloned gene into its normal chromosomal location. The first children were treated by transfer of the adenosine deaminase gene in 1991. The procedure has been partially successful. By 1999 more than 400 clinical gene therapy trials were planned or in progress. Nevertheless, development of suitable vehicles for gene delivery has been slow.

Uncertainty about the safety of adenovirus vectors is one problem. Poor efficiency of gene transfer is another. A glycogen storage disease of knockout mice has been cured by transfer of human α-glucosidase using an adenovirus vector. Mice have also been used in developing gene therapy for hemophilia, sickle-cell disease, and aspartylglycosaminuria. However, gene targeting in animals other than the mouse has been difficult. Genetic therapy may be most effective when the gene is transferred into stem cells, which can then take up residence within the body. Young stem cells, which can be obtained from unbilical cord blood at birth, can be used.

For many diseases therapy will probably be best soon after birth, or even prior to birth during the second trimester of pregnancy. An alternative approach is to synthesize highly specific hydrolysis-resistant DNA analogs that can form triple helical structures with DNA. If these can be made specific enough they might bind to a targeted DNA site, such as the sickle cell anemia locus and induce a "back mutation" from the faulty A•T base pair to a T•A pair in at least some of the hematopoietic cells that give rise to hemoglobin. Another possibility is to make oligonucleotide analogs that serve as mimics of *antisense RNA*. This could impede translation of bad mRNAs, such as that giving rise to polyglutamine chains in the triple repeat diseases. At this time no effort is being made to alter the DNA in human germ cells. It seems undesirable to experiment with such changes. However, as methods are developed for genetic therapy of somatic cells, we will rear more and more healthy carriers of serious genetic defects. Eventually we may need to develop therapy for germ cells.

Genetic Engineering of Bacteria, Plants, and Animals

Many improvements in bacteria used in industrial fermentations have been made. The number of copies of a useful gene may be increased, and repressed genes may be made more active by deletion mutations in regulatory genes. New genes are being transferred between bacteria and into plants and animals. For example, Bacillus *thuringensis* produces crystalline protein toxins, which are harmless to mammals but active against many insects. The Bt gene for this toxin has already been transferred into many different plants and is having a major effect on agriculture in allowing for decreased use of insecticides. Genetic engineering may allow the toxin to be made more specific for particular insect species. Genetic engineering of plant genes may improve the quality of storage proteins in cereal grains and the flavors of fruits, provide more drought-resistant plants, and offer increased resistance of crop plants to particular herbicides and to viruses.

Many things can be done to enhance the nutritional qualities of foods. For example, rice has been engineered to produce and store β-carotene in the grain, a development that may benefit 400 million people in the world deficient in vitamin A and often suffering from infections and blindness. Specialty products such as poly-β-hydroxybutyrate and many others can also be produced in plants. Genetic engineering of animals has produced not only knockout mice but the possibility for nutritional improvement and for production of useful products such as vaccines in milk.

Ethical Problems

With a host of new medicines and agricultural products coming and with the ability to alter genomes at will, we face new ethical problems. For example, who owns the human genome?

Should patenting of human genes be allowed? How private are genetic data? Who has access to human DNA data? Insurance companies? Should routine screening with new genetic tests be used (as is now the case for phenylketonuria, diabetes, and others), even if no effective treatment is available? Ethical questions arise in application of experimental forms of genetic therapy to patients dying of inherited disorders. What if the therapy simply provides a longer life of suffering to patients? Some biologists foresee a future in which human beings learn to control their own genes and prevent genetic deterioration resulting from the accumulation of harmful mutations. They find it exciting to think that we can, in an intelligent way, elect to continue our evolution in a desired direction.

Others caution that our present knowledge is such that attempts to eliminate all "bad" genes from the population might be disastrous. They point to the hemoglobin *S* gene and the role it once had in preserving life in a malaria-infested environment and urge that at this stage we allow maximum heterogeneity of genetic types. There are other dangers in allowing genetics to control human lives. In the past "eugenic" doctrines have been used to justify racist laws and (in Nazi Germany) genocide. Problems have resulted from the availability of hormones and other products of genetic engineering. For example, should parents be allowed to request human growth hormone for short children who do not have a pituitary deficiency? Should it be available to athletes and to aging people?

Use of stem cells in genetic therapy raises another set of questions. Present guidelines of the National Institutes of Health (NIH), which supports much of the gene therapy research, allows use of stem cells in therapy but doesn't allow their use in reproductive cloning or the combining of human stem cells with animal embryos. Dangers from retroviruses may lurk in attempts to transplant animal organs (*e.g.*, pig hearts) into people. In attempting to improve crop plants, proteins causing allergic reactions in some individuals may be transferred. This is a compelling reason that mandatory monitoring and labeling of genetigelally modified foods is recommended. The danger is emphasized by the highly popularized contamination of foods with "starlink" maize. Should we rush commercial developments of genetic engineering to increase food supply at a time when there is an excess worldwide? Should we engineer increased herbicide-resistance into plants when drinking water supplies are being contaminated by herbicides already in use? On the other hand, less toxic herbicides effective at low levels can be used on plants engineered to resist them. Will we generate superweeds resistant to herbicides by gene transfer? With so many genes being engineered, are there other dangers in their release into the environment? A *terminator* technology that produces plants with sterile seeds has been rejected because it would force poor farmers to purchase seeds year after year from multinational corporations. Yet, it could eventually have beneficial consequences in preventing cross-fertilization with genetically modified plants.

Since plants with the *Bt* gene kill insects, will their widespread use decimate populations of butterflies or of various beneficial insects or adversely affect soil organisms in the *rhizosphere*? The darkest prospect may be possibility that military organizations will develop new biological warfare weapons using recombinant DNA methods. An international treaty bans such weapons. However, governments often respond with the argument "We must do it because we know that 'they' are doing it." Couldn't such an attitude bring on unparalleled disaster? Could new types of viruses spread throughout the world and literaly tear the human genome to bits? Does any kind of imagined danger justify preparation to attack populations with new viruses or new toxins?

What about the smallpox virus? Should remaining stocks in the United States and Russia be destroyed?

Cystic Fibrosis

One of the commonest of genetic diseases, cystic fibrosis affects persons all over the world. The incidence is unusually high in persons of northern European descent, about one in 2500 children being born with the defect. The inheritance pattern showed that cystic fibrosis is recessive and is caused by a single-gene defect that is carried by almost 5% of white Americans. In the United States there are ~30,000 persons with the disease. Many die in early childhood and even with careful treatment only 50% live into their late twenties or beyond. Through extensive linkage analysis the cystic fibrosis gene was mapped to chromosome 7, and its location was narrowed further to a 1600-kbp region between the oncogene *met* and another marker designated J3.11. Random searching located closer markers, and "chromosome walking and jumping" led to identification and characterization of the gene in 1989. The large 250-kbp gene contains 27 exons. The transcribed mRNA is 6129 bp in length, and the gene product is a 1480-residue amino acid protein which is known as the *cystic fibrosis transmembrane conductance regulator* (CFTR).

Children with cystic fibrosis lose excessive amounts of salt in perspiration and become dehydrated readily. A salty taste of the skin and an elevated chloride concentration of sweat are traditional diagnostic symptoms. More serious problems arise from progressive respiratory failure and inadequate pancreatic secretion. Lung infections with *Pseudomonas neruginosa* are the major cause of death. The CFTR gene is expressed in many tissues, especially those of the mucous membranes. An alternatively spliced isoform may form chloride channels in heart. As mentioned, CFTR protein is a member of the ATP-dependent ABC transporter family. However, it is atypical because it also contains a regulated chloride channel. In secretory epithelia of intestines, pancreas, lungs, sweat glands, and kidneys Cl^- enters epithelial cells through their basolateral surfaces using an Na^+ + K^+ + 2 Cl^- cotransporter and exits the cells through their apical surfaces using the CFTR channel. Absorptive epithelia also contain both the cotransporter and the CFTR channel, but Cl^- flows into the cells from the exterior surface, and the distribution of the cotransporter and CFTR between basolateral and apical surfaces is opposite to that in secretory cells.

From the amino acid sequence the CFTR protein is predicted to form two 6-helix transmembrane domains, two ~240-residue cytosolic ATP-binding domains, and a cytosolic regulatory (R) domain that contains at least five serine residues that may be phosphorylated by the cAMP-dependent *protein kinase A*. See adjacent scheme. The two ATP-binding domains resemble those of myosin and other ATP-hydrolyzing proteins. The chloride channel, which is probably formed by helices M1, M3, and M6, remains closed unless serine residues of the regulatory domain are phosphorylated. However, opening of the channels also requires binding of ATP in the nucleotide-binding domains. Binding of either vanadate or BeF_x stabilizes the open state of the channels. There are four small 55- to 65-residue cytosolic loops, labeled 1- 4 in the diagram. Transmembrane loop 3 also appears to function in regulation of the channel, which involves a complex regulatory mechanism. Two sites of *N*-linked glycosylation are present in the short extracellular loop between helices M7 and M8.

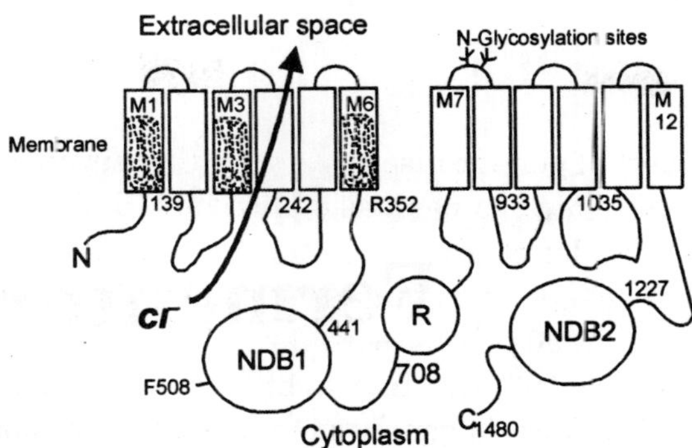

The CFTR protein can undergo endocytosis into clathrin-coated vesicles as part of its regulatory mechanism. Since HCO_3^- is usually exchanged for Cl^- in epithelial ion transport, regulation of HCO_3^- uptake is also a significant aspect of CFTR function.

Mutations that cause cystic fibrosis are found at many locations in the gene. However, -70% of the mutations are caused by the absence of phenylalanine 508, as a result of a three-nucleoticie deletion, in the first nucleotide-binding domain. This deletion causes misfolding of the CFTR to give an inactive protein. Hundreds of other mutations, some in the regulatory domain and some in the cytosolic loops, also cause the disease. Cystic fibrosis is one of the diseases for which targeted gene transfer may become an effective treatment. Cystic fibrosis induced in mice by targeted disruption of the CFTR gene has been successfully treated by gene therapy. However, the gene is large and efficiency of transfer into animals is low. Nevertheless, human gene therapy for cystic fibrosis is being pursued cautiously.

The CFTR protein has additional medical significance. Its stimulation by bacterial toxins is responsible for *secretary diarrhea*, which kills 3 million young children annually.

CHAPTER

4 Fermentors

BATCH CULTURE FERMENTATION

The purpose of this chapter is to give an introduction to the methodology of batch fermentation using a small, autoclavable, benchtop fermentor system. Specifically, this chapter should provide an understanding of the following:

(*i*) the engineering and biological concepts of fermentors;

(*ii*) the principles behind the instrumentation used with a modern fermentor;

(*iii*) the assembly and preparation of a fermentor vessel and anciliary equipment for autoclaving;

(*iv*) how samples are taken from the fermentor vessel during operation;

(*v*) how simple continuous culture can be accomplished.

THE MAIN COMPONENTS OF A FERMENTOR

The main subdivisions of a bench-scale fermentor are as follows:

(*i*) base components including drive motor, heaters, pumps, gas control, etc.;

(*ii*) vessel and accessories;

(*iii*) peripheral equipment such as reagent bottles;

(*iv*) instrumentation and sensors.

These components combine to perform the following functions:

(*i*) provide operation free from contamination;

(*ii*) maintain a specific-temperature;

(*iii*) provide adequate mixing and aeration;

(*iv*) control the pH of the culture;

(*v*) allow monitoring and/or control of dissolved oxygen;

(*vi*) allow feeding of nutrient solutions and reagents;

(*vii*) provide access points for inoculation and sampling;

(*viii*) use fittings and geometry relevant to scale-up;

(*ix*) minimize liquid loss from the vessel;

(*x*) facilitate the growth of a wide range of organisms.

The example used to illustrate these points is the smallest, simplest system which displays all these components and attributes, *i.e.* a small, autoclavable, bench-top fermentor. Mention is made of larger, *in situ* sterilizable units and special features which can be added to achieve certain objectives such as containment work. Also, a range of basic applications are illustrated, e.g. mammalian cell culture as a starting point for anyone contemplating the use of small-scale fermentors for these purposes. The specific information provided should be treated with a great deal of circumspection as the parameter values required for any particular microbes or cell line could be completely different.

COMPONENT PARTS OF A 'TYPICAL' VESSEL

The vessel is constructed either as a single-walled cylinder of borosilicate glass with a flat bottom or as a glass-jacketed system which typically has a round bottom. The top plate is made from '316' stainless steel and is compressed onto the vessel flange by an easily released clamping system. A seal separates the vessel glass from the top plate. Port fittings of various sizes are provided for insertion of probes, inlet pipes, exit gas cooler, cold fingers, sample pipes, etc. These work by compressing the sides of the probe/pipe against an O-ring seal. A special inoculation port will have a membrane seal held in place with a collar. Culture can be withdrawn into a sampling device or a reservoir bottle via a sample pipe situated in the bulk of the fermentor fluid. A gas sparger is also fixed into the top plate and this terminates in a special assembly which ensures that incoming air is dispersed efficiently within the culture by the flat-bladed 'Rushton-type' impellers fixed to the drive shaft.

A drive motor provides stirring power to the drive shaft and is usually fitted directly to the drive hub on the vessel top plate. An exit gas cooler works like a simple Liebig condenser to remove as much moisture as possible from the gas leaving the fermentor to prevent excessive liquid losses during the fermentation and wetting of the exit air filter. A narrow platinum resistance (Pt-100) temperature sensor completes the list of minimum essential fittings. Temperature control is either by direct heating using a heater pad or by circulating warm water around the vessel jacket.

If direct heating is used, a cold finger is used to cool the vessel contents. The sensors which fit into the vessel do so either by direct coupling with a thread on the body of the electrode, as is the case in the gel-filled type of pH electrode, or by a special fitting on the vessel top plate, as in the case of the dissolved oxygen electrode (polarographic type). Another system involves the use of a simple compression fitting which holds the body of the electrode, as with the foam probe. In this case, the height is variable and the tip of the probe is above the culture fluid. Illustrates all the main components of a fermentor.

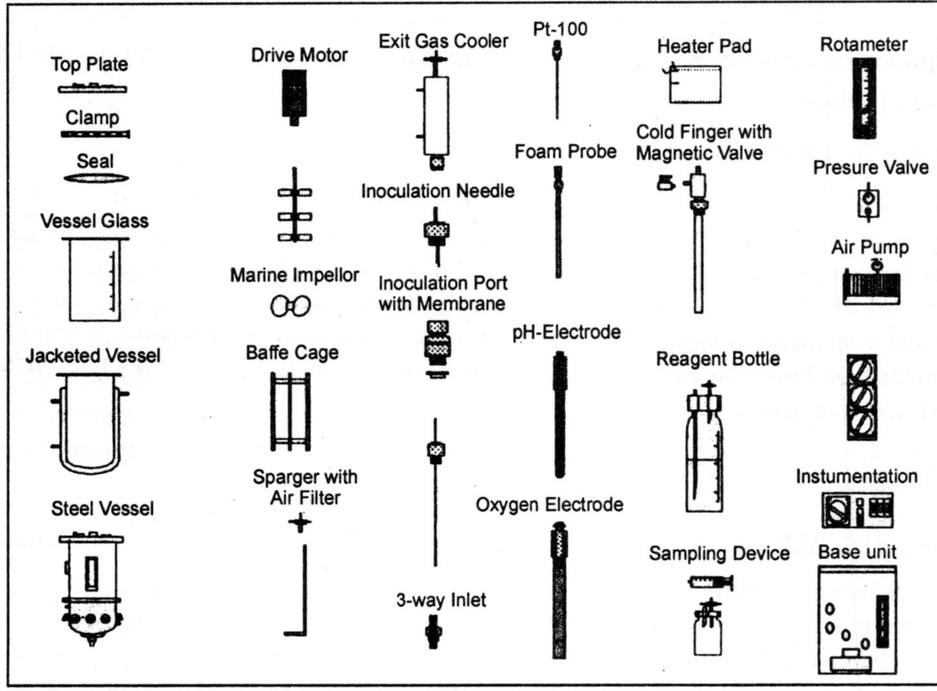

Fig. 4.1. Major Components of a Fermentor.

PERIPHERAL PARTS AND ACCESSORIES

Reagent Pumps

Pumps are normally part of the instrumentation system for pH and antifoam and control. Peristaltic pumps are used and the flow rate is usually fixed with a timed 'shot and delay' feed system of control. Flow rates will depend on the bore of the tubing used and it is a good idea to move the tubing every few days so that one piece does not become worn out by the pump rollers. The peristaltic tubing links the reservoir bottles with the vessel multi-way inlet. The tubing is clamped shut during autoclaving, already connected to the vessel, and opened for active addition of reagent. Alternatively, the bottles can be connected following autoclaving by using an aseptic coupling to join the tubing between the pump and the multi-way inlet. A dip tube goes to the bottom of the bottle. A filter in the top of the reservoir bottle is kept open during autoclaving to prevent a build-up of pressure. For extremely high accuracy of addition, the reservoir bottles are placed on analytical balances which can be used to determine how much reagent has been pumped in a given time.

Medium Feed Pumps and Reservoir Bottles

Medium feed pumps are often variable speed to give the maximum possible range of feed rates. The speed of operation of the pump can be set manually or the whole system put under computer control. The reservoir bottles are usually larger, *e.g.* 5-20 l, but are prepared in the same way as normal reagent bottles. These bottles may have to be changed several times during a long continuous culture experiment, so it is usual to fit an aseptic coupling in the tubing. An effluent pump is often used to remove culture fluid from the fermentor vessel into a storage reservoir bottle.

Rotameter/gas supply

Gas input, usually air, can be provided by use of a laboratory air supply or from a separate pump (which *must* be oil-free). A variable area flow meter or rotameter is used to control the air flow rate into the fermentor vessel. A pressure regulator valve before the rotameter ensures safer operation. A sterile filter (usually 0.22 μm) is used as a bridge between the tubing from the rotameter and that connected to the air sparger of the fermentor. A second filter on the exit gas cooler stops microbes being released into the laboratory air as the gas leaves the fermentor under a slight positive pressure.

Sampling Device

This allows culture fluid to be removed aseptically during the fermentation at intervals decided by the user. The frequency of sampling and the size of each sample is determined empirically according to the needs of the experiment.

Alternative Vessel Designs

There are alternatives to using a conventional, 'stirred tank' fermentor. Usually, alternative vessel designs are tried when the standard vessel configuration does not allow adequate growth of the organism (*e.g.* animal cells are often disrupted by shear forces in fermentors with turbine impellors) or the scale-up criteria require a different design of bioreactor (*e.g.* production of large quantities of single-cell protein is cheaper on a large scale if air lift fermentors are used to eliminate energy costs associated with a drive system). A number of these special designs are available at the bench/pilot scale of operation to allow small-scale research into the suitability of a particular method.

Air Lift

This vessel design eliminates the need for a stirrer system. A tall, thin vessel is the best shape with an aspect ratio of around 10:1 (height to base diameter). Sometimes a 'conical' section is used in the top part of the vessel to give the widest possible area for gas exchange. Sensors can be mounted in a steel base section, from a collar at the base of the conical section or from the vessel top plate. The culture fluid is both mixed and aerated by a stream of air which enters near the base of the vessel. A hollow pipe or draft tube in the centre of the vessel provides a 'riser' for the air (which is full of bubbles) to move upwards to the top of the vessel. If a very large vessel is used, the hydrostatic head of the fluid provides a pressurizing effect to the lowest region of the culture where the air enters and so increases the dissolved oxygen concentration. The draft tube is usually double-walled to allow heating and cooling using a thermocirculator system. When the aerated culture fluid reaches the top of the draft tube, it 'spills over' and begins to fall towards the bottom of the vessel via the space between the outer wall of the draft tube and the inner wall of the vessel.

A large head space above the top of the draft tube allows for easy gas transfer from the liquid to the gas phase which causes the density/specific gravity of the liquid to increase and so it descends to the bottom of the vessel. The descending liquid returns to the base of the vessel where it is regassed and begins to rise again. A common use of air lift fermentors is the growth of shear-sensitive cells such as plant and animal cultures. Also, the design has been used for the production of large amounts of biomass as single-cell protein.

Fluidized Bed

The microbes/cells are trapped in a physical medium (*e.g.* alginate beads) and held in the vessel by a mesh. Medium is recirculated via a pump and this can be easily adapted to give a continuous/semi-continuous flow to allow the trapped cells to effect chemical changes on constituents of the medium without being washed out along with spent medium. This system is well suited to growth of animal cells on the smaller scale and has large-scale application in effluent/decontamination treatment plants.

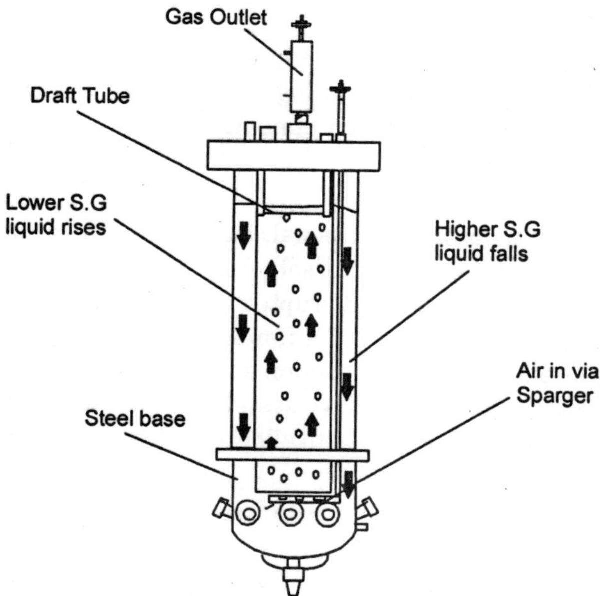

Fig. 4.2. Air lift fermentor.

Hollow Fibre

This is a similar idea but the cells are now embedded in fibres contained in a cartridge which is bathed in circulating culture medium. This is often used for mammalian cell culture, where anchorage-dependent cell lines can be perfused with oxygenated medium. An extension of this method is to use cartridges containing two different bundles of fibres as a separation 'membrane' between a pair of fermentor vessels and allow dissolved gases and metabolites to be exchanged without cells crossing the barrier to allow, for example, toxicity or interactive studies.

In Situ Sterilizable Fermentors

The use of autoclavable vessels of greater than 5-10 l working volume quickly becomes impractical. For safety and insurance considerations, vessels above this size are usually made of stainless steel (316 L) and are designed to be sterilized *in situ* using house steam or an electrical steam generator (which may be built into the base unit). The heating for the vessel is normally provided via a double jacket which can either be the full length of the vessel or cover just the bottom third. The bottom section contains large (25 mm) port fittings for electrodes and usually has some kind of steam-sterilizable sampling device/harvest valve. The mechanical seal and drive shaft usually enter from the bottom on this size of vessel. As the vessel body is steel, a sight window and a light have to be fitted in order to see the culture. The vessel top plate has port fittings which use a membrane seal and port closure.

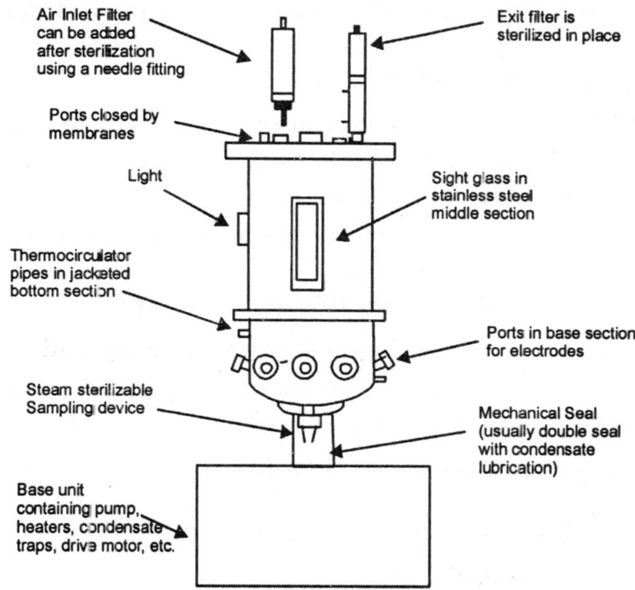

Fig. 4.3. Major components of an *in situ* sterilizable fermentor vessel.

Unlike the bench units, *anything* which dips into the culture liquid or is used as an addition port must be autoclaved separately and pushed through a membrane using aseptic techniques after the vessel has been autoclaved. In some fermentors, this even includes the air inlet filter and connecting pipe to the air sparger. The steam for sterilization is either supplied from an in-house source and used to heat the vessel jacket or is raised electrically from a separate steam generator. The medium in the vessel is heated to 121°C and often supplies the steam for sterilization of the exit gas filter.

Containment

An *in situ* sterilizable fermentor may have to be altered in certain ways if the organism to be cultured is pathogenic or genetically modified. The alterations are designed to contain any release of microbes into the environment by using features such as a double mechanical seal, magnetic coupling, additional air filters, extra foam control systems and special sampling devices. More elaborate precautions include steam traces to all vessel fittings and discharge of any released liquid directly to a tank of disinfectant. There are several different categories of containment (1-3 for pathogens and 1-4 for genetically modified organisms). Requirements differ at each stage and relate to building and working protocols as much as the design of the fermentor. Applications for this sort of technology are medical research and vaccine manufacture.

DIFFERENT TYPES OF INSTRUMENTATION

Fermentor instrumentation can range from simple analogue control modules arranged as a 'stack', to powerful, embedded microprocessors which operate as on-board computers to directly operate the heaters, pumps, valves and other control actuators of a fermentor.

Analogue Controllers - Rack System

Here the individual modules for the measurement and control of each parameter are made to plug into a rack system for compact, flexible instrumentation. A bus system in the racking transmits signals from one module to another to allow measured values to be passed to controllers and for one controller to influence another, *e.g.* oxygen levels altering speed.

Analogue Controllers - Separate Modules in Housings

Each measurement/control module has a separate housing and essentially operates in isolation. The modules may be stacked one on top of another and can be supplied with 'pass through' power connections so that only one power socket is needed. This system allows for gradual building up or swapping of instrument modules between several fermentors.

Digital Controllers - Embedded Microprocessor

The measurement sensors link directly to the single control module and several parameters are displayed immediately on a single screen. Control is by direct action on heaters, valves, etc. (Direct Digital Control or DDC). The microprocessor is permanently embedded in the instrumentation and may even be a single chip. Operation is usually via a simple menu system.

Digital Controllers - Process Controllers

A complete process controller (usually from a production control environment) is added to a housing containing all the signal processing and control actuators for the fermentors. It exerts control in the same way as an embedded controller but is essentially a 'plug-in' component. The controller is usually 'programmed' using simple commands to input set-point values, etc.

Digital Controllers - Direct Computer Control

In this case, there is no external instrumentation or processor between the actuators and the computer. A printed circuit board with operational amplifiers for the probe input signals is the only electronic part which may be present in the fermentor base unit. A special Input/Output (I/O) card is needed for the computer which allows the input values to be accessed by the measurement and control software and sends signals out to operate relays, *e.g.* to turn a heater on or off.

The processing power and speed of modern PCs make them more than capable of replacing separate PLCs or embedded controllers. The advantage with this system is that the computer display and control software is integrated totally with the fermentor. Fewer components means that these systems are usually less expensive. However, the fermentor cannot be used without the computer and there is no backup for the control systems should the computer develop a fault and 'hang'.

COMMON MEASUREMENT AND CONTROL SYSTEMS

Speed Control

Speed control relies on the feedback from a tachometer located within the drive motor determining the power delivered by the speed controller to maintain the speed set-point value set by the user. A digital display shows the actual speed in rpm, as determined by the tachometer signal. A power meter is sometimes included which indicates how hard the motor has to work to maintain the set speed and thereby, indirectly, the viscosity or 'density' of the culture fluid. A DC, low voltage (24-50 V) motor is often used for safety reasons. Speed range is typically from 50 to 1500 rpm (exceptionally up to 2000 rpm).

Where speed is used to control the level of dissolved oxygen, an external signal from the oxygen controller can have an effect on stirrer speed. In this case, an absolute maximum and minimum value for speed can be set on the speed control module to limit the effects of the oxygen controller (which could set either too low a speed and impair mixing or too high a speed and cause excessive foaming)

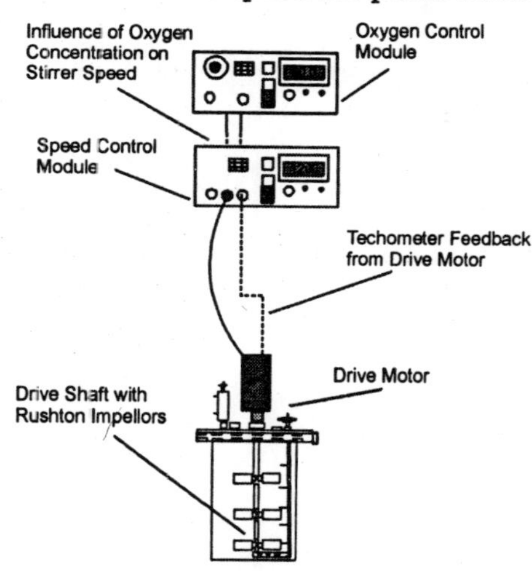

Fig. 4.4. Speed control system.

Temperature Control

A thermocirculation system around a vessel jacket has been chosen as an example here because it is the most complex of all the methods of temperature control. For simple direct systems such as a heater pad, it is simply a matter of fixing the heater, setting the desired temperature and switching on. Cooling is normally via a cold finger and flow of cooling water is controlled via the action of a solenoid valve. The Pt-100 sensor provides the feedback signal which causes the controller to take one of the following actions:

- heat at full power as the actual temperature is some way below set point;
- pulse the heater power as the actual temperature is close to set point;
- turn on the cooling valve as the actual temperature is above set point.

Indicator lights usually show which action the controller is taking at any given moment. A low-voltage DC heater system, *e.g.* 50 W, is often chosen for safety reasons. A circulation pump and pipework are added to the system for water circulation and any heating is indirect, *i.e.* on the water circulating in the vessel jacket. In this case, the connections to cold tap water must be made (securely, using jubilee clips or cable ties) and a drain pipe provided from the overflow point to a sink with a clear fall to the drain, *i.e.* the sink must be the lowest point for the whole length of this pipe. The water should be delivered from the mains at a minimum pressure of 2 bar and a flow rate of greater than 51 min^{-1}. The water hardness should be no more than 50 ppm suspended solids to protect the heating elements from 'furring'.

The vessel must be connected to the circulation loop (normally by rapid coupling connectors and flexible pressure tubing). Water is first supplied by opening a manual valve until the jacket is filled. The manual valve is then used to maintain just a trickle of water going through the jacket to the overflow to compensate for evaporation losses, *e.g.* 30-50 drops per minute. The heating and cooling is controlled in exactly the same way as a directly heated system but only the water in the jacket is affected. The jacket provides a large surface area in contact with the vessel wall for heat exchange.

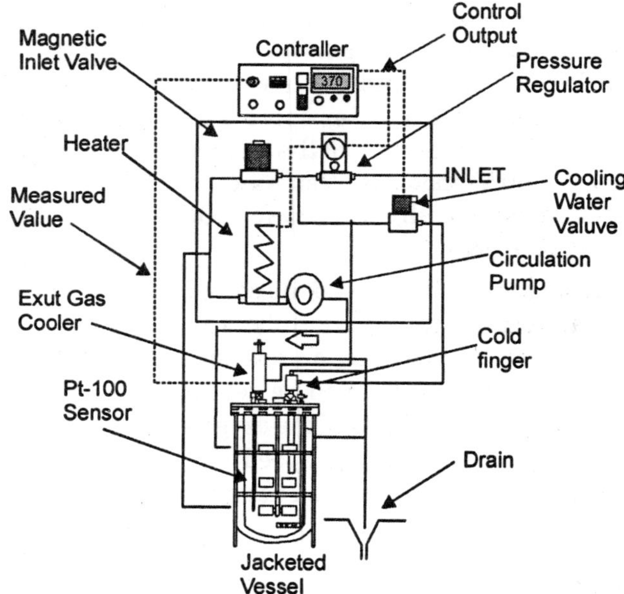

Fig. 4.5. Temperature control system using water circulation.

Good temperature control can be achieved from approximately 5-8°C above the ambient temperature or 5°C above the temperature of the cooling water. Counter-cooling with water ensures stable temperature control when operating near ambient temperatures. Measured range is typically from 0 to 60°C (exceptionally up to 90°C).

Control of Gas Supply

The supply of gas (normally air) to the fermentor vessel is provided from a compressed air supply (oil-free) which may provide air for more than one vessel depending upon its size and output. A pressure regulation valve ensures that air reaches the vessel at a maximum of 1.5 bar. The rotameter controls the actual flow rate of air through the fermentor. This should not exceed 1.5 vessel volumes per minute or droplets of water may become entrained in the stream of gas leaving the fermentor and so wet the exit gas filter, causing it to block.

A valve at the bottom of the rotameter is turned and the indicator 'ball' in the rotameter tube rises or falls in proportion to the valve position. A scale on the tube gives flow rates in ml min^{-1} or h^{-1}. The air passes through the inlet air filter which prevents any microbes from entering the vessel via this path. The end of the sparger is typically a ring with small holes through which the air is forced. The bubbles are immediately broken up and dispersed by the impellors on the drive shaft and the baffles which can be fitted near the wall of the vessel. The use of several impellors ensures all regions of the vessel receive good aeration.

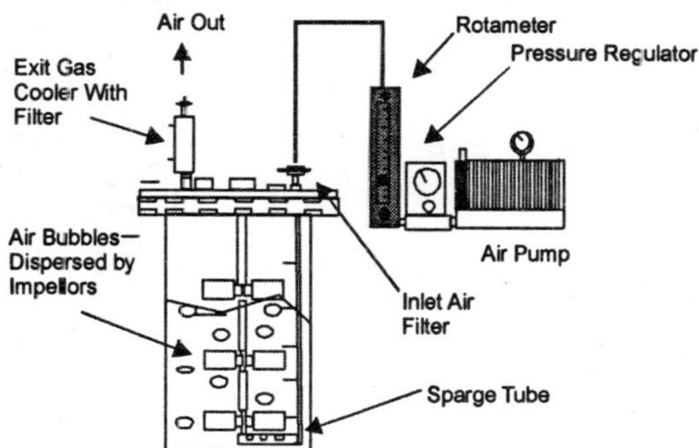

Fig. 4.6. Gas supply system.

A 'head space' of around 20% is normally left between the culture level and the vessel top plate. Sometimes, gas can also be introduced into this region via a short pipe in the fermentor top plate. For certain types of fermentation, *e.g.* mammalian cell culture, a gas mixing station can be used to pre-mix several gases before they are introduced into the fermentor.

Control of pH

pH control is achieved by the addition of either acid or alkali to correct changes in the pH of the culture during growth. The controller uses a pH electrode (either glass or gel type) to sense these pH changes and provide a feedback signal. The pH measurement system is identical to a bench pH meter and is calibrated in the same way *before* the electrode is autoclaved. Temperature compensation during operation may have to be set manually or may be automatic. The calibration normally requires that the temperature is manually set to equal the temperature of the buffer solutions.

Autoclavable electrodes have a limited life (20-50 sterilizations typically). The pumps used to supply the acid and alkali are normally built into the control module or the base part of the fermentor. Measured range is typically pH 2-12. The reagent bottles are connected to the fermentor via transfer lines of silicone tubing. The bore of tubing selected will determine the volume of acid or base added by each activation of the relevant pump by the controller. Selecting the concentration of the acid or alkali will determine how much effect each dose has on the vessel contents.

Normally, solutions of between 0.2 and 0.5 M of acid and base are used as a starting point. Using an ammonium salt for alkali addition can provide an extra nitrogen source for the growing culture. A set-point value is set on the controller and/or an upper and lower limit to provide a 'dead band' range in which the controller is not active. This band should normally be around 0.5 pH units each side of the desired value to prevent overdosing of acid and alkali sequentially. A proportional band adjustment may be present to widen or tighten the range of pH value over which the controller acts and even a switch to give a non-linear control if good control cannot be achieved by any normal adjustments.

Control of Dissolved Oxygen

Dissolved oxygen is one of the most important and yet one of the most difficult parameters to control properly during a fermentation. The electrodes used to measure dissolved oxygen are of two distinct types:

1. Galvanic electrodes, which are generally simple, cheap, slow and have a limited life.
2. Polarographic electrodes, which are generally more complex, accurate, rapid and robust but expensive.

The polarographic electrode is used more commonly and will be the example provided. The key point with this type of electrode is that it requires a voltage to polarize the anode and cathode of the detecting cell. This polarization can take between 2 and 6 hours to complete (in an emergency, 40 min is just sufficient for most tasks). During this time, the electrode must be connected to its relevant module which in turn must be switched on. The membrane of the polarographic electrode needs periodic replacement and a special cartridge kit is available from the manufacturer to make this a simple task. The electrode can be 'zeroed' prior to autoclaving using a special zeroing gel provided by the manufacturer.

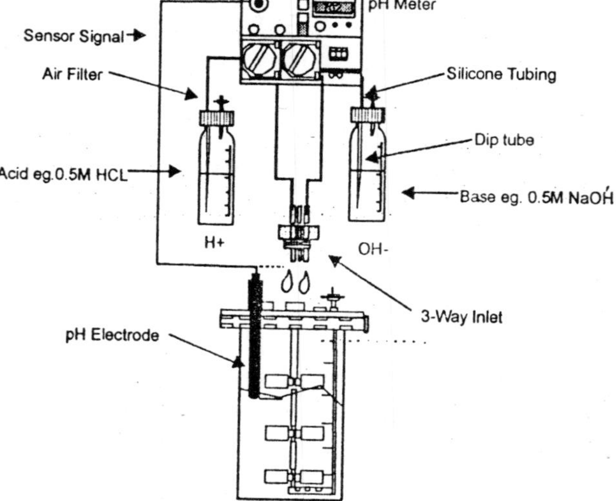

Fig. 4.7. pH measurement and control.

The electrode is simply inserted into a sachet of gel, left for a short time and then the displayed value adjusted until it reads zero. Alternatively, the electrode can be zeroed after autoclaving by passing nitrogen gas through the vessel for some minutes and then adjusting the zero point. The 100% value is a relative setting made after autoclaving and polarization of the electrode. The air flow and stirrer speed are turned to the maximum needed for the fermentation and the vessel left for some minutes. The display is now adjusted until it reads 100%. Control of dissolved oxygen can be purely by speed control, purely by control of a proportional air valve, or by a combination of both.

Speed adjustment may interfere with good mixing or increase the degree of foaming. Adjusting the air flow rate may also affect levels of foam. The most accurate form of flow control is to use a thermal mass flow control valve which measures and controls air flow based on the cooling effect the gas exerts when passed over a heated element. A simple option is the use of a solenoid valve with a manually set 'bleed' valve to simply vary air supply between a maximum and a minimum flow rate. Range is typically 0-120%.

Antifoam Control

The control of foam is based on its detection by a conductance probe in the vessel head space, which leads to the controller

delivering a dose of liquid antifoam reagent via a peristaltic pump. A delay timer ensures the antifoam reagent has adequate time to reduce the foam level before another 'shot' of antifoam is added. The sensitivity of the probe to foam can be adjusted by the user to suit the conditions prevailing in the fermentor. A sheath of inert material around the probe prevents splashes of foam thrown up from the culture from giving 'false positives.

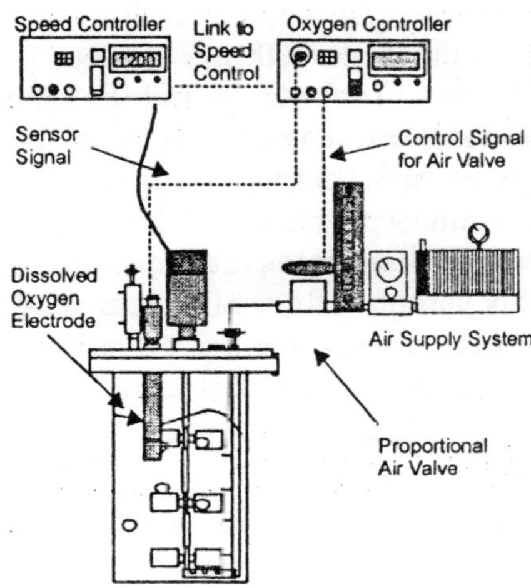

Fig. 4.8. Dissolved oxygen measurement and control.

Normally, the metal top plate is used to provide the electrical circuit for the probe to operate so a flying lead is provided which fits into a socket somewhere on the top plate. Antifoam reagents can be mineral oils, vegetable oils or certain alcohols. Commercial preparations are available for use with cultures which ultimately provide materials for pharmaceutical manufacture. The key thing with using oils is that they can form a 'skin' on the surface of the culture and interfere with gas transfer at the liquid/ air interface. If foam is allowed to build up unchecked it can eventually reach the exit gas filter so blocking it and providing a path for contamination.

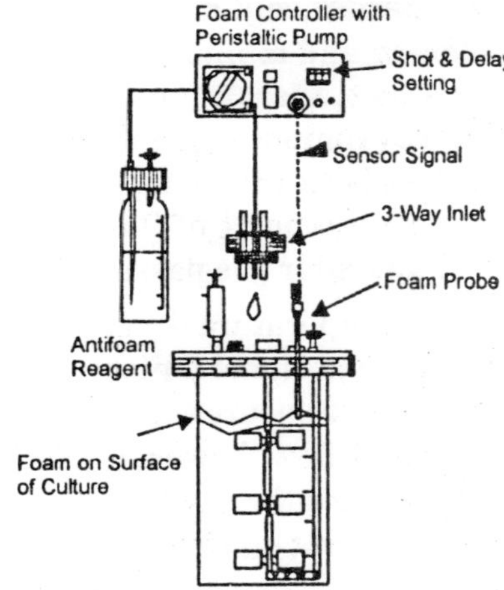

Fig. 4.9. Foam control system.

ADDITIONAL SENSORS

There are several additional process values which can be measured and controlled within a fermentor. The most common are listed below:

Redox

This refers to the reduction/oxidation (redox) potential of a system and can be applied to biological cultures as well as chemical reactions. Redox values are given in millivolts and usually refer to the oxidation or reduction potential of a given chemical relative to a standard hydrogen half-cell (which is defined as having a redox potential of 0.00 V). A half-cell which is more reducing than the hydrogen standard has a negative value and a stronger oxidizing agent will have a positive value. Reducing agents donate an electron to oxidizing agents.

Depending on conditions (the pH can affect redox potential), this reaction is reversible. The complexity of a typical fermentation broth (in terms of the sheer number of separate reactions going on at any one moment) means that the actual measured value is only of limited importance. However, for anaerobic cultures, the redox electrode provides a very sensitive 'watchdog' for the presence of oxygen. Tiny amounts of oxygen will have a relatively large effect on the redox value. A redox value below -200 mV is a good indicator of anaerobic conditions. The redox electrode is very similar to a pH electrode and the conditions for handling and care are almost identical. An electrical zero can usually be set and shorting out the electrode connections should give a reading close to 0 mV. There is no calibration procedure as such. A check on the electrode can be made using ordinary pH buffers which have been saturated with an excess of solid quinhydrone. The quinhydrone dissociates into equal quantities of benzoquinone and benzoquinol below pH 8.0. The redox potential for this system at pH 7 should be around +285 mV and changes by 59 mV for each pH unit, e.g. pH 9 = 403 mV. Deviations of ±5 mV are acceptable. If an analogue meter is used, a variable offset is usually provided for output to a chart recorder.

Air Flow

This has been mentioned previously in connection with dissolved oxygen control. A simple system would use a magnetic sensor and a variable area flow meter (rotameter) to measure the current air flow and adjust a motorized valve between 0 and 100% open-or closed to control the flow. Alternatively, a solenoid valve could pulse the air flow very rapidly on and off between a maximum and minimum value set manually by the rotameter and a 'bleed valve' respectively. In this case, there is no displayed value for flow rate which could be used for data-logging or control by computer. A more sophisticated system uses a thermal mass flow control valve which combines both the measurement element and the control valve in one compact unit. In this case, measurement of flow is by detecting the change in temperature of a heating element before and after gas passes over it.

The temperature difference is proportional to flow rate and this can be processed to give a measurement signal for logging and/or control. The practical points to consider when using this system are to ensure the valve is the right way round (an arrow on the metal section shows the direction of flow, however, this is sometimes obscured by the valve mounting) and that the control valve either bypasses the rotameter or the rotameter has a maximum value set to act as a physical limit on the amount of gas flow into the fermentor.

Of course, the flow control systems can measure any gas flow, not just air. They can be used in combination under computer control to give a precise gas mix for any total flow rate using one mass control valve per gas and post-blending the mixture. However, this tends to be an expensive option. A simpler system uses solenoid valves to introduce each gas into a pre-mixing chamber at manually set flow rates and then controls the final gas flow into the fermentor with a single mass flow control valve.

Weight

This parameter is useful for continuous culture or fed-batch work where the rate of addition of feed must be known and controlled very accurately. It is possible to mount a whole fermentor on a balance and 'tare' out its weight. However, a more precise approach is to use a system with a small load cell mounted in such a way that only the vessel and its contents are measured. The 'deformation' of the load cell provides an output which can be transduced into an electrical signal. Control is normally by setting a maximum and/or minimum weight such that a feed pump operates whenever the value is less than the maximum.

At the maximum value, a second pump operates to remove culture until the minimum weight is reached. A variant of this system can be used to measure very accurately the amount of reagent, *e.g.* alkali, added to a fermentor. In this case, the reagent bottle is placed on an analytical balance which has a computer output (normally RS232) and the decrease in weight used as a measure of the amount of reagent added.

Pressure

The use of pressure as a control parameter is more common in larger, *in situ* sterilizable fermentors, for the obvious reasons that the vessel is made of steel and is already adapted to working at pressures of several bar. Also, it is typically in the larger vessel that maintaining the dissolved oxygen concentration at a high level presents a problem (an increased pressure increases the amount of oxygen which can be dissolved in the fermentation broth). However, even a small glass vessel can often operate at some overpressure (*e.g.* 1-1.5 bar) but it is vital that this is first confirmed by the manufacturer.

The measurement and control is the same whatever the size and type of vessel, except that the size of the control valve would alter. Measurement is normally provided by a piezo-electric sensor which can be mounted into a port closure and fitted to the vessel top plate. As with a load cell, it is the deformation of the crystal by the internal vessel pressure which gives a proportional electrical signal, typically in the range of 0-2 bar. The control element is a proportional valve which restricts the flow of gas out of the fermentor and thereby creates a 'back-pressure'. A mechanical overpressure valve or burst-disc is an essential safety requirement if overpressure is to be used.

On-line Measurement of Biomass

A direct measurement of the number of organisms in a culture is clearly desirable for the control of feed rate, oxygenation and general process optimization. Normally, measurements of total cell numbers, wet weight, dry weight and viable cell counts would be made by taking samples of the culture at set times and performing the relevant laboratory analysis. The results would be entered some minutes, hours or even days later, when they would be of little value for real-time control. Fluorescence microscopy has also been used as a measure of viable cell density either with a probe or externally through the vessel glass. Two probe-based systems which are reasonably accurate and can be used with bench-scale vessels will be described by way of examples below.

Mettler-Toledo *FCS* Turbidity System

This is actually a nephelometer as it measures the scattering of light by particles in the culture (*i.e.* microbes). The advantage over a standard nephelometer is that the sensor is directly mounted in the vessel and connected to the electronics via a fibre optic cable. A two-point calibration procedure is used and results can be expressed in a variety of units. One of the restrictions of such a system is that if the culture is very thin, then shiny, reflective components close to the sensor could affect the readings. Conversely, too thick a culture would not allow the light scattering to be detected properly. However, a range up to 250 g l^{-1} dry weight is measurable. The probe is resistant to fouling over time.

Capacitance/conductance-based Biomass Monitor

This uses a totally novel measurement system to give viable cell numbers for all types of organisms (bacteria, yeast, filamentous fungi and animal cells). A probe in the vessel uses a radio-frequency electrical field to measure the natural capacitance of living cells with an intact plasma membrane. This build-up of charge is proportional to the cell concentration and specific to different types of cell. The signal is processed and can be ex-pressed as dry weight or concentration in cells ml^{-1}.

An anti-fouling system is used to maintain the ability of the probe to give accurate readings over long time periods. This type of instrument makes possible the control of feed pumps by biomass concentration, and could form the basis for an automated transfer system for delivering a seed culture to a larger vessel after an inoculum of sufficient concentration has been produced in a smaller fermentor.

SIMPLE CONTINUOUS CULTURE

There are a number of different ways of setting up a continuous culture. The method of choice has the virtue of being able to be added to any fermentor vessel without modification or the addition of extra controllers beyond the acquisition of a feed pump and a harvest pump. The feed pump is variable speed but the harvest pump can be fixed speed as long as its speed is greater than that of the feed pump. The harvest pipe, being sited at the surface of the culture, has the disadvantage of taking off only the top of the culture (which may not be representative of the bulk culture) but the advantage of helping to prevent build-up of foam. If foam is a problem then antifoam reagent can be added to the medium feed at a background level, *e.g.* 1 part in 20 000. Samples can be taken by putting a 'Y' piece in the take-off tubing and placing a sampling device in one arm of the 'Y'.

ADDITIONAL ACCESSORIES AND PERIPHERALS

Feed Pumps

A separate feed pump is necessary for fed-batch or continuous applications. The type of pump used will depend on the size of the vessel, the required flow rates, the nature of the additive (*e.g.* viscous, corrosive, etc.) and the speed at which the addition should be made. Often, a feed pump is included as part of the basic fermentor or provision is made for its inclusion as an optional accessory.

Peristaltic Pump

This is the most common type of addition pump used in fermentor applications and is often found in a fixed-speed version for addition of reagents such as acid and alkali. The pump uses rollers to occlude sections of silicone tubing sequentially. This action creates a 'peristaltic' flow of liquid through the tubing (analogous to the movement of food through the body by waves of peristalsis). The more rollers a peristaltic pump has, the smoother the flow. The pumps used for medium feed are typically variable speed units which can accept signals from an external source such as a computer or process controller to allow adjustment of feed rates based on, for example, respiratory quotient (RQ) or optical density (OD). The key advantages of peristaltic pumps are as follows:

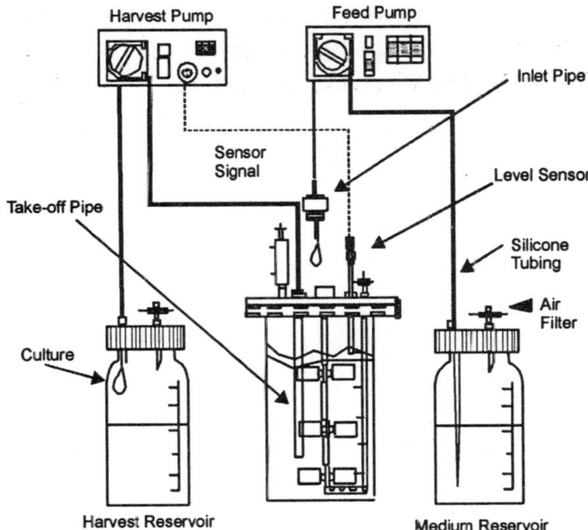

Fig. 4.10. Simple continuous culture.

(*i*) The liquid being pumped is not contaminated by any part of the pump.

(*ii*) The pump is not contaminated by the liquid being pumped.

(*iii*) A wide range of flow rates is possible using tubing of different bores.

(*iv*) The accuracy of delivery is good over a wide range of flow rates.

(*v*) The pump can run safely even when the tubing is empty.

The disadvantages are:

(*i*) The flow is pulsed and cannot 'push' against high backpressure.

(*ii*) The tubing wears over time and flattens, reducing accuracy and increasing the risk of bursting.

(*iii*) Silicone tubing is not suitable for all liquids and alternative tubing material may be too rigid to allow correct peristaltic action.

The problem with tubing wear can be partly overcome by leaving extra length in the tubing when connecting it to the fermentor vessel and moving a new length into the pump head every few days to stop wear in just one small section.

Syringe Pump

This type of pump works on a completely different principle and is useful for fed-batch applications rather than continuous medium feed. In this case, a syringe (of various sizes from µl to 100 ml, typically) is filled with the liquid to be added and secured onto the syringe pump. This is done in such a way that the syringe plunger is linked to a movable piston and the syringe barrel is held firmly to the main body of the pump. A flow rate is set (in mm min^{-1} or directly in ml min^{-1} or ml h^{-1} if the pump can be calibrated).

The piston is then moved at the desired speed by the action of toothed gears linked to a stepper motor engaging with 'teeth' on the underside of the piston. When the syringe is empty, an alarm condition is indicated and the pump automatically stops or, on very sophisticated pumps, a signal can be sent which allows automatic refilling from a reservoir. This type of pump is typically used in medical applications for the precise delivery of drugs. Most now allow for computer control. The advantages of syringe pumps are:

(*i*) Very precise and accurate delivery; non-pulsile flow.

(*ii*) Very low flow rates possible.

(*iii*) Can work against back-pressure.

(*iv*) Can pump very viscous fluids, *e.g.* oils.

The disadvantages are:

(*i*) Must be replaced or refilled regularly.

(*ii*) Range of flow rates not as wide as with peristaltic pumps.

(*iii*) Usually more expensive.

Diaphragm/Positive Displacement Pump

This type of pump is a hybrid of the two previous types in that it uses a flexible membrane and soft tubing to displace the liquid and control its flow but also employs a piston and pump heads of fixed volume. A stainless steel pump head is connected to soft tubing and autoclaved. It is then located onto the pump and held securely. A central piston and two small 'pinch valve' pistons move in and out to push liquid out of the central chamber of the pump head and allow fresh liquid to refill the chamber on each half of the pump stroke. The flow rate is controlled by the frequency of the pump stroke and the size of the pump chamber (which can usually be exchanged). A return spring returns the diaphragm to the correct position after each inward stroke.

A variant of this is to use a 'captive syringe' where the piston moves in and out of the chamber, emptying it fully then drawing back to refill on each stroke.

Advantages of this type of pump include:

(*i*) Accurate delivery.

(*ii*) More choice in type of tubing for corrosive liquids, for example.

Disadvantages are:

(*i*) Less flexibility on flow rate ranges for each pump head.

(*ii*) Stainless steel chamber and diaphragm may not be suitable for some liquids.

(*iii*) Pulsed flow even more pronounced than with a peristaltic pump.

Exit Gas Analysis

Measurement of the amounts of different gases leaving a fermentor vessel can provide valuable information about the metabolic processes taking place under the conditions in which the culture is growing. For example, the ratio of oxygen and carbon dioxide entering and leaving the vessel, or respiratory quotient (RQ), can be used to determine whether a yeast is producing biomass or alcohol. The RQ can be calculated by most fermentation software packages and can then be used as part of a control algorithm so that the flow rate of a feed pump can be altered accordingly. The entry gas does not need to be analysed providing it is air, because the amounts of oxygen and carbon dioxide will be those of the atmosphere. The flow rate of the air into the vessel will need to be accurately measured - normally by using a thermal mass flow controller. The exit gas may need to be conditioned (*e.g.* moisture removed) before going into the analyser, depending on the type of instrument used.

Infrared Carbon Dioxide Analyser

A hot wire is used to generate a source of infra-red radiation which passes through the gas which comes from the fermentor into the sample chamber. An infra-red detector measures the amount of radiation reaching it after some is absorbed by the gas. An optical filter is used to make sure the detector only responds to the gas of interest.

A 'chopper' or rotating shutter is used to allow the detector to see a reference source at regular intervals. This type of detector allows continuous measurement with good sensitivity, accuracy and selectivity. Output signals are normally provided either as analogue signals (*e.g.* 0-10 V) or a string of ASCII characters for printing or transfer to computer software.

Paramagnetic Oxygen Analyser

Oxygen has a particular physical property which this type of analyser can utilize, *i.e.* that it is more susceptible to a magnetic field than other gases. This property is measured using a finely balanced test apparatus suspended in the test chamber of the analyser. A dumb-bell of gas-filled spheres is linked to a support mechanism suspended in a magnetic field created by permanent magnets. If oxygen is present in the test gas, its attraction to the magnetic field will cause the dumb-bell to be displaced. The movement is detected using a mirror in the centre of the balance system which displaces a beam of light shone via the mirror onto a photocell.

The signal from the photocell will be proportional to the amount of oxygen present in the sample. A low flow rate is needed for this system to work properly, so inlet gas is typically pumped into the detector cell with most being discarded through bypass pipework. Once again, suitable analogue and computer outputs are normally provided.

Both infra-red detectors and paramagnetic oxygen analysers need to undergo a calibration procedure before use.

Mass Spectrometer

This represents a step upwards in versatility, capability and also cost! A wide range of gases can be analysed (both in a gas stream and as dissolved gases in the vessel), *e.g.* oxygen, carbon dioxide, argon, nitrogen, ammonia, hydrogen, methanol, ethanol and several other organic volatiles.

The analysis is rapid, so allowing a multi-inlet system to be used which 'shares' the analyser between separate vessels within a bank of fermentors. Mass spectrometers can be magnetic sector, multi-collector or quadropole. Each has its own particular advantages but the magnetic sector can be used to illustrate the principles upon which they all work.

The physical principle involved is the ionization of gas molecules by an electron source, usually a hot filament, followed by their separation in a magnetic field (the quadropole analyser uses a combination of RF and DC electrical fields to do this job) according to their mass/ charge ratio. This separation takes place in a virtual vacuum to minimize collisions prior to sorting. The magnetic field is 'tuned' so that only the ions of interest will be focused onto the detector system - a Faraday Cup. This is a metal plate which generates a tiny electric current whenever a gas ion strikes it. This signal can be amplified and displayed. By re-tuning the magnetic field very quickly (milliseconds) different ions can be detected and measured in a sequential way which is so fast it appears to give an almost simultaneous measurement.

FERMENTOR PREPARATION AND USE

Disassembly of the Vessel

The fermentation is shut down from the control unit and transfer lines plus cable connections removed. Following a fermentation, the vessel should be re-autoclaved, taking care that inlets and outlets are properly prepared. The culture should be disposed of as laid down by departmental safety procedures. The clip/clamps/bolts which retain the vessel top plate are undone until the whole assembly springs free. The top plate can now be lifted upwards away from the glass section of the vessel, taking care that the air sparger, drive shaft/impellors and Pt-100 temperature probe completely clear the vessel. The flanged top section of the glass vessel and seal can now be seen.

Cleaning

The pH and dissolved oxygen electrodes should be removed and stored in suitable reagents according to the manufacturer's instructions. Periodic cleaning and regeneration of the electrodes are also covered by these instructions. Doing this maintenance is very cost-effective. The vessel should be rinsed several times in distilled water to remove any loose culture residues. Cleaning of growths of culture on the vessel walls may require disassembly and light brushing of the glass. At this point, an examination of any chips or cracks in the vessel glass can be carried out and a replacement made if necessary.

Vessels must be stored clean and dry. In use, any spillages of reagents or medium should be wiped up immediately with a damp cloth and not be allowed to dry out. Contact between the top plate and liquids with a high chloride ion content (*e.g.* common salt solutions, hydrochloric acid) should be avoided to prevent corrosion. The pump heads and covers must be thoroughly cleaned if a tube breaks and reagent leaks out.

Preparations for Autoclaving

The vessel seal should be removed and lightly greased with a suitable silicone grease. On replacing the seal, it must be correctly located so that there is no chance of any part lifting or kinking. At this point, the vessel can be filled with medium to a maximum of 80% full (if active aeration is to be used this space is vital for gas exchange). The minimum medium volume is the amount needed to cover the electrodes adequately. The vessel top plate can now be replaced and any clamping ring or bolts tightened firmly. The ports for electrodes have O-ring seals which should be lightly greased with a silicone preparation before autoclaving.

Electrodes normally push directly into the port fitting and the collar is tightened down to compress the O-ring seal. All other fittings such as pipes are fitted in the same way. Ports not in use have 'stoppers' fitted, and their O-ring seals should have a light coating of silicone grease applied. The pH electrode should be calibrated in appropriate buffers *i.e.* pH 7, then either pH 4

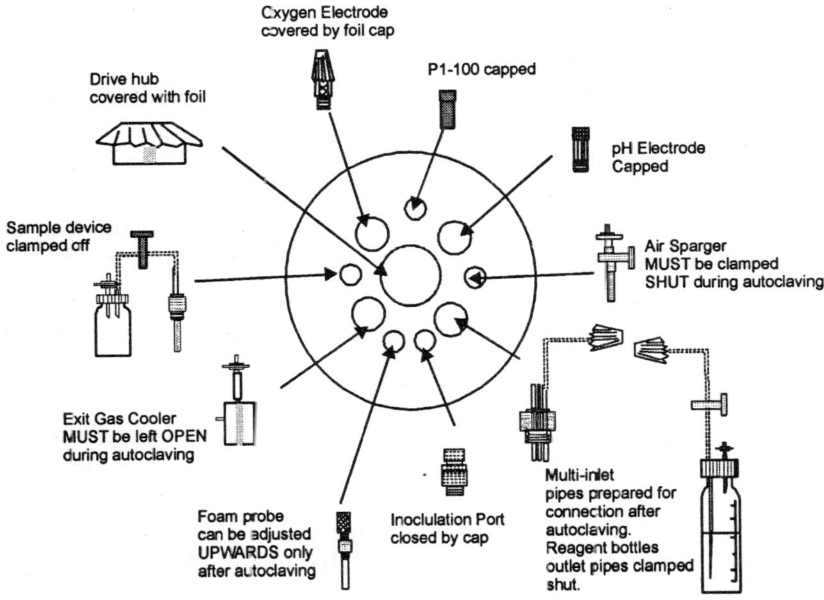

Fig. 4.11. Vessel and accessories prepared for autoclaving.

or pH 9 (or just 4 and 9 for a 'high reference-low reference' calibration of a DDC controller channel). The pH and dissolved oxygen electrodes should be fitted, taking care not to damage them by careless insertion into the port. Both must be tightly capped to prevent moisture getting into the electrical contacts.) For the dissolved oxygen electrode, a cap may have to be improvized from aluminium foil. The Pt-100 temperature sensor must be fitted and capped. If used, the foam probe is fitted so that it is above the liquid level in the head space. Reagent bottles are prepared in a similar way to the fermentor vessel. A membrane is placed over the mouth of the filled bottle and an open-topped cap screwed into place. A short needle is connected to a disposable filter which is then used to pierce the membrane in the bottle. The needle must not dip into the liquid and nothing must block the free passage of air through the filter.

A long piercing needle/pipe is used to puncture the membrane and push it into the liquid as far as possible. This pipe should now have a length of silicone tubing fitted to it which is long enough to reach the peristaltic pump heads of the fermentor instrumentation for which it is intended. The tubing is clamped so that no liquid can escape during autoclaving. A similar procedure is used for sampling and/or harvest bottles except that two short pipes are used so neither dips into the collected culture.

Some preparation is needed at the autoclaving stage for the aseptic coupling of silicone tubing ready for operation. The exit gas cooler should be fitted to one of the wider ports. A short length of silicone tubing should be attached to the top of the air outlet and a small 0.22 mm filter fitted. The air outlet line must be kept open during autoclaving. A short length of silicone tubing must be fitted to the air sparger inlet pipe with a 0.2 μm disposable filter mounted on top. The tubing between the sparger pipe and the filter must be clamped shut during autoclaving. If a port is to be used for inoculation or piercing with a needle, a silicone membrane must be fitted into the empty port and a clamping collar/cap used to hold it in place.

Autoclaving

The vessel and any reagent/sampling bottles already connected by silicone tubing are assembled together on a steel tray or in an autoclave basket. A final check should be made that at least one route is available for air to enter and leave the vessel(s) and that all lines dipping into liquid are clamped closed. If the vessel has top drive and a mechanical seal, the seal must be lubricated (normally with glycerine). It the medium cannot be autoclaved, a suitable volume of distilled water should be used and removed via a sample line before the actual medium and inoculum are aseptically transferred.

A quantity of liquid is certain to be lost during autoclaving (approximately 10%) so either the medium is over-diluted to compensate for this or sterile, distilled water is added afterwards to restore the volume. Some form of indicator such as autoclave tape should be included to provide a warning if the correct sterilization procedure has not been carried out. Autoclaving at 121°C for 30 min is normally considered adequate for vessel sterilization to minimize damage to the constituent chemicals of the medium. However, some work may require temperatures of 134°C for several hours to ensure sterility. If in doubt, your safety committee should be consulted. Also, the autoclave used must have good pressure equalization during the cooling down phase of operation to prevent medium being boiled off. The vessel and any accessory bottles must be allowed to cool completely before handling.

Set-up Following Autoclaving

Firstly, the silicone tubing prepared prior to autoclaving must be connected to the fermentor vessel (if the bottles they are connected to were autoclaved separately). The air sparger is connected to the rotameter by a piece of silicone tubing from the top of the filter to the air outlet of the rotameter. The air sparger line is undipped between the metal pipe and the air filter. The exit gas cooler is connected to the water supply, either directly or via the fermentor base unit.

The tubing for water in, out and drain is connected to the vessel jacket for a water system, and the water turned on so that the vessel jacket is filled. Alternatively, any pads or heater cartridges are connected to the base unit or temperature control module; the cold finger is connected to the water supply. The tubing from the reagent bottles is connected to the multi-way inlet (if necessary), and the silicone tubing from the reagent bottles is located in the relevant peristaltic pump. Any aseptic connections must be made first if the reagent bottles were autoclaved separately from the vessel. The clamps are removed so liquid can flow freely.

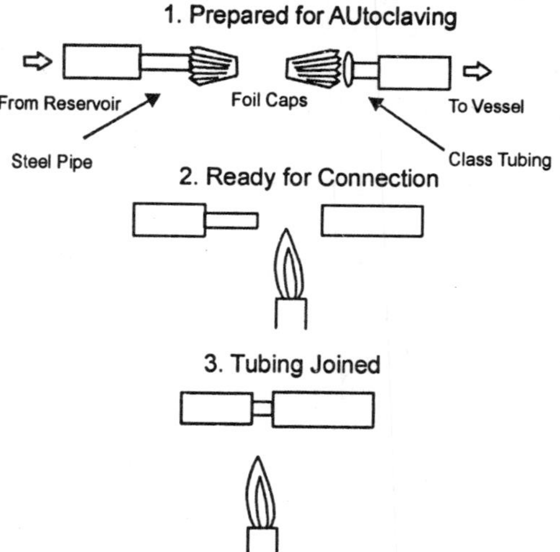

Fig. 4.12. Connecting tubing aseptically.

A manual switch is often fitted which allows the pumps to be primed with liquid prior to use. The drive motor is located onto the top plate (if appropriate), ensuring a good connection is made to the drive shaft. The Pt-100 temperature sensor is connected to the control module and the pH electrode by removing the shorting cap and screwing in the cable. The dissolved oxygen electrode is connected to the appropriate cable (this requires some care, but the connector should lock firmly when it is correctly positioned). Connections to the foam probe are made, usually one wire into the electrode and one on the vessel top plate to make a circuit. The mains electricity to the instrumentation modules is switched on, allowing some hours (minimum of 2, preferably 24) for the dissolved oxygen electrode to polarize properly. This delay is not needed for galvanic electrodes.

Setting the temperature control at this stage will ensure the fermentor is ready to inoculate after calibration of the dissolved oxygen electrode. After polarization the dissolved oxygen electrode is calibrated and the air supply turned on. The maximum stirrer speed to be used and the maximum air flow required on the rotameter are set. After leaving for about 5 min, the 100% level is set. The fermentor is now ready to inoculate.

Inoculation of a Fermentor Vessel

This section assumes that all the set-up procedures listed above have been carried out. The simplest way to inoculate is to have a dedicated port fitted with a membrane which is capped off before autoclaving. The inoculum (which should normally be no more than 5-10% of the total culture volume) is aseptically transferred to a sterile, disposable syringe of a suitable size. The port fitting is removed and held vertically to prevent contamination of the bottom end. The syringe needle is quickly pushed through the membrane and the inoculum transferred into the vessel. The vessel may be actively aerated during this procedure to minimize the risk of a contaminant getting into the vessel (safety considerations permitting).

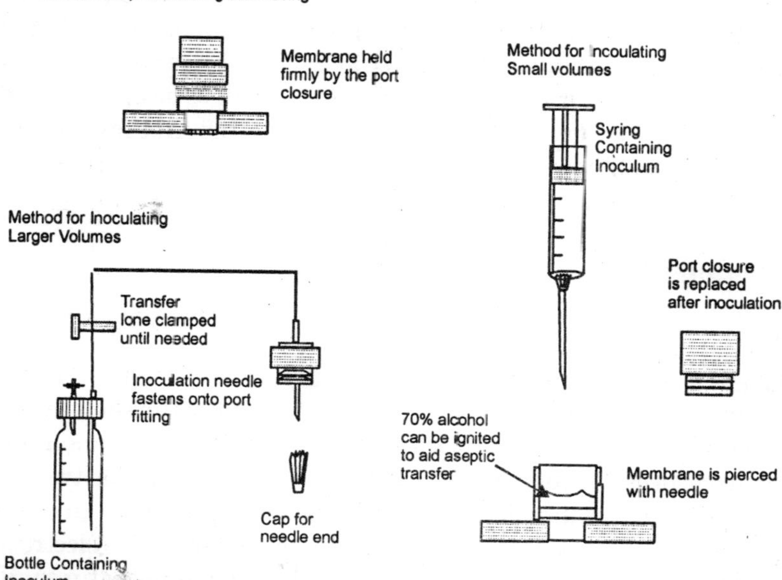

Fig. 4.13. Inoculation of a fermentor vessel.

The syringe needle is quickly withdrawn and the silicone membrane reseals. The port fitting is now replaced. For added security, a couple of drops of 70% ethanol can be placed on the membrane surface before piercing and ignited to provide a thermal barrier. Alternatively, an 'aseptic connection' can be made to an inlet pipe. If the line connected has a 'Y' coupling in it, then the same aseptic connection could be used to introduce medium. This technique is useful if a dedicated inoculation port cannot be provided.

Sampling from a Fermentor Vessel

All sampling starts with a sample pipe, which should dip into the bulk of the culture liquid. The simplest system is to have a length of silicone tubing connected to the sample pipe (and clamped off until a sample is to be taken) which is plugged and covered at the open end. After autoclaving, the tubing is quickly unwrapped, the plug removed and the open end dipped in a container of 70% ethanol. When a sample is to be taken, a sterile syringe is quickly coupled to the tubing and the sample withdrawn into the syringe. Aeration is stopped during this process to prevent 'surging' of culture into the syringe. Any culture in the transfer line is discarded into a 'kill jar' and the tube replaced in the ethanol. Clearly, this method can be prone to contamination and is not suitable for organisms with any risk to operators. A better approach is to use a sampling device. At its simplest, this is a bottle which has two metal needles/pipes permanently fixed through both the metal and rubber seals of its cap. One pipe has a 0.22 µm air filter connected and the second is linked by silicone tubing (clamped off until a sample is needed) to the sample pipe.

A syringe is fitted to the air filter after autoclaving. The sample device is usually attached to the vessel top plate so that the glass bottle hangs down vertically beneath the cap and a supply of bottles of the same size are autoclaved ready for use. When a sample is to be taken, the clamp on the sample pipe tubing is released, and the syringe pulled back to create a partial vacuum in the sample bottle. Culture will flow into the sample bottle. When the required volume of sample has been taken (never more than 75% of the total bottle volume) the syringe is pushed in to clear the line and the sample tubing is clamped shut again. The sample bottle with the culture is quickly removed and capped with a sterile cap from a new glass bottle which replaces it under the sampling device. It is often preferable simply to discard the first few millilitres of the next sample by using it as a 'wash' for the transfer line, putting a fresh bottle under the sampling device to actually take the sample.

Routine Maintenance of Fermentor Components

pH Electrodes

pH electrodes are a semi-consumable item. Typically, an electrode will need replacing after 20-50 autoclave cycles. This broad range reflects the different results sometimes found between different brands and types of electrode and the harshness of the environment in which they are used, *e.g.* high temperatures, frequency of autoclaving, etc. Ageing of electrodes can show up as a sluggish response time, and a smaller slope on calibration, especially in the alkaline region. The ageing process is speeded up by high temperatures, so an autoclavable electrode will invariably have a shorter working life if repeatedly heated to sterilization temperatures. However, during its working life, the pH electrode may need attention to ensure optimal performance. In this respect, the gel-filled pH electrodes are normally less demanding and are physically more robust than the glass, liquid-filled variety.

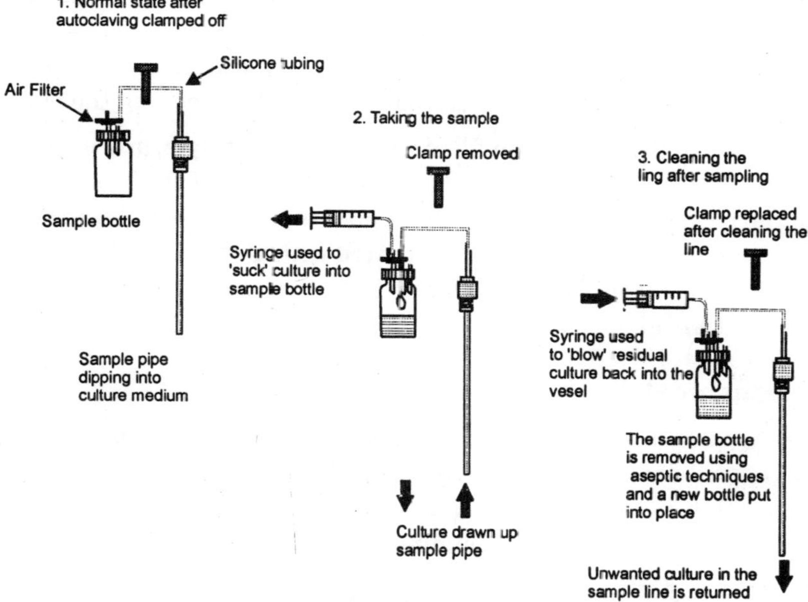

Fig. 4.14. Sampling.

Dissolved Oxygen Electrodes

Galvanic Electrode : This is a simple, self-powered system which relies on a lead anode and a silver cathode which are shorted together. Oxygen is reduced at the silver cathode and an electrical current flows in proportion to its partial pressure. The response time is slower than a polarographic electrode but the initial cost is far lower. The electrode is conditioned for the first time by the autoclaving process. The probe is relatively robust except if the lead anode is exposed to air due to a drop in the level of electrolyte in the probe. Therefore, all air bubbles must be removed when electrolyte is added and probes must always be stored upright with an adequate level of electrolyte and with the electrode shorted out (cap on).

Periodically, the electrode membrane will need replacing. This involves soaking a section of new silicone rubber tubing in chloroform to swell it (safety hazard!). While this is happening, the old tubing and membrane is cut away. A new membrane is stretched tightly across the electrode tip and the silicone tubing placed over it to hold it in place without any wrinkling. The end away from the tip is trimmed off and secured with a little Araldite glue and the electrode tapped gently to remove air bubbles.

Polarographic Electrode

These electrodes work on a different principle (the Clark Cell). The electrode consists of a silver anode/ reference electrode with a platinum cathode encased in a glass rod. The end of the glass rod is shaped to accept a replaceable membrane cartridge. A temperature sensor is included for automatic compensation. The whole assembly is usually encased in a stainless steel housing. A chloride solution is used as an electrolyte. A polarization voltage must be applied to the electrode which can then reduce the oxygen passing through the membrane and generate a current. For this reason, the electrode must be autoclaved shorted out and be connected to a source of polarizing voltage for a minimum of 2 h before use. The membrane must be replaced periodically by unscrewing the bottom section of the electrode and gently removing the old cartridge.

A new cartridge is fitted, having been filled with fresh electrolyte, and pushed onto the inner glass rod. If necessary, the small O-ring in the bottom section of the housing can be renewed before replacing the cartridge. A functional check on the electrode can be performed by setting the reading of a polarized electrode in water saturated with air to 100%. The air is replaced with nitrogen gas and the value should fall to less than 1.5% after about 5 min. If the electrode is now removed into the air, the value should return to 98%+ within 45-90 s.

Changing a Mechanical Seal

Different arrangements of mechanical seal are used for different makes of fermentor. The type described is a good example as it shows many of the features used with the mechanical seals employed on both bench-scale and *in situ* sterilizable fermentors. Seals should be replaced regularly (at least every 6-12 months, depending on use).

MAJOR TYPES OF ORGANISMS USED IN FERMENTATION

A wide range of organisms can be used in fermentation for the production of various chemicals and antibiotics.

Bacteria/yeast/fungi

Either a natural isolate of the culture is made or an uncontaminated 'master' culture is revived from a 'freeze-dried' state or from liquid nitrogen storage. Bacteria, yeasts and filamentous fungi are the usual starting materials. The starting cultures are grown in petri dishes on agar containing the required nutrients and incubated at an optimum temperature, humidity and gaseous atmosphere. If necessary, shake-flask cultures are prepared in an incubator shaker to produce larger volumes of culture for analysis or for use as an inoculum for a fermentor. Flasks are typically filled to 20-25% to provide a large area for gas exchange and are 25 ml to 5 1 in total capacity. An inoculum is prepared either from the shake-flask culture or by suspending colonies from growth on solid media in a nutrient/buffer solution. This is aseptically transferred to a fermentor vessel. The inoculum is typically 5-10% of the volume of medium in the fermentor and is often diluted in order to introduce a known number of cells/ml in this volume.

A laboratory scale fermentor can be used to provide larger volumes (typically 1-51 of culture). Growth is normally as a batch fermentation with the following parameters measured and controlled as necessary:

Temperature typically 20—40°C (4-90°C at the extremes) Speed typically 150-1500 rpm (20-2000 rpm at the extremes) pH typically 4-8 (2-12 at the extremes) Dissolved oxygen typically 40-80% (0-100% at the extremes)

In some cases, a continuous culture system may be used to draw off culture of a known cell concentration for long periods. A concentration of a growth-limiting metabolite (chemostat) or an indirect estimate of cell numbers (*e.g.* turbidity, turbidostat) is used to keep the culture density constant for many days/weeks. Samples are taken from the fermentor for counting, culturing on solid medium or biochemical assay for a particular metabolite, sugar utilization, etc. On completion of the culture, the whole culture can be centrifuged, *e.g.* 4000 × g for 15 min to separate. whole cells from the culture supernatant fluid.

Alternatively, a membrane or hollow-fibre system can be used for this initial separation. The separated cells can be washed in buffer, resuspended, and can then be disrupted using a variety of methods, *e.g.* ultrasonic disruption, freeze-thaw, shaking with beads, etc. Supernatant fluids can be further separated and analysed using separation columns, fraction collectors and spectrophotometry. Gel electrophoresis, ELISA assay, etc. are also used. Applications are almost too numerous to mention but obvious categories include brewing (yeasts), antibiotic production (fungi), enzyme production (bacteria) and food manufacture such as yoghurt (bacteria). Environmental uses include removal of toxic wastes, biological control of crop pests and soil regeneration.

Plant Cells

The whole plant is grown by conventional means. A sterile knife is used to cut a small section of root, stern or leaf which is transferred onto solid medium to produce a ball of cells (a thallus). The thallus is separated with a sterile blade and the cells grown in shake-flask culture in an incubator shaker at 20-30°C and 100-200 rpm. The cells are transferred to a fermentor under the following conditions: *e.g.*

- stirrer speed = 100-150 rpm;
- temperature 20-30°C (cooling needed if lights used);
- pH not generally controlled as it tends to cycle naturally;
- dissolved oxygen rarely controlled as it is not generally a limiting factor;
- foaming can be a problem unless controlled.

Cells may be cycled between the shaker and fermentor several times to build up shear resistance. The cell growth is normally too slow for effective continuous culture. For material inside the plant cell a 'Potter' type homogenizer can be used for gentle disruption. Centrifugation at approx. 1000 rpm for whole cells will separate them from the culture supernatant. Higher speeds will be needed to deposit cell fragments. The cellular pellets are discarded or extracted depending on where the required component is located. Plant cell cultures are often used for production of specific biochemicals such as flavinoids and alkaloids used in foods and chemical production. Culture of plant cells for agro-chemical uses is another obvious application.

Mammalian Cell Culture

There are many different cell lines of human, mammalian, avian and even insect origin. For simplicity, a single type, hybridomas, will be used as an example here. Hybridoma cell lines are prepared using cells from mouse spleen which secrete desired antibodies crossed with a continuous cell line grown *in vitro.* The cells are placed in suspension culture in an incubator shaker at 37°C and 60-100 rpm speed. The cells are transferred to a specially adapted fermentor. Large, broad-bladed impellers are used to minimize shear forces and typical stirrer speeds are 20-100 rpm. A gas mixing station can provide an air/carbon dioxide mixture and CO_2 gas can be used for pH control via a solenoid valve as replacement for a peristaltic pump. Oxygen and nitrogen can be introduced for dissolved oxygen control. For many cell lines, the control of foam is vital as its presence can contribute to cell damage.

The cells can sometimes grow freely in suspension but more often are anchored to a mesh, alginate beads or a hollow-fibre system. A special mesh can be connected to the drive shaft to provide a 'spin-filter', or to the sample pipe for taking samples of culture fluid relatively cell-free. Some alternative approaches to growth of mammalian cells are as follows:

(*i*) Air lift. There is no drive motor and it is the pumping of liquid around the draught tube which provides for mixing and gas exchange.

(*ii*) Hollow-fibre cartridges. They remain outside the fermentor vessel but culture fluids can circulate between the fibres and the vessel. All mixing and gas exchange is done in the vessel and the enriched medium passes to the cells anchored to the hollow fibres via a pump.

(*iii*) A gassing basket inside a stirred fermentor. Where cells need good levels of dissolved oxygen but are very sensitive to shear, a large amount of thin-walled silicone tubing supported on a frame can be placed in the vessel to provide a large area for gas exchange. Air is bubbled through the pores in the silicone tubing to give good gas transfer rates without forming large bubbles.

Applications of mammalian cell culture include monoclonal antibody production, culture of viruses for vaccine manufacture and large quantities of specific cells for medical research.

Algae

This category covers a diverse range of microorganisms which differ vastly in their growth conditions and metabolic requirements. Some are halophiles (salt-loving), barophiles (survive under high pressure), thermophiles (grow at high temperatures) and psychrophiles (grow at low temperatures). Most need special culture conditons and so it is best to seek out published literature with specific details. Applications for the use of algae include food additives and chemical production by biotransformation.

MICROBIOLOGY OF INDUSTRIAL FERMENTATION

Microorganisms play a central role in the production of a wide range of industrial chemicals, enzymes and antibiotics. The diversity of microbial processes may be attributed to many factors, including the high surface-to-volume ratio and adaptability to a wide spectrum of carbon and nitrogen sources. The high surface-to-volume ratio supports a very high rate of metabolic turnover, *e.g.* the yeast *Saccharomyces cereviciae* has been reported to be able to synthesize protein several orders of magnitudes higher than plants. On the other hand, the ability of microorganisms to adapt to different metabolic environments makes them capable of utilizing inexpensive renewable resources such as wastes and by-products of the farming and petrochemical industries as the primary carbon source.

Industrially important microorganisms include bacteria, yeasts, moulds and actinomycetes. The rate of product formation in a given industrial process, a significant parameter, is directly related to the rate of biomass formation which, in turn, is influenced directly or indirectly by a whole host of different environmental factors, *e.g.* oxygen supply, pH, temperature, accumulation of inhibitory intermediates, etc. It is, therefore, important that we are able to describe growth and production in quantitative terms. The study of growth kinetics and growth dynamics involves the formulation and use of differential equations and while the mathematical derivation of these equations is beyond the scope of this chapter, it is important that we understand how such equations can be used to further our understanding of microbial growth in general and its impact on product formation (yield) and recovery in particular.

THE GROWTH CYCLE

Now let us consider the basic equation used to describe microbial growth:

$$\frac{dx}{dt} = \frac{ax}{b} \qquad \text{...(4.1)}$$

Equation (4.1) implies that the rate of biomass (x) formation changes as a function of time (t) and that the rate of change is directly proportional to the concentration of a particular factor (a) such as growth substrate or temperature but is inversely proportional to the concentration of another factor (b) such as inhibitors. In the equation above a and b are independent of time t, and the proportionality factor in equation (4.1) can in effect be ignored.

In the early stages of any fermentation process, the increase in biomass is unrestricted and, as such, the pattern of growth follows an autocatalytic first-order reaction (autocatalytic growth) up to a point where either side of equation (4.1) becomes negative, resulting in autocatalytic death. During batch fermentation, a typical pattern of growth curve, otherwise known as the growth cycle, is observed. Clearly a number of different phases of the growth cycle can be differentiated. These are:

- I Lag phase
- II Acceleration phase
- III Exponential (logarithmic) phase
- IV Deceleration phase
- V Stationary phase
- VI Accelerated death phase
- VII Exponential death phase
- VIII Survival phase

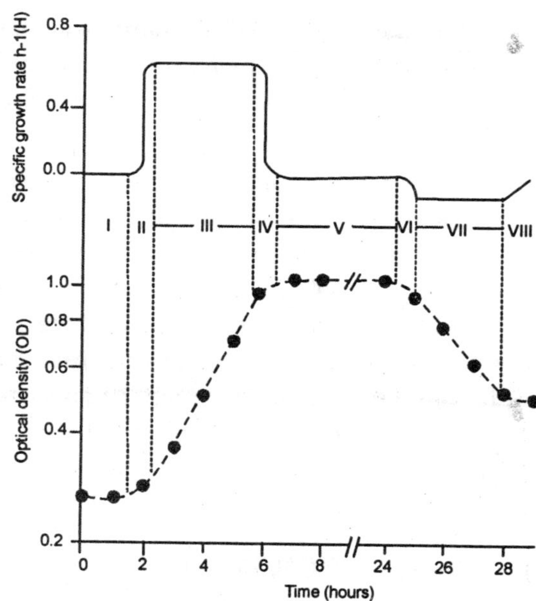

Fig. 4.15. Typical pattern of growth cycle during batch fermentation. The vertical dotted lines and the Roman numerals indicate the changes in specific growth rate (m) throughout the cycle.

The changes in the specific growth rate (μ) as the organism progresses through the growth cycle seen in Fig. 4.15.

We shall now describe the metabolic events and their implications in as far as growth, survival and productivity are concerned. Naturally the scenario begins with the first phase of the growth cycle.

The Lag Phase

In this phase the organism is simply faced with the challenge of adapting to the new environment. While adaptation to glucose as a sole source of carbon appears to be relatively simple, competition with other carbon sources, although complex, is resolved in favour of glucose through the operation of two different mechanisms, namely catabolite inhibition and catabolite repression. Adaptation to other carbon sources, however, may require the induction of a particular set of enzymes which are specifically required to catalyse transport and hydrolysis of the substrate, *e.g.* adaptation to lactose, or to fulfil anaplerotic as well as regulatory functions as is the case in adaptation to acetate as the sole source of carbon and energy.

Irrespective of the mechanisms employed for adaptation, the net outcome at the end of the lag phase is a cell that is biochemically vibrant, *i.e.* capable of transforming chemicals to biomass, *i.e.* growth. Entering a lag phase during the course of industrial fermentation is not desirable as it is very costly and as such should be avoided. The question of whether a particular organism has entered a lag phase in a given fermentation process can be determined graphically by simply plotting log *n* (biomass) as a function of time. Note that the transition from the lag phase to the exponential phase is generally interrupted by an acceleration phase, as described earlier. This difficulty canbe easily overcome by extrapolating the lag phase sideways and the exponential phase downwards as shown, with the point of interception *(L)* taken as the time at which the lag phase ended. What is also interesting about the graph in Fig. 4.16 that if one continues to extrapolate the exponential phase downwards, then the point at which the ordinate is intersected gives the number of cells that were viable and metabolically active at the point of inoculation.

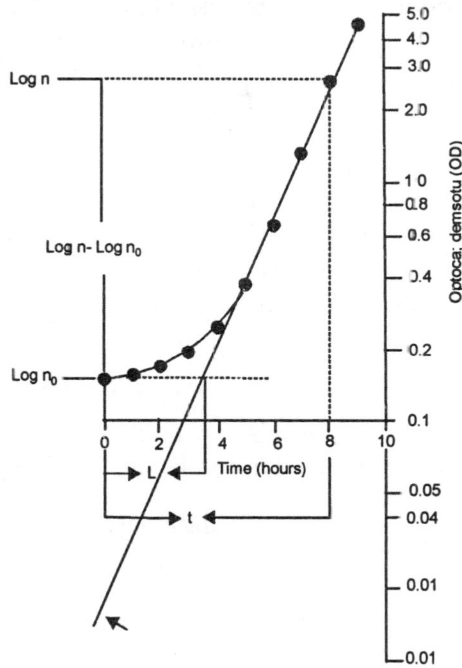

Fig. 4.16. Graphical determination of the lag phase and the number of viable cells at the onset of batch fermentation. As the exponential phase is extrapolated downwards, it intercepts the extrapolated line of the lag phase and the ordinate respectively (see text for details).

The question of whether a lag *(L)* has occurred during the course of the fermentation process and for how long can be easily determined from the following equation:

$$\frac{\log n - \log n_0}{t - L} = \frac{\log 2}{T} \qquad ...(4.2)$$

where n is the total number of cells after a given time *(t)* since the start of fermentation, n_0 is the number of cells at the beginning of fermentation and T is the organism's mean generation time (doubling time). Equation (4.2) describes the exponential growth, taking into consideration a lag phase in the process. In the following section we shall describe the exponential phase in general and the derivation of equation (4.2) inparticular.

The Exponential Phase

In this phase, each cell increases in size, and providing that conditions are favourable, divides into two which, in turn, grow and divide and the cycle continues. During this phase, the cells are capable of transforming the primary carbon source into biosynthetic precursors, reducing power (NAD(P)H, $FADH_2$, HS-CoA, etc.) and energy which is generally trapped in the form of ATP, PEP and proton gradients. Such a complex function is achieved through the activities of various enzymes of central metabolism. The biosynthetic precursors thus generated are then channelled through various biosynthetic pathways for the biosynthesis of various monomers (amino acids, nucleotides, fatty acids, sugars) which, in turn, are polymerized to give the required polymers (proteins, nucleic acids, ribonucleic acids, lipids). Finally, these polymers are assembled in a precise way and the cell divides to give the new biomass characteristic of each organism. The time span of each cycle (cell division) is known as generation time or doubling time but because we generally deal with many millions of cells in bacterial cultures, the term mean generation time is more widely used to reflect the average generation times of all cells in the culture. Such a rate, providing conditions are favourable, is fairly constant. If a given number of cells (n_0) are inoculated into a suitable medium and the organism was allowed to grow exponentially, then the number of cells after one generation is $2n_0$; at the end of two generations, the number of cells becomes $4n_0$ (or 2^2n_0). It follows therefore that at the end of a certain number of generations (Z), the total number of cells equals 2^Zn_0. If the total number of cells, or its log value, at the end of Z generations is known, then

$$n = n_0 2^z \qquad ...(4.3)$$

$$\log n = \log n_0 + Z \log 2 \qquad ...(4.4)$$

To determine the number of cell divisions, *i.e.* number generations (Z) which have taken place during fermentation, the above equation can be modified to give:

$$Z = \frac{\log n - \log n_0}{\log 2} \qquad ...(4.5)$$

If T is the mean generation time required for the cells to double in number and t is the time span over which the population has increased exponentially from n_0 to n, then

$$Z = \frac{t}{T} = \frac{\log n - \log n_0}{\log 2} \qquad ...(4.6)$$

If during the course of a particular fermentation, a lag time has been demonstrated, equation (4.6) can be modified to take account of this observation. The modified equation is as follows:

$$Z = \frac{t - L}{T} = \frac{\log n - \log n_0}{\log 2} \qquad ...(4.7)$$

The above can be rearranged to give the familiar equations governing the determination of the mean generation time (T) as follows:

$$\frac{\log n - \log n_0}{t} = \frac{\log 2}{T} \qquad ...(4.8)$$

$$\frac{\log n - \log n_0}{t - L} = \frac{\log 2}{T} \qquad ...(4.9)$$

While equations (4.8) describes the exponential phase of growth in pure terms, takes into consideration the existence of a lag phase in the process.

The exponential pattern (*i.e.* 2-4-8-16, and so on) demonstrated in Fig. 4.17 obeys equations (4.8) and (4.9) for logarithmic growth. Note the different position of $\log n_0$ to that cited in Fig. 4.16.

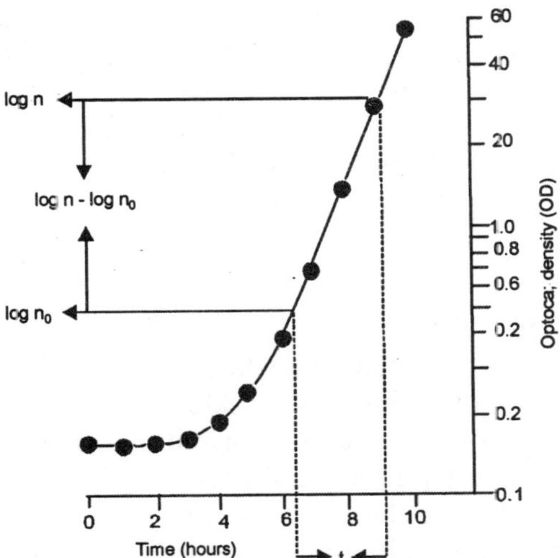

Fig. 4.17. Graphical determination of the mean generation time (T) during batch fermentation. Note the different position of log n_0 with respect to that in Fig. 4.16.

Although the use of log to the base 2 has the added advantage of being able to determine the number of generations relatively easily, because an increase of one unit in log $2n$ corresponds to one generation, the majority of researchers continue to use log to the base 10. In this case, the slope of the line. The relationship between the mean generation time (T) and the specific growth rate (μ) can be described mathematically by the following equation:

$$\ln 2 = \mu T \quad ...(4.10)$$

or

$$\mu = \ln 2/T \quad ...(4.11)$$

Because ln 2 = 0.693, either of the above equations can be rearranged to give

$$T = \frac{0.683}{\mu} \quad ...(4.12)$$

During the course of exponential growth, the culture reaches a steady state and as such the intracellular concentrations of all enzymes, cofactors and substrates are considered to be constant. During this phase, one can therefore safely assume that all bacterial cells are identical and that the doubling time is constant with no loss in cell numbers due to cell death. The rate of growth of a given population (N) represents, therefore, the rate of growth of each individual cell in the population multiplied by the total number of cells. Such a rate can be described mathematically by the following differential equation:

$$\frac{dN}{dt} = \mu N \quad ...(4.13)$$

This equation implies that the rate of new biomass formation is directly proportional to the specific growth rate (μ) of the organism under investigation and the number of cells (N). This pattern of growth may be referred to as ***autocatalytic***: a term which has been described earlier. If the above equation describes the exponential phase correctly, then a straight line should be obtained when ln N is plotted as a function of time (t). During the exponential phase, it is generally assumed that all cells are identical and as such the specific growth rate (μ) of individual cells equals that of the whole population.

Therefore eqn, 4.13 can be rearranged to take account of the fact that dN/dt is proportional to the number of cells (N) and that the specific growth rate (μ) is a proportionality factor to give:

$$\mu = \frac{1}{N}\frac{dN}{dt} \qquad ...(4.14)$$

The specific growth rate (μ) is usually expressed in terms of units per hour (h^{-1}).

Although describe the exponential phase satisfactorily, in some fermentations, as is the case during growth of *E. coli* on sodium acetate (M. El-Mansi, unpublished results), the organism fails to maintain a steady state for any length of time and so μ. falls progressively with time until the organism reaches the stationary phase. Such a drop in growth rate (μ) can be accounted for by a whole host of different factors including nutrient limitations, accumulation of inhibitors and/or a crowding factor, *i.e.* as the population increases in size μ decreases.

Metabolic Interrelationship

To account for the effect of nutrient limitations on growth rate (μ), Monod modified the above equation so that the effect of substrate concentration (limitations) on μ could be assessed quantitatively. Monod's equation is as follows:

$$\mu = \frac{\mu_m S}{K_S + S} \qquad ...(4.15)$$

where S is the concentration of the limiting substrate, μm is the maximum specific growth rate and K_S is the saturation constant, i.e. when S = K_S then μ = u' m/2

During growth in a steady state, the following equations may be used to describe growth using biomass (X) rather than number of cells (N) as a measure of growth

$$\frac{dX}{dt} = \mu X \qquad ...(4.16)$$

$$\frac{dS}{dt} = -\mu\frac{X}{Y} \qquad ...(4.17)$$

where X is the biomass concentration, Y is the growth yield (*g* biomass generated per gram substrate utilized). It is noteworthy, however, that during steady state growth, *i.e.* in a chemostat or a turbidostat, the terms used to describe the bacterial numbers (*N*) and biomass (*X*) in equations identical. This is not necessarily the case in batch cultures because the size and shape of bacterial cells vary from one stage of growth to another.

While eqn. (4.15) addresses the relative change in biomass (the first variable) with respect to time, eqn. (4.16) addresses the relative change in substrate concentration (the second variable) as a function of time. Note that following the exhaustion of substrate concentration dX/dt = dS/dt = 0. While either equation (4.16) and (4.17) can surely predict the deceleration and stationary phases of growth, it should be remembered that nutrient limitation is not the only reason for the deceleration of growth and subsequent entry into the stationary phase. In addition to environmental factors, crowding is reported to have an adverse effect on growth rate (μ), *i.e.* growth rate diminishes as the size of the population increases. Although the equation describing the effect of crowding on growth rate (Verhulst-Pearl logistic equation) predicts a sigmoidal pattern of growth and fits rather well with the growth curve for populations of higher organisms, yeasts as well as bacteria.

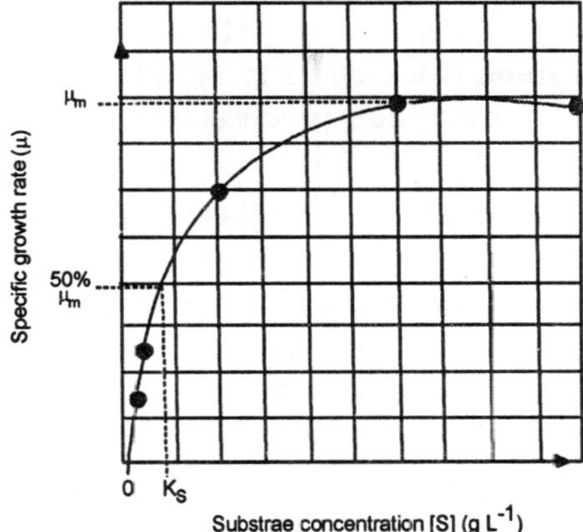

Fig. 4.18. Graphical determination of the maximum specific growth rate (μ) and the saturation constant (K_s). Note that this graph was constructed without taking maintenance energy into consideration. With maintenance in mind the curve should slide sideways to the right in direct proportion to the fraction of carbon diverted towards maintenance.

Stationary Phase and Cell Death

As the exponential phase draws to an end, the organism enters the stationary phase and then the death phase. The equations used earlier to describe growth do not adequately describe cell death. However, if we assume that the kinetics of death are similar to that of growth, then the specific rate of cell death (λ) can be described mathematically by the following equation:

$$\frac{dN}{dt} = (\mu - \lambda)N \qquad ...(4.18)$$

Equation (3.18) clearly indicates that if μ is greater than λ, then the organism will grow at a rate equal to μ - λ. If on the other hand μ is less than λ, then the population dies at a rate equal to λ – μ. Under conditions where μ equals λ, neither growth nor death is observed: a situation which is thought to prevail throughout the course of the stationary phase. As the energy supply continues to fall, the equilibrium between λ and μ will finally shift in favour of λ and consequently cell death begins. Microorganisms respond differently to nutrient limitations during the stationary phase.

For example, while *Bacillus subtilis* and other Gram-positive, spore-forming organisms respond by sporulation, *E. coli* and other Gram-negative, non-spore-forming bacteria cannot respond in the same way and as such other mechanisms must have evolved. Although a fraction of the bacterial population die during this phase of growth, a relatively large number remains viable for a long time despite starvation. The ability of cells to remain viable despite a prolonged period of starvation is advantageous, as most microorganisms in Nature are subject to nutrient limitations in one form or another. Our understanding of the molecular mechanisms employed by microorganisms for survival during this phase of growth is rather limited.

However, contrary to the notion that microorganisms enter a logarithmic death phase soon after the onset of the stationary phase, recent investigations have revealed that some microorganisms such as *Saccharomyces cerevisiae* and *E. coli* adapt to starvation, presumably through mutations and induction mechanisms very effectively. The ability of some cells to survive prolonged starvation inspired some researchers to pursue the question of whether such cells are biochemically distinguishable from their predecessors which were actively growing or had just entered the stationary phase.

In the case of *Saccharomyces cerevisiae,* analysis of mRNAs which are specifically expressed following the onset of the stationary phase revealed the presence of a new family of genes. This family is referred to as SNZ (short for snooze) and sequence analysis revealed that they are highly conserved. The functional role of each member of this family, however, remains to be determined. It is interesting that some researchers use the term Viable But Non-Culturable (VBNC) to describe cells in the stationary phase. However, as colony-forming ability is our only means of assessing whether a particular cell is alive or not, other microbiologists argue the case for reversibility, *i.e.* the cell's ability to transform itself from being dormant to being metabolically active.

Recent investigations have revealed that resuscitation of stationary phase cells may be aided by the excretion of pheromones; this has recently been demonstrated in *Micrococcus luteus.* In the case of *E. coli,* an attempt was made to identify the genes which are uniquely turned on in the stationary phase in response to starvation, *i.e.* not expressed during the exponential phase, with the aid of random transposon mutagenesis.

A number of mutants which were unable to survive in the stationary phase were isolated and, subsequently designated as survival negative mutants (Sur$^-$) and the genes involved were designated as *sur* genes. While some genes, *e.g. sur A,* are required for survival during 'famine' (starvation), others, *e.g. surB,* enable the organism to exit the stationary phase as conditions change from 'famine' to 'feast'. Recent studies have also revealed that starvation induces the production of a stationary phase-specific transcriptional activator (sigma factor) that is essential for the transcription of *sur* genes by RNA polymerase. The physiological function of the *sur* gene products is to downshift the metabolic demands made on central and intermediary metabolism for maintenance energy.

Maintenance and Survival

Maintenance can be defined as the minimal rate of energy supply that is required to maintain the viability of a particular organism without contributing to biosynthesis. The fraction of carbon oxidized in this way is, therefore, expected to end up in the form of carbon dioxide (CO_2). The need for maintenance energy is obvious as the 'living-state' of any organism, including ourselves, is remote from equilibrium and as such demands energy expenditure. Moreover, in addition to the carbon processed for maintenance, another fraction of the carbon source may be wasted through excretion, *e.g.* acetate, a-ketoglutarate, succinate, lactate, etc. In this context, excretion of metabolites ought to be seen as an accidental consequence of central metabolism rather than by design to fulfil certain metabolic functions. The fraction of carbon required for maintenance differs from one organism to another and from one substrate to another.

For example, maintenance requirements for the lactose phenotype of *E. coli* are greater than those observed for the glucose phenotype as the *lac*-permease is much more difficult to maintain than the uptake system employed for the transport of glucose, *i.e.* the phosphotransferase system (PTS). The metabolic interrelationship between maintenance and growth was first described by Pirt (1965). In his theoretical treatment of this aspect, he formulated a maintenance coefficient (m) based on the earlier work of Monod, and defined the coefficient as the amount of substrate consumed per unit mass of organism per unit time (*e.g.* g substrate per g biomass per h). If s represents the energy source (substrate), then the rate of substrate consumption for maintenance is governed by the following equation:

$$-(ds/dt)_M = mX \qquad ...(4.19)$$

where m is the maintenance coefficient. The yield can therefore be related to maintenance by the following equation:

$$Y = \frac{\Delta x}{(\Delta s)_G + (\Delta s)_M} \qquad ...(4.20)$$

where Y is the growth yield, Δx is the amount of biomass generated, $(\Delta s)_G$ is the amount of substrate consumed in biosynthesis and $(\Delta s)_M$ is the amount of substrate consumed for mainteneace of cell viability.

If maintenance energy $(\Delta s)_M$ was determined and found to be zero, then equation reduces to:

$$Y_G = \frac{\Delta x}{(\Delta s)_G} \qquad ...(4.21)$$

Eqn, (4.21) clearly means that growth yield (Y) is a direct function of the amount of substrate utilized. The growth yield (Y, g dry weight per g substrate) obtained in this case may be referred to as the true growth yield to distinguish it from the Y_G where maintenance $(\Delta s)_M$ is involved. However, if a specific fraction of carbon is diverted towards maintenance energy or survival, then one might conclude that the slower the growth rate (μ), the higher the percentage of carbon that is diverted towards maintenance, the less the growth yield.

On the other hand, however, during unrestricted growth, *i.e.* no substrate limitation, the interrelationship between substrate concentration and maintenance can be assessed from the following:

$$ds/dt = (ds/dt)_M + (ds/dt)_G \qquad ...(4.22)$$

In this case eqa. (4.22) implies that the overall rate of substrate utilization-equals the rates of substrate utilization for both maintenance and growth. With growth rate expressed in the usual way, *i.e.* $dx/dt = \mu x$, and assuming that Y equals Y_G, can be rearranged to give:

$$1/Y = m/\mu + 1/Y_G \qquad ...(4.23)$$

According to eqa. (4.23) if m and Y_G are constants, then the plot of $1/Y$ against $1/\mu$ should give a straight line, the slope of which is maintenance (m) and the point at which the ordinate is intercepted is equal to $1/Y_G$.

During the course of fermentation, the minimum rate of energy supply may fall below maintenance requirements due to a shortfall in the supply of phosphorylated intermediates, ATP and/or reducing powers (NAD(P)H, $FADH_2$, etc.). Although such a drop in energy supply may not be fatal, it is likely to have an adverse effect on the fitness of the organism, *i.e.* the cells become less capable of taking advantage of favourable changes in the environment and consequently cells resume growth after a lag period. On the other hand, if the cells were able to grow immediately following inoculation or transfer into another medium, *i.e.* without lag, then the cells must have been left in optimal conditions.

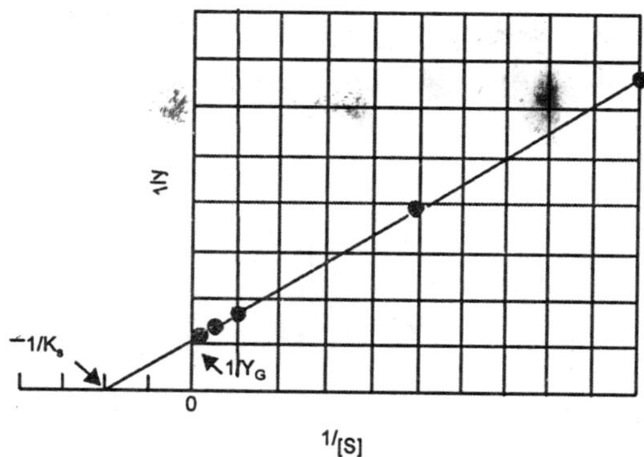

Fig. 4.19. Graphical determination of maintenance (m) and the saturation constant (*Ks*).

However, if the cells were to enter the lag phase prior to growth then it is fair to suggest that the organism was left in a sub-optimal state, *i.e.* the organism must have suffered a drop in energy supply below its maintenance requirements. It follows, therefore, that the physiological state of the cell at the time when energy supply falls below maintenance is very important and that the deeper the shortfall in energy supply, the longer the lag period. A lag period due to the above condition is surely different to that observed when the cells are faced with the challenge of changing phenotype, as is the case when glucose phenotype of *Escherichia coli* is forced to change to lactose phenotype). Individual cells in a given population may respond differently when a given drop of energy supply is exerted. The ability of a given population to respond successfully to 'shift-up' or to survive a 'shift-down' in nutrients appears to be directly related to the intracellular concentration of ribosomes. For example, in comparison with cells grown in rich medium, restricted growth of *Salmonella typhimurium* was accompanied by a 35-fold drop in the concentration of ribosomes. In the yeast *Saccbaromyces cerevisiae*, a drop in growth rate from 0.40 h^{-1} to 0.10 h^{-1} was accompanied by a drop in the cellular concentrations of RNA, DNA and proteins from 12.1%, 0.6% and 60.1% to 6.3%, 0.4% and 45% respectively. Furthermore, it was demonstrated that the cells were capable of 'scaling up' or 'scaling down' the intracellular level of ribosomes in response to the 'shift-up' or the 'shift-down' in nutrients respectively. Recent investigations have revealed that such ability to modulate the cellular content of ribosomes is achieved through a mechanism of autogenous regulation.

To account for such a phenomenon, Nomura *et al.* (1984) proposed a new hypothesis, *translational couplings*, which simply implies that binding to and initiation of translation at the first ribosome binding site of the operon is essential in exposing the ribosome binding sites of all other cistrons and that inhibition or prevention of translation at the first binding site means that all other sites within the polycistronic message will remain unexposed and as such untranslated. The argument over the question of whether smaller cells are less able to cope with environmental changes, *i.e.* nutrient limitations and the drop in energy supply below maintenance, because they contain less ribosomes can now be answered to the satisfaction of everyone as recent research revealed that it is not the number of ribosomes that matter but rather their concentration inside the cell.

Furthermore, the need for the cell size to be relatively large during growth under no limitations is not a reflection of high concentration of ribosomes but rather the need to attain a certain mass before replication of the chromosome can be initiated. The essence of Donachie's discovery is that if initiation of replication is triggered in response to a certain volume of cell mass, then the faster the cell divides, the faster the growth rate, the larger the size of the cell.

DIAUXIC GROWTH

The ability of microorganisms to display biphasic (diauxic) growth patterns is well documented. For example, *E. coli* and *Aerobacter aerogenes* display such a pattern when faced with two different substrates, glucose and lactose in the case of the former and glucose and citrate in the case of the latter. This is generally due to the preferential utilization of glucose to either of the competing substrates. This can be explained on the grounds that glucose has been found to exert catabolite inhibition, *i.e.* inhibition of other sugars or organic acids uptake systems, and catabolite repression of the enzymes required for the transport as well as hydrolysis of other sugars, as hydrolysis of other sugars, as is the case in lactose metabolism.

GROWTH YIELD

The relative contribution of each substrate to total biomass formation can be assessed with a good deal of precision providing that growth limitation is only enforced as a result of exhaustion of the substrate in question rather than the accumulation of toxic end-products or adverse changes in pH. The relationship between total biomass formation and the concentration of any given substance can therefore be determined from the plot of log *n* (biomass) as a function of substrate concentration and providing that all other components are in excess, the organism will enter the stationary phase upon depletion of the substrate or the substance in question. This experiment should obviously be done over a wide range of substrate concentrations and from the slope obtained, the yield constant or yield coefficient per unit of the substance in question can be determined. If the unit is, say, one mole of a substrate, then the yield coefficient obtained is referred to as the molar growth yield coefficient. This method is used widely as a biological assay for the determination of vitamins, amino acids, purines and pyridines.

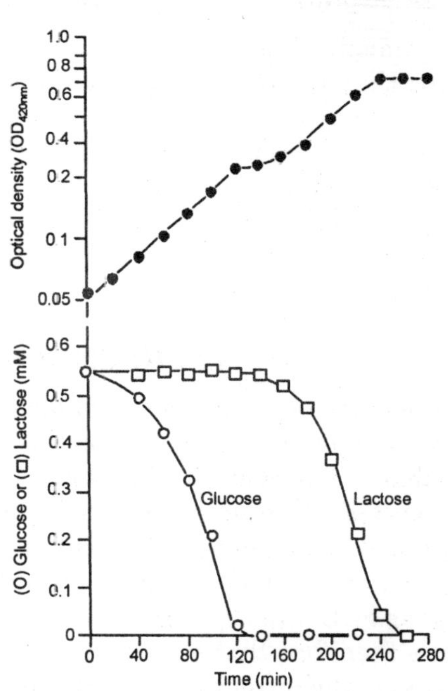

Fig. 4.20. Diauxic pattern of growth and the pattern of sugar utilization during growth of *E. coli* ML30 on equimolar concentrations (0.55 mM) of glucose and lactose.

FERMENTATION BALANCES

Carbon Balance

Apart from the fraction of carbon used for biomass formation, the remainder is partitioned between products and by-products, including carbon dioxide. The ratio of recovered carbon to that present at the onset of fermentation is referred to as the carbon balance or carbon recovery index. Such a balance or an index is a measure of efficiency; *i.e.* the higher the index, the higher the carbon recovery, the more efficient the fermentation process. The carbon balance is generally calculated by working out the number of moles (or millimoles) produced of a given product per 100 moles (or millimoles) of substrate utilized. The values obtained can then be multiplied by the number of carbon atoms in each respective molecule. The resulting values for the products in question can then be totalled and compared with that of the substrate. If the values are equal, *i.e.* 1 : 1 ratio, then a complete recovery of carbon into product formation has been achieved.

Although this is theoretically possible, our experience indicates otherwise as part of this carbon is used for maintenance and assimilation to support growth, however slow. A complete carbon balance can, therefore, be calculated for any given fermentation if the fraction of carbon diverted towards biosynthesis and maintenance is determined. In addition to the carbon balance, some fermentations involve calculation of the redox balance.

Redox Balance

Fermentations of sugars and other primary carbon sources give rise to a whole host of different intermediates, some of which are phosphorylated (energy rich) while others are not. While biosynthetic intermediates are utilized for the biosynthesis of monomers, other intermediates are produced in excess and this is balanced by their excretion into the medium. The stoichiometry of product and by-product formations of any given fermentation process can be ascertained by carefully analysing the culture filtrates at different stages. From a physiological standpoint and in order for fermentation to go to completion, the redox balance must be maintained.

EFFICIENCY OF CENTRAL METABOLISM

The efficiency of carbon conversion to biomass and desirable end-products is influenced by many different factors (see Chapters 4, 6 and 7 for more details) of which futile cycling is a major contributor. For example, *Klebsiella aerogenes* has been shown to have an NAD^+-dependent pyruvate reductase, an enzyme which catalyses the formation of D-lactate, and an FAD-dependent D-lactate dehydrogenase which catalyses the conversion of D-lactate to pyruvate. The operation of these two enzymes provides a futile cycle in which NADH,H^+ is oxidized at the expense of reducing FAD, thus bypassing the first phosphorylation site in the electron transport chain. It follows therefore that any changes in the metabolic environment which lead to a transient increase in the intracellular level of pyruvate may lead to the operation of this futile cycle and, in turn, lower the efficiency of central metabolism.

The presence and function of glutamine synthetase and glutammase in ammonia-limited cultures represents yet another possible futile cycle in which the synthetase generates glutamate while the other affects its hydrolysis with a net loss of one molecule of ATP per turn. In addition to futile cycling, flux to metabolite excretion in general and acetate in particular is another significant factor in influencing the efficiency of central metabolism (El-Mansi and Holms, 1989).

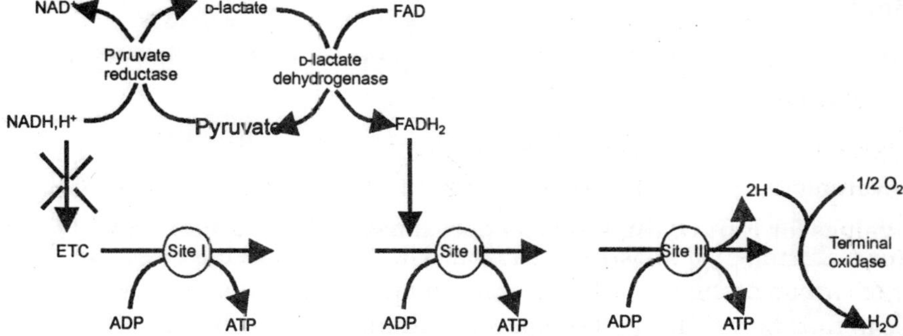

Fig. 4.21. The role of pyruvate reductase and D-lactate dehydrogenase in bypassing the first phosphorylation site of the electron transport chain (ETC) in the oxidation of NADH, H^+.

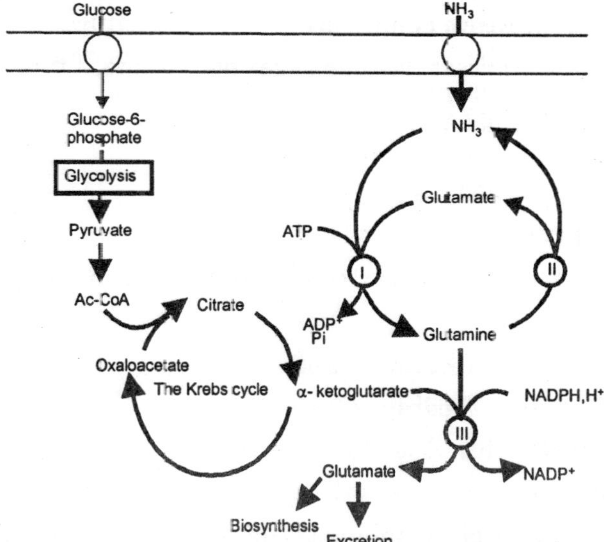

Fig. 4.22. The role of glutamine synthetase and glutaminase in energy dissipation through the formation of a futile cycle.

Furthermore, because acetate, particularly in its undissociated form, has proved to be a potent uncoupler of oxidative phosphorylation, any conditions which promote flux to acetate excretion will, in turn, diminish flux to product formation due to the diversion of a much larger fraction of carbon towards high maintenance energy requirements (El-Mansi, unpublished results).

Calculation of RQ

When data are obtained from an exit gas analysis, it is generally required that the parameter RQ is calculated in order to gain an insight into the metabolism of the organism under investigation. This is normally done by programming a definition into a computer software package for datalogging and control. Here is a generalized outline of the steps needed:

RQ = CPR (Carbon dioxide production rate)/ OUR (Oxygen uptake rate)

CPR = [Flow CO_2 out] × [Conc'n CO_2 out] miny [Flow CO_2 in] × [Conc'n CO_2 in] gmol/O_2/h

OUR = [Flow O_2 in] × [Conc'n O_2 in]-[Flow O_2 out] × [Conc'n O_2 out] gmol/O_2/h

Concentrations of gases in inlet gas (AIR) O_2 = 20.95% CO_2 = 0.03%

The working volume of the fermentor (WV) and the gas flow rate (FR) must be known to make the calculation:

VUO_2 (Volumetric uptake O_2) = (FR × (0.2095 – (Exit O_2/100))) WV

$VPCO_2$ (Volumetric production CO_2) = (FR × ((Exit CO_2/100) – 0.0003))) WV

Typical values for Exit O_2 could reach approximately 16% and for Exit CO_2 around 5%.

$$RQ = \frac{VPCO_2}{VUO_2}$$

Values of RQ for yeast biomass fermentation, for instance, would need to be kept around 1 for optimal biomass production.

A worked example of the above definitions:

WV = 1 l. FR = 1 l min^{-1} (1 vessel volume of gas per minute, a typical flow rate).

Exit O_2 = 16% and Exit CO_2 = 5%.

VUO_2 = 10/10 × (0.2095 – 0.16) = 0.0494

$VPCO_2$ = 10/10 × (0.05 – 0.0003) = 0.497

RQ = 0.497/0.494 = 1.006

To achieve control, a sequence along the lines of the following would have to be set up:

if RQ < 0.95 then feedrate = feedrate × 1.05

if RQ > 1.05 then feedrate = feedrate × 0.95

Preparation of Tubing for Aseptic Connection

Any silicone tubing to be joined needs the two ends to be prepared as follows:

1. The open (female) end of the tubing is simply covered in aluminium foil held in place with autoclave tape. For added security, the open end can be sealed with a short piece of glass rod flattened at one end to make a 'bung'.
2. The other (male) end has a short length of stainless steel pipe pushed into the silicone tubing. The exposed length of pipe is covered with foil which is taped closed.
3. Silicone tubing is connected to any of the inlet pipes intended for use for reagent addition. The tubing must stretch to the peristaltic pump heads when the vessel is in place.
4. The tubing is closed with clamps or any other closure system which will withstand autoclaving (the arterial clamps used in medicine are often employed where a connection must be opened and closed many times, e.g. to a sampling device). The tubing which goes to the sampling device from the top plate must be clamped shut during autoclaving.

Routine Maintenance of pH Electrodes

1. Always store an electrode which is not in use in liquid – preferably 3 M KCl solution,

although pH 7 buffer or even fermentation medium will do (in descending order of preference).

2. An electrode which has completely dried out will need regenerating—a procedure which normally involves a brief immersion in a hydrofluoric acid solution (typically supplied by the manufacturer) followed by prolonged soaking in storage solution. This procedure may also regenerate 'aged' electodes for a time. (Note that glass vessels cannot be used for this task due to the effects of the hydrofluoric acid).
3. A build-up of protein material on the electrode membrane can be dealt with by soaking the electrode in a pepsin/hydrochloric acid solution for about an hour.
4. If a discolouration (blackening) of the diaphragm has been caused by sulphides in the medium, soaking for several hours in an HCl/thiourea solution will recover its shiny surface.
5. Never wipe the tip of the electrode with a paper towel as this causes scratching.
6. Glass/liquid electrodes should have the reference electrolyte topped up periodically.
7. To test if the pH meter/cable are functioning correctly, a pH simulator can be used. If such a device is not at hand, a paper clip can be used to short the center pin of the electrode plug and the outside metal ring of the cap. The pH reading should be very close to 7.0.

Replacing a Mechanical Seal

The procedure for replacing a typical single mechanical seal is as follows:

1. Unscrew the top part of the drive shaft and remove. Measure the distance between the bottom of the retaining ring and the inner surface of the top plate. This allows the new seal to be correctly tensioned by setting the same gap at the end of the procedure.
2. Remove any retaining ring by loosening the small screw(s) holding it in place. Disassemble the retaining ring, hard seal, soft seal and bottom O-ring in order. Examine the faces of the static and rotating seals. Scoring of the surface indicates the seal has worn out. Never try to replace the old seal if it looks undamaged - it cannot be reused.
3. Insert a new O-ring into the 'hole' in the vessel plate which leads to the drive shaft bearings. This may be a tight fit but it must seat properly to prevent leaks. Replace the static seal by pushing it into the O-ring to obtain a snug fit. The protective rubber cover over the top face must be removed at this point.
4. Fit the rotating seal onto the drive shaft. Ensure the inner O-ring and metal washer are firmly pushed down into place as the tensioning spring will rest on these. Fit the new tensioning spring into place then push onto this firmly with the retaining ring until the spring is tightly compressed. Measure the gap with a ruler/calipers and only tighten the screw(s) on the retaining ring when the correct distance has been achieved.
5. Replace the drive shaft and gently turn the shaft by hand a small distance to ensure it moves freely. Lubricate with glycerine.

CHAPTER

5

Sterilisation of Disinfection

Sterilisation is the action of eliminating microorganisms from a medium. Sterility is the absence of any detectable and viable microbes in a culture medium or in the gas phase. Sterilisation is a process that destroys all living organisms, spores and viruses in a pressurised vessel at high temperature. In the food and dairy industries, sterilisation is commonly used to preserve food products. At the laboratory scale, huge steel vessels with live stream at 105 kPa (15 psig) are commonly used for 20-30 minutes. This is a closed system known as an autoclave; therefore it is batch sterilisation. The temperature is raised to 121 °C when air is initially flashed out and all live streams are replaced. Wet steam is usually used for effective autoclaving. The high temperature and long duration may kill all living microorganisms, spores and viruses.

When media is prepared, to eliminate bacterial and fungal contaminants it must be heat treated at high pressure. Even at high temperatures the fungal spores may survive if only heat is used. Therefore, media is autoclaved at 121 °C and 105 kPa. In fact, an autoclave is used like a pressure cooker, which is convenient and economic autoclaving equipment. Overheatmg the prepared media may have negative impact, causing the media pH to be unstable.

Acid pH is very sensitive to overheating. Overheating in media containing sugars cuses the carbohydrate to be caramelised. This media may have reverse impact, reducing bacteriological performance. Gelatinous media or any nutrient agar at acidic pH is hydrolysed. This is due to the acidic condition of catalytic activities or excess amounts of protons breaking down the solid media and extra sugars forming. This may cause substrate inhibition on growth of organisms on solid slant media.

BATCH STERILISATION

Batch sterilisation uses steam to eliminate living organisms. Heat losses, heating and cooling are major steps, and it is a time-consuming process. It requires air to be evacuated replaced with steam. At first the chamber is flashed with pressurised steam. With this technique we should get rid of the existing air. Steam sterilisation is performed in a jacketed vessel by supplying steam and maintaining the set pressure and temperature at a constant level for a fixed duration. Batch sterilisation wastes energy and overcooks the medium. There is no conservation of energy; therefore the process may not be economical for implementation at a large scale. Batch sterilisation is common at bench and laboratory scales. Batch media sterilisation is done in an autoclave.

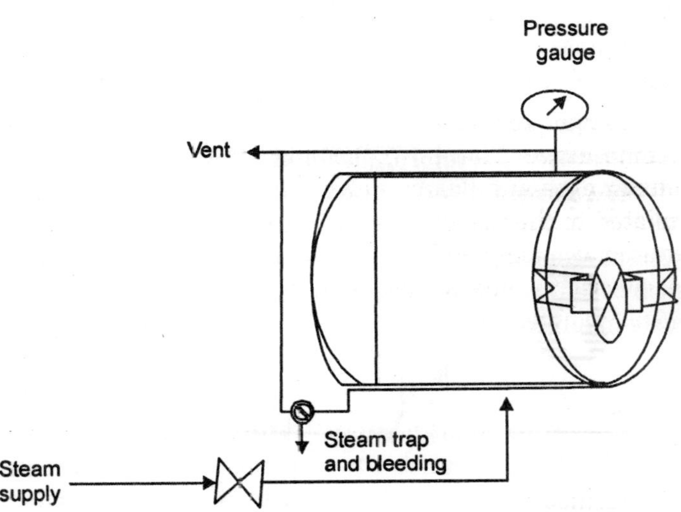

Fig. 5.1. Pressure steam sterilizer (autoclave).

Basically, it is a huge steam cooker. Steam enters the jacket of the surrounding chamber. When the pressure from the steam has been built up in the jacket, a venting valve for the outlet of chamber air is closed and the inlet valve allows the steam to enter the chamber. The pressure of the chamber is increased to 105 kPa. (15 P_{sig}) At this point the sterilisation time begins to count down. Usually 15-30 minutes are used, depending on the type and volume of media. For instance, sterilisation of one litre of media at 121 °C requires 20 minutes. Larger volumes may require longer retention times for sterilisation. There are recommended holding times of 20, 10 and 3 minutes for 121, 126 and 134 °C, respectively. The high pressure in a closed container allows the temperature to exceed 121 °C.

CONTINUOUS STERILISATION

One of the methods for continuous sterilisation of medium for fermentation is the direct use of live steam by injection of steam into the medium. The heat exchanger is eliminated. The medium stays in a loop for a predetermined holding time until the entire medium is sterile. The problem with directly injecting steam is dilution of media since it is initially cold. However, it has better heat economy, since it comes from substituting heat exchangers for direct steam injection. To utilise all the supplied heat, preheaters or heat economisers are used. Instead of having a cold water stream to cool the sterile media, the lower temperature, unsterile media stream takes the primary heat from the warm stream, cooling the sterile media.

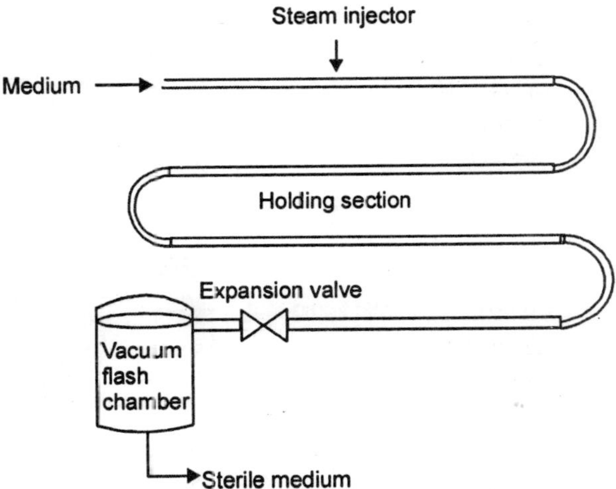

Fig. 5.2. Continuous sterilization.

Continuous sterilisation has a holding coil for detention long enough to kill all the microorganisms. The medium from a make-up vessel flowing through the exchanger is held in the coil, and then goes through the heat exchanger, heating more unsterile medium while become cool itself, as it is collected in a sterile fermenter Fig. 5.2, presents the sequence of piping media goes through, along with steam. Finally, the media is collected in a vacuum flash drum. Heat economisers are important for large-scale plant unit in continuous sterilisation. For a small-scale bioreactor use of direct steam injection is much simpler to operate.

A heat exchanger is then needed with cooling water to bring the medium back to ambient temperature. Therefore, jacketed vessels are commonly used with cooling water. There are advantages and disadvantages to batch and continuous sterilisation. The energy savings are related to the use of an economiser and direct stream or indirect sterilisation. In a largescale system economy and role decide which one is the most suitable process to be implemented. For each case the specific design needs to be evaluated in terms of the thermal efficiency and fixed costs involved.

HOT PLATES

In directing steam heating, continuous sterilisation is preferable. The medium passed through La preheater may reach about 90 °C; then quite fast sterilisation may take at 140 °C. Fig. 5.3 shows a counter-current hot-plate heat exchanger is normally used for continuous sterilisation. Hot-plate heat exchangers are extremely efficient and easily maintained if any fouling or scale deposition takes place. Normally fouling is serious problem, occuring by deposition of proteins on the hot surface of the exchanger, which causes reduction in the overall heat transfer coefficient. Plate exchangers are easily separated and cleaned. Special care must be taken when the media contains any agglomerated particles. Correlation is required to calculate the suitable residence time for sterilisation.

The choice of a suitable sterilisation process is based on economics and cost of the heat exchangers to reduce energy consumption. In a suitable exchanger, the larger the heat transfer coefficient, the greater the energy recovery.

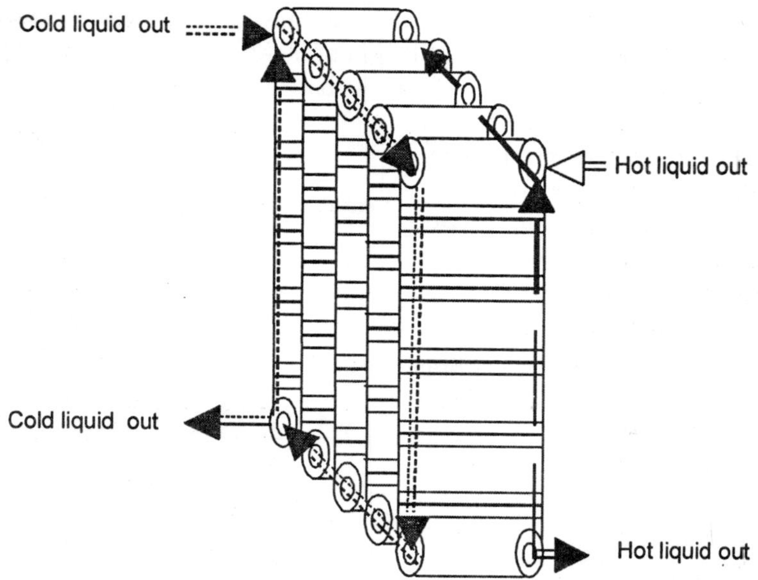

Fig. 5.3. Counter current hot plate heat exchanger.

HIGH TEMPERATURE STERILISATION

Fast sterilisation is performed at high temperature. High-temperature sterilisation requires short holding times. This technique is used in the fast preparation of nutrient media for industrial bioprocesses and in pasteurising milk. A short retention time may favour media with heat-sensitive proteins and cause less damage to the biochemical composition of the media than more prolonged times at lower temperatures. The ability of high temperatures to perform rapid sterilisation is related to activation energies. That is affected by how fast bacteria are killed in an elevated temperature.

On the other hand, long sterilisation at high temperatures may destroy the protein and biochemical composition of the media. Short-duration sterilisation at high-temperatures are more lethal to organisms and less chemically damaging than are longer sterilisation processes at lower temperatures. Sterilisation at high temperature is recommended for 3 minutes at 134 °C. It is preferable to 20 minutes at 115 °C in conventional operation.

STERILISED MEDIA FOR MICROBIOLOGY

Sterile media are used for pure culture. The media used to culture microorganisms depend on the living conditions of the microorganisms. Media compositions must be identified based on the needs of the organisms for carbon, nitrogen and phosphorus. Nutrients such as protein and sugars are added at wide ranges of pH. The acidic, neutral and alkaline conditions are defined based on differentiation of microbes in their biochemical reactions. Colour indicators are used to observe any transition of pH. Buffer is used in the media to maintain constant pH. Growth factors and growth stimulants are also used to accelerate microbial growth and to maintain exponential growth.

The most common media in the liquid phase is nutrient broth; in the solid phase, nutrient agar is used to propagate microorganisms. When microbiological media is prepared, it has to be sterilised before any inoculation of organisms. Sterilisation is used to eliminate any microbial contamination that may originate from air, glassware or hands. Microbes propagate quite quickely: within a few hours there will be thousands of organisms reproduced in the media. To control the growth of unwanted organisms, the media has to be sterilised quickly before any microbes start to utilise the nutrients.

The sterilisation process makes the media absolutely free of contaminants and guarantees that the medium will stay sterile, so that is used only for the desired organisms. The kinetics of culture media sterilisation describe the rate of destruction of microorganisms by steam using a first-order chemical reaction rate model. As the population of microorganisms (N) decreases with time, the rate is defined by the following equation:

$$-\frac{dN}{dt} = k_d N \qquad ...(5.1)$$

where N_0 is the number of viable organisms present in the culture media, t is the retention time or sterilisation time, and k_d is the reaction rate constant as it is known for a specific death rate. Using separation of variables and integrating with initial condition, the following useful expression is obtained:

$$N(t) = N_0 e^{-k_d t} \qquad ...(5.2)$$

where N_0 is the number of viable organisms present in the media at the starting point before sterilisation. Now take the natural log; it is reduced to a linear model:

$$\ln \frac{N(t)}{N_0} = -k_d t \qquad ...(5.3)$$

The graphical presentation of the equation shows a straight line with a negative slope for k_d. As the death rate constant follows Arrhenius' law, the death rate constant is temperature dependent. The value of k_d is about 0.02 min^{-1} at 100 °C, the death rate constant increases by 10-fold at 110 °C and 100-fold at 120 °C.[2]

$$k_d = k_0 e^{-E/RT}$$

where k_0 is the death rate constant at a reference temperature also known as Arrhenius' constant, R is the gas constant, T is the absolute temperature, E is the activation energy, which is 60-70 kcal per mole for microorganisms, 100-150 kcal per mole for spores, and 30-40 kcal per mole for media with vitamins and protein solution.[2]

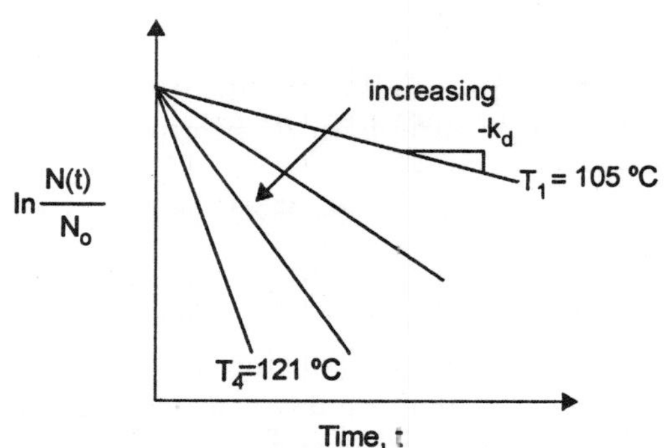

Fig. 5.4. The loss of cell viability at various temperatures.

Figure 5.4 shows the linear model for the loss of cell viability at various temperatures. As the temperature increases from 105 to 121 °C, the value for the slope of the line increases. This means that the number of viable cells at a fixed time of sterilisation will drastically decrease as the temperature increases by 16 °C.

Sterilisation of Media for Stoke Cultures

To clear slant tubes, culture agar solution is prepared. To have clear agar solution, it has to boil. Agar is a polysaccharide. It is not soluble in cold water, so heating helps to dissolve it in water. Once the solution is dispensed in the tube, it goes for sterilisation. Agar is in the solid state at room temperature, about 35 °C.

Sterilisation of Bacterial Media

Often, sterilisation of culture media is done in an autoclave at temperatures between 121 and 134 °C. It is good to know the damage caused to the medium by heating it. Heating the culture media, which contains peptides, sugars, vitamins, minerals and metals, results in nutrient destruction, either thermal degradation or reaction between the components of the media. Protein in the media is denatured by high temperature, or sugars are easily caramelised. During the heat treatment, toxic compounds are formed, which may retard or inhibit the microbial growth. For minimal damage to the ingredients of the media, it is important to optimise and minimise the heating and holding time of the sterilisation process, respectively.

Sterilise Petri Dishes

There are several ways to handle Petri dishes. The standard dishes are 15-20 mm deep and 12-15 cm in diameter. They are normally available in glass and also transparent plastic. Petri dishes are individually wrapped. The media is separately sterilised while we add the sterilised media to the Petri dishes in front of a flame. It is recommended to use 20-25 ml of medium in each Petri dish. These are all called culture dishes. The media contains about 1.5-2% agar. After the media cools off it solidifies. The Petri dishes with the nutrient agar should be sterilized and stored in a refrigerator. It is common to sterilise the media and Petri dishes separately. When the medium is cooled to about 55 °C, in front of a flame or in a laminar flow chamber, lift the lid of the dish enough to pour about 25 ml of the medium to the desired depth and lower the lid in place.

It is best to gently move the Petri dish in way that spreads a thin layer of agar uniformly without any air bubbles. Distribution of media in the Petri dishes should be done in front of a flame. Most plastic Petri dishes are made of polystyrene and are not autoclaveable. Plastic Petri dishes are easily deformed during sterilisation at high temperature. Some plastic dishes can be autoclaved, but they are more expensive. Please follow the instructions given by the manufacturer or obtain information from catalogues. It is recommended to autoclave glass dishes and medium separately. If your dishes are autoclaveable, you may dispense the agar into the dish and then autoclave it. In this case it is best to cool the dishes in the autoclave or in the pressure cooker to reduce the amount of water that will condense on the cover of the dish.

DRY HEAT STERILISATION

Dry heat sterilisation is used for equipment that can withstand high temperature and dry heat but cannot withstand wet or steam autoclave. This method is often used for glassware as it dries and sterilises in one operation. The pipets must be wrapped in dustproof aluminum foil or placed in metal pipette cans. The can lids are removed during heating and replaced after sterilisation, that is before any dust can get in the can. Disposable items are not recommended for dry heat sterilisation. This method may only be good for permanent reusable glass pipettes. Normal laboratory glassware must first be washed and cleaned. It has to be rinsed with deionised water.

The clean glassware is sterilised in an oven set at 200 °C for 1–4 hours. It is suitable to cover glassware with aluminum foil to maintain aseptic conditions after removing the glassware from the oven. If aluminum foil is not available, special heat-resistant wrap paper can be used. The sterile glassware must be protected from the air, which has micro-flora, or any contaminants. Avoid the use of any plastic caps and papers. Detach any labelling tape or other flammable materials, as they are fire hazards.

STERILISATION WITH FILTRATION

Certain media components are susceptible to heat; it is going to be denatured if it is heated. Therefore they must be added to the media after autoclaving. To do so, it is necessary to carry out filtration, the components using a 0.22 µm pore size filter that is appropriate to the solvent used. Filters are normally available from Whatman, or Fisher Scientific. It is recommended you consult filter experts or suppliers about the solvent and the special application you are intending to implement for the desired filters for the process. Normally filtration of media is used instead of sterilised media with an autoclave. To filter sterilised media, filter the water using a 0.45 µm pore size filter, before using a 0.22 µm pore size filter for final sterilisation.

MICROWAVE STERILISATION

Rapid sterilisation of media is achieved by using microwave ovens. Most plant tissue culture media can be sterilised using a microwave, although it may not be suitable with a medium containing complex additives like oatmeal.

ELECTRON BEAM STERILISATION

Electron beam sterilisation is a high-voltage potential established between a cathode and an anode in an evacuated tube. The cathode emits electrons, as a cathodic ray or electron beam. A high intensity of electrons is produced. These electrons are accelerated to extremely high velocities. These accelerated electron intensities have great potential as a bacteriocide. Most electron beams operate in a vacuum. As a result the unwanted organisms in the media vanish and the media is sterilized. The electron accelerator equipment producing the high-voltage beam also has various applications in the medical field and surgical supplies. Electron beam sterilisation is a successful technology used for sterilisation of disposable medical appliances and devices with a wide range of densities. The electron beam inactivates microorganisms that cause destruction of biomolecules, and results in death of microbes by an indirect chemical reaction. Irradiation by gamma rays is a similar mechanism. Advanced electronics precisely control the use of electron beams in the sterilisation of medical devices. Electron beam sterilisation results in less material degradation than with gamma irradiation. Medical products are sterilised in the original shipping containers, saving lots of time and maintaining integrity of the original package.

CHEMICAL STERILISATION

Chemical agents are used to sterilise heat-sensitive equipment. Chemical solutions are used as a suitable method for sterilising long pipettes and glassware. Normally at pilot scale in the absence of life, steam and chemical agents are often used. Application of an oxidising agent such as 10% chlorox for 20 minutes or longer proves the system operates without any contamination. Excess amounts of chemical agent have to be removed; otherwise organisms are able to grow in a toxic environment. The bleach (sodium hypochlorite) then needs to be removed by rinsing with sterile water or rubbing with ethyl alcohol, 70% ethanol-water (pH = 2) or an alcohol solution of 70% *iso*-propanol without recontaminating the glassware or tools. Ethanol is commonly used for cleaning surfaces. Use of bleach on metal devices is not suitable as it corrodes metals rapidly.

Ethylene oxide in the gas phase is commonly used. This is a special chemical effectively used for column bioreactors. It is a volatile compound and strong oxidizing agent. It boils at ambient temperature, therefore the solution of ethylene oxide (liquid phase) must be stored in a refrigerator (4°C). An excellent oxidizing agent such as a 3% sodium hypochlorite is used for chemical sterilization of equipment.

Tyndallization

John Tyndall, a famous British scientist, noticed that a period of about 24 hours at ordinary room temperatures is usually sufficient to enable dormant and heat-resistant spores to germinate and grow into the vegetative state in which they are sensitive to heat. He devised a process of sterilization based on these observations. It consists in steaming for a few minutes at 100° C, on three or four successive occasions, separated by 24-hour intervals at a temperature favorable to spore germination. The intervals permit the dormant, resistant spores to become active, vulnerable vegetative cells, readily killed by 100° C. This process renders an infusion sterile, whereas one single continuous boiling for one hour may not, since many spores remain in their dormant and resistant state during this time. This intermittent process is sometimes put to practical use in the home canning of foods. A disadvantage of the method is the time required to achieve complete sterilization; an advantage is that it requires no special apparatus. In some fluids, such as water, spores may not grow out promptly. Also, if the material is freely exposed to air, anaerobic spores may not germinate and may survive the process. If not freely exposed to air, aerobic spores will not grow out freely.

Table 5.1. Boiling point at various altitiudes

Location	*Altitude**	*Boiling Point***
New York City	0	100
Chicago, Ill	589	98.9
Denver, Col	5,280	94.3
Fort Laramie, Wyo	7,390	92.2
Tahoe, Nev	10,000	89.1
*Feet above sea level.		
Degrees C.		

* Feet above sea level.

** Degrees C.

Compressed Steam: Autoclaving

Anyone familiar with the operation of a home pressure cooker is familiar with an *autoclave,* because the cooker is a simple form of autoclave. In the autoclave, be it a small and simple home pressure cooker or an apparatus large enough to fill a room and fitted with various gauges, pipes, valves, clocks and wheels, the object of both is alike: to heat the articles to be sterilized by means of steam under considerable pressure.

Steam under pressure is hotter than boiling water or free-flowing steam such as is used in tyndallization. The higher the steam pressure, the higher the temperature will be.

It must be remembered that it is the compressed steam (moisture, hydration) that sterilizes and not compressed air (dry and usually not as hot as steam).

For example, pure steam at 15 pounds pressure has a temperature of 121° C. If the steam is mixed with an equal amount of air, at the same pressure the temperature is only 110° C, while if the mixture is two thirds air the temperature is only 109° C.

Steam *hydrates* and thus promotes coagulation; air does neither. Steam, being water vapor, also produces *hydrolysis* at autoclave temperatures. Dry air cannot do this at any pressure or temperature. In autoclaving, therefore, as in using a pressure cooker, a valve is left open for the escape of all air before the steam pressure is allowed to rise.

The actual amount of water present as steam in an autoclave is small, and articles soon dry off after removal, especially if removed from the autoclave while hot.

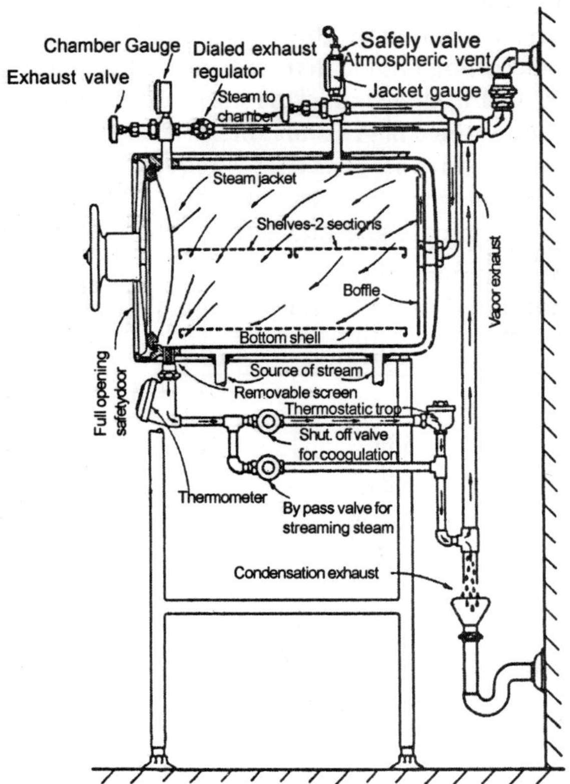

Fig. 5.5. Diagrammatic illustration of steam jacketed autoclave. Steam enters the *jacket,* a double-walled shell, at two places just beneath the cylinder. It passes out the top through a pipe to which are attached: a wheel valve admitting the steam to the inner *chamber,* a *safety valve,* a gauge showing *pressure in the jacket.* The steam enters the inner chamber at the right of the diagram, filling the upper portion. Its pressure registers on the *chamber gauge.* It may be allowed to escape rapidly by the *exhaust valve.* If this is closed, steam pushes the cooler air in the lower portion out at the bottom (left), where the *thermometer* registers proper temperature only when the air is gone and is followed by the hot steam. The escaping steam may be allowed to flow out without building up any pressure if the *by-pass valve* is fully opened. If the bypass valve is closed and the *shut-off valve* is opened, steam passes through the *thermostatic trap* where the heat shuts off all but a pinhole opening. This causes pressure to build up in the chamber, yet prevents stagnation by permitting a constant minute flow of steam through the apparatus.

The *thermometer* on the autoclave is the important guide to the process of autoclaving, not the pressure gauge. However, the latter, as well as a steam-escape or safety valve, is essential to safety.

The common practice in autoclaving fluids or freely exposed surfaces such as those of dishes and instruments is to apply 115° to 125° C (10 to 20 pounds' pressure) for 20 minutes. The pressure must be allowed to subside *slowly* after the heating is over or *superheated* fluids in open (cotton-plugged) vessels will boil over. Any large, solid masses must be heated a longer time to allow for heat penetration. Packages must be spaced so as to allow free circulation of steam. Substances such as mineral oil or petrolatum, sand, or any dry objects in tight jars, or substances that are impervious to moisture, cannot be satisfactorily sterilized in the autoclave. The temperature may rise as high as 125° C, but in the absence of moisture this is ineffective. Such materials are more effectively sterilized by dry heat in an oven.

Acid materials, such as canned tomatoes, acid fruits, pickles or sauerkraut (pH more acid than 4.5), require much shorter periods or lower temperatures for sterilization than low acid foods (pH 6.5 to 7) such as milk, corn and meats. This is because of the synergistic action between pH and temperature.

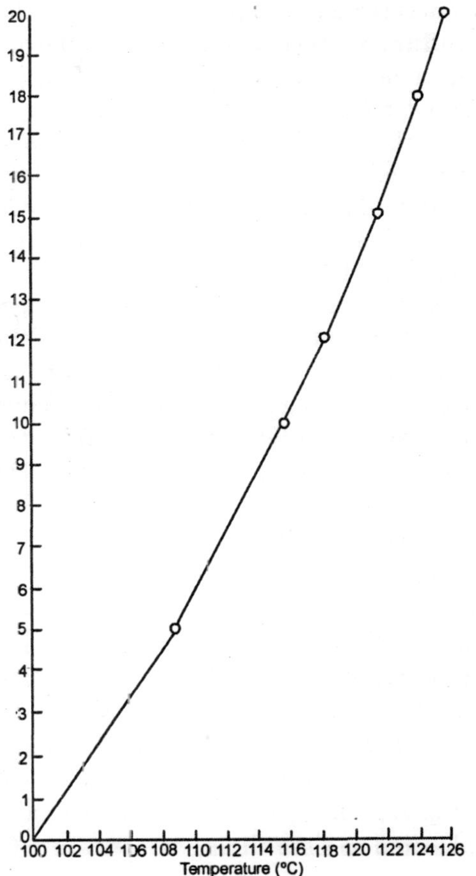

Fig. 5.6. Relation between temperature and pressure of pure steam.

DRY HEAT

Dry heat is used in oven sterilization. It is necessary to bear in mind that significant coagulation does not occur when moisture is not present.

Articles in ovens are very dry and, therefore, in order to be freed of live spores, must reach a very high temperature (165° to 170° C; 329° to 338° F). It is customary to apply 165° C for a period of two hours. This accomplishes no coagulation but, what is more effective, slight charring.

A home oven can easily be used for sterilization. A "moderate" temperature (330° F) is satisfactory, and the heating should be allowed to proceed for two hours *after* reaching that temperature. Paper wrappings should be slightly browned but not brittle; muslin or string should be faintly yellow due to the heat.

Only dry articles not injured by baking (glassware, bandages, instruments, mineral oils, petrolatum, talcum powder and the like) may be thus sterilized. Solutions containing water, alcohol or other volatile substances will, of course, boil away and be ruined.

TIME AND CONTROLS

Thorough Heating Necessary

In any process of disinfection or sterilization by heat it is absolutely essential that the object be heated through and that the center of the object be held at a killing temperature long enough to destroy the bacteria.

Thus, in home canning, a quart jar of spinach may be held in free steam (100° C) for five to ten minutes and when grasped with the hand will feel very hot, yet the center may be well below a microbicidal temperature. Large masses of nonfluid materials like quart jars of canned vegetables and roast meat, in which the contents cannot circulate, require a long time (one and one-half to two hours), even in the autoclave, to be heated so thoroughly that the center reaches a sporicidal temperature. Pieces of meat or vegetables to be sterilized in jars by heating should be loosely packed, allowing space for circulation of the fluid in the jar. Air pockets in the jars should be carefully removed. Penetration of the heat is facilitated if the cans or jars or pieces of roasting meat are small and not packed too closely together, promoting free circulation of steam or hot air around and between them.

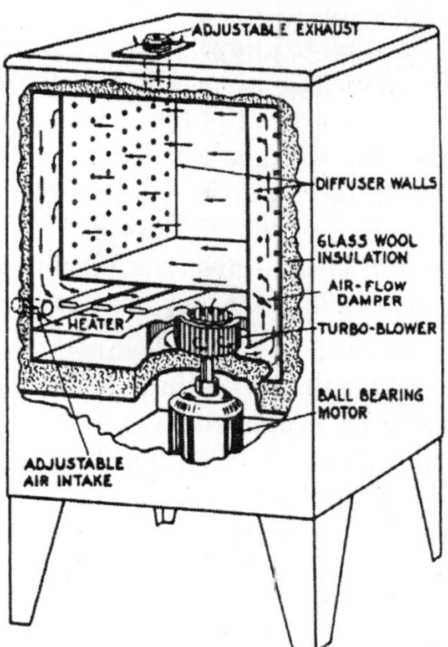

Fig. 5.7. A type of sterilizing oven for microbiology. The motor at the bottom of the oven forces streams of hot dry air through the chamber of the oven, as indicated by the arrows.

Sterility Check

For many purposes—industrial, hospital, research—it is important to have some absolute means of determining after a heat sterilizing process that even the most heat-resistant spores have been killed. Two general types of check are commonly used. One consists of pieces of paper, gauze or thread to which are applied spots of some thermolabile dye. Inserted in the materials being sterilized, the dye turns a distinctive color if the contents of the sterilizer have been subjected to a sterilizing temperature for a sufficient time.

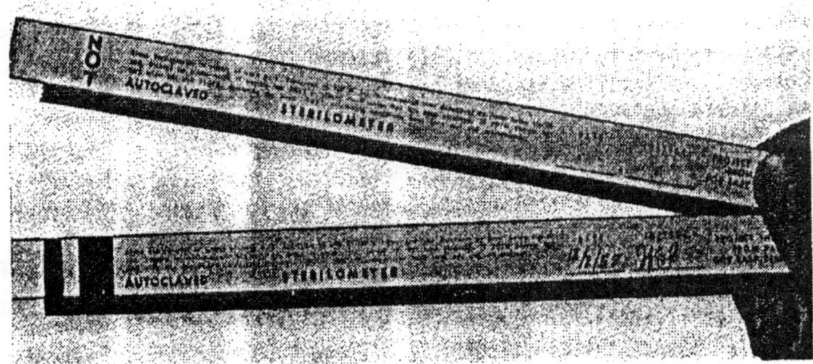

Fig. 5.8. One type of device for indicating proper autoclaving. Before heating the word "NOT" appears clearly as on the upper indicator. After partial heating the leftmost black bar appears as seen on the lower indicator. Not until autoclaving is complete both as to time and temperature is the word "NOT" obliterated by a second dark bar on the lower indicator.

A more direct check (though more time-consuming) consists of strips of filter paper having on them the very heat-resistant spores of *Bacillus subtilis*. This is a harmless aerobic bacterium. The spore-impregnated strips are inserted (enclosed in sterile envelopes) inside materials being sterilized. Afterward the strips are removed to tubes of culture medium and incubated. If growth occurs in the tubes within seven days then it is evident that the sterilization process in which they were involved was inadequate to kill the spores. In a modification of this process, sealed ampules (commercially available) containing culture medium with dormant spores of harmless *Bacillus stearothermophilus* already in it are inserted in the materials being sterilized. After sterilization the complete ampules are incubated at 55° C (since *B. stearothermophilus* is a strict thermophile). If no growth occurs in seven days it is proof of effective sterilization, since spores of *B. stearothermophilus* are among the most heat-resistant of known living things.

SOME USEFUL DISINFECTANTS

There are hundreds of disinfectants in the market and it would obviously be impossible to discuss even a small part of them here. A few commonly used representatives of several classes will illustrate general principles and usages. For convenience we may group the disinfectants as follows: halogens and halogen compounds, compounds of heavy metals, phenol and its derivatives, alcohols, detergent disinfectants, microbicidal gases and radiations.

HALOGENS AND HALOGEN COMPOUNDS

Iodine and chlorine are the most widely used of this group. Chlorine gas is used to disinfect filtered water at all municipal water-purification plants and many sewage-disposal plants. It is usually handled in tanks or tubes like oxygen. It is applied to drinking water in a final concentration of about 1.0 part per million.

A more convenient form of chlorine for the individual and household user is calcium or sodium hypochlorite. Solutions (5 per cent) of sodium hypochlorite are purchasable in all grocery stores under various names (*e.g.*, Clorox). They have a multitude of household and sanitary uses. They depend for their effectiveness probably on their liberation of free chlorine. Directions for use of these solutions are on each bottle. Four points need to be remembered in their use: the chlorine tends to evaporate from the solution; they give a bad odor to the hands; chlorine is poisonous; and it is readily inactivated by all organic matter.

Chloride of Lime

Chloride of lime (0.5 to 5 per cent aqueous solution) is excellent for similar purposes and is inexpensive. The ordinary chloride of lime of commerce is unstable and soon loses most of its free chlorine. Many organic chlorine compounds which liberate their chlorine more slowly are very effective. *Azochloramid* is one of these; *dichloramine toluol* is another. The odor of chlorine may be objectionable.

In addition to the above substances, there are many new proprietary chlorine compounds on the market based on principles given above.

Lodine

The most actively antimicrobial of the halogens is iodine. However, it is not commonly used for the same large-scale purposes as chlorine because of its physical properties and its cost. Formerly used widely as a local antiseptic in a 7 per cent tincture (alcoholic solution), this strength is far too caustic and poisonous for local application and does more harm than good. The 2 per cent tincture is much better and is the most generally useful for small cuts and abrasions. If the alcohol causes too much pain, the tincture may be diluted with two to four volumes of water.

An aqueous solution containing approximately 1.85 per cent of iodine and 2.2 per cent of potassium iodide is excellent but has a high surface tension.

Lodophors

Nonirritating, nonstaining, virtually odorless and very effective solutions of organic compounds of iodine in which the iodine is loosely combined with a surfactant organic *wdophor* (Gr. *phor* = bearer or carrier) are available commercially. The iodophor releases the iodine slowly and also lowers the surface tension of the solution. Some idea of the amount of iodine remaining in such solutions is readily gained by comparing the diminishing brown color with that of the original full-strength solution. This is not possible with colorless chlorine solutions. Examples of such iodine products are Wescodyne, Iodide and Betadine. In such forms iodine may be used for a variety of purposes, from clinical uses to sanitization of dishes, with aqueous solutions containing 200 parts per million. Under certain conditions of concentration and temperature such solutions may be sporicidal. They appear to kill bacteria, molds and some viruses quickly. Iodine is a much more effective germicide than chlorine.

Iodine, like chlorine, combines indiscriminately with organic matter; its solutions are unstable and, like chlorine, it is poisonous.

COMPOUNDS OF HEAVY METALS

The first scientifically designed, specifically acting drug was based on the heavy metal arsenic. Arsenic was compounded with an organic group to diminish its poisonous properties. Prior to 1907, Paul Ehrlich and his coworkers had synthesized and tried many such compounds in the treatment of spirochetal diseases, notably syphilis, without success. The 606th compound ("SOS" for "606" or Salvarsan or arsphenamine), discovered in 1907, was very effective and, with various modifications such as neoarsphenamine, was used for years in the treatment of syphilis and other spirochetoses. It fell into disuse after 1943 when it was found by Mahoney et al., in the U. S. Public Health Service, that penicillin is much more active against the same organisms and almost nontoxic.

Other metals in common use as microbicides today are iron and zinc, previously discussed, and mercury, silver and copper. All act mainly by coagulative action except in very low concentrations (*e.g.,* 1: 10,000 and lower) in which they appear to have specific molecular toxic effects (see *oligodynamic action* and *ionic effects*).

Mercury compounds were first used in the fifteenth century specifically to treat syphilis. More recently zinc in a chelated form called *Zinc Omadine* has been shown to have potent microbicidal properties, as well as the sodium salt of *Omadine.*

Bichloride of Mercury

Long used as a general disinfectant and as an antiseptic, bichloride of mercury (mercuric chloride) is now much less used than formerly. For general disinfection it may be prepared in dilutions of 1: 1000 to 1:5000. However, it is very toxic to tissues, very poisonous internally and corrosive to materials, especially metals (hence its older name of *corrosive sublimate).* In many situations it fails because it forms a thick coating of coagulum on the outer surface particles of mucus, tissues and feces, and thus protects the microorganisms within. As described previously, in dilutions of 1:10,000 or more its action is more bacteriostatic than coagulative. This action may be reversed by sulfur compounds unless coagulation has occurred.

Organic Molecules

Attempts have been made, following Ehrlich's tead with arsenic, to decrease the toxic, corrosive and irritating qualities of mercuric disinfectants by incorporating mercury in complex organic molecules. This has yielded a number of products that are less toxic and irritating than bichloride of mercury, among them *Mercurochrome, Merthiolate* and *Metaphen*. In the presence of organic matter they are more effective than bichloride of mercury. Mercurochrome is often used for local, superficial application for disinfecting cuts, wounds and skin. Some of these organic mercury compounds are used as preservatives in biological materials. Other useful organic mercurials are *phenylmercuric nitrate,* very effective and of sufficiently low toxicity for external use, and *ammoniated mercury,* long used as a 10 per cent ointment for external wounds and fungal infections of the skin. These mercurials all mainly act microbistatically, and their action, like that of bichloride of mercury, is reversed by thioglycollate and hydrosulfuric acid. All mercury compounds are poisonous internally to some degree, and mercury is absorbed through the skin.

Silver Nitrate

Silver nitrate (1 per cent aqueous solution) is used principally for application to infants' eyes at birth to prevent infection. Silver is also used as a local disinfectant in the form of organic colloidal preparations such as Argyrol or Protargol, in which form it is nonirritating and may be used in adult eyes at 20 per cent concentration. It is not for use in infants' eyes in place of silver nitrate. The use of silver nitrate (1 per cent aqueous) or an equivalent, approved antibacterial in the eyes of infants at birth is required by law in most states as a safeguard against blindness of the newborn *(ophthalmia neonatorum)*. However, antibiotic treatment may be satisfactory in some situations.

Copper Sulfate

Copper sulfate is used chiefly to control the growth of algae in open water reservoirs and as a fungicide in garden sprays *(Bordeaux mixture)*. Organic compounds of iron are also used as fungicides in garden sprays.

PHENOL AND ITS DERIVATIVES

Phenol (C_6H_5OH) (carbolic acid) in the pure state is not commonly used as a disinfectant because it is expensive and because there are more than a dozen derivatives that are not only less costly but more effective. All act similarly to phenol (*i.e.,* mainly by coagulation, partly by little understood lytic and toxic effects). Unlike the halogens, which are indiscriminately effective against a wide range of microorganisms, susceptibility to phenolic disinfectants varies greatly. Some organisms use phenol as a food.

Surface (Residual) Disinfection

Phenol and, more especially, many of its derivatives are surface tension reducers and are commonly used in solutions containing additional surfactants to improve wetting properties.

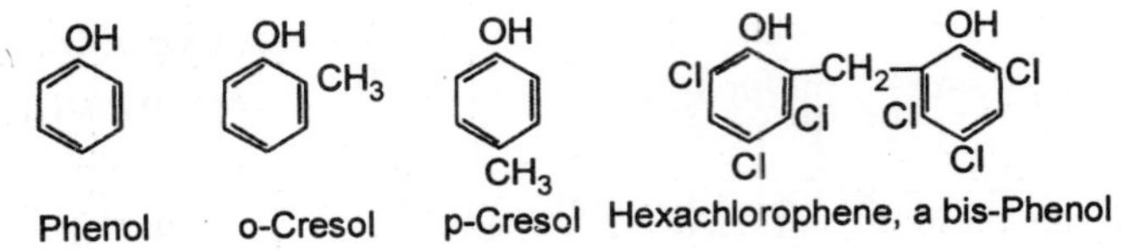

Since they are surfactants themselves, the phenolic disinfectants tend to be adsorbed in thin, and more or less durable, films on inert surfaces to which they are applied, resulting in what is termed *surface,* or *residual,* disinfection. As compared with the volatile halogens or alcohols their action is thus prolonged, often for many hours. This is an especial value of certain preparations of phenol derivatives for surgical and hospital hand washes.

Fig. 5.9. Structures of phenol, representative cresols and one commonly used bisphenol (hexa-chlorophene).

Types of Phenolic Disinfectants

Phenolic disinfectants are of many types, common among which are: the *cresols,* the molecules of which are like phenol but have methyl groups attached; and the *bis-phenols,* the molecules of which consist of two phenolic groups joined directly or through some other radicle.

Crude Cresols

Crude cresols form colloidal (milky) suspensions in water and are therefore especially effective as disinfectants, because each colloidal droplet consists of *concentrated* cresol. This is true of many colloidal disinfectants. There are many excellent cresol preparations on the market. Some have a clinging odor. Most of them contain a surface tension reducer in addition to the disinfectant. They are commonly used for disinfecting floors, furniture, barns and stables in from 1 to 5 per cent concentrations. Some may also be used in temporary contact with human tissues in 0.5 to 1.0 per cent strengths. Good examples are Lysol and saponated cresol solution, *U.S.P.* The preparation of such solutions is an important industry.

Bis-Phenols

Like the cresols, bis-phenols have surfactant properties. In order to increase their solubility in water they are commonly used in solutions with other surfactants or detergents. In general they are less toxic than the cresols. Among well-known bis-phenols is one combined with chlorine (one of the class of *halogenated bis-phenols),* called Hexachlorophene (G 11). Because it is a surfactant it provides good residual surface disinfection of the skin, and it is not irritating if used properly. It is incorporated especially in surgical soaps and hand washes (Gamophen, pHisohex and Hex-O-San).

Orthophenylphenol

A similar compound orthophenylphenol, **is used** in mixtures such as *O-syl. Chlorothymol* and *Chlorohexidine* are other halogenated bis-phenols which are similar in general properties. All are broadly germicidal, and each has distinctive properties that adapt it to particular situations. Some can be used effectively as irrigations in surgical wounds in 1 :1000 to 1:10,000 concentrations.

The activity of all phenolics is markedly depressed in the presence of extraneous organic matter, and they are highly microbistatic.

ALCOHOLS

There are many substances that chemically are alcohols: many of these are effective microbicides. Like various organic disinfectants, their microbicidal activity is directly related to increased length of their carbon chain and molecular weight; beyond their critical point (different for each type): microbicidal activity decreases. The two alcohols most commonly used for disinfection or sanitization are ethyl ("grain") alcohol and isopropyl ("rubbing") alcohol. They are mild disinfectants and are nontoxic on external application. Methyl or "wood" alcohol and its vapor are extremely poisonous and should never be used in any bodily contact whatever.

Ethyl alcohol is a surface tension reducer, a lipid solvent and a coagulating agent, properties that should contribute to microbicidal potency. However, it is also a potent dehydrating agent, a property that interferes greatly with its coagulating power and probably also with its other antimicrobial properties. Because of its dehydrating power it is best used as a 70 per cent aqueous solution. Stronger concentrations (95 or 100 per cent "absolute") are much less effective. As a coagulating agent it tends to combine nonspecifically with extraneous organic matter and can produce a thick, protective coating of coagulum around microorganisms. As a lipid solvent it is useful as a *cleansing* agent ("disinfectant"?) for skin prior to hypodermic injections. It adversely affects viruses and cells with envelope lipids. As a surface tension reducer it is a potent adjuvant of other disinfectants such as phenolics and quaternaries. Similar considerations apply to isopropyl alcohol.

DETERGENT DISINFECTANTS

As just mentioned, not all detergents are disinfectants. However, several compounds combine the properties of detergent, surface tension reducer and disinfectant.

Soaps

Among these versatile compounds are the common soaps; potassium or sodium esters of higher fatty acids (stearic or glyceric). They are actively lethal for certain species of microorganisms: *Treponema pallidum,* the syphilis spirochete; *Diplococcus pneumoniae,* a cause of pneumonia; some streptococci; and a few other organisms. Unfortunately, with these few exceptions, soaps are relatively ineffective as general disinfectants. However, soaps are potent surface tension reducers and consequently are good emulsifying and foaming or "sudsing" agents. Foaming results from the fact that, since the surface tension is greatly weakened, the surface can be extended almost indefinitely as foam or as droplets in an emulsion; compare with mercury. When soapy water is applied to skin or instruments, microorganisms are carried into the emulsified grease droplets and removed by rinsing. Detergency is the chief virtue of neutral soaps. Soluble soaps are precipitated by acids, ions of heavy metals, and by Ca^{++}. Hand soap is often used as a carrier for phenolic disinfectants, as previously discussed.

Synthetic Detergents

As generally understood, this term comprises a group of potent surface tension reducers that are not typical soaps. They are synthesized commercially and much used for household and commercial cleansing and laundry purposes. They are available in "large, economy" packages in all grocery stores. Most of them are long-chain, alkyl-benzyl-sulfonates. Few are markedly microbicidal. They are *anionic* compounds (see following paragraphs).

Another group of synthetic compounds, classed as detergents because they are strong surface tension reducers, are much more important as disinfectants and sanitizing agents. The first was discovered in 1936 by Domagk (Nobel Prize winner), who described the properties of dodecyl-dimethyl-benzyl-ammonium chloride, now familiar as *Roccal* or *Zephiran*. This is a quaternary alkyl ammonium salt, and is representative of a number of similar compounds that have since been developed commercially as disinfectants or detergents. As a group they are commonly called *quaternaries* or "quats." Basically, they are ammonium halides (for example, ammonium chloride) in which the hydrogen atoms have been replaced by various alkyl groups.

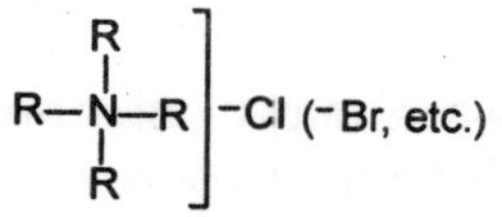

$$\left[R-\overset{R}{\underset{R}{N}}-R\right]-Cl\ (-Br,\ etc.)$$

A. General formula of alkyl ammonium halides

$$\left[CH_3-(CH_2)_{11}-CH_2-N^+(CH_3)_2-CH_2-C_6H_5\right]-Cl$$

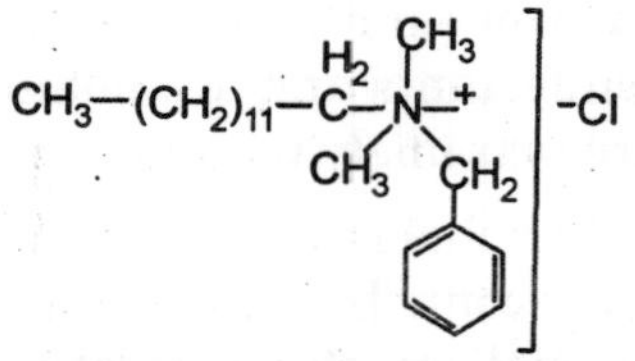

B. Alkyl-dimethyl-benzyl-ammonium chloride; a cationic, microhicidal detergent

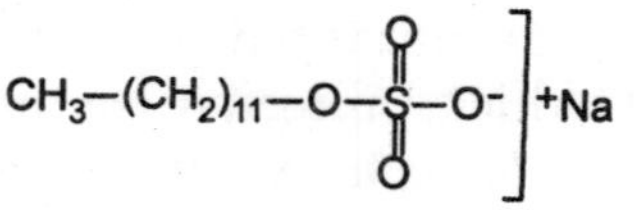

$$\left[CH_3-(CH_2)_{11}-O-\overset{O}{\underset{O}{S}}-O^-\right]+Na$$

C. Sodium lauryl sulfate; an anionic mirobicidal detergent.

Fig. 5.10. Forms of microbicidal detergents.

Chemical Structure and Activity

In general, the more effective of these detergent-disinfectant compounds are those in which at least one of the alkyl groups contains a chain of 12 to 16 carbon atoms, whereas the less effective are those of lower or higher molecular weight. This represents a common observation previously mentioned; *i.e.*, the germicidal efficacy of many organic compounds is greatly affected by chemical structure, especially length of carbon chains.

Cationics, Anionics and Nonionics

There are three types of microbicidal detergents: (1) those in which the surfactant portion is the *cation* (*e.g.*, quaternary alkyl ammonium halides those in which the surfactant portion is the *anion* (*e.g.*, complex soaps like sodium lauryl sulfate; and (3) those that do not ionize. These compounds are classified as *cationic, anionic* and *nonionic* detergents, respectively. The cationic compounds appear to be the most generally effective as disinfectants and are what is generally meant by the term "quaternary." Soaps are anionic detergents. The nonionic compounds are relatively weak disinfectants. They are sometimes used with anionic compounds in household detergents as adjuvants.

Mechanism of Action

Since detergents like the quaternaries and sodium lauryl sulfate are generally alkyl compounds of chlorine or sodium, they are bipolar, *i.e.*, they exhibit both fat-soluble (lipophilic) and water-soluble (hydrophilic) sides. When dissolved in aqueous culture media, the longer alkyl chains tend to be oriented toward, and to dissolve in, the cell lipids. The negatively charged residue (chlorine or sodium ion) is oriented away from the cell and is drawn toward the surrounding aqueous fluid. The surface tension reducer molecules are systematically arranged side by side at the cell-aqueous medium interface, as closely placed as possible.

Since the reducer molecules at any surface have a lower mutual attraction than the water molecules in which they are dissolved, the surface tension of their solution is lowered. Lipids upon which detergents in aqueous solution become oriented tend to become divided into smaller and smaller particles, each coated with the surface tension reducer (*i.e.*, to be *emulsified*); detergents thus characteristically "cut" (remove) grease. The lipophilic property of quaternaries and alkyl sulfates is thought to disrupt the lipids, and therefore the whole structures, of bacterial cell walls and membranes, especially of those rich in lipids like many gram-negative bacteria. Disruption of the cell walls and membranes would permit outward passage of vital intracellular components and passage inward of the surface tension reducer. Electron micrographs show the destruction of bacterial cell walls and inner structures by antimicrobial substances which reduce surface tension. Probably these detergents also act in part by occluding the bacterial surfaces by being adsorbed there. In use, soaps and other anionic compounds should not be mixed with cationic compounds, because the oppositely charged molecular complexes neutralize each other and the agents are precipitated or inactivated. For the same reasons hard, acid and iron-rich waters tend to inactivate them.

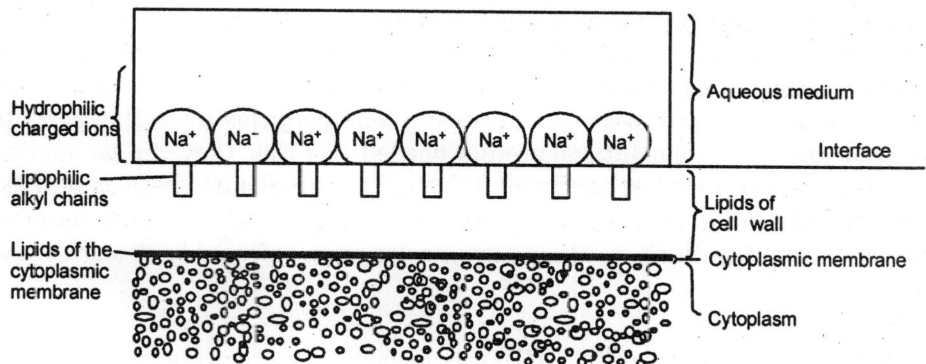

Fig. 5.11. Orientation of polar groups of an *anionic*, surface-active detergent, in this case sodium lauryl sulfate [$Na^+(C_{12}H_{25}OSO_{,1})^-$], in relation to the surrounding aqueous medium and a cell having lipids in its cell wall or cytoplasmic membrane. The hydrophilic group (here represented by Na^+) orients at the fluid-cell interface; the hydrophobic group (here represented by the lauryl sulfate group$^-$) is oriented in the lipids of the cell wall or membrane.

At the air-water interface the hydrophilic groups orient themselves within the surface water while the hydrophobic alkyl groups are at the air-water interface, weakening the intermolecular attraction between water molecules at the surface; *i.e.*, *lowering the surface tension* of the water. Accumulation of molecules of a surface-active, polarized agent at any interface between solids and fluids or between immiscible fluids is called *adsorption*.

A very similar relationship would be that between a cell and a *cationic* detergent of the class of quaternary ammonium compounds; *e.g.*, cetyl pyridinium chloride (Ceepryn):

$$\left[C_5H_5N—C_{16}H_{33} \right]^+ Cl^-.$$

Unlike many other disinfectants, quaternaries retain much of their activity in the presence of considerable organic matter. The quaternaries tend to be microbistatic or slowly lethal rather than to kill instantly. When the effectiveness of quaternaries as disinfectants is being measured, their marked microbistatic action must be offset by the use of special inactivators to allow the microorganisms to grow in the subculture. *Tinctures* of quaternaries appear to be very effective indeed against tubercle bacilli, and some are effective against molds, but they are of limited value against certain viruses.

Since they are emulsifying agents for lipids quaternaries should be effective against viruses that have lipid-rich components, *e.g.,* envelopes, and there are data in support of this view. Opinions differ concerning other viruses. Pi-cornaviruses are resistant. Quaternaries are not sporicidal. They are much used for disinfection of skin and for purposes of sanitization in restaurants and hospitals.

MICROBICIDAL GASES

Attempts to cure and prevent disease with vapors and smokes date from the earliest human records, when "aromatics," smudges, perfumes and the like were endowed by popular supposition with almost miraculous, and generally quite factitious, powers to drive away evil spirits and "noxious humors."

More objective studies in the last few decades have yielded several exceedingly valuable results in the form of definitely sporicidal, and therefore *sterilizing,* gases: formaldehyde, ethylene oxide and beta-propiolactone. Chlorine and sulfur dioxide, formerly used, are too destructive and poisonous for general purposes. As sterilizing agents, gases have the advantages that destructive heat and aqueous solutions are not needed and that they can be applied in large volumes in buildings. On the other hand they affect only exposed surfaces, except those made of porous or permeable materials, and they must be applied under conditions of controlled temperature and humidity. Such sterilizing agents are often referred to as *sterilants.*

Formaldehyde

This very irritating gas (formula: HCHO) is generated by heating para-formaldehyde or concentrated solutions of formaldehyde. Unless very carefully applied at raised temperatures, formaldehyde tends to polymerize as paraformaldehyde or *paraldehyde* in a thin white film over all surfaces, and hence has definite limitations to places not containing valuable furniture or equipment. Its action is to some extent microbistatic and reversible with sulfite ions. In some solutions it is definitely sporicidal and therefore sterilizing but, like bichloride of mercury, it tends to form protective coatings of coagulum around microorganisms if extraneous organic matter is present. For best results a relative humidity of about 70 per cent and a temperature of about 22° C, are requisite. Formaldehyde vapor has low penetrability into such materials as blankets. It is much less used now than formerly.

Ethylene Oxide

This substance (formula: $CH_2 \cdot O \cdot CH_2$) is liquid at temperatures below 10.8° C (51.4° F), its boiling point. It is relatively inexpensive and is handled in metal tubes or bottles like other compressed gases. In the pure state it is very toxic, irritating and explosive. For use it is mixed with other substances: ethylene oxide, 10 per cent and carbon dioxide, 90 per cent (sold as Carboxide); ethylene oxide, 20 per cent and carbon dioxide, 80 per cent (sold as Oxyfume); ethylene oxide, 11 per cent and halogenated petroleum, 89 per cent (sold as Cryoxcide). Each preparation has special advantages in certain situations.

The desired mixture is generally applied in autoclaves or other closed containers where humidity, temperature and pressure may be controlled. The concentration of the gas is generally expressed in milligrams of pure ethylene oxide per liter of space. A typical use situation is 500 mg. of gas per liter at about 58° C (130° F) and a relative humidity of 40 per cent for four hours. If one factor is varied the others must be varied accordingly. For example, if the concentration is doubled, the time may be halved. At a relative humidity of 30 per cent the action of the gas is about 10 times as fast as at 95 per cent relative humidity.

In using ethylene oxide in an autoclave, a typical procedure involves seven steps, all of which can be carried out by fully automatic equipment now available for the purpose:

(1) Withdraw almost all air with a vacuum pump.

(2) Admit a controlled volume of steam or water vapor.

(3) Admit ethylene oxide in required concentration.

(4) Adjust the temperature as indicated and hold constant.

(5) After the required time turn off heat.

(6) Take out all gas with vacuum pump.

(7) Admit sterilized air to normal pressure.

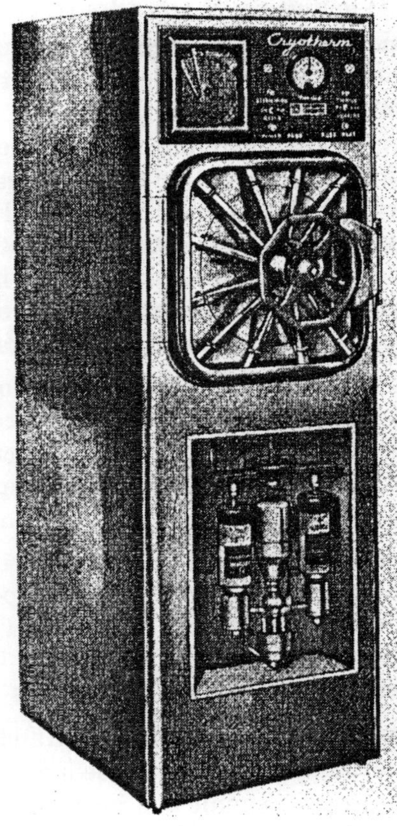

Fig. 5.12. Autoclave arranged for sterilization by sporicidal ethylene oxide gas (Cryoxcide). Below the steel door are seen the containers of gas and a water inlet so arranged that the vapors can be mixed and humidified (much as gasoline, air and water vapor are mixed in a modern automobile carburetor) before admission of the sterilizing vapor mixture to the chamber of the autoclave. Above the door are the dials controlling sterilizing temperature and time, vacuum pump, gas and humidity inlet and a maximum and minimum recording thermometer. Operation is almost completely automatic.

Ethylene oxide is a very effective general microbicide and kills bacterial endospores. It is therefore a sterilizing agent. It acts by alkylation or introduction of an organic group in place of a hydrogen atom, in various compounds in protoplasm. It does not injure even delicate organic materials that are injured by heat and aqueous solutions. It penetrates well, does not condense on objects and soon dissipates from all materials.

It can also be used in the liquid state in 1 per cent concentration to sterilize unstable organic fluids, such as bacteriological culture media, by mixing fluid ethylene oxide with them at 10° C. After a few hours the ethylene oxide is driven off by a short incubation at 37° C.

Beta-Propiolactone (BPL)

In pure form at 20° C this is a liquid with a sweetish but very irritating odour [formula: $(CH_2)_2 \cdot CO$]. Unlike ethylene oxide, it is not explosive. It may be stored at 4° C, but is more stable in sealed glass containers at –25° C. Its aqueous solutions are very unstable. It kills most microorganisms, including bacterial spores, and is therefore a sterilizing agent. It probably acts by alkylation, as does ethylene oxide. In fluid form it has been used to sterilize vaccines, tissues used for grafting and other delicate biological materials that could not be sterilized in any other way. Like ethylene oxide, it must be used in closed chambers where temperature and humidity can be accurately controlled. A relative humidity of 70 to 80 per cent and a temperature of about 25° C are optimal, with a final concentration of beta-propiolactone of about 2 to 4 mg. per liter of air and a sterilizing time of 2 to 3 hours.

When properly applied, beta-propiolactone is much more active than either formaldehyde (HCHO) or ethylene oxide. It is noncorrosive, and does not condense on surfaces as does formaldehyde. However, beta-propiolactone lacks the penetrating power of ethylene oxide and is more like formaldehyde in this respect. It also requires higher relative humidities. It is generally recommended to replace formaldehyde for use in buildings.

Since they are so effective, gaseous disinfectants such as ethylene oxide, formaldehyde, and beta-propiolactone have tremendous potential value in combating massive and widespread dissemination of infectious agents such as might be employed in biological warfare.

Toxic and Destructive Properties

Experience shows that prolonged and repeated exposure to any chemically active disinfectant, or any sterilant, may cause tissue irritation or destruction, and may damage articles containing certain unstable organic and inorganic substances such as magnesium (instrument housings) and some plastics. Users must always be alert to this possibility.

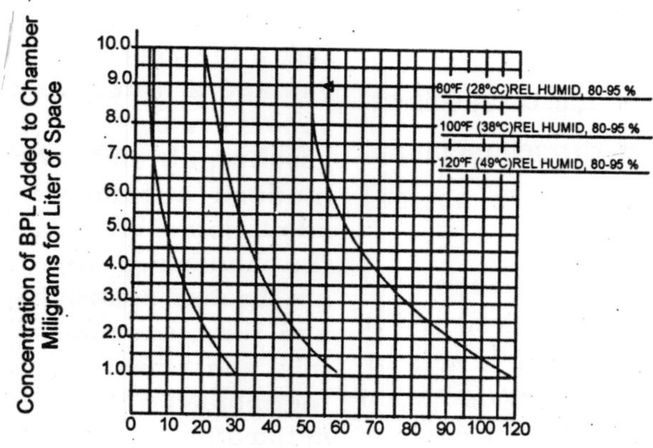

Fig. 5.13. Relation between concentration of beta-propiolactone and time required to kill spores of *Bacillus globigii* on filter paper strips. Curves are shown for three different temperatures at a relative humidity of 80 to 95 per cent.

They must also be alert to toxic properties, especially dangerous potentialities of newly developed disinfectants. For example, with continued study of beta-propiolactone, data have appeared suggesting that it may be carcinogenic as well as irritating. The toxic or carcinogenic properties of a disinfectant do not necessarily militate against its use; otherwise, such substances as chlorine, formaldehyde, bichloride of mercury and phenol, and radiations such as ultraviolet, x-rays and gamma rays would be banned. However, proper precautions must be taken in handling such agents to prevent ingestion or undue contact.

RADIATIONS

Ultraviolet Light

As described, ultraviolet light is actively microbicidal when applied properly. It has very slight powers of penetration. Exposure must be direct, intense and sufficiently prolonged. Eyes must be protected, as they can be severely damaged. It is sometimes used in hospitals, especially in operating rooms, reportedly with good results, to kill microorganisms on instrument tables, in the air and on floors and walls. Ultraviolet light is used on a large scale to prevent growth of molds in meatpacking houses, bakeries and other commercial and industrial establishments.

Another application is in the treatment of biological fluids, such as blood plasma and vaccines, to kill contaminating viruses. The rays are directed on fluid that is made to flow in a micro-thin film that ultraviolet rays can penetrate. The wave length commonly used in disinfectant applications of ultraviolet light is 2537 A, because it is most practical. Shorter waves, such as 1849 A, are much more lethal but are difficult to handle and apply. Ultraviolet kills spores as readily as vegetative cells.

Cathode Rays

A recent development in radiation sterilization, cathode rays (electron beams) are expensive to apply, but are being developed on a commercial scale. They can be made to penetrate thin metal, paper or plastic sheets and have been used successfully to sterilize sealed packages for surgical use, packaged meats, vegetables and other foods.

INHIBITION OF MICROORGANISMS (MICROBISTASIS)

As previously indicated, microbistasis denotes a condition of microorganisms in which, although they are alive (or *viable),* they do not multiply. Their enzymic or other vital functions are in a static condition.

In one sense the dormant ascospores of yeasts, conidia or molds, cysts of some protozoa and the endospores of bacteria represent microbistasis as a part of their normal life cycles. However, microbistasis is usually thought of as affecting vegetative cells and as being produced artificially. Microbistasis is exceedingly important in all aspects of microbiology. There is even published discussion of preserving important(?) people in “deep-freeze” (“Homostasis”?).

Physical Methods

As mentioned elsewhere, there are several physical methods of inducing microbistasis. Most of them are fundamentally methods of depriving the cell of liquid water: desiccation; immobilizing intracellular water by changing it into ice; drawing most of the water out of the cell by immersion in a fluid of high osmotic pressure; and combinations of these.

In addition, mild irradiations with ultraviolet light and x-rays are often microbistatic instead of lethal. Their microbistatic action is not clearly understood but it is not related to dehydration. Apparently, oxidations are among the most important reactions that occur.

Reversibility

The action of microbistatic agents is *characteristically reversible;* that is, the organisms can be reactivated if the inhibiting action has not been so prolonged that the cells have gradually died off. We have seen in foregoing chapters how the microbistatic effect of irradiations may be reversed by visible light and certain chemicals, and how bichloride of mercury poisoning may be "cured" by glutathione. Tubercle bacilli dried in a vacuum were found fully viable and virulent after 17 years; diphtheria bacilli and scarlet fever streptococci, after 25 years. Microbial spores (natural microbistasis) will survive for decades. Viruses are commonly held unaltered for months or years frozen in carbon dioxide ice at –76° C. Cattle breeders and others preserve sperm by freezing. In all cases the totally inactive cells can be reactivated.

MICROBISTASIS AND DISINFECTION

The distinction between chemical microbistasis and disinfection is interesting but entirely arbitrary; a matter of semantics. The differentiation depends on what one means by killing "quickly" (*i.e.,* disinfection) and what one may mean by killing "slowly" (*i.e.,* microbistasis). Even under the most effective conditions of microbistasis, some of the weaker cells die fairly soon; presumably the others are not actually immortal and die eventually, even though only after decades or centuries. As applied to bacteria, one could say that death within a time less than the maximal for one generation (*i.e.,* one cell-fission) under optimal growth conditions might be regarded as more than mere prevention of multiplication and, therefore, true killing or disinfectant action. If death were delayed beyond one generation-time, one might say that multiplication had been delayed or that the organisms had been in a static condition for at least a short time. This last situation could represent true bacteriostasis. There are many obvious difficulties in making such a distinction. The term bacteriostasis (or microbistasis) is, however, widely used and usually means "long" delayed death (*i.e.,* the cells die only after some hours, days or decades), rather than in, say, 10 to 30 minutes.

CHEMICAL MICROBISTASIS

We may note three general types of chemical microbistatic agents: certain aniline dyes; sulfonamide and similarly acting drugs; and antibiotics.

Aniline Dyes

We have already referred to selectively bacteriostatic culture media containing dyes. A representative dye used in such media is crystal violet, commonly used in the Gram stain. Since it is a basic dye, it has a marked affinity for acidic constituents of the bacterial cell, notably nucleic acids. With these the dye forms unstable salts that (in the absence of iodine) soon dissociate, and are either disposed of by the cell or can be removed by washing the cell with a harmless solvent such as water. Cells stained in this way with crystal violet, especially gram-positive cells, are in a static condition. They will not grow in media containing such dyes. They can be quickly reactivated by washing.

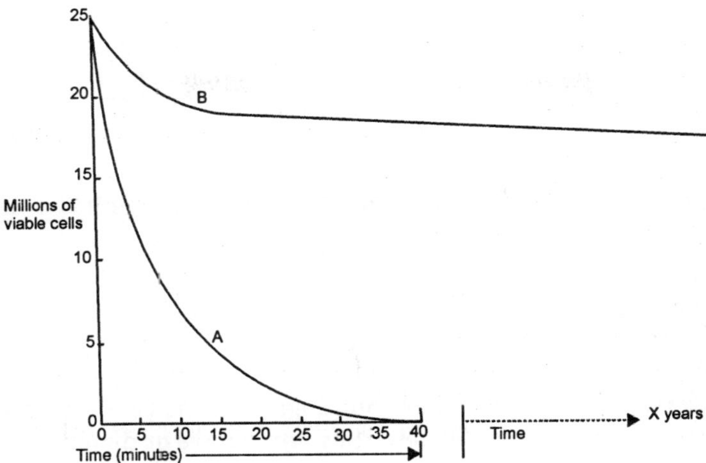

Fig. 5.14. Comparison of disinfectant action with microbistasis. The microorganism concerned has a generation time of 40 minutes at 37° C in broth at pH 7.2. Curve *A* shows complete sterilization within one generation time resulting from action of an applied microbicidal agent such as heat or disinfectant. Curve *B* shows initial rapid diminution of numbers caused merely by the action of the microbistatic agent such as freeze-drying. Bacteriostatic action is characterized by very slow decline in numbers.

On the other hand, gram-negative cells in contact with crystal violet are not stained and generally grow well on media containing such dyes as crystal violet, basic fuchsin or methylene blue-eosin. Such dye-containing media are therefore commonly used for *selective* bacteriostasis; especially to isolate gram-negative bacteria from materials contaminated with gram-positive bacteria.

Sulfonamide Drugs

The sulfonamides are excellent examples of bacteriostatic agents. They were discovered by Domagk (Nobel Prize winner) about 1936. Sulfonamide drugs are derived from the same basic material (coal tar) as aniline dyes. All are derivatives of sulfanilamide like basic dyes they are effective mainly (with a few notable exceptions: *Nesseria, Shigella, Proteus, Klebsiella* and *Escherichia)* against gram-positive bacteria.

Sulfonamide Drugs and Metabolite Antagonism

The true sulfonamide drugs are 'those that readily release the sulfanilamide residue in the living cell. Sulfanilamide so closely resembles paraaminobenzoic acid (PABA) that it can displace PABA in the absolutely essential vitamin, folic acid. This important vitamin is necessary to a co-enzyme (CoA) involved in synthesis of amino acids and hence of proteins. Such absolutely essential specific substances, necessary to the completion of a particular functional part of the cell, but not used as sources of energy or bulk, are sometimes called *essential metabolites* or *micronutrients.* In the present discussion PABA is considered to be an essential metabolite. When we compare the diagrams of the molecules of sulfanilamide and PABA, the resemblance between the major parts of the two is made clear.

As shown in the diagram, the folic acid molecule consists of three functional groups: pterin, PABA and glutamic acid. If any form of a sulfonamide drug is present in a sufficient amount, part of it goes into the folic acid complex in place of PABA. It is said to be a *metabolite antagonist.* In this case it *antagonizes,* or is *competitive* with, the metabolite PABA. The sulfonamide is therefore often called a *competitive inhibitor.* Because the structure of the sulfonamide is analogous to that of PABA, the drug is also called a *metabolite analogue.* But no form of sulfonamide is capable of carrying on the functions of PABA. The sulfonamide is, in effect, a molecular imposter.

The enzymic functions of the microbial cell depending on folic acid are therefore promptly stopped by sulfonamides. Since all of the functions of the cell are closely interdependent, the whole cell mechanism stops also. The cell is then in a state of microbistasis. The production of microbistasis in this way by chemicals or drugs is the basis of modern *chemotherapy*. There are many substances besides sulfonamides (*e.g.,* amino acids, certain ions [see ion antagonism], organic acids and esters, purines, pyrimidines and sugar derivatives) that, because of similarity of structure, can act as metabolite analogues. They antagonize or displace ions or molecules of vital cell structures such as proteins, enzymes and coenzymes or nucleic acids. In the nucleic acids nucleotide analogues can produce mutations.

Reversal of Metabolite Antagonism

It is important to note that, like most other microbistatic reactions, the enzyme blockage produced by sulfonamide is not fatal and may be reversed in several ways—for example, by adding enough PABA to displace and exclude the sulfonamide; by supplying new folic acid ready-made in the medium; by supplying, in the culture medium, the product that would have been formed by the blocked enzyme (or any subsequent enzyme in the chain); or by supplying the finished product of the entire enzymic series, in this case certain amino acids.

DRUG RESISTANCE

The cell itself may be capable of bypassing or overcoming the effects of metabolite antagonists. For example, let us suppose that in a parallel series of enzyme reactions enzyme D is blocked by some microbistatic drug. However, let us assume that some of the substance usually made by the blocked enzyme D is produced as a side reaction or intermediate stage in the function of another enzyme, H. This product becomes available to E and thus the blockage at D is bypassed. Such a cell is more or less *drug-resistant,* depending on how much of the product needed by E is available from H.

Such enzymic rearrangements and bypassings *(alternate metabolic pathways)* are quite common. From the clinical standpoint, sulfonamide resistance of an infecting organism means that the patient must receive some other drug.

Resistant Mutants

It has already been shown that among any large number of microbial cells in a given pure culture, mutant forms may be found. Among the large numbers of bacterial cells of a given species infecting a patient there are, therefore, likely to be mutants with enzymic arrangements that will be unaffected by a microbistatic drug such as sulfonamide. In the presence of the drug these mutants will thrive, while all the other microbial cells are killed off. We thus have bred up a sulfonamide-resistant strain of the infecting organism which is dangerous to the patient and to those who contract the resistant organism from him.

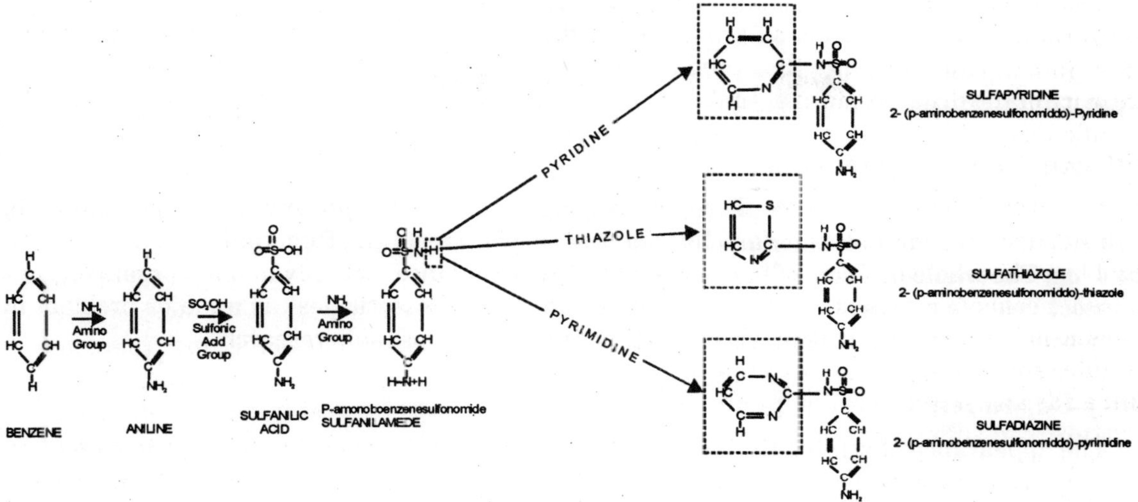

Fig. 5.15. How the sulfonamide drugs are derived from benzene. First, an amino group is added, yielding aniline (a base from which dyes are synthesized). The addition of a sulfonic acid group to aniline yields sulfanilic acid. The addition of another amino group yields sulfanilamide. The addition to this, in place of one hydrogen atom, of various other chemical groups yields other sulfanilamide drugs, three of which are shown in the diagram.

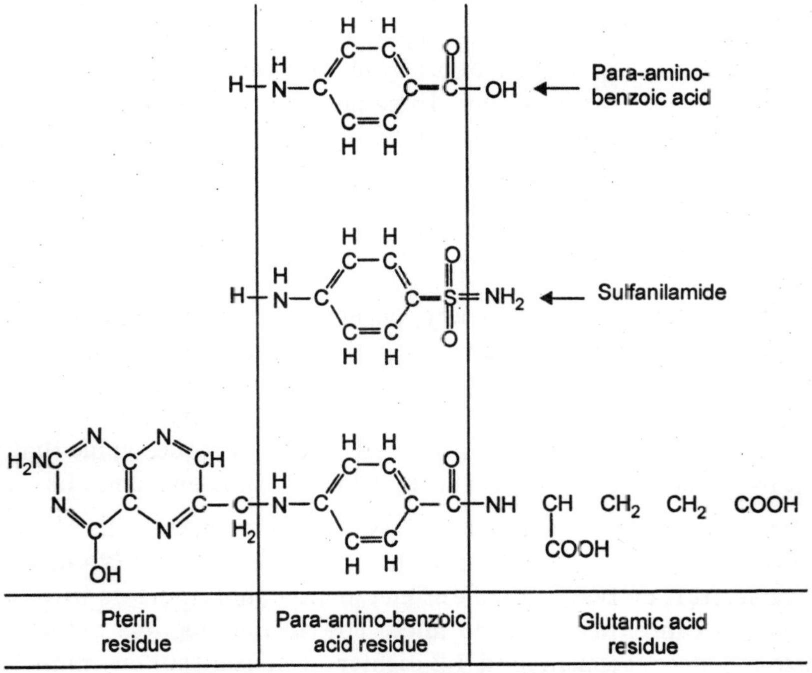

Para-amino-benzoic acid

Sulfanilamide

Pterin residue	Para-amino-benzoic acid residue	Glutamic acid residue

Folic (pteroyl monoglutamic) acid = vitamin B_c

Fig. 5.16. Structural resemblance between molecules of paraaminobenzoic acid (PABA) and sulfanilamide (above), and the antagonism of PABA by sulfanilamide in the folic acid molecule (below).

In some microorganisms that are ordinarily susceptible to penicillin, notably *Staphylococcus,* penicillin-resistance depends on production of an extracellular enzyme *(ptenicillinase)* which destroys the drug. Mutants that produce such enzymes are able to grow unhampered in the presence of the drug, to the eventual exclusion of the nonenzyme-producing sensitive mutants.

Induced Drug Resistance

In addition to the occurrence of drug-destroying enzymes that appear as a result of mutation, such enzymes appear also as *induced* or *adaptive enzymes* resulting from contact of the cell with the drug. The whole problem of mechanisms of drug resistance, not only in microorganisms but in insect pests (*e.g.,* resistance of mosquitoes and flies to insecticides), is a major problem in chemotherapy and pest control, and an excellent field for profitable research.

Drug Dependence

The appearance of strains that are not merely resistant to drugs but wholly *dependent* on them seems as strange as the emergence of resistant strains. The two are probably not related except that they are two manifestations of the phenomenon of mutation. Drug-dependent mutants obviously could not survive and develop significantly except in an environment containing the drug on which they are dependent. It is only because we have tried to cultivate microorganisms in contact with antimicrobial drugs that these curious mutant forms have been revealed.

CHEMOTHERAPY

Chemotherapy, broadly speaking, is therapy by means of chemicals. The term is generally restricted, however, to the use of drugs that act by metabolite antagonism or enzyme blockage. In order to be truly chemotherapeutic such drugs must act on the parasite in concentrations that have no effect (or at least no serious "side" effects) on the host. Some valuable chemotherapeutic drugs are toxic to the patient as well as to the parasite and so must be used with great care and under medical supervision.

Chemotherapy and Cancer

A very interesting aspect of chemotherapy concerns the possibility of selectively poisoning the cells of *neoplasms* (Gr. *neo* = new; *plasm* = form) (*e.g.,* cancers). Neoplasms are thought to have metabolic mechanisms different from those of normal cells. The obvious possibility presents itself of finding some metabolite antagonist that will poison an enzymic mechanism peculiar to the neoplasm cell and not present in the normal cell. Thus, it is conceivable that we could cure cancer overnight by means of a few tablets or injections of some antibiotic-like drug.

This rosy dream may not be too far from reality. For example, the cells of certain cancers of mice differ from normal mouse cells by synthesizing cyanocobalamin (vitamin B_{12}), a cobalt-containing coenzyme. At least two types of antimetabolites that interfere with synthesis of B_{12} produced cures of the cancer when fed to mice. The antimetabolites are 1,2-dimethyl-4(P-carboxyphenylazo)5-hydroxybenzene and 1,2-dichloro-4-benzenesulfonamide-5-nitrobenzene. Several other drugs have been found that, if not the whole answer, at least point definitely to the possibilities for chemotherapy in research on cancer metabolism. Honor, Fame and Fortune are to be won in this field!

CHAPTER

6 Foods and Dairy Products

The modern human dietary includes a wide variety of substances from many sources. Much of it is subject to natural contamination by many different kinds of microorganisms, including some that are dangerous pathogens, and it is modified with many additives (antimicrobial preservatives, hygroscopic agents, colors, flavors) and by various preparative processes involving many species of microorganisms; consequently, the microbiology of foods is an exceedingly complex subject.

CLASSES OF FOODS

Since water and dairy products are dealt with elsewhere, they will not be included here. Other common foods may be grouped as follows:

1. Fresh foods (*e.g.*, meats, vegetables, fruits and fish)
2. Foods preserved by:
   a. Drying
   b. Canning
   c. Pickling, brining, salting or fermenting
   d. Low temperatures, especially rapid freezing
   e. Antimicrobial substances
3. Breads
4. Eggs

For purposes of this discussion fresh foods may be defined as those recently harvested or prepared, that are in their natural or original state, not affected by any means of artificial preservation excepting refrigeration (not freezing) for limited periods, and unchanged by effects of holding for sale or use beyond *slight* wilting or drying. Foods still edible but held so long that perceptibly undesirable changes in volume, weight, color, flavor, odor, appearance or other properties of fresh products have occurred may be classed as stale. An *accurate* definition of *fresh* and *stale* is very difficult indeed, as are chemical or bacteriological determinations of these conditions. We shall use the terms in their commonly accepted meanings.

Foods may also be classified on the basis of stability:

1. *Perishable* foods such as meat and fish
2. *Semiperishable* foods such as potatoes
3. *Stable* foods such as cereals, flour and sugar.

Of course, any stable or semistable foods that become over-moist or water-soaked are no longer stable.

We may divide microorganisms of foods into three general groups on a functional basis:

1. Those causing spoilage or *undesirable* changes in the food
2. Those producing *desirable* changes
3. Those producing *disease*

Attempts to combat undesirable microorganisms or to encourage desirable microorganisms in food are the basis of industries and research in this country involving billions of dollars annually.

Autolytic Enzymes

In any discussion of the stability of foods, autolysis must be considered. When cells die, certain incompletely understood intracellular enzymic processes act within a few hours to disintegrate the cells. Such self-disintegration is called *autolysis.* Autolysis proceeds best under conditions optimal for enzymic action: pH, temperature and so on.

Many meats undergo autolysis. Venison and poultry are often "hung" until tender (*i.e.,* until some autolysis has taken place). Beef is more tender after a ripening period in refrigerators, partly due to autolysis. Autolysis may cause the uninitiated housewife to wonder why the pound of liver she bought in the butchershop seemed to melt away to a half pound as she drove home on a warm summer day. Since conditions for autolysis and microbial growth are parallel to a great extent, autolysis under market conditions is usually accompanied by microbial action unless foods are sterile.

Autolysis will occur even in sterile foods if the autolytic enzymes have not been destroyed by heat. Leafy vegetables (lettuce, spinach, endive), bananas and other fruits are made very soft during autolysis, especially in warm weather. Loss of weight, color, flavor and nutritive value result during excessive autolysis. Decomposition by yeasts, molds and bacteria can advance rapidly when over-autolysis has prepared the way. Hence in preserving foods, efforts are made to (1) prevent autolysis beyond a certain desirable *ripening* point; (2) to prevent decomposition by microorganisms. Usually the same preservative measures are effective against both. For example, in preparation of vegetables for preservation by freezing, autolytic enzymes, as well as some of the superficial microorganisms, are partly destroyed by *blanching* or scalding for a few minutes before packing.

FRESH FOODS

MEAT

When an animal dies of natural causes (disease or age), before death there is a short period (the agonal or moribund state) during which there is a collapse of the defensive mechanisms which normally prevent invasion of the blood (and thence of the tissues) by microorganisms in the gastrointestinal tract, respiratory tract, skin and other body surfaces. Examination of muscle, liver and other tissues of such an animal (and especially lymph nodes which arrest bacteria coming from such sources) immediately after death reveals the presence, often in considerable numbers, of microorganisms characteristic of the intestine, respiratory tract and so on. These include species of *Clostridium,* Enterobactcriaceae, *Micrococcus,* fecal and respiratory streptococci; *Proteus, Pseudomonas,* and the like. Meat from such an animal will spoil quickly. If a healthy animal is *suddenly* killed, as in the abattoir by a blow on the head, relatively little postmortem invasion of the blood and tissues occurs. Cutting a large vessel with a sharp, clean knife, as in killing hogs, probably introduces very few microorganisms. If the animal is then immediately dismembered in a cleanly manner, as is done in well-conducted abattoirs, relatively few organisms are to be found in the depths of solid tissues; spoilage must proceed mainly on the surface.

Surface Flora of Meat

Meat becomes contaminated immediately upon exposure in the abattoir. Dust from hides and hair, bacteria on gloves, hands and cutting and handling instruments, all contribute to this. The principal damage done by such organisms is to cause decomposition of organic substances in the surface tissues unless preventive and preservative measures include prompt hanging in refrigerators under ultraviolet light.

The microorganisms involved in surface decomposition and spoilage of meat and other protein foods include saprophytes of the genera *Pseudomonas, Bacillus* and *Micrococcus,* various Actinomycetales, yeasts and molds, *Achromobacter, Proteus,* enterococci, *Clostridium, Corynebacterium, Escherichia* and *Aerobacter.* Some psychrophilic microorganisms cause unsightly but harmless blackening, greening, or other discoloration during refrigeration and may produce undesirable tastes and odors, such as rancidity owing to decomposition of fats.

Antibiotics in Meat

Spoilage of fresh meat may be prevented for days and even weeks by intravenous injection of antibiotics just before or after killing. Broad-spectrum antibiotics such as chloramphenicol and the tetracycline group appear to give good results. The antibiotic is distributed throughout the tissues by the blood before the animal is killed. The method has proven of value in the tropics and in situations in which refrigeration is not feasible. Such use of antibiotics, however, is a source of some controversy. Probably *legal* residues of tetracyclines are not harmful. The use of penicillin is generally frowned upon because it is highly allergenic. Antibiotics in *any* materials can interfere greatly with bacteriological examinations. Such use of antibiotics is properly restricted to very special situations.

Antibiotic-resistant Organisms in Foods

A serious problem in the use of antibiotics for preservation of foods of any sort concerns the development of antibiotic-resistant species. Antibiotic-resistant pathogens (*e.g., Salmonella*) capable of infecting man as well as animals are developed by the continuous use of antibiotics; the various spoilage saprophytes also become resistant.

Ground Meat

The flora of freshly ground meat is much richer in numbers and types of microorganisms than that of large pieces of meat such as roasts, because in the ordinary processes of preparing ground meat, the meat is cut into small pieces, scraps are used and the surface contamination is thoroughly mixed with the meat as it is ground. It is also usually warmed somewhat. The microorganisms present in the product are those found initially on the surface and, in addition, molds, yeasts, sporeformers, Actinomycetales and others from the grinding machine, hands, implements and dust. Unless effectively refrigerated the microorganisms grow throughout the whole mass and the meat spoils rapidly.

Bacterial Count

Bacteriological standards for ground meat are difficult to establish and to evaluate. According to some workers, using a medium like that used for enumerating bacteria in milk, incubated at 32° C, an aerobic plate count of 10 million organisms per gram of meat is a reasonable maximum. Occasional counts in ground meat of market quality may run into the billions per gram, and this is exclusive of anaerobes, molds, and other organisms that do not grow under the cultural conditions provided.

BACTERIA IN OTHER COMMINUTED FOODS

The basic principles that apply to ground meats hold true also for products such as separated, fresh, crab meat; "shucked" shellfish; flaked, fresh fish; ground horseradish; chopped spinach; salad mixtures; cole-slaw, and the like. The more foods are handled and the more they are chopped, ground or grated *(comminuted),* the more the surface microorganisms are mixed with the interior and with richly nutritive sap or juice, and the heavier the inoculation. Such foods must be kept refrigerated at all times, and never sold if more than 24 hours old.

As in preservation of ground meats, admixture of antibiotics to initially reasonably fresh, clean products holds spoilage in check for many days, even without refrigeration. The acidity of products such as slaw and horseradish, mixed with vinegar or citric acid (lemon juice), tends to retard spoilage.

POULTRY, FISH AND SHELLFISH

Poultry

In 1955 the Food and Drug Administration (FDA) approved the use of not over 10 parts per million (ppm) of chlortetracycline (CTC) or oxytetracycline (OTC) in the ice-slush used for chilling eviscerated birds. No part of the finished product may contain over 7 ppm. If the birds are fresh-killed, clean and promptly eviscerated, the *"shelf life"* of the antibiotic-treated products is considerably extended. The advisability of the use of antibiotics for such purposes has been questioned and is now very carefully controlled.

Standards of hygiene and cleanliness surrounding the processing of antibiotic-treated poultry are stringent and are prescribed by the FDA and the Poultry Inspection Branch, U.S. Department of Agriculture. These requirements are necessary to prevent the substitution of cheap and easy antibiotic treatment for expensive and laborious cleanliness and sanitation.

Fish

As is true of the flesh of mammals, the deep tissues of freshly killed, healthy fish are sterile or nearly so. Autolysis is slow in fish muscle. It is common practice on commercial fish boats or landing places to gut fish immediately and pack them in ice. The surfaces become contaminated with a variety of bacteria from the skin, scales and intestines of the fish, the decks and cutting boards, and other environmental sources including ice and wash water in tanks.

Influence of Refrigeration Temperatures

Failure or delay of refrigeration is one of the commonest causes of spoilage of any nonsterile food. Another is overlong holding at refrigerator temperatures several degrees *above* zero instead of *at* or *below* 0°. Bacterial spoilage of many refrigerated foods may occur twice as rapidly at 5° C as at — 1° C. At temperatures above 5° C psychrophiles grow well. Many microorganisms not definitely classed as psychrophiles can also grow to a significant extent at low temperatures. Maximum numbers of live bacterial cells accumulate *slowly* at the lower temperature. When the refrigerated products are afterward gradually warmed for use, these enormous active populations (along with their accumulated enzymes, and other growth products) act very rapidly indeed to effect spoilage.

Psychrophilic organisms commonly associated with spoilage of recently caught fish include species of *Pseudomonas, Achromobacter,* coli-forms, and sometimes *Clostridium.* Usually present but less active in spoilage are *Micrococcus, Flavobacterium* and *Corynebacteriwn.*

Antibiotics in Fish

Chlorinated water, ice with calcium chloride, and sodium nitrite have been much used to preserve fresh fish. Admixture of antibiotics to the flaked ice in which freshly caught fish are packed keeps the product, if cleanly prepared, in a perfectly fresh state, even unrefrigerated, for days. In 1955 the FDA permitted the use of chlortetracycline in raw, whole or gutted fish, shucked scallops and unpeeled shrimp, to a maximum concentration of 5 ppm in any part of the product. The same remarks apply to the use of antibiotics in fish as to their uses in meat and poultry.

Shellfish

Under market conditions shellfish are often rather heavily contaminated with dirt from the shells and from benches at which *shucking* (removal of shells) is done. The flora includes marine species similar to those found on fish, organisms from soil and from equipment used in handling. If the water in which they have grown has a rich bacterial flora, the shell liquor may contain considerable numbers of various harmless bacteria. Oysters, mussels and some other shellfish fatten on sewage, and have long been known as vectors of typhoid fever if taken from polluted waters.

Tyhioid bacilli may live for two weeks or longer in live and in shucked oysters. Formerly common, typhoid due to shellfish is now very rare. Uncooked shellfish are vectors of enteric viruses, especially that of epidemic hepatitis, and they should be regarded with suspicion in respect to polio and other viruses. In the United States, state and government supervision of the shellfish beds has virtually removed infected shellfish from the market. It has recently been shown that fluids inside the shells of steamed clams may not always reach temperatures high enough to inactivate the heat-stable virus of epidemic hepatitis.

Among the bacteria that may be present in shellfish are the coliform group. These are not harmful per se, but represent potentially dangerous sewage pollution. Tests for the coliform group of bacteria in shell liquor, or in the liquid around shucked shellfish, are made in much the same manner as tests are made for this group in drinking water. The results are expressed in a similar manner. However, coli-forms in shellfish may be entirely nonsewage in origin if the shellfish are taken from clean beds.

Oysters to be offered for sale raw are generally *floated* (allowed to remain for some hours or days) in clean chlorinated water. In the process they pass a large volume of the clean water through and around their bodies and thus greatly reduce their bacterial content.

FRUITS AND VEGETABLES

The principles underlying the microbiology of fresh meat and fish products apply equally to fresh vegetables and fruits. That is to say, the internal tissues of whole, healthy plants and fruits contain very few bacteria, but contamination of the exterior surfaces by microorganisms from the soil, hands and packages occurs. Vegetables growing in the soil, such as root crops, have adhering to them soil saprophytes such as species of *Bacillus, Pseudomonas* and many others, depending on the nature of the soil. These soil organisms are usually held in check by washing, drying, exposure of the surfaces of the vegetables to sunlight, and by refrigeration.

Intact vegetable skins free from superficial water resist invasion of microorganisms for considerable periods, especially if kept in a cool, dry place. Nonsucculent vegetables with whole skins, such as turnips and potatoes, will stand storage better than soft, succulent, nonacid vegetables and fruits like lettuce, asparagus, spinach and ripe peaches, the juices of which offer good pabulum to microorganisms. Such soft vegetables soon autolyze and then decay. Invasion and decomposition are retarded by gentle handling and storage under cool, dry conditions.

EGGS AND EGG PRODUCTS

These may be considered under two general headings: (a) shell eggs (fresh or storage); (b) egg products such as liquid eggs; eggs that are frozen, dried or whole; egg whites and yolks; and eggs that are salted, plain, sugared, desugared, or fermented.

Shell Eggs

As soon as an egg is laid, the outer surface becomes contaminated. If, as is the practice in many large-scale poultry plants, the eggs are laid by clean, healthy hens on a clean wire frame and automatically collected immediately, contamination is at a minimum. The sanitary and keeping qualities of such eggs (and of commercial egg products derived from them) are clearly superior to those of eggs laid in damp, mud-and-feces-fouled straw nests and collected perhaps once a day (or every two or three days!. Surface microorganisms are prevented from entering the egg for some days largely by the dried, mucilaginous surface coat, a sort of natural varnish. Molds and some bacteria can grow on the outer mucinous coating if eggs are stored in humid atmospheres (above 70 per cent saturation) at ordinary climatic temperatures. The microorganisms eventually penetrate the shell and contaminate the interior. These give the eggs "off" odors and tastes and unsightly appearance. Eventually the mucinous protective film is entirely decomposed and the interior of the eggs is overwhelmingly invaded, with consequent decay. The flora of stale or bad eggs is largely of fecal and soil origin and includes the various saprophytes listed elsewhere.

Pseudomonas and Pyoverdine

Early spoilage of eggs stored at 15° C is often due to *Pseudomonas,* especially *P. fluorescens* and *P. ovalis* from soil or dirty wash water. These bacteria impart "musty" tastes and odors to so-called *fresh* eggs long before overt, late-stage decomposition develops. These bacteria produce a fluorescent pigment called *pyoverdine* that can readily be detected by the ingenious expedient of examining eggs, shell or liquid, or extracts of liquid products, in ultraviolet light. Higher than normal fluorescence indicates heavy growth of *Pseudomonas* and usually other bacteria, incipient spoilage and low commercial quality.

Infection of Eggs

Eggs may contain pathogenic bacteria when laid, even though the hen *seems* healthy. *Salmonella* species are commonly found. These bacteria have at times caused serious outbreaks of gastroenteritis, *Salmonella* may be detected by means of selective cultivation and fluorescent-antibody staining.

EGG PRODUCTS

In commerce, the best (freshest) eggs are sold in the shell, while second or lower grade eggs are used for liquid, frozen or dried products. Plate or direct microscopic counts of such egg products may run from a million to a billion or more per gram. This does not necessarily make the eggs unfit for use in cooking, any more than high bacterial counts destroy the commercial value of lower grades of milk. The problem is one of profit and loss. Egg products are a picnic for microorganisms, just as are milk, crabmeat and ground meat. If the egg products are not promptly frozen, refrigerated or dried, the heavy initial contamination quickly ruins them.

Liquid Eggs

Egg whites are often sold separately. To improve their keeping quality they (and some other food products) are often heavily inoculated with the nonproteolytic *Escherichia coli* or, better, with *Saccharomyces cerevisiae.* The whites are held at about 28° C for 24 to 48 hours. Glucose in the whites, main source of energy for putrefactive organisms, is thus promptly removed by fermentation with nonputrefactive *E. coli* or yeast. The whites must be initially clean or they will spoil during the intentional fermentation.

BREAD

The better grades of bakers' bread are produced by allowing bakers' yeast (varieties of *Saccharomyces cerevisiae)* to ferment sugars (glucose, maltose, sucrose) in a mixture *(dough)* consisting mainly of flour and water (or milk), with some salt. In commercial baking softening and hygroscopic agents, and flavoring, raisins, caraway seeds, vitamins and shortening are commonly added. Small amounts of cane sugar are often included in home bread making, both for flavor and to stimulate fermentation. The dough is a soggy, plastic mass at first. *S, cerevisiae* does not attack the starch in flour.

However, the necessary amylases and proteinases from the grains are present in the flour. These enzymes hydrolyze the starch and proteins of the dough. Sometimes *malt* is added to aid this process. The products of the hydrolysis support growth of the yeast. Early growth of lactobacilli, derived from the grains, gives the dough an initial acidity which helps suppress undesirable organisms and favors yeast. The lactobacilli and yeast also contribute to the flavor and aroma of the bread.

The yeast, or *leaven* (L. *levare* = to raise), produces carbon dioxide, water and ethanol. The gas causes the bread to rise and gives it its foamy texture. Baking drives off the alcohol and partly dries and *sets* (firms) the bread.

Leavens

An important phase of the work of the microbiologist in some bakeries is the preparation of leavens. Many bakeries maintain their own leavens as trade secrets. The leaven may consist of a pure culture of a selected strain of yeast cultivated massively in aerated wort made of malted grains or other starch derivatives. Many tons of yeast in dried, compressed cakes are prepared in commerce daily for use by brewers, bakers and in the home. Leaven may also consist of a mixture of pure cultures of *aerogenic* (gas-producing) bacteria: *Aerobacter cloacae* (subgroup A), the hetero-fermentative *Lactobacillus brevis* and *Leuconostoc,* as well as yeasts. *Clostridium* species are undesirable in leaven because they produce rancid flavors. Bacterial leavens are used to prepare bread made of sour dough (sauerteig), and "self-rising" or salt-rising breads. Such leavens often consist of previous lots of sour dough and are fortuitous mixtures; the standby of the old-time prospector, or "sourdough," of Alaska.

Microorganisms in Bread

Most organisms in bread, aside from those added as leaven, appear to come originally from the flour or meal and include the familiar list of the environmental saprophytes. Bacterial counts of flour range from a few hundred thousands to several millions per gram. These organisms, especially molds, can cause spoilage in the stored product, especially under humid conditions.

During ordinary baking most of the vegetative forms of molds, yeasts and bacteria are killed. Heat-resistant bacterial spores and conidia of molds may survive. Temperatures inside the loaf of baking bread rarely rise much above 80° C.

Ropy Bread

If bread is made with ingredients containing large numbers of spores of slime-forming species of *Bacillus B. polymyxa* and *B. pumilis,* some of these can survive baking and may grow in the bread, producing a mucinous slime. When the bread is broken apart this slime is drawn into long threads, resulting in a defective product called *ropy bread.* These difficulties with bacterial spores may be largely eliminated by cleanliness and modern methods of inducing germination of spores before baking.

Moldy bread is usually the result of extraneous contamination of cut surfaces or crust by hands, dust and knives, followed by holding under humid conditions at household temperatures. Slight surface growth of mold gives bread a musty odor and taste; some species of mold produce toxins (aflatoxin).

Bread in the Kitchen

Since bread is not sterile it should be cooled promptly after baking. The wise housewife keeps bread in the refrigerator in warm, humid weather. Wrapping bread in waxed paper helps keep the bread clean and prevents drying out, but conserves a humid atmosphere and favors growth of molds and bacteria, even if refrigerated.

Nonsporeforming pathogenic bacteria cannot survive proper baking of bread but may be introduced by unsanitary handling after baking.

SOME FERMENTED FOODS

As noted elsewhere, soil and plants harbor, among many other microorganisms, numerous species of lactic-acid bacteria such as *Lactobacillus, Leuconostoc* and *Pediococcus.* These can ferment the carbohydrates and metabolize the other nutrients in the tissues, sap and juices of green plants. This is the basis of the making of several kinds of fermented plant foods: ensilage for cattle; sauerkraut and pickles for humans. The methods of preparing each kind of food differ somewhat in form, but the principles and microorganisms are identical, or very similar, in all three processes. The preparation of ensilage for cattle is representative.

Ensilage

Finely chopped, partly mature plants such as corn stalks or alfalfa are tightly packed in tall cylindrical tanks *(silos).* Microorganisms of many kinds start to grow in the plant juices and ferment the carbohydrates. As fermentation proceeds, the material becomes warm and acid. The heat can be reduced if the rate of oxidation is decreased (exclusion of free oxygen) by tight packing. Oxygen is used up rapidly so that molds and other strict aerobes soon cease growth.

Only facultative and strict anaerobes continue. In the first stages probably the Enterobacteriaceae and other rapidly growing nonaciduric but fermentative microorganisms predominate. These are undesirable since many produce gas and unpleasant flavors. As acidity increases they subside, and the more aciduric homofermentative lactic-acid bacteria predominate. These produce more lactic acid with small amounts of other products of fermentation (diacetyl and other volatile substances) which give an aroma and flavor to ensilage that is relished by cattle. The last stages of the fermentation and final increase in acidity are caused by the very acidophilic and aciduric lactobacilli, *e.g., L. delbrueckii* and *L. plantarum.* After three or four weeks, the process slows and the fermented mass gradually cools.

Carbon dioxide is produced during the fermentation process and often settles in the lower part of silos, so that a person ignorant of this may die if he stays in the depths of a poorly ventilated silo.

It has been suggested that a desirable type of fermentation in silos may be facilitated by introducing cultures of various fermentative bacteria, such as *Streptococcus lactis* or *Lactobacillus* sp., as the material is packed.

Under natural conditions various other organisms are doubtless involved, including the bacteria of the soil. In some sections of the country molasses is added to promote fermentation by the acid-formers, and to improve palatability.

If too much soil is introduced with the fodder, undesirable and excessive putrefactive processes may spoil the product. For example, butyric-acid organisms such as *Clostridium butyricum* get in and ruin the silage by producing butyric acid, which makes it rancid. The action of such organisms constitutes a "disease" of silage. *Cl. botulinum,* a soil anaerobe forming a deadly exotoxin, has also at times caused much damage to livestock by its growth in silos.

Souerkrout

In the production of this savory delicacy pure cultures of lactobacilli are sometimes used to aid the process. Commonly, however, the fermentation is allowed to proceed naturally. Salt is placed between layers of shredded cabbage as it is packed in large crocks or barrels. The salt inhibits undesirable bacteria and draws out the juices of the cabbage. Wooden frames are placed on the cabbage to keep it submerged (anaerobic). Except for the salt, sauerkraut is analogous to silage. Only facultative, anaerobic, aciduric and acidophilic and thermoduric forms not sensitive to salinities of two per cent or higher can survive. During the first two to five days species of *Leuconostoc,* espedially *L. mesenteroides,* are common. Acids, esters and diacetyl give pleasant aromas and flavors. Temperatures near 70° F favor the best fermentations. Fermentation is complete in two to three months.

Pickles

Fermentation by mixtures of organisms normally present, in a manner analogous to the manufacture, of ensilage and sauerkraut, is part of the processing of pickles, ripe olives and the like. The process is made very selective by progressively increasing salt concentrations up to 30 per cent. The action of the lactic acid, brine and microbial metabolism change the color, consistency and flavor of the cucumbers, olives, etc.

In making pickles, after 8 to 10 weeks the vat is emptied and the pickles, now called *salt stock,* are packed in fluid containing vinegar, sugar and various flavorings, spices, dill, and so on. Slimeforming organisms such as *Leuconostoc mesenteroides;* sporeforming bacilli; pectinaseformers that destroy vegetable tissues; organisms that destroy lactic acid; molds and other hydrolytic spectes may ruin the pickles if temperatures remain long near 20° C or if salinity is too low. Stock to be used for dill pickles (low in acid and salt) is particularly liable to such spoilage.

PRESERVATION OF FOODS

Methods of food preservation are of two general classes: (1) *Bactericidal* or *sterilizing* methods, involving canning and such processes as making jams, jellies, and preserves (*i.e.,* heating processes). The use of highly penetrating, microbicidal, ionizing radiations will also probably take its place among these methods in the future. (2) *Bacteriostatic* methods include: drying, freezing, refrigeration at temperatures slightly above freezing, various types of pickling, brining, salting, smoking, the use of antibiotics as already described, and other preservatives such as various organic acids.

Modern home or commercial canners have to consider not only the killing of bacteria likely to cause spoilage or disease but also the effect of the processing on palatability and appearance of the food. If canned foods are processed long enough they can be absolutely sterilized. This may require such prolonged heating in steam under pressure (autoclaving) (especially of non-acid foods) that the foods become mushy and discolored and have bitter flavors. Prolonged heating also adds to the cost of canning.

The aim of the canner, then, is to heat as little as possible, consistent with safety from food poisoning and loss from spoilage. From the standpoint of food poisoning the only organism likely to resist autoclaving is *Clostridium botulinum.* Even the most resistant spores of this species are killed by heating at 121° C for 20 minutes at pH 7.0. Acidity greatly reduces the time and temperature necessary to preserve foods by heat even though they may contain resistant spores.

Canned tomatoes, rhubarb and acid fruits (pH 3.5 to 4.5), for example, require only a few minutes at 100° C to preserve them. On the contrary, nearly neutral materials such as meats, corn, spinach and beans require much longer periods depending on solidity and size of packs, and preheating. The cans are often preheated to 100° C and sealed in a vacuum.

Table 6.1. Acidity (pH) Range of Some Representative Foods

Range of acidity	*pH*	*Examples in order of acidity*
Very acid	2.5 to 3.5	Lemons
		Cranberries
		Rhubarb
		Grapes
		Pineapples
Moderately acid	3.6 to 4.5	Oranges
		Apples
		Tomatoes
		Pears
Slightly acid	4.6 to 6.8	Carrots
		Squash
		Spinach
		Fish
		Beef
		Poultry and eggs
		Corn
		Shellfish

The effectiveness of expert canning is well illustrated by the history of several cans of meat prepared in London several years after the process of canning was devised by Appert, in 1805 in Paris, at the request by Napoleon Bonaparte for a means of preserving foods for his troops. A can of roast veal prepared in London and dated 1823 was carried the next year on the expedition of H.M.S. *Fury* (Sir Edward Parry, R.N., Captain) in search of a Northwest Passage. The *Fury* was crushed in the ice and was abandoned, with stores including the can in question, for four years in the arctic. The *Fury* was visited by another ship in 1829 and the stores were brought back to London. When the can was opened, in 1958, after 125 years, the meat was found in good, unspoiled condition although the fat had emulsified somewhat. Also in 1958, a can of plum pudding prepared in London in 1900 and kept in South Africa was found on opening to be in excellent, unspoiled condition.

Commercially Sterile Foods

Canned foods may contain spores of organisms that are viable but fail to grow under ordinary conditions of storage. These foods may be said to be *virtually sterile* or *preserved by heating* or *commercially sterile.* For example, spores of nonpathogenic *Bacillus stearothermophilus* and *Clostridium thermosaccharofyticum* generally withstand ordinary processing. The spores of these species are among the most heat-resistant of all. They are often used to test the efficacy of heating. Both can cause sour spoilage of canned goods. However, being obligate thermophils, they do not grow unless the food is stored in a *very* warm place (55° to 70° C). Under ordinary conditions of storage the food is, from a practical standpoint, sterile. In any case, care must be taken to eliminate the spores of *Clostridium botulinum,* cause of deadly botulism.

PRESERVATION BY FREEZING

Frozen foods represent one of the greatest money-exchange items in the American dietary. These foods may be classified for microbiological purposes as (1) uncooked and (2) precooked.

Uncooked Frozen Foods

These are fresh foods to be eaten raw (*e.g.,* frozen berries) or to be cooked (*e.g.,* frozen steaks). At the time of freezing, fresh foods have on or in them the original microbial flora plus any that are added during handling, packaging or processing. Factors affecting the microbial content of uncooked frozen foods include: freshness, condition at time of freezing, pH, preliminary processing, packaging, time and temperature of frozen storage, kind of food (protein, starch, comminuted). If foods are dirty or have been held long at room temperatures prior to freezing, they may be on the verge of spoilage at the time of freezing and the microbial counts may be very high. However, since freezing and storage at — 25 ° C soon destroys many bacteria, low (deceptively favorable) plate counts may be obtained with long-frozen food that was initially of low quality. In general, storage at higher temperatures (0° C to —10° C) is more bactericidal, especially in acid foods such as orange juice, than storage at lower temperatures: –25° C in freezer lockers to -75° C in carbon dioxide ice.

Vacuum drying after nearly instantaneous freezing ("freeze-drying") (analogous to *Lyophilizatian* used to preserve viruses, bacteria, appears to offer practical usefulness in the preservation of foods.

As with milk, a direct microscopic count on frozen foods reveals much concerning initial numbers of now-dead bacteria. Microorganisms that will not grow in ordinary plate-count media are revealed. The same is applicable to foods preserved by any of the methods described farther on. High microscopic count and low plate count suggests food of initially poor bacterial quality.

Precooked Frozen Food

Some precooked frozen foods, such as certain "TV dinners," are mixtures of cooked or partly cooked components. that are not further cooked before packaging and freezing. Some of these foods are meant to be oven-baked before serving. Others, such as meat pies, are said to be ready to eat after mere warming. Bacterial counts on cleanly produced, promptly frozen, cooked foods are relatively low. Excessively high bacterial counts, especially high direct microscopic counts, indicate staleness or dirtiness of ingredients or processing.

Baking meat pies (a representative precooked food) does not necessarily sterilize them. Commercial-size meat pies held at 425° F for 40 minutes have very low counts but the same temperature for 20 minutes is wholly inadequate and may, by merely warming the interior, actually incubate the pathogens in the pies.

Sanitation in Frozen Foods

Virtually all frozen foods are subject to some sort of contamination before freezing. The usual environmental saprophytes are found. Since many foods contain animal products, and are handled by people, fecal streptococci, coliforms and staphylococci are frequently found. The full sanitary significance of these indicator organisms in frozen foods is not yet fully clarified, but they are generally regarded with suspicion.

As indicators, enterococci tend to persist in contaminated frozen foods, while the coliforms tend to diminish in numbers and disappear with continued frozen storage.

Thawing of Frozen Food

Freezing during preparation and thawing for use should be completed within two to four hours under ordinary conditions. Most frozen foods remain wholesome so long as they remain frozen. On thawing, frozen foods undergo rapid deterioration because the natural resistance of fully active tissues is reduced or absent. Autolysis occurs and microbial decomposition can set in with little delay at room temperature. Pathogens can grow at summer temperatures (70° to 100° F)

Interruptions to electric current supplying frozen-food lockers may cause serious spoilage if long enough for frozen foods to thaw. Indeed, food refrozen after being thawed without the consumer's knowledge may be a very dangerous product. Large masses of ice in the bottom of the package suggest that thawing and refreezing have occurred.

General Rule

A sure means of making any food, canned, frozen or otherwise, safe from infectious organisms and from botulinal (but *not* staphylococcal) toxin is to heat all of it to 100° C for at least 15 minutes just before eating.

As a matter of experience of over three decades, food poisoning (botulinal or staphylococcal) and food infection (salmonellosis, shigellosis) due to *commercially* prepared foods are rare (but by up means unknown!) in the United States.

CHEMICAL PRESERVATIVES

Several organic acids and their salts: acetic, lactic, benzoic, salicylic, butyric, caproic, sorbic and others, have marked microbistatic and microbicidal action. As previously indicated, the effectiveness of these acids is dependent mainly on the toxic action of the *undissociated* acid or salt. Two factors that diminish dissociation of these acids or their salts are salinity and low pH. The higher the salinity and the lower the pH the more effective these organic acids are as preservatives. Undesirable *yeasts* and *molds* in brined or fermented pickles, in sauerkraut, and in unfermented acid products, such as fresh apple juice and tomato catsup, are well controlled by the addition of 0.05 per cent to 0.10 per cent sodium benzoate or sorbic acid (or sorbate). *Bacterial* growth, desirable or undesirable, is not significantly affected by sorbate.

Sodium nitrate or nitrite, or mixtures of these, are commonly added to sodium chloride in mixtures for curing meats (corned beef, ham). Whether the nitrates or nitrites have any bactericidal or bacteriostatic action per se seems to be undecided, but meat packers find less spoilage when these salts are used. They appear to have an adjuvant action in acid solutions (most cured meats are acid, pH near 5.8) and on the preservative effect of sodium chloride. Nitrites and nitrates are particularly desired because they give good red color to meat.

Sulfur dioxide has a bleaching effect desired in some fruits and also suppresses growth of yeasts and molds. It is applied as a gas to treat drying fruits and is also used in molasses.

OTHER METHODS OF PRESERVATION

High Osmotic Pressures

The salting and brining of fish, corning of beef and sugarcuring of hams and brining of green olives are examples of the use of solutions of high osmotic pressure for food preservation. Salt concentrations of 20 to 30 per cent are commonly used in brines; 50 to 70 per cent of sugar, in sugar syrups. Their preservative action depends almost entirely on their withdrawal of water from microorganisms. Other uses of sugar are in the making of jams and jellies. Heating of jams and jellies greatly reduces the lnitial count of microorganisms and enhances the effectiveness of sugar.

Dry salt and sugar-cured products will spoil if allowed to stand in very humid atmospheres, since they are not sterile and since sugar and salt are strongly hygroscopic. Spoilage of salted products is sometimes caused by halophilic microorganisms in the brines or salt. Clean ingredients and equipment help prevent losses due to spoilage. Molds and yeasts are causes of early spoilage of sugared and salted products since these fungi grow well in solutions of high osmotic pressure.

Drying

Drying of foods is another time-tested means of withdrawing water from spoilage microorganisms and their environment. Drying is largely microbistatic in effect; therefore, dried foods are not sterile and will decompose promptly if kept warm in humid atmospheres.

Smoking

Meats and fish to be smoked are usually first salt-cured to prevent rapid deterioration, since smoke curing is slow. Modern "streamlined" injection processes preserve and flavor more rapidly than salt or smoke curing. The preservative factors in wood smoke are various cresols, a mixture of formic, acetic and other organic acids and alcohol ("pyroligneous" acid), and formaldehyde, all of which are bactericidal or bacteriostatic substances. These are gradually absorbed in small quantities by the tissues exposed to the smoke. The meat is thus cured and rendered impervious to the action of most microorganisms, as well as given an agreeable aromatic flavor. Kept dry, the meat is preserved almost indefinitely. In warm, humid weather it may become moldy and rancid, especially on the surface.

Radiations

As pointed out elsewhere, streams of electrons (cathode rays) under sufficient potential (1 million volts or more) are potentially lethal ionizing radiations. Beta rays and gamma rays from radioactive cobalt and other wastes of atomic research have similar properties. Fresh perishable foods (as well as culture media, surgical supplies and drugs) in packages, *e.g.*, plastics, that can be readily penetrated by such rays can be sterilized *without heat* and kept in their moist, "fresh" condition with *no refrigeration*.

The possibility of keeping foods fresh without refrigeration, whether by antibiotics or irradiation or other means, obviously offers enormous advantages. The necessary amounts of radiation are expensive and produce undesirable changes in flavor, color, consistency, and possibly in vitamin and other nutrient content in many foods. However, nonsterilizing amounts of radiation can be used economically as a preliminary treatment. Spores are sensitized to heat by radiation. The use of irradiation for pasteurizing (not sterilizing) as a means of reducing spoilage is still in the experimental stage.

MICROBIOLOGICAL EXAMINATION OF FOODS

Quantitative Microbiology of Food

Estimating the total numbers of bacteria in milk. water, ground meat, vegetables, frozen foods and the like involves basically the same procedures. The main differences lie in methods of collecting, preparing and measuring the sample. These steps are guided by the nature of the material. Fluids are diluted and plated directly, as is milk. In examining solid foods such as meat or vegetables, washings, swabbings or scrapings from measured surfaces may be collected and shaken in measured diluting fluid. Deep samples of solid foods are cut from within after sterilizing the surface with a hot spatula.

Colony Counts

Any solid material is first weighed and then comminuted in a Waring blendor or similar device to get the microorganisms of the sample of food *into a fluid suspension*. This can then be diluted and put into Petri plates with "Standard Methods" agar for colony counting. The diluted sample must not be too cloudy or contain visible particles since particles confuse colony enumeration. For inoculation into tubes of broth, clarity of inoculum is not so essential. Highly acid products should be neutralized.

Direct Microscopic Count

The direct microscopic examination is very useful, both quantitatively and qualitatively.

It is made in much the same manner as the Breed count for milk. High microscopic counts and low plate counts on a food generally indicate that an initially poor product has been treated in some microbicidal manner or that the organisms seen will not grow in the medium provided for the plate count. Numbers of yeasts and molds and of other special morphological types that do not grow on the routine media may also be estimated in the direct count.

Qualitative Microbiology of Food

Methods of examining water or milk for the coliform group or for fecal streptococci typify *qualitative* procedures in examining foods for these groups. Tubes of selective broths (for determining MPN) or of selective plating medium (for colony counts) may be used. The same sorts of media and procedures are used, with minor modifications, as for milk. In several dried and frozen foods, fecal streptococci appear to be a more dependable indicator of fecal contamination than coliforms.

The coliform organisms are so ubiquitous that their presence in many foods does not accurately indicate fecal pollution at all. Further, they do not survive in preserved foods, especially acid foods, as do the enterococci. In adapting microbiological procedures to the examination of foods the nature of the medium used, the temperature (thermophilic, mesophilic or psychrophilic) and conditions of incubation (aerobic or anaerobic), and other factors (acidity, salinity) must be modified to suit the flora. For example, if a plate count of yeasts and molds is desired, plate with acidified potato-dextrose agar (pH 3.5) or with malt agar (pH 3.5). For lactic acid bacteria, use orange-juice agar (pH 5.5). Other special media for coliforms and staphylococci are suggested in the literature cited. Numbers of bacterial spores may be estimated (not accurately counted) by heating the material to 90° C for 10 minutes before plating. Other selective procedures will occur to the ingenious microbiologist and can be adapted to various problems.

FOOD SPOILAGE: DOMESTIC

The two principal aims of the domestic kitchen director are much the same as those of the commercial food handler: (a) prevention of spoilage; (b) prevention of food infection and food poisoning.

Prevention of Spoilage

The housewife relies mainly on low temperatures, osmotic pressure (*e.g.*, in jams or jellies), and on heating for preservation of foods. As a domestic economist she is interested in saving from spoilage those foods bought in quantities at lower-than-usual prices, or bought only during weekly or monthly trips to food stores. The culinary expert is also interested in "left-overs." The surest mark of high achievement in the art and economics of the kitchen is the skillful and appetizing use of foods that are kept from one meal to another—the pot of good soup (excellent bacteriological medium!) that is saved for tomorrow's dinner, the left-over porridge or bread for pudding, and so on. All are valuable assets to the resourceful domestic economist. But they are also subject to spoilage by microorganisms. The household refrigerators or freezing cabinets offer the readiest means for saving foods.

When these places are overcrowded or unavailable, spoilage can be retarded by reheating contaminated foods in covered vessels and *not openling* the vessels until more of the food is needed. For example, the pot of soup or other nonacid food may be brought to a boil for a few minutes in a vessel with a good dust-tight cover. The heat kills any vegetative forms of microorganisms and the spores of wild yeasts and molds that might have gotten in while it was cool and uncovered.

A good many bacterial spores will be killed also, but some will survive. However, these probably will not grow sufficiently to cause spoilage for 18 hours or more. If the soup should be brought to a boil the next day, the cover not having been removed in the meanwhile, virtual fractional sterilization has been accomplished and the food should keep for some time. With solid food not immersed in water, such as a roast of meat, spoilage tends to occur mainly on the surfaces. Much of this is introduced by utensils.

Clean utensils introduce less contamination than soiled ones. If the cooled roast is reheated in a very hot oven for a few minutes after the meal, so as to cook only the exterior, and is left in the oven without opening the door, thus avoiding recontamination from the air, surface growth is definitely checked and the meat may remain good for some days, depending on the extent and depth to which initial contamination was introduced. In reheating food in this manner it is

clean handling and routine *effective* sanitization of milk-handling equipment. Their development can be retarded by prompt refrigeration and by pasteurization.

Pasteurization

Pasteurization consists in holding the milk in tanks at 145° F (63° C) for 30 minutes ("low-temperature holding" or LTH) and immediately refrigerating. *Disinfection* of the milk is accomplished; *not* sterilization. In many dairies the same result is achieved by heating the milk rapidly in a tube or in thin layers between metal plates, to 71.6° to 80° C and holding at that temperature for 15 to 30 seconds, then cooling. These high-temperature-short-time (HTST *or flash)* methods save time and money and are effective so far as sanitation of milk is concerned.

Fig. 6.2. Bottling milk in a sanitary dairy. All of the piping can be demounted in a few minutes for steam sterilization and the floors and walls hosed down. In some plants light, transparent, plastic tubing is used. The bottles in this picture have been steam sterilized just before filling.

CHANGES IN FLORA OF MILK

Since milk is an excellent medium for bacterial growth, the numbers of bacteria in it increase steadily the longer it stands, even if pasteurized and refrigerated. Even if milk is refrigerated so that growth of thermophiles and heat-resistant sporeformers is retarded, psychrophilic species will grow. Under storage at 3° C some of these can soon cause discoloring, ropiness, off-flavors and other undesirable condition. Common dairy psychrophiles are *Pseudomonas, Achromobacter, Alcaligenes, Flavobacterium* and *Micrococcus* species.

Not all are killed by pasteurization and the survivors can multiply. It is important that refrigeration be at very near 0° C. Much commercial refrigeration is at about 10° C, a temperature ideal for many psychrophiles. The temperature of the average household refrigerator (not freezer) is about 5° C. At best, such refrigeration is effective for not much over 24 hours.

Fig. 6.3. Conditions contributing to high bacterial counts in milk. Milk cooler open to dust and dirt; not readily cleaned.

Row Milk

If allowed to stand at about 22° C, the flora in raw market milk rapidly undergo a series of changes. Numbers of bacteria increase to almost astronomical figures within 24 hours. Enterobacteriaceae, lactic streptococci (*e.g.*, 5. *latis), Micrococcus,* some sporeformers such as *Bacillus polymyxa* and other species of saprophytes that thrive at a pH near neutrality grow rapidly and dominate the picture at first. The lactose is fermented. As acidity increases, these species are inhibited. The aciduric lactic organisms then gain the ascendancy, especially *Lactobacillus* and species of *Leuconostoc.*

Many *Clostridium* species will also ferment the lactose, but the presence of large numbers of *Clostridium* spores in fresh milk is not usual and indicates excessive contamination of the milk with soil or dung. When the acidity reaches a pH of about 4.7, curdling occurs. The curd shrinks and settles out. Eventually, organisms capable of attacking lactic acid develop, especially aciduric yeasts and molds growing on the surface. These lower the acidity of the milk by destroying the acid and by producing alkaline products of protein decomposition: amines, ammonia and the like.

Since the carbohydrates (lactose) have been decomposed by this time, fermentation does not reoccur. Organism capable of hydrolyzing the fat and casein now thrive: eucaryotic fungi, Bacillaceae, both aerobic and anaerobic, Pseu-domonadaceae and many other lipolytic and proteolytic saprophytes. As the oxidation-reduction potential of the milk is reduced, species of *Clostridium* and other anaerobes, both obligate and facultative, gain the ascendancy and the odors (ammonia, odoriferous amines, mercaptans, hydrogen sulfide, rancid odors) and effects, of putrefaction become evident.

The casein is hydrolyzed; the milk is darkened. After the situation has somewhat stabilized, a more prolonged decomposition continues, mainly by eucaryotic fungi and various microbial enzymes.

Pasteurized Milk

Pasteurized milk does not promptly undergo souring because many of the lactose-fermenting species, being non-sporeformers, are killed by the heat of the process. The milk may then undergo *sweet curdling* caused by rennet formation by bacteria, especially proteolytic streptococci and aerobic spore-formers *(Bacillus).* Often pasteurized milk does not sour, but the casein undergoes digestion and, later, putrefaction by the proteolytic enterococci, sporeformers and other thermoduric, proteolytic saprophytes that survive pasteurization.

Significance of Coliform Organisms in Milk

Coliform organisms are always present in market milk *before* pasteurization. They are derived from hay, soil, dust, dung and utensils. They do not necessarily indicate human fecal contamination. However, these organisms do not survive in milk in significant numbers if pasteurization is properly carried out. It is possible to detect coliforms in milk by methods similar to those used for determining coliforms in water. Plating the milk in desoxycholate-lactose agar or violet-red bile agar (both selective media) are especially recommended. Red colonies may be counted as coliforms and transferred for *completed test* if desired.

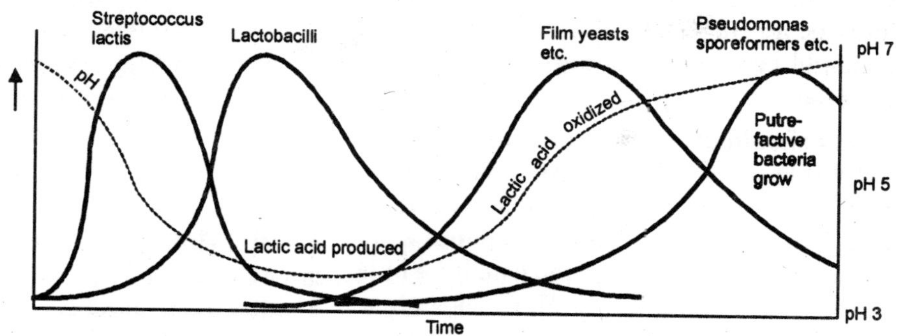

Fig. 6.4. Changes in numbers and types of microorganisms in raw milk held at summer temperature. For explanation see text. Note the initial decline in pH (increase in acidity) during fermentation of the lactose, with later rise in pH as the resulting lactic acid is metabolized and alkaline products of casein putrefaction accumulate.

The details for all important laboratory procedures in connection with dairy products are to be found in "Standard Methods for the Examination of Dairy Products" (see supplementary reading). Unless almost-surgical precautions are used, a few coliform organisms gain entrance to milk *after* pasteurization, during cooling or bottling. Small numbers are of little significance and are not inimical to health, but their numbers give a good indication as to post-pasteurization cleanliness and refrigeration or staleness.

Coliform organisms are particularly undesirable in milk to be used for cheese, since they cause rapid souring, gassy fermentation, and undesirable odors and flavors in cheese and other products made from the milk. If considerable numbers of coliforms (more than about 1 to 5 per milliliter) are found in pasteurized milk it is evident that: (a) the milk has not been properly pasteurized; or (b) it has been excessively contaminated after pasteurization by unclean conditions, possibly by sewage, feces or dung; or (c) the milk has been held unduly long above about 15° C after pasteurization; or (d) raw milk was mixed with it after pasteurization.

The Phosphatase Test

The presence of excessive numbers of coliform bacteria in pasteurized milk suggests staleness or improper pasteurization or illegal adulteration with raw milk. Concerning pasteurization, more definite and accurate information can rapidly be obtained by means of the *phosphatase test.* This test is based on the obstruction by pasteurization of the heat-sensitive enzyme, *phosphatase,* normally present in fresh milk.

The test for presence or absence of phosphatase is based on the power of phosphatase to liberate phenol from phosphoric-phenyl ester added to a sample of the milk. In this test, 0.5 ml. of milk sample is added to 5.0 ml. of *buffered substrate* (disodium phenyl phosphate buffered with $NaHCO_3 \cdot Na_2CO_3\text{-}2H_2O$) and held at 40° C for 15 minutes. If the milk is unpasteurized or insufficiently pasteurized, the phosphatase enzyme normally present in the milk will be active and will decompose the added phenyl phosphate, liberating phenol. The phenol turns blue if 2,6-dichloroquinone-chloroimide (CQC), with $CuSO_4$ as catalyst, is added to the mixture.

The appearance of blue color indicates the presence of free phenol liberated by undestroyed phosphatase in the milk. The color of the tested sample of milk is then compared with the color of standards containing known amounts of phenol and treated with the same reagent. When milk is pasteurized at 143° F for 30 minutes, 96 per cent of the enzyme is destroyed. Only a trace of blue color should appear. Heating above 145° F for 30 minutes insures complete inactivation of the phosphatase.

No blue color should develop. When milk has been underheated (in respect to either temperature or time) or when there is an admixture of raw milk afterward, phosphatase will be present in larger amounts than when the milk is properly processed, and a definite blue color appears in the phosphatase test. The phosphatase test can be made quantitative by comparing color with known standard. This test will detect 0.5 per cent raw milk mixed with pasteurized milk, or one degree below standard temperature, or five minutes of underheating during pasteurization. Color values (in the Sharer rapid method) greater than 0.5 µg of phenol per ml. of milk indicate progressive degrees of improper handling of milk.

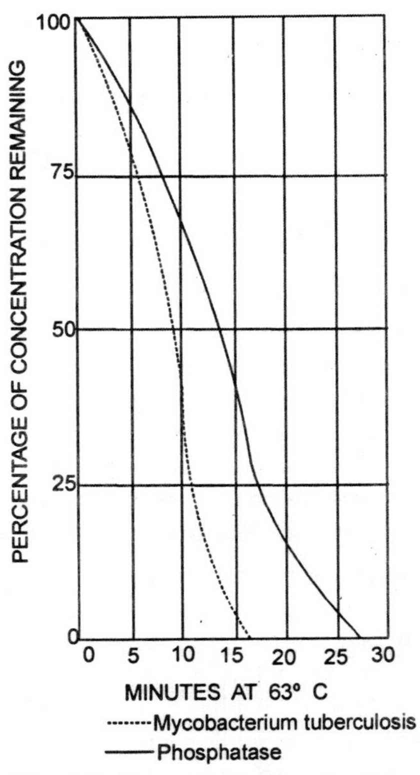

Fig. 6.5. Representative comparative time-concentration curves for destruction of pathogenic bacteria where represented by *Mycobacterium tuberculosis*) (———) and phosphatase (———) in milk during pasteurization at 63°C. Ordinates show approximate percentages of original concentrations or numbers remaining at 63° C after minutes numbered on abscissa. Note that at all times destruction of the bacteria is more rapid and extensive than destruction of the phoshatase.

Sanitary Significance of Phosphatase Test

In pasteurization, *Mycobacterium tuberculosis* (the most resistant of the nonsporeforming pathogens commonly found in milk) is destroyed more quickly than phosphatase. Therefore a heat treatment adequate to inactivate the enzyme likewise kills this organism and all other common pathogenic microorganism. A sample of milk that does not have more phosphatase present than the standard allows can be regarded as both safely pasteurized and free from subsequent contamination with raw milk.

Sources of Error

In some cases the phosphatase seems to become reactivated after proper pasteurization.

Certain bacteria can give falsely positive results in properly pasteurized milk because they produce a thermostable phosphatase before pasteurization. This remains active even after proper pasteurization.

ENUMERATION OF BACTERIA IN MILK

In order to have some measure of the conditions under which milk has been produced and handled and to have a legal control over its sanitary quality, health departments and dairymen have set up various standards by which to judge milk. Important among these standards is the number of bacteria present.

Plate Colony Count

Numbers may be determined by one or more of several methods, legally recognized procedures for several methods, legally recognized procedures for which are detailed in literature cited. One of these, the plate (colony) count, is closely analogous to the platecount procedure.

For counting bacteria in milk the plates may be incubated at 32° or 35° C for 48 hours for routine work, or at 7° C for 7 to 10 days to enumerate psychrophiles; at 55° C for 48 hours to enumerate thermophiles. The plating medium officially recognized for all of these, as well for enumerating the bacteria in drinking water and widely used for counting bacteria in other materials (*e.g.*, foods), contains: agar, 1.5 per cent; yeast extract, 2.5 per ceut; protein digest, 5.0 per cent; glucose, 1 per cent. This medium is generally called *Standard Methods Agar* and must conform in bacteriological quality to that of a standard lot of medium specified by the American Pubic Health Association in 1963.

Direct (Breed) Count

Bacteria may also be enumerated by the direct microscopic examination of milk in a smear. The smear is prepared by spreading exactly 0.01 ml of the sample on a slide over an area of exactly one square centimeter (100 mm^2). After staining with specially prepared methylene-blue solutions designed to remove fat globules, the smear is examined by means of a microscope calibrated with a stage micrometer so that the field is exactly 0.206 mm. in diameter. Such a field represents approximately 1/ 300,000 ml of milk. The number of bacteria seen in it must therefore be multiplied by the reciprocal of this fraction (the *microscopic factor* or MF) to determine the numbers of bacteria per milliliter:

$$MF = \frac{10,000}{\pi r^2}$$

Numerous large *clumps* of bacteria indicate unclean utensils. Many *pus cells* indicate infected udders. Streptococci and staphylococci indicate mastitis. It is common practice to report results in terms of *clump counts* and, in reporting, to translate the microscopic observations directly into such terms as "good," "bad" and "mastitis." The method is quick and inexpensive. It is applied mainly to unpasteurized milk, since heating causes many of the bacteria to the lose their staining properties.

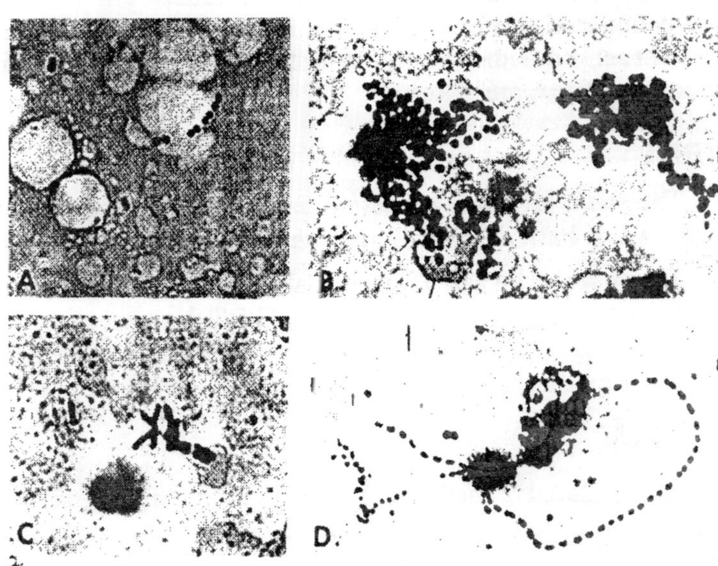

Fig. 6.6. Examples and probable sources of microorganisms detectable by direct microscopic examination of milk: *A*, *cocci*, milk probably improperly refrigerated; *B*, large masses of bacteria, improperly cleaned milk-handling utensils; *C*, spore-forming bacilli, probably from soil and dust contamination; *D*, pyogenic streptococci and a pus cell from cow with mastitis.

Numerical Relationship Between Counts

The *total* direct microscopic count is usually five to ten or more times as high as the plate count. This is because the total direct count enumerates individual cells, even those in clumps, and also dead bacteria. As in the plate count applied to water and soil, the plate count applied to milk enumerates only live bacteria capable of developing in the medium and under the environmental conditions used. In the plate count each clump, even though it may contain scores of live cells, counts only as a single bacterium since each clump forms only one colony. The plate count is usually closer to the clump count than to the total direct count.

QUALITY AND OXIDATION-REDUCTION POTENTIAL

The Reduction Test

Most actively growing bacteria cause a lowered oxidation-reduction (O-R) potential in their medium. The presence in milk of large enough numbers of growing bacteria to produce a significantly lowered O-R potential can be detected by the use of methylene blue because, with a lowering of the O-R potential to about-0.01 volt, this dye becomes colorless. The reduction (so-called "*reductase*") test is used principally with raw milk, and furnishes a rough but useful approximation of the number and kinds of living bacteria present. In performing the test, 10 ml. of milk sample are pipetted into a sterile tube and 1 ml. of a standard methylene blue solution (final concentration about 1:250,000) is added. The tube is closed with a rubber stopper and slowly inverted three times to mix. It is placed at 35.0° C in the water bath immediately. At the end of each hour during the test the tube is inverted once. Observations are made after 30 minutes, 1 hour, and later.

MBRT

The *methylene-blue reduction time* (MBRT) is the interval between the placing of the tubes in the water bath at 35.0° C and the disappearance of the blue color from the milk.

The shorter the MBRT, the greater the number of active bacteria in the milk and the lower its bacteriological quality. However, MBRT does not give an accurate *count* of bacteria present; only an overall measure of its bacteriological quality. Milk with an MBRT of 30 minutes is of very poor quality; an MBRT of six hours is very good.

Resazurin

A dye related to methylene blue, resazurin undergoes a *series* of color changes depending on O-R potential changes, whereas methylene blue changes only from blue to colorless. Resazurin, therefore, permits readings of *degrees* of reduction at shorter intervals than does methylene blue. In one procedure (the *l-hour test),* color of the milk, initially blue, is compared after one hour at 37° C with several exactly described *(Munsell)* color standards designated as 5P7/4 (mauve), PBP7/5.5 (purple), and so on. Milk showing no greater change during the hour than from blue to PBP7/5.5 is grade 1; to PRP7/8 (lavender), grade 2; and so on to grade 4, complete decolorization. In the *triple-reading* or *three-hour* test, three successive readings are made at one-hour intervals to see how long it takes to reach the color 5P7/4. High grade (acceptable) milk requires at least three hours (resazurin reduction time, or RRT, three hours).

FACTORS AFFECTING BACTERIOLOGICAL QUALITY

Bacteriophages in Milk

A factor of importance, especially to manufacturers of dairy products that depend on early, rapid and luxuriant growth of certain bacteria such as *Streptococcus lactis,* is the presence of bacteriophage lytic for that species. These phages are common in dust in and around dairies. They are known to interfere with many sorts of dairy work (cheese-making) that is dependent on bacterial growth. Very rigid aseptic technique in preparing and handling the pure *starter* cultures is necessary to eliminate the phages. Steel filters with triple layers of fiberglass have been successfully used to remove phages from air of dairy laboratories and work rooms.

Antibiotics and Disinfectants

Another important factor in bacteriological studies of milk is the possibility that the results have been influenced by (1) preservatives illegally added; (2) residues of disinfectants used to sanitize the dairy equipment; (3) antibiotics or other drugs used to control udder infections, and otherwise administered to cattle. Antibiotics, disinfectants and preservatives can interfere with growth of the bacteria used in the manufacture of cheese and butter, cultured milks (*e.g.,* yoghurt), and in such bacteriological controls as reduction tests and plate counts. In most manufacturing procedures, uninhibited growth of S. *cremoris, Leuconostoc citrovorum, Streptococcus lactis, Lactobacillus* species and other bacteria is essential.

Penicillin is especially undesirable in milk because it can produce severe allergic reactions in persons hypersensitive to it.

GRADES OF MILK

The actual numbers of bacteria permissible in milks of various grades vary in different cities. A good guide is the standard milk ordinance and code of the Public Health Service of the U. S. Department of Health, Education and Welfare. Various localities may have somewhat different standards.

Grode A Row Milk for Pasteurization

Grade A raw milk for pasteurization is raw milk from properly supervised producer dairies conforming to standards of sanitation of workers, cattle, premises and equipment as prescribed in the ordinance. Cattle and personnel must be free from diseases transmissible in milk. The bacterial plate count or the microscopic clump count of the milk, as delivered to the pasteurizing plant for pasteurization, shall not exceed 200,000 organisms per milliliter as determined by standard methods of the American Public Health Association or must have a MBRT of less than 5.5 hours or RRT of less than 2.75 hours.

Grade A Pasteurized Milk

In all cases Grade A pasteurized milk shall show efficient pasteurization as evidenced by satisfactory phosphatase test, and at no time after pasteurization and before delivery shall the milk have a bacterial plate count exceeding 30,000 per milliliter, or a coliform count exceeding 10 per milliliter, as determined by Standard Methods of the American Public Health Association.

Grade B Pasteurized Milk

Grade B pasteurized milk is pasteurized milk that does not meet the bacterial-count standard for grade A pasteurized milk, and certain other sanitary requirements. Such milk may be used in some commercial processes.

Most communities now permit the sale only of grade A milk, pasteurized.

Criteria of Good Milk

Qualify tests such as bacterial enumerations and reduction tests, as well as others, have a distinct value and usefulness from the standpoint of *cleanliness* but not necessarily with regard to *infection.* The present status of microbiology of milk was well summarized by Robertson, who said:

> None of the routine laboratory procedures for estimating the number of bacteria in milk will determine whether or not infectious bacteria are present. The best assurance of freedom from infectious bacteria is that provided by proper pasteurization of the milk. The best assurance of pasteurization is that demonstrated by a satisfactory phosphatase test on the bottled pasteurized milk. The best assurance of freedom from recontamination in freshly bottled milk after pasteurization is a satisfactory coliform test in 1.0 ml. portions of the bottled product.

Certified Milk

If milk is to be offered for sale unpasteurized, it is often required that it be produced only under very carefully supervised conditions. The American Association of Medical Milk Commissions has established rules and regulations concerning veterinary inspection of cows, especially for tuberculosis, brucellosis and infectious mastitis, and sanitation of barns and utensils. These very rigid regulations are often used by health departments and milk dealers in certifying qualified farms to produce such milk. It is usually called *Certified milk* or *baby milk.* The use of certified milk has much to recommend it, especially its cleanliness. The coliform standard for Certified milk—raw, is 10 per milliliter; for certified, pasteurized, 1 per milliliter. It is said to contain a larger proportion of certain vitamins essential for infants than milk which has been heated.

Most cities and states, as well as the A.A.M.M.C., require that all persons occupied in preparing Certified milk, or, indeed, any food for the public, be examined periodically for typhoid, paratyphoid and dysentery bacilli. Examinations for presence of organisms that cause diphtheria, tuberculosis, scarlet fever and other transmissible diseases are also required for Certified milk handlers.

CONCENTRATED AND DRIED MILKS

Milks from which part or all of the water has been withdrawn are termed *concentrated* or *dried.* Assuming that approved standards of cleanliness, freshness, sanitation and chemical content (fat, solids and so on) have been met in selecting the milk to be dehydrated, the microbiological quality of the finished product is determined largely by: (a) temperaturess and time of storage (if any) prior to processing; (b) times and temperatures of processing; (c) cleanliness of the apparatus and final containers; (d) time and temperature of final processing of canned milks.

Evaporated Milk

The raw milk is first cooled and may be clarified. Fat and solids contents are "adjusted" to meet required standards. The milk is then heated to boiling or nearly so (94 to 100° C) for about 20 minutes. This kills all but the most heat-resistant microorganisms. Water is driven off *in vacua* at about 55° C, a temperature which favors development of undesirable thermophiles, *e.g., Bacillus stearothermophilus* and which therefore requires careful bacteriological control.

After homogenization, the product is cooled and canned. It is "commercially sterilized" at about 117° C for about 15 minutes. Spoilage problems can result from inadequate heating and subsequent storage at unduly high temperatures that favor growth of spores of thermophilic molds and Bacillaceae.

Sweetened Condensed Milk

Milk can be doubly preserved by the addition of about 20 per cent of sucrose or glucose, or both, to whole milk, and by subsequent heating at temperatures near boiling and by partial dehydration as with evaporated milk.

Nonfat Dried Milk

Nonfat dried milk is prepared by preliminary steps similar to those used for evaporated milk. The fat is removed by high-speed cream separators. The water is partly removed by preliminary heating (about 85° C) and then by: (a) spraying the milk as a mist into a current of hot (about 120° C) air in a closed chamber; or (b) by spreading the milk as a thin film on hot rollers or drums—at about 145° C if no vacuum is used, or at less than 100° C in a vacuum chamber.

The dried *flakes* from hot drums or the *powder* from sprayed milk are packaged to prevent access of moisture. Dried milks should be kept dry and cool at all times or they may spoil since they are not sterile. As previously pointed out, dry heat is an inefficient sterilant. Microorganisms surviving the process are mainly Bacillaceae and thermoduric streptococci, lacto-bacilli, micrococci and species of *Microbacterium*. Nonsporing pathogens are eliminated. The presence of coliform bacteria or of pathogenic streptococci or other heat-sensitive microorganisms is of the same sinister significance that it is in pasteurized whole milk.

Reconstitution

In reconstituting dried or evaporated milk it is clearly desirable to use clean, cool and hygienically acceptable water and utensils. The reconstituted milk should be kept and handled under the same conditions of sanitation and refrigeration as are recommended for whole milk.

Standards of Quality

Quality standards for dewatered milks are similar to those for whole milk, bearing in mind the changes in flora caused by heating. The standards are established by the U. S. Department of Agriculture, The American Dry Milk Institute and the Evaporated Milk Association (Evaporated Milk Industry Sanitary Standards Code, Chicago, 1ll). Standards for various grades of dried milk, for example, are 50,000 per gram for "extra" quality and 100,000 per gram for "standard" quality. Counting procedures are the same as for fresh milk after the dried milk is reconstituted with sterile water.

SOME MANUFACTURED DAIRY PRODUCTS

Market milk contains numerous species of microorganisms in varying numbers whose uncontrolled action is too unreliable to serve as a basis for commercial operations requiring uniformity of product. Pure-culture inocula (called *mother* or *stock cultures)* of constant properties are essential to continued success in this highly competitive field. For these purposes *Lyophilized* ("freezedried"), or other pure, stock cultures of desired organisms may be maintained in the dairy if a competent bacteriologist and adequate laboratory facilities are available. Otherwise it is best to obtain stock cultures from dairy-supply houses. Cultures of species of lactic streptococci, *Leuconostoc* and *Lactobacillus* are used especially.

Starter Cultures

In practice mother cultures in about 2 per cent volume are added as nearly aseptically as possible to about 600 ml. of sterile or very-low-count milk (previously heated 30 minutes at about 88 °C and cooled to 21 °C, and incubated. The lactic organisms soon outgrow other species, if any are present. This culture is called a *starter*. It may be used to inoculate a tank-size batch of milk or cream for butter or cheese, or to inoculate a still larger lot of starter.

Fresh, high-quality (low-count) pasteurized milk is brought *quickly* to the desired incubating temperature. A large, *preemptive,* virtually pure, starter inoculum of vigorously growing young cells of the desired lactic organism is added and thoroughly mixed with the milk or cream. Before the other bacteria in the milk have time to recover from their previous refrigeration or pasteurization and overcome their lag phase, the acidity quickly produced by the actively metabolizing added lactic starter suppresses them.

BUTTER

Butter is generally made by churning cream that has been soured by lactic acid bacteria.

Two species of bacteria, each with a distinct function, are added together to the cream simultaneously. *Leuconostoc citrovorum* is depended on for flavor, *Streptococcus cremoris* or S. *lactis* is selected primarily for *rapid,* initial lactic acid production. If high acidity (pH 4.3) is not produced promptly, numerous undesirable contaminants may grow excessively. When the pH reaches about 4.3 *Leuconostoc* ceases growth, but its enzymes attack the citrates in the milk and produce diacetyl. This substance gives butter and similar products their characteristic buttery flavor and aroma. Neither *S. cremoris* nor *Leuconostoc* alone can produce the desired result in commercial practice.

CHEESE

Cheeses may be divided into three general types: (a) soft- or cottage-type cheese, and cream cheese (these are eaten in a fresh or unripened state); (b) hard- or rennet-curd cheese, including Roquefort, American cheddar-type ("rat-bait"), Edam, and Swiss (these are *ripened* by the enzymes and slow growth of bacteria or molds or both, which cause some, but not extensive proteolysis); (c) soft or semisoft rennet-curd cheese, of which Camembert, Limburger and Liederkranz are types (these are ripened by proteolytic and lipolytic organisms which soften the curd and give it flavors). The hardness of cheese depends to some extent on moisture and fat content as well as on heating and acidity of the curd, draining, salting and conditions of storage.

Soft Acid-Curd Cheese

In making cottage cheese, starters containing mixtures of *Leuconostoc citrovorum, L. dextranicum, S. lactis* and the like are added to pasteurized milk. These ferment the lactose, *Leuconostoc* adding flavor and aroma. The lactic acid coagulates the casein. Rennet may be added to hasten the coagulation and make the curd firmer. The curd is cut into small cubes. To firm the curd and separate it from the whey, the mass is heated slowly to about 50 °C and held so for 30 minutes. Water is added; the curd settles. The water, with the whey, is drained off and the curd is pushed into heaps to drain. It is washed a second time with water and drained. About 0.5 per cent salt is added. Just before packaging many manufactures add a little cream.

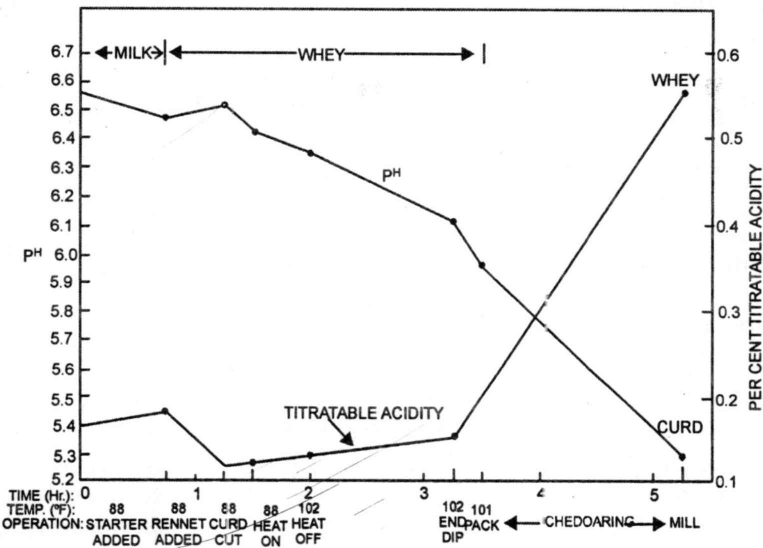

Fig. 6.7. Cheddar cheese: changes in temperature and in acidity as measured by determination of pH and by titration of total acid, during successive early phases, from addition of starter (low left) to milling (lower right); about 5 hours in all. At the time starter is added at 88° F, the acidity is low (pH 6.58; titrable acidity 0.18 per cent). As lactose is fermented acidity quickly increases. At pH about 6.46, rennet is added and the curd forms. It is soon cut, heated, and the whey is drawn off (dipped). Cheddaring and milling then proceed. Further increases in acidity occur during the ripening process.

Table 6.2. Types of Natural* Cheese

Representative Cheeses	*Distinctive Organisms in Ripening Flora*
Soft:	
Cottage	not ripened
Cream	
Liederkranz	*Strep. Liquefaciens, Brevibacterium*
Camembert	*Penicillium camemberti, Brevibacterium*
Semisoft:	
Blue (or Bleu)	
Roquefort	*Penicillium* strains such as
Gorgonzola	*P. roqueforti*
Hard	
Swiss	*Propionibacter species*
Cheddar	Lactic group, *Geotrichum*
Parmesan	Lactic group (brine cured)

* Not processed artificially. (Pasteurized and processed cheeses and cheese spreads are not included since they are made almost entirely from the natural cheeses such as those listed above.)

Hard-Curd Cheese

In the preliminary stages, nearly all natural (*i.e.,* not "processed") cheeses are much alike. Differences result from different methods of treating the curd: degree of acidity, addition of different amounts of salt, special ripening microorganisms, moisture, temperature and humidity of ripening and other factors.

For yellow cheeses of the Cheddar type, color is added. After a slight acidity has developed, rennet is added to make an elastic, rubbery curd which is later cut into pieces about one inch in diameter warmed to about 35° C. The curd becomes firmer and the whey separates and is drained off and may be used for stock feed. The clumped masses of firm curd are chopped *(milled)* again, and piled up to press out whey. This is called *cheddaring* in Cheddar cheese making. The curd is again milled, and then is salted, drained and pressed in hoops to cure. Curing of Cheddar cheese proceeds at about 15° C. It becomes "sharper" with aging.

Curing or Ripening of Hard Cheeses

During the curing process of Cheddar and other hard cheeses, various microorganisms, the varieties depending on the kind of cheese, continue a slow fermentative, lipolytic and proteolytic action, the products of these processes yielding the characteristic flavors, textures and aromas of various cheeses. Prominent among these flavors are diacetyl, lactic, butyric, caproic and acetic acids, and various amines, as well as various esters such as those that give flavors to ripe fruit juices. In addition, since many of these organisms synthesize vitamins, especially nicotinic acid and vitamins of the B complex, the nutritive value of the cured (or ripened) cheeses is increased.

Gas formers such as species of *Clostridium* and Escherichiae are undesirable because they produce gassy cheeses off flavors; they may be especially active in the early stages. They generally occur in milk of poor quality.

Swiss cheese is heated to 50° C after cutting the curd. Starters therefore usually contain the *thermoduric* lactics: *Streptococcus thermophilis* and *Lacfobacillus bulgaricus* or *L. lactis,* as well as *Propionibacterium shermanii* and *P. freudenreichii.*

The cheese is soaked in 23 per cent brine for some days at 13° C. Propionibacteria are then favored by incubation at 22° C. Later the cheese is ripened for months at about 13° C.

The "eyes" in Swiss cheese are a result of the production of carbon dioxide by species of *Propionibacterium,* while its bitter-sweet flavor is caused in part by the formation of glycerol, propionic and succinic acids by *Propionibactenrim* species while ripening.

Semisoft cheese, such as Roquefort, Gorgonzola or Blue (or *Bleu*), contains much fat and as high as 5 per cent salt and relatively little moisture. The high salt content prevents continued growth of most bacteria, as does the low ripening temperature (7° to 8° C) and humidity (60 per cent). A species of mold *(Penicillium roqueforti)* is introduced by the inoculation of spores into the milk or into the curd as it is put into hoops for ripening. The mold grows in the interior, producing the masses of blue-green conidia and the sharp flavor so characteristic of this type of cheese. As the mold is aerobic, perforations are made in the cheese to aerate the interior.

Soft Cheeses

Limburger, Liederkranz and Camembert cheeses are cured mainly by the growth of organisms in a red-orange, slimy coating on the outer surface. Numbers of microorganisms in this slime sometimes exceed 10 billion per gram. In Limburger cheese, yeasts *(Geotrichum* species) begin to grow on the surface after subsidence of the initial acidity caused by lactic organisms. They persist in the surface slime for about a week. *Brevibacterium linens* and *B. crythrogenes* then grow all over the surface, forming a reddish brown coating, or *"smear,"* commonly seen on soft and semisoft cheese (Brick, Camembert). Camembert cheese is sometimes inoculated on the outer surface (or before curd-formation) with a pure culture of the mold, *P. camemberti.* The enzymes of the various micrococci, yeasts or molds in the slime penetrate into the interior of the cheese, producing the flavors, softening, and the famous aroma of Limburger and similar cheese.

FERMENTED MILK BEVERAGES

In certain countries lactobacilli have been used for centuries in combination with certain yeasts and streptococci to produce foods of fermented milk. The *yoghurt* of eastern central Europe (now available in all grocery stores), the *busa* of Turkestan, the *kefir* of the Cossacks, the *koumiss* of central Asia and the *leben* of Egypt are examples of these. Formerly, the microbial nature of these processes was unknown. In all of these fermented milks lactobacilli act in company with other microorganisms : yeasts, lactic streptococci and various rods. For example, kefir, made from milk of various domestic animals, is prepared by putting *kefir grams* (small, cauliflower-like masses) into the milk. These grains consist of dried masses of lactobacilli *(L. brevis),* yeasts *(Sacch. delbruckii), Streptococcus lactis* and probably other lactic organisms held together in a matrix of coagulated casein and bacterial polysaccharide gum. The kefir grains increase in size and break apart as the fermentation proceeds. The combined growth of the mixed flora yields a characteristically flavored, soured milk containing small amounts of alcohol. The kefir grains are found in the bottom of the vessels of fermented milk.

Yoghurt

In the United States this is made of pasteurized milk thickened by rennet or addition of dried milk, and soured by species of *Lactobacillus.*

Acidophilus Milk

Milk soured by *Lactabacillus acidophilus* is thought by some to have medicinal properties in the intestinal tract. *Bulgarin buttermilk,* a similar beverage, is prepared with pure-culture starters of *Lactobacillus bulgaricus.*

Buttermilk

Much of the product commonly sold in the United States as buttermilk is in reality pasteurized skim or whole milk soured mainly by Streptococcus cremoris with Leuconostoc citrovorum and then beaten so as to produce a smooth, creamy beverage. It is a pleasant, nourishing drink. Addition of 0.15 per cent of citric acid to the milk result in formation of increased flavor because diacetyl is produced by *Leuconostoc citrovorum.*

CHAPTER

7 Biochemistry of Parasites

The youngest basic-science diseipline applied to the study of parasite and parasitic diseases is molecular biology. The ability to clone parasite genes and express them in *Escherichia coli* or yeast has provided the first glimmer of hope for the development of vaccines against some of the parasites. Before the availability of recombinant DNA technology, there was no conceivable way in which sufficient antigenic material could be produced from any of the medically important parasites for use in vaccines. In only a few years the very rapid application of the new methods led to cloning of plasmodial genes, large-scale antigen production, and phase I and phase II clinical trials of an antisporozoite vaccine. Although recombinant vaccines against other plasmodial stages and other parasites have not yet progressed this far, their development is underway. From experience with the sporozoite vaccine, it is clear that more than cloning and expressing genes of interest may be required to obtain a successful vaccine. Yet the availability of large amounts of antigenic material and the primary sequence data that can be derived by application of molecular techniques have revolutionized the field.

Perhaps the only alternatives to this approach in vaccine development, given the paucity of antigenic material available directly from the parasite, are antiidiotype vaccines (such vaccines employ antibodies [antiidiotypic antibodies] that mimic the structural conformation of an epitope.) An additional practical application of molecular technology has been in the area of rapid diagnosis. With specific cDN A probes, it is now possible not only to determine the presence of leishmania in skin biopsies, but also to speciate them. The effectiveness of this diagnostic method is facilitated by the recognition that the extrachromosomal DNA of the kinetoplast contains so-called minicircles, which are highly divergent within the genus, allowing for species discrimination by kinetoplast DNA (kDNA) hybridization techniques. This achievement is an important practical development because conventional diagnostic methods, which require parasite identification by microscopy or isolation of parasites in culture of biopsy material, are cumbersome and often unreliable. Plasmodia also can be detected in blood in this manner.

Although the current level of sensitivity of the molecular method of diagnosis is no greater than that achievable by microscopic examination of blood smears, modifications (that might include amplification by the polymerase chain reaction method) may increase sensitivity. The application of molecular techniques to malaria diagnosis presumably will have greater application in epidemiological surveillance efforts than in clinical practice in areas where malaria is endemic.

Application of molecular biology approaches to the pharmacology of certain parasitic diseases has spurred great excitement and optimism for the idea of developing new antimicrobials by rational design rather than by the traditional and inefficient approach of drug screening. The strategy of this approach is to identify critical enzymes of the parasite that are distinctive from that of the host. Isolating the gene encoding the enzyme, expressing it in *E. coli*, and gleaning the fine structure of the protein by X-ray crystalography—especially its substrate-binding domain—provide the data base.

With this information, drugs might then be designed that can bind avidly to this domain and competitively block access by the substrate. Insights into the genetic basis of drug resistance also have been revealing. For example, resistance to folate antimetabolites in *Leishmania* species has been found to result from extraordinary amplification of the gene that encodes the enzymes that are the targets of such antimicrobial agents. Molecular pharmacology, as this new discipline is called, could hold great promise for winning the race of medicine against parasites such as *Plasmodia,* which can develop multidrug resistance at a rate that exceeds our ability, by conventional means, to find effective alternative agents. One intriguing strategy is to prepare derivatized oligonucleotides that aneal with specific parasite mRNA, prevent translation, and thereby incapacitate parasite metabolism. Although practical application of the new techniques has understandably received the greatest attention, emphasis on basic aspects of the molecular biology of certain parasites has uncovered some unexpected findings of broad biological and evolutionary significance.

For example, comparisons of primary ribosomal RNA sequences suggest that the plasmodia causing human malaria are more closely related, in evolutionary terms, to rodent and avian parasites than to parasites causing malaria in other primates. Similar ribosomal RNA analysis in Kinetoplastida (such as African trypanosomes and *Leishmania)* suggests that these eukaryotes may be among the most distantly divergent, in evolutionary terms, perhaps having branched even before plants. Another striking finding has been the unique manner in which these organisms transcribe genetic information: by a noncontiguous construction of mRNA. In contrast to the conventional cis-splicing of most eukaryotes, these protozoa can carry out trans-splicing, a process whereby mRNA transcribed from genome on remote sites on a chromosome—or even on different chromosomes—forms a hybrid mRNA that then is translated. Many interesting biological questions about specific parasites have yet to be subjected to molecular scrutiny. One of the most obvious is gene regulation in the transformation of the parasite from one stage to another. Efforts in this direction have been initiated.

For example, expression of the gene for α- and β-tubulin has been found to be expressed as leishmania transform from the intracellular (aflagellate) amastigote stage to the extracellular (flagellate) promastigote stage. Tubulin is the major protein in promastigotes, perhaps not surprisingly so in view of the presence of an extensive subpellicular microtubular system that includes the flagellum. What triggers gene tubulin gene expression now awaits definition. Along similar lines, an important challenge to the biology of malaria is to determine the molecular basis of gametocytogenesis (the emergence from plasmodial merozoites of gametocytes, the sexual stages) Since gametocytes are responsible for initiating transmission in the mosquito, knowledge that could lead to strategies for preventing their genesis in the—human host could have distinct practical value. A molecular understanding of the manner whereby specific environmental factors trigger interstage transformation in schistosomes is but one example of a worthy goal in molecular helminthology. Indeed one particularly exciting aspect of parasite molecular biology is that contemplation of parasite cell biology can engender numerous interesting questions that are now, or soon will be, amenable to molecular biological investigation.

Furthermore, the parasitic relationship may provide a fertile ground for gaining insights into fundamental evolutionary aspects of gene structure and regulation. As the molecular biology of parasites matures, it is likely that many important discoveries will be made as a result of asking the right biological questions—with the assistance of some serendipity. Inasmuch as none of the disciplines in modern parasite biology stands alone, new knowledge in the cellular biology of parasites and the immunology of parasitic diseases will no doubt influence perspectives and research directions in molecular biology studies. By the same token, these other areas can be expected to be influenced by progress in basic molecular parasitology.

BIOCHEMISTRY OF PLASMODIUM

Protozoan parasites of the genus *Plasmodium* are the causative agent of malaria, a disease that today affects more than 200 million people in the tropics. Of the four human malarias, those due to *Plasmodium falciparum* and *Plasmodium vivax* are the most common. This chapter will focus on the role of molecular biology in the efforts to understand and eradicate malaria and discuss the molecular biology relating to diagnosis, taxonomy, genetics, gene analysis, and vaccine production.

Evolution

Members of the genus *Plasmodium* have historically been grouped according to the hosts that they infect, then subdivided according to morphological and biological characteristics. Thus, for example, the species are classified as primate, rodent, avian, and lizard malarias. This classification scheme implies that the parasites have evolved with their hosts and that there is a greater relatedness among parasites in related hosts than among those greatly separated in evolution. A major exception to this scheme is the human malarial organism P. *falciparum;* this parasite species appears to be more closely related to parasites that cause avian and rodent malarias than to those that cause primate malaria. This exception was suggested on the basis of the similarities in the structure and composition of the genomes of *Plasmodia* causing avian and rodent malaria with those of *P. falciparum.* More recently these relationships have been confirmed by comparing the primary sequence of ribosomal RNAs from the different species of malarial parasite. Ribosomal RNA sequence comparisons constitute a generally accepted measure of evolutionary relationship and thereby add weight to these conclusions.

Diagnosis of Malaria

In the standard, time-consuming diagnosis of malaria, a skilled technician does a microscopic examination of blood smears. In many areas diagnosis is omitted altogether; people with malaria-like symptoms are simply treated with antimalarial drugs empirically. The detection of malarial infections in patients by the development and use of a DNA hybridization assay could be a valuable tool in the control of the disease. The success of these assays depends on finding DNA probes that hybridize to the DNA of a particular malarial parasite in a sample of blood and not to the DNA of the host or of other microbes that might be in the same sample. In comparison to molecular diagnosis, microscopy by a skilled technician can detect a concentration of about 20 parasites/1 microliter (µl) of blood. In addition, he or she can determine by microscopy the species of the parasite and degree of anemia. Refining of a molecular diagnosis for malaria to obtain this information will be difficult but clearly possible.

The degree of sensitivity that has already been reported for molecular probes approaches 50 parasites per microliter, and increased sensitivity in assays of this type can be expected in the future. Species-specific probes certainly will be available soon; it is clear that potential diagnostic probes include ones that detect the highly repetitive sequences of *P. folciparum* not found in the other species. It is even possible that similar, although less sensitive, molecular assays may be developed for detecting such other characteristics of a parasite as drug resistance. Perhaps the strength of molecular diagnosis is not in performing better than a skilled microscopist but in allowing a large number of assays to be done with great consistency. A perfected molecular form of diagnosis should allow relatively untrained personnel to do a large number of assays every day with great accuracy.

Plasmodium Genes

This chapter focuses on the three *Plasmodium* genes that have received the greatest attention and discusses reasons for interest in a particular gene as well as the technology involved in its study. An overlying theme in this discussion is how gene structure relates to rapid changes in immunodominant portions of a protein.

The Circumsporozoite Protein

Interest in the circumsporozoite protein (CSP) of malarial parasites is related mainly to the need for a malarial vaccine. The important sporozoite antigen, being studied for its ability to elicit a protective response, is a single protein, the circumsporozoite (CS) protein. It covers the entire surface of the sporozoite. The reaction of a fluorescein-tagged anti-CS protein antibody with the sporozoite surface is shown in Fig. 7.1. The dominant epitope recognized by antibodies to the CS protein and the one involved in the protective response is included in a single immunodominant region of the protein.

The fact that a single antigen could elicit a protective response suggested that a molecular biological approach could provide a source of material for a vaccine. The important gene would be cloned and the product produced by microbes in the laboratory.

Cloning of the Circumsporozoite Gene

Cloning of the CS protein gene presented a particular problem to molecular biologists because of the lack of availability of the sporozoite stage of the parasite. The first CS protein gene was cloned from the simian malaria *P. knowlesi*. To do this, a large number of sporozoites were collected from the salivary gland of infected mosquitoes. *P. knowlesi* was chosen because in the laboratory it produces approximately 10 times more sporozoites in an infected mosquito than any of the human malarial species. From these sporozoites, mRNA was isolated and used to make a cDNA library in *E. coli* that contained copies of sporozoite-specific messages. The cDNA containing a fragment of the CS protein gene was identified with a monoclonal antibody to the CS protein. This gene segment was then used to identify the entire CS protein gene. Unfortunately, the P. *knowlesi* gene was of little use in identifying CS genes of other species of malaria, because there were no apparent sequence similarities among genes carried by different species of parasites.

Fig. 7.1. Immunofluorescent microscopy of a sporozoite with the use of anticircumsporozoite antibody. The distribution of the fluorescinated antibody reveals that the circumsporozoite antigen is distributed over the surface of the parasite. Monoclonal antibodies to the circumsporozoite protein identified this surface component as the immunodominant epitope and facilitated cloning of the gene that encodes the protein.

It therefore appeared that the most direct approach to isolating genes from different species was to start from the beginning and collect sporozoites from each of the parasites of interest. Asexual parasites were grown in culture and their DNA was used to prepare an expression library containing intact genes, including those from all stages of the parasite's life cycle. This work was made possible by an earlier discovery indicating that the mung bean nuclease cuts *Plasmodium* DNA in areas flanking genes, thereby reducing the genome to fragments containing intact genes. Bacterial clones containing the CS protein genes were detected by anti-CS protein antibody and standard screening techniques. The clones detected with this approach contained the complete CS protein gene of *P. falciparum.*

Sequence Evolution of the Circumsporozoite Protein

The cloning and DNA sequencing of the *P. falciparum* and *P. vivax* CS protein gene allowed comparison of these sequences with that of the *P. knowmlesi* gene. Although there were only two small regions of DNA sequence homology among the three, there did appear to be a common structural relationship. Evolutionary conservation of structure at some level was expected among analogous proteins that are presumably responsible for the same functions. For example, concentrations of hydrophobic amino acids at the amino and carboxy terminus of the protein probably serve as signal and anchor regions for the CS proteins. Regions of the three CS proteins that contain a concentration of charged amino acids can also be aligned and probably result from structural restraints on the molecule. The two regions of sequence homology between the *P. falciparum* gene and the other two CS genes are small and flank the repeated or immunodominant portion of the molecule. It seems possible that the strong conservation of sequence in these two areas reflected their involvement in some essential parasite function such as reception for liver invasion.

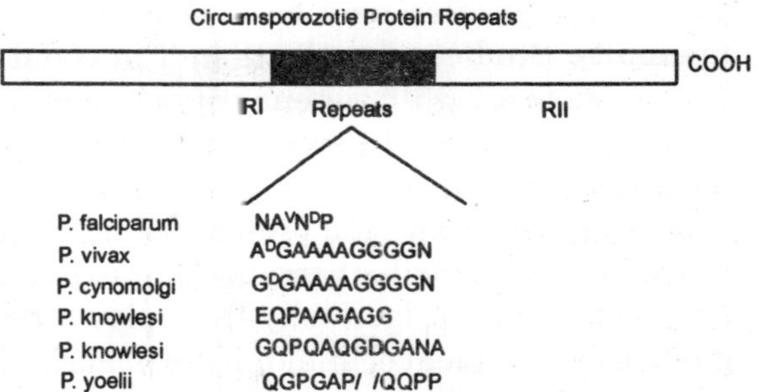

Fig. 7.2. The immunodominant repetitive epitopes in the circumsporozoite protein of several different species of *Plasmodia*. In contrast to the species specificity of the repeating units, the regions (RI and RII) that flank them are conserved in the genus.

The sequence homology between the *P. vivax* and *P. knowlesi* genes is much more extensive and extends over the molecule, except for the central repeating area. In general, at least small regions of conserved sequence in each of the CSP genes flank a species-specific repeating epitope. In schematic of the CS gene shows the position of the highly conserved regions (regions I and II) surrounding the species-specific repeats. This repeated area in the center of the molecule is the immunodominant area of the protein and the subject of the most intensive investigation regarding vaccines. Some typical DNA sequencing data from the analysis of a CS protein gene are shown in Fig. 7.3. The data shown in the figure have been read from the arrow on the left in are ascending order to the arrow on the right. The translation of raw data to the gene sequence and then to the protein sequence of the CS repeating epitope is shown in the Fig. 7.3. Notice that although the primary DNA sequence varies somewhat from repeat to repeat, only one nucleotide change in the five segments results in an amino acid change.

The other changes are in silent third positions, which do not result in the amino acid change. The extent and rate of variability in the repeat region are important questions in vaccine research. Changes in the immunodominant target would nullify the effectiveness of a vaccine. Perhaps much can be learned from the work done on *P. falciparum* S-antigen, discussed later, which has a repeating immunodominant region similar to that found in the CS protein. Clearly the repeat area of S-antigen varies so rapidly that there are antigenically distinct populations of *P. falciparum*. The CS proteins do not appear to vary as rapidly as the S-antigens, but there are indications of a similar type of drift. The analysis of CS-protein genes from malarial parasites that infect monkeys has shown intraspecies variation of the immunodominant region of the protein. Although to date no variation has been seen in isolates of *P. vivax* or *P. falciparum,* this work raises at least two questions. Are there reasons to believe that the *P. falciparum* CS protein gene will be more stable than the CS protein of corresponding monkey malarias? What effect would small subpopulations containing variant CS protein epitopes have on the effectiveness of a vaccine?

The S-Antigen

The S-antigen is a soluble, heat stable protein that is released from the malarial infected erythrocyte as it ruptures. S-antigens circulate in the blood of infected individuals, and the serum of some immune individuals recognize these molecules. Although the expression of a single type of S-antigen is a stable genetic characteristic of a parasite, there are probably a large number of immunologically distinct types. Apparently these immunologically distinct types of S-antigen have a worldwide distribution, since antibody to a mixture of types can be found in a single immune individual and a given type of S-antigen can be found in endemic areas throughout the world. If, as evidence suggests, the frequency of a single type of S-antigen in a particular area varies greatly with time, the dominant form in a single isolated area may vary from year.

The S-antigen also may be a target for protective immunity. Studies of patients in The Gambia. West Africa, show a high frequency of different S-antigen serotypes occurring in sequential natural infection, which could indicate that immunity to parasites expressing one S-antigen protect the individual from reinfection by parasites carrying that particular type. Antibodies to S-antigen may then play a role in the gradual development of immunity to malaria. This type of immunity would not extend to protection from the parasites expressing the S-antigen types that the host immune system has not seen before. It may also be that the frequency of individual S-antigens oscillates in a particular area in response to the development of immunity to specific S-antigens by members of the host population. The general approach that resulted in the isolation of the S-antigen gene was to isolate any gene that produced a product recognized by antibodies of people who are immune to malaria.

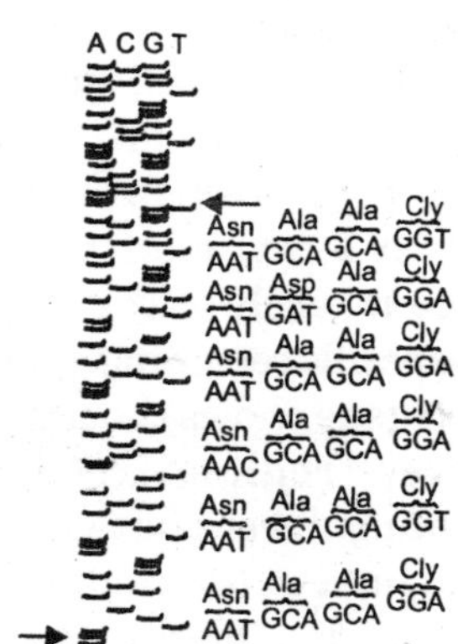

Fig. 7.3. Representative DNA sequence analysis of a circumsporozoite protein gene. The sequence has been read in ascending order from the arrow on the left to the arrow on the right. Translation of the codons into the amino acid sequence is shown. This type of analysis permits derivation of the primary peptide structure from knowledge of the gene structure.

Individual genes were then characterized and identified by assessing the properties of the protein product of the gene. The S-antigen gene was identified in this manner because the protein produced in the parasite had previously been extensively characterized. The series of events leading to the cloning of the S-antigen gene proceeded in the following manner. A cDNA library was made with mRNA from a mixture of the asexual stages of *P. falciparum.* The library vector, bacteriophage lambda-gt 11, allowed the expression of *P. falciparum* cDNAs as peptides after infection in *E. coli.* Bacterial colonies expressing the *P. falciparum* proteins encoded on the lambda vector were detected with serum from humans that were immune to *P. falciparum* infection. This procedure allowed these researchers to isolate a number of bacterial colonies that contained clones of individual *P. falciparum* genes whose proteins were being recognized by the host immune system. Immunologically specific serum recognizing only a single *P. falciparum* product could then be produced by injecting the bacterially produced product into mice. Having monospecific serum allowed researchers to localize the position of the unknown *P. falciparum* protein on the malarial parasite by an immunofluorescence assay and to characterize its time of expression. It was clear from the immunofluorescence study that one clone from the library produced a product that was synthesized in schizonts, that seemed to be released upon rupture of the infected erythrocyte, and that was isolate-specific.

All these features indicated that the clone carried an S-antigen gene. They then used their monospecific serum to identify *P. falciparum* proteins on Western blots. They founi that it recognized single proteins of variable size in several cloned lines of parasites. These parasite lines produced S-antigens that were antigenically similar but physically distinguishable by size. By correlating specificity and size they were able to match the serum's specificity to the S-antigen protein. This series of experiments selected for clones carrying and expressing the most immunogenic portion of the S-antigen molecule. When the cDNA clone of S-antigen was analyzed it was shown to contain a DNA sequence that encoded a sequence of 11 amino acids repeated 23 times. The only nucleotide sequence variation that occurred in the repeating element was located in the third position of codons and did not alter the encoded repeating peptide unit.

Although it was clear from this study that there were other areas of the S-antigen molecule that were not composed of this repeat, it was suggested that the repeats were the immunodominant portion of the protein. This work also showed that the nucleotide sequence was isolate-specific. It could be found only in strains of *P. falciparum* expressing one particular antigen type, which indicated that the difference between S-antigen genes was not simply the result of alteration of an existing sequence by point mutation insertions or deletions. The comparison of the gene sequences of antigenically distinct S-antigens of *P. falciparum* has been detailed. These studies showed that the repeating element of S-antigen was bounded on either side by a nonrepeating sequence. Although the repeating elements of the two genes were different, the nonrepeating sequences were very similar. This pattern is similar to that found in comparisons of CSP genes from different species of *Plasmodium*. Hence, S-antigen genes are variable between strains and are therefore varying faster than the species-specific CSP genes.

The mode of change, however, appears to be similar. Variation in the S-antigen form does not appear to be occurring by rearrangement of sequences, in that the repeated sequence of one gene cannot be found in the genome of another line of parasite expressing a different S-antigen. Although the precise mechanism that generates diverse forms of S-antigen is not understood, it seems to involve simple sequence alterations as one repeating unit of a gene is spread to other units. For example, any changes to the common theme or base pattern of a repeating element are generally reproduced exactly in several other units in the string of repeats. Since it is unlikely to alter each of several units identically by chance, a mechanism of spreading of single sequence alterations has been proposed. Sequences surrounding the coding portions of the genes were homologous; which indicates that the genes are allelic and that polymorphism exists, because the repeated areas can change more rapidly than their surroundings. The study of the S-antigen genes has revealed several important tacts. (1) The structure of the immunodominant epitope has been elucidated. This information has allowed researchers to compare the S-antigen gene structure with other DNA structures that are constructed similarly. (2) Variations in types of S-antigen are not linked to rearrangement and expression of preexisting sequences, as is the case with mammalian immunoglobulin genes or the trypanosome variant surface antigen.

A single parasite therefore does not have an infinite repertoire of changes with regard to the S-antigen. (3) The definition of differences between antigenically distinct types indicates that different S-antigens consist of alleles of a single gene rather than a family of different genes. (4) The mechanism that creates diversity has been partially revealed. We know that the repeating immunodominant epitope is changing faster than the rest of the molecule. The mechanism apparently disseminates local changes in one repeat unit to neighboring units. The mechanism and rate of change of the S-antigen may give us a valuable lesson by providing an extreme example of the way in which other repetitive antigens can vary.

It is unlikely that the accumulation of point mutations in the repeating element of S-antigen occurs any more rapidly than it does in any other repeating element. The question is, Why is the result of this spreading of the initial base change through the S-antigen repeats apparently more frequent? In a population of parasites a great variety of amino acid sequences may be acceptable in S-antigen, whereas sequence variation in the circumsporozoite gene, for example, may be selected against. It also may be that the rate of change is inversely proportional to the amount of host immune pressure. In any case, the different repetitive antigens of *Plasmodium* probably change by similar mechanisms.

Histidine-Rich Proteins

Proteins of unknown function in several species of malarial parasites are characterized by an astonishingly high content of histidine. *Plasmodium lophurae* (an avian parasite) synthesizes one such protein (HRP), which has a histidine content of 73 percent. It may constitute as much as 50 percent of the total protein in asexual forms of the parasite. The protein is found in cytoplasmic granules. One of the things about the HRP that has particularly interested researchers is the report that this antigen protects ducklings against infection with P. *lophurae.* Another parasite, P. *falciparum,* has at least three histidine-rich proteins. One of these proteins (PfHRP-I) is associated with the formation or function of the knoblike protrusions on *P. falciparum* -infected erythrocytes. An example of localization of PfHRP-I on these knoblike protrusions by immunoelectron microscopy is shown in Fig. 7.4.

In panel A, an antibody specific for the PfHRP-I protein has been labeled with an opaque tag, allowed to react with an intact parasite, and the antibody reactivity localized at the knob is visualized by electron microscopy. Panel B shows a similar experiment using normal mouse sera as antibody; the resulting microscopy shows no localization of antibody over knobs as expected. Panel C repeats the experimental protocol used in panel A, except that the parasite used does not have knobs and hence there is no reactivity.

The association between knobs and PfHRP-I is of interest because the knobs mediate cytoadherence of infected erythrocytes to endothelium. Another histidine-rich protein from P. *falciparum* is designated PfHRP-II. It is synthesized throughout the asexual cycle and is released as a soluble protein into culture supernatants from intact cells. There is at least one other histidine-rich protein from P. *falciparum* that remains uncharacterized. The three histidine-rich protein genes of *P. falciparum* have been isolated. An interesting comparison can be made between the gene sequences of PfHRP-II and PfHRP-III. The sequence of the intron and sequence beyond the carboxy terminus of the protein are the same in both genes. Both genes also contain a similar hexapeptide repeat. The hexapeptide repeats vary between genes in only one of six positions.

Presumably, one set of repeats arose from the other. Divergence of one region of hexapeptide repeats from the other seems to occur in a uniform way: each repeat unit of the series is altered in the same way. Another set of repeats that are unrelated to the hexapeptide repeats occurs in PfHRP-III. The implication is that both genes were derived from a common precursor by a gene duplication and then diverged independently and assumed different functions. This process, at the very least, establishes the event of gene duplication in *Plasmodium.* It also outlines a mechanism whereby two proteins can share a common immunodominant epitope.

Merozoite Surface Antigens

The merozoite stage of *Plasmodium* invades host erythrocytes to start a new round of asexual replication. The process may involve several independent steps, including (1) initial attachment of the merozoite to the erythrocyte surface; (2) reorientation of the attachment, so that the apical end of the merozoite contacts the surface of the erythrocyte; (3) release of a substance into the erythrocyte from an apical organelle of the merozoite (at this point the shape of the erythrocyte is altered and a vacuole forms opposite the apical end of the merozoite); and (4) entry of the merozoite into the vacuole. Antibody against epitopes of important antigens involved in any of these steps could be a target for protective immunity. The genes of three *P. falciparum* antigens that are potentially important for vaccine development and are associated with the merozoite stage of development have been cloned. All three can serve as immunogens and give a protective response that blocks merozoite invasion when administered with Freund's complete adjuvant. This does not mean that they would necessarily be useful candidates for a vaccine. It is not known in any of these cases if the protective response is purely antibody mediated, which would be the ideal situation in terms of a vaccine.

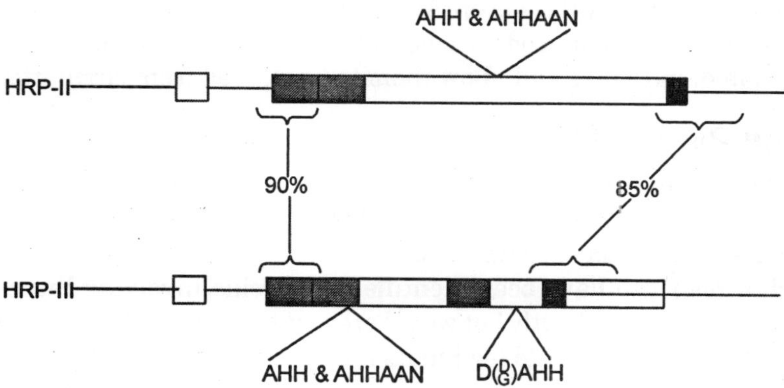

Fig. 7.4. A comparison of two histidine-rich protein (HRP) genes from *P. falciparum*. It is believed that the two genes were derived from a common precursor by gene duplication and then diverged independently, leading to the differences in the sequence of one region of the hexapeptide repeats shown.

Furthermore, not all antibody responses would be useful in a vaccine. Ideally, a neutralizing antibody that can block invasion at low titers is needed. Glycophorin A appears to be a merozoite receptor molecule on the surface of the erythrocyte. A glycophorin A binding protein (GBP) of *P. falciparum* has been identified which is probably involved in the initial binding of the merozoite to the red-cell surface. This protein is on the surface of the merozoite, binds glycophorin, and appears to recognize the surface of the erythrocyte. Antibodies to this protein block invasion of erythrocytes; however, the involvement of this particular protein in reception is unproved. The gene for the GBP has been cloned and the DNA sequenced. The protein contains a 50-amino-acid repeating segment. Preliminary evidence suggests that the structure of the repeat does not vary among isolates of *P. falciparum*. A merozoite-associated protein has been identified that may be localized to one of the apical organelles involved in erythrocyte invasion.

Antibody against this protein stained the surface of ring-infected erythrocytes, but the stain disappeared as the parasite matured. Again, antibodies against the protein block invasion. A protein of molecular weight (M_r) 195,000 that is synthesized in schizonts is the precursor to three major surface proteins of the merozoite. The function of the proteins is unknown. Preparations of the precursor molecule can induce protective immunity against blood-stage parasites. One of the three proteins (M_r 83,000) is shed when the merozoite invades the red cell.

This molecule contains a repeating DNA sequence that may represent an isolate-specific epitope. Two other proteins remain in associated with the ring after invasion. The particular antigenic sites that induce protection have not been defined. Whether or not isolate-specific sites are involved in the protective response also has not been clarified.

Asexual Erythrocyte Parasites

The proteins that are produced by the malarial parasite but reside on the surface of the infected erythrocyte may be an important target of protective immunity. For example, parasite-encoded proteins may be localized in knoblike protrusions on the surface of infected cells that bind to host endothelial cells (a process called cytoadherence). Cytoadherence prevents erythrocytes containing mature parasites from circulating in the peripheral blood and thereby protects them from filtration by the spleen. Analysis of genes whose products are directly involved in this binding might provide a good vaccine target. Antibodies to one of these proteins might block sequestration of parasites and allow them to be cleared from circulation by the spleen.

Sexual-Stage Parasites

Other possible vaccine targets are proteins associated with the sexual cycle of the parasite. It has been shown that the transmission of malaria through the mosquito can be blocked by antisera directed against malarial gametes and zygotes. The specific malarial proteins involved in eliciting such a response have been identified. A vaccine involving these protein products would not prevent the disease directly but would prevent its spread. The hope is that the reduction of transmission of the disease would lead to its eradication.

Genetics of Malaria

The production of an appropriate antigen is one step in producing an effective vaccine. Presently four different sources of antigen are being investigated: chemical synthesis, microbial production, vaccinia vaccine, and avirulent *Salmonella* vaccine. Many of the advantages and disadvantages of each approach relate directly to the biology of the vaccine target. The specific example used here is the production of antigen for a circumsporozoite-derived vaccine, and the relationship of approach to CS protein biology. The specific considerations involved in the production of an antigen will of course vary with different vaccine targets. Chemical synthesis of the repeating epitope of the CS protein appears to be the most direct approach to producing a vaccine antigen. One major objection to this approach, based on work in other systems, is that short, chemically synthesized peptides often do not display an antigenic site in the manner that a complete folded protein would. Initial experiments with chemically synthesized peptides from the repeating epitope of the CS protein indicated that as few as two repeats containing seven amino acids weakly mimicked the antibody-binding site of the complete CS protein.

Three repeats or twelve amino acids appear to constitute a complete epitope. This suggested that overall configuration of the CS protein was not necessary to display the appropriate antigenic site and that short, chemically synthesized molecules might be useful in a vaccine. This will not necessarily be the case with other antigens. Microbial production of antigen based on a recombinant DNA approach has been used by two groups. The P. *knowlesi* CS protein has been produced in yeast cells. A protein that is approximately the size of the precursor CS protein was detected.

These researchers report production of 1 microgram (μg) of stable CS protein per 20 milligrams (mg) of total protein. The bacterial fermentation approach is also being investigated.

The complete CS protein could not be synthesized in bacteria as it was in the yeast system. Clearly products were rapidly broken down in vivo. But when a smaller segment of the immunodominant region was expressed in the bacterial vector, the products were stable and the yields were nearly 5 percent of total protein, or several thousand times better than the yeast system. These researchers went on to show that antisera raised to their product block hepatoma cell invasion by sporozoites in vitro. Work directed toward a vaccinia-derived sporozoite vaccine is also proceeding. The gene coding for the CS protein of *P. knowlesi* has been inserted into the *Vaccinia* genome. Cells infected with these recombinant viruses produced a protein that reacted with monoclonal antibodies against the CS protein. Also, rabbits infected with virus produced antibodies against sporozoites.

To date, studies of successful vaccinations have not been reported. Some success in producing recombinant vaccinia vaccines for other diseases makes this system look like a promising approach for a malarial vaccine. Another potential "live vaccine" involves the use of nonvirulent, nonreverting strains of *Salmonella* to deliver antigens. These strains of bacteria could be genetically engineered with any antigen or group of antigens. They grow in the laboratory but cannot grow in the host, because they depend on media supplements that do not exist in host tissue. They apparently exist long enough to elicit an immunogenic response in the host but do not grow and divide.

Vaccines potentially could be delivered orally, and there has already been a great deal of testing of the system in the veterinary field. A live bacterial vaccine, however, is another level of increased complexity over the vaccinia system. The fear of loss of control over a "live vaccine" therefore increases. Although antigen for a sporozoite vaccine can be produced in a number of ways, host immune responses to varied products will not be identical. Thus, a few considerations with regard to antigen production for a sporozoite vaccine relate to the epidemiology of malaria. The first consideration is the dynamics of infection. The sporozoites that are injected into the blood reach the liver within minutes, where they may be protected from the host immune response. If only a single parasite escapes the immune response, the infected individual will contract malaria.

Subsequent stages of the parasite's life cycle would not be affected by antibodies directed against sporozoites. Antibody tilers must then be held at a very high level to be effective. The immune system must be more than primed to respond: it must be continually producing specific antibodies. The hope for effective long-term protection is that vaccination will produce a lasting response and that there will be natural boosting of this response by the bite of infected mosquitoes. Clearly, the possibility exists that the vaccine response to constructs consisting of less than the entire CS protein may not be boosted by sporozoites contacted in the field. Even the structure of various CS protein may be sufficiently different so that sporozoites from one strain of *P. falciparum* will not boost the response to a CS protein from another strain of *P. falciparum.* In terms of antigen production, these facts favor the live vaccines that use complete gene sequences. The host responses to live vaccines also tend to be more enduring.

Another factor that relates to a sporozoite vaccine is that a purely antibody-mediated response can be protective. This does not exclude the possibility that cell-mediated immunity may be much more effective in protection.

Many of the standard arguments against the clean, inexpensive peptide vaccines therefore may not apply in this situation. For example, if only an antibody response is necessary, then the peptide vaccines become attractive even though they would not elicit certain types of T-cell responses. A final consideration relates to the fact that the group at highest risk is children. Even a short lived vaccine that had to be continually boosted would be of great value in saving lives in this age range. There appears to be no simple definition of the ideal vaccine. Both the live and peptide vaccines have advantages, and a sense of urgency may dictate the parallel development of both.

Drug Resistance

Little work in molecular biology has been done on the origin of drug resistance in malaria. There are two likely candidates for work in this area, pyrimethamine resistance and chloroquine resistance. Pyrimethamine is a chemotherapeutic agent that is used alone in the prevention of malaria and in combination with sulfonamides in the treatment of malaria. This drug inhibits the dihydrofolate reductase (DHFR) of the blood schizonts of avian, rodent, and primate malarias at concentrations that have little effect on the corresponding host enzyme. Some drug-resistant parasites can tolerate levels of the drug at 1000 times this concentration. Although DHFR is a classic example of an amplifiable gene, restriction analysis of DNA from pyrimethamine-susceptible or resistant lines of parasites showed no obvious gene amplification.

Such analysis indicates that resistance is likely to be the result not of an increase in gene copy number but of altered properties of the protein conferring both a different affinity for the drug and an increased stability. Chloroquine is another commonly used drug in the prevention and treatment of malaria. Most resistant forms that have been studied carry only a two or threefold level of resistance to the drug, although some greatly increased amounts of resistance have been reported lately. The mechanism of resistance is unknown. Drug resistance seems to be acquired gradually in response to drug pressure. It is assumed that increases in resistance are the result of multiple mutations.

The origins of chloroquine resistance may be much more difficult to find than those of pyrimethamine resistance but are of great practical importance. Contributions of molecular biology to the discovery of new drugs may well be significant in the future.

Analysis of proteins such as tubulin may help us focus on central differences between comparable host and parasite enzymes. Potentially, defining these differences will help researchers identify drugs that will selectively kill parasites. Another possibility for the creation of new drugs rests with the use of derivatized oligonucleotides as drugs that will selectively anneal with parasite messenger RNAs and prevent their translation. Neither of these approaches has yielded a new drug to date, but both may have future potential.

Inheritance Techniques

The techniques commonly used to study the inheritence of free-living organisms are difficult to apply to malarial parasites, which have a complex cycle involving two hosts; therefore, the amount of classical genetics done in malarial parasites is limited. The number of classical genetic markers is small, and no markers have been reported to be linked. Selectable genetic markers, which would make the recovery of recombinant clones easier, are mainly limited to chloroquine and pyrimethamine resistance. Each generic cross is an arduous task. Parents with readily definable markers must be cloned and then mated in the mosquito. Progeny of each mating must then be cloned, propagated and analyzed. Even in the face of these difficulties, several important facts have been revealed by classical genetics in malaria.

Several markers have been shown to segregate in a normal Mendelian fashion. These studies have also indicated that the blood-stage parasites are haploid and that the diploid stage of the parasite resides in the mosquito after fusion of the gametes to form a zygote. The use of pulsed-field gradient gel electrophoresis will help resolve some of the genetic questions in the field of malaria. This technique allows researchers to electro-phoretically resolve the chromosomes of *P. falciparum* and analyze them individually. Two advances have come from this work to date. A large number of genetically linked markers have been established, and it has been shown that the chromosome sizes of different geographical isolates can vary. The latter may allow the firm demarcation of various taxonomic categories, such as an isolate, strain, and subspecies. Analysis of the parents and progeny of mated parasites will also yield significant information about the genetics of *P. falciparum.*

Coda

Plasmodium is not an especially good model system for studying molecular biology. The complete life cycle of the parasite is difficult to maintain, classical genetic techniques are difficult to apply to the system, and no procedure for transformation has been developed. The research effort has been directed toward understanding and controlling malaria rather than understanding molecular mechanisms in general. The link between understanding molecular mechanisms and understanding the biology of the parasite should, however, not be underestimated. Of considerable importance in this regard is understanding more fully the way in which this parasite evades the host immune system, the amount of natural genetic diversity of the parasites and how it arose, and the evolutionary relationship among the *Plasmodium* species. Most of the genes that have been studied encode antigens that are on the surface of the parasite or are actually released from the parasite.

The unique features of these *Plasmodium* genes seem to be the ubiquity of the short repeating epitopes and the frequent sharing of cross-reacting epitopes among different antigens. Clarifying the reasons for these unusual features of *Plasmodium* genes may enhance our understanding of the way that these parasites evade host immunity and may contribute to the production of an effective vaccine.

BIOCHEMISTRY OF TRYPANOSOMES

Through the application of recombinant DNA technology, we have recently gained some insight into the inner workings of the trypanosomes. Few, if any, other biological systems have revealed such a remarkable collection of extraordinary phenomena in such a short period of time. This is a testimony not only to the power of the techniques being used but also to the unusual position these organisms occupy on the evolutionary tree. This chapter will present an overview of the areas where some or the more interesting findings have been made. As there are many recent reviews on antigenic variation in the African trypanosomes, this topic will receive proportionately less attention. Instead, areas that are just as unusual and interesting but that have not received the same coverage will be emphasized. The chapter is divided into six parts, each representing a functional area of study: genome and genetics, transcription, translation, antigenic variation, mitochondria and respiration, and glycolysis. Most of the work described has been obtained from studies of *Trypanosoma brucei,* but some of the results extend to other species in the order Kinetoplastida.

Genome and Genetics

Like most other eukaryotes, all trypanosomes have DNA within two organelles, the nucleus and mitochondrion. The mitochonrial DNA is highly unusual and will be discussed in detail in the section entitled Mitochondria and Respiration. The genome of *T. brucei* is essentially diploid, with a total complexity of about 4.3×10^7 base pairs (bp). Unfortunately, the chromosomes of trypanosomes do not visibly condense during mitosis (possibly owing to an apparent lack of histone Hl); their analysis therefore has awaited the arrival of technologies that do not depend on cytology. Such procedures, termed pulse-field gradient gel electrophoresis, or orthogonal field alternating gel electrophoresis, have recently been developed. These techniques are similar to ordinary agarose gel electrophoresis but are capable of resolving much larger DNA pieces.

The method exploits the fact that the time taken for large DNA to reorient in an electrical field that has been rotated through 90 degrees depends on the length of the molecule (although the analogy is biophysically inaccurate, this is akin to the molecules having to repeatedly turn a tight corner). Thus, by switching the orientation of the electrical field from north-south to east-west in repeated pulses DNA of up to several million base pairs can be resolved.

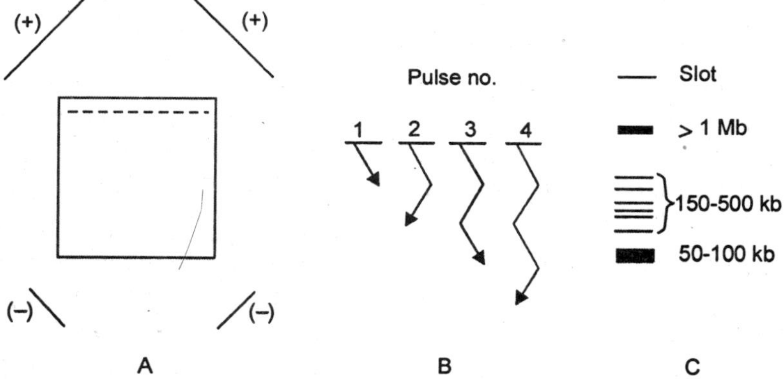

Fig. 7.5. Chromosome configuration. Part A shows the principle behind orthogonal field alternating gel electrophoresis (OFAGE). A horizontal agarose gel is placed between four electrodes, as shown. Each pair of electrodes (+ and -) is connected to a voltage source in alternating sequence. Under pulse times of a few seconds to many minutes, the DNA slowly migrates into the gel. Ultimately, conventional bands of discretely sized DNA are produced without enormous distortion of a given lane because of the vector sum of the voltage fields, as shown in part B. Part *C* shows a schematic of trypanosome chromosomes analyzed in this way.

With the use of these techniques, the chromosomes of trypanosomes have been found to fall into four general size classes. The largest is well over 10^6 bp (perhaps even over 10^7 bp) and comprises about 60 percent of the genome. The second set of about 3-6 chromosomes is $1 - 2 \times 10^6$ bp. Most, if not all, of these two size classes are apparently present at two copies per nucleus. The third class of chromosomes is about $2 - 7 \times 10^5$ bp. In T. *brucei,* these number about six and are apparently present at only one copy per nucleus. The mini-chromosomes form the fourth and smallest class, with an average size of about $0.5 - 1.5 \times 10^5$ bp. This class, which is unique to the African trypanosomes, is comprised of the most members, totaling about 100 chromosomes per nucleus (or about 15 percent of the total nuclear DNA).

The concept of ploidy is somewhat misleading when applied to members of the mini-chromosome class: they are highly redundant and repetitive, carrying related surface antigen genes, as discussed later. The chromatin of trypanosomes is apparently conventionally organized in nucleosomes comprised of a histone octamer and about 200 bp of DNA. As yet, however, no H1 histone or equivalent cross-linker has been identified. The chromosome ends (referred to as telomeres) also appear to be conventional, possessing over the last few hundred base pairs a repeated hexanucleotide with occasional nicks.

The sequence of this hexanucleotide is similar but not identical to that seen in other protozoa. Preceding this is a subtelomeric sequence that is apparently conserved within the species. Only a few genes have been mapped to a given chromosome or even class of chromosomes. However, even at this early stage, a trend is clearly emerging, so that "housekeeping genes," or genes involved in routine cellular processes common to most eukaryotic cells, are located on the largest class of chromosomes. These genes are therefore diploid in the trypanosomes. The surface-antigen genes involved in the phenomenon of antigenic variation, on the other hand, are found on all classes of chromosomes, even the mini-chromosomes.

Indeed, these smallest chromosomes appears to function exclusively as vehicles for the many surface antigen genes necessary for antigenic variation. This fact also serves to explain their lack in Kinetoplastida species such as *T. cruzi* which do not undergo extensive antigenic variation. Little is known about the sexual cycle of *T. brucei,* but it is now clear that such a cycle does occur during transmission by the insect vector, the tsetse fly. On the basis of findings to date, it appears that the entire sexual cycle, from meiosis and gametogenesis to zygote formation and subsequent asexual expansion of the population, occurs within the fly. The details of the ploidy of the different stages in this process is still a matter of debate. During meiosis, it may be that the mini-

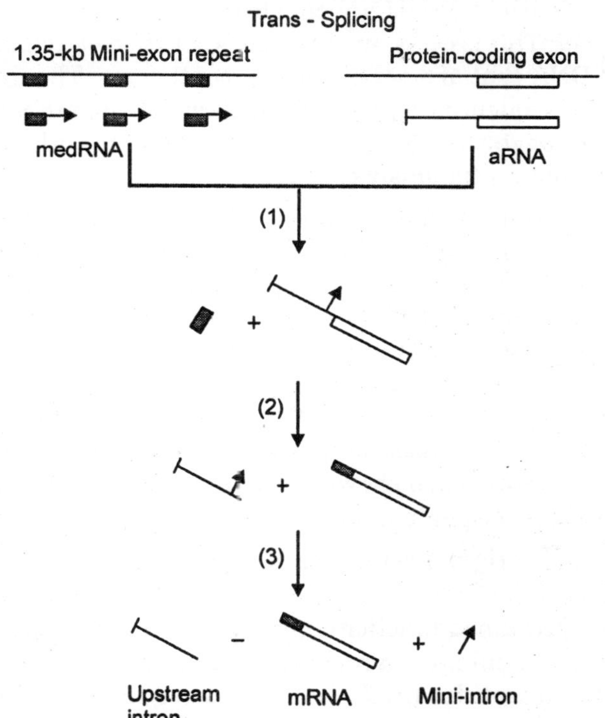

Fig. 7.6. Model of trans-splicing. The mini-exon repeats direct the synthesis of mini-exon -derived RNA (medRNA); the protein-coding exon directs the transcription of the acceptor RNA (aRNA). The first step in trans-splicing involves formation of abranched species (comprising the intron portion of medRNA (minRNA) and aRNA and free miniexon. In the second step, the two exons are joined so that the two introns are released as a discrete, branched species. The third step comprises the debranching of the branched species into the two component introns. This reaction yields three final products: the upstream intron of aRNA, minRNA, and the mature mRNA.

chromosomes are not evenly segregated (owing to their haploid state in an otherwise diploid nucleus).

However, this would not impair the viability of a daughter cell so long as it received sufficient genes to enable it to undergo several cycles of antigenic variation in the mammalian host and thereby survive long enough to be picked up by another fly. Subsequent genetic exchange during the sexual cycle in the fly would enable it to collect a new set of surface antigen genes when the pool is reasserted. There is no definitive evidence for a sexual cycle in the other Kinetoplastida so far studied.

Transcription

One of the most unusual features of the trypanosomes is their mode of transcribing nuclear genes. This process involves the production of a chimeric mRNA by transcription of two segments of DNA that are not physically contiguous. The first indications of this unusual process came from sequence analysis of the 5′-end of the mRNAs coding for surface antigens. This showed that roughly the 5′-most 35 nucleotides of the mRNAs encoding different antigens were identical and not contiguously encoded with the protein-coding portion of these genes.

Synthetic oligonucleotide probes were constructed and used to determine where the mini-exon coding for this spliced leader was located. The results indicated that the mini-exon is part of a 1.35-kilobase (kb) unit (the mini-exon repeat), which is tandemly repeated about 200 times in the genome and is unlinked to the protein-coding exon for these surface antigen genes. In some cases, the mini-exon repeats are on different chromosomes from the surface antigen genes. It is now known that the mini-exon sequence is present at the 5′-end of most, probably all, mRNAs in all stages of all Kinetoplastida. Although the mini-exon repeat may vary in size and overall composition, the mini-exon sequence is highly conserved. Its exact size in *T. brucei* has recently been shown to be 39 nucleotides plus the usual eukaryotic 7-methyl guanosine in 5′-5′ tri-phosphate linkage.

The original estimates of 3 5 nucleotides were incorrect as a result of the highly modified state of the first four nucleotides, which probably affected the assays used. Given their abundance and presumed functioning in the generation of so many mRNAs, it was not surprising to find that each mini-exon repeat is independently transcribed to give a short, discrete RNA of about 140 nucleotides (in *T. brucei,* shorter in other Kinetoplastida species), the first 39 of which correspond to the mini-exon sequence. Based on these observations, two basic schemes for the synthesis of chimeric mRNAs were originally proposed. The first involved the use of the med (mini-exon-derived) RNA as a primer in a transcriptional reinitiation event upstream of the protein-coding exon followed by cis-splicing of the resulting molecule. The second model supposed that some sort of bimolecular mechanism operated to combine the mini-exon portion of medRNA with the exon portion of an independently generated transcript from protein-coding genes. Some sort of splicing must be involved because only 39 nucleotides of the medRNA are ultimately found on the mature mRNA.

This splicing apparently uses much of the same basic machinery used by higher eukaryotes. This fact is apparent from looting at the sequences located at the exon - intron boundary in the medRNA and intron-exon boundary in the coding gene. These conform to the absolute consensus sequences found in nuclear-encoded genes of all other eukaryotes. Moreover, it has recently been found that the Kinetoplastida have several of the mRNAs involved in cis-splicing in higher eukaryotes. Recent evidence strongly suggests that the second model, trans-splicing, is correct.

This model supposes that the medRNA and an acceptor RNA are spliced by a pathway similar to the more conventional cis-splicing, except that the two RNAs are never found as a single, covalently linked precursor. Evidence for this model came from the ability of a debranching activity of HeLa cells to release the tail portion of medRNA (the mini-intron sequence) from higher-molecular-weight RNA. It is now clear that the analogy to cis-splicing is accurate because the linkage between the miniintron and the acceptor intron is a 2′-5′-phosphodiester bond and the trypanosomes have a debranching enzyme with the same activity as the HeLa enzyme. The unusual process of trans-splicing in Kinetoplastida is critically important to our understanding of the way trypanosome genes are regulated.

Trans-splicing may have equally important significance, at least to the molecular biologist, in telling us something of the evolution of RNA splicing, an exciting and contentious subject. This suggestion stems from the fact that trypanosomes are also unusual in having no known conventional (cis) introns in the many genes now studied; that is, no protein-coding exons are interrupted by introns. This total lack is without precedent in eukaryotes and begs the question of whether trypanosomes have somehow removed their introns. An alternative suggestion is that they never had them. If true, this idea argues that trans-splicing predates cis-splicing and that the cis-machinery adapted that of the trans-process for its own purposes. What is the function of trans-splicing and what is the role of the mini-exon sequence itself? One way in which trans-splicing is apparently used to advantage is in the generation of multiple, capped mRNAs from a single, polycistronic precursor transcript. This process would explain the high frequency of trypanosome genes found tandemly arranged on the chromosome, sometimes with the same gene(s) repeated several times.

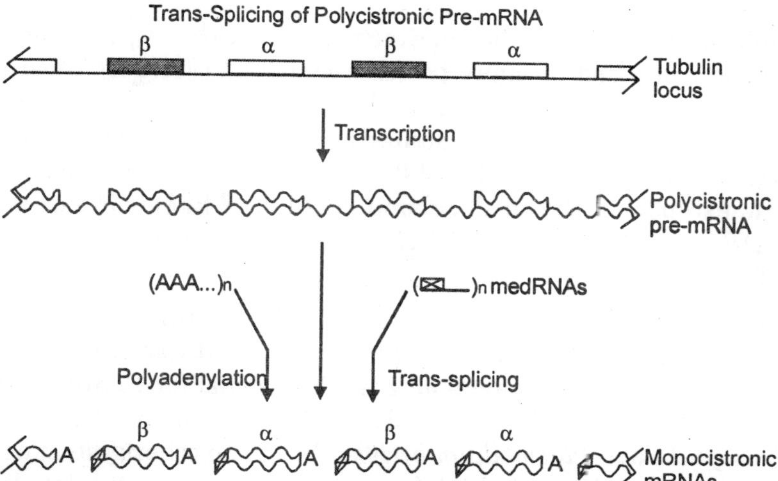

Fig. 7.7. Model of polycistronic precursor transcripts. The model proposes that a polycistronic RNA is transcribed from a locus composed of multiple coding regions (in this case, the tubulin locus of *T. brucei* encoding both α– and β-exons). Multiple monocistronic mRNAs are generated from this transcript by trans-splicing to yield a capped (medRNA derived) mini-exon sequence at the 5′-end, and by cleavage and polyadenylation to yield the polyadenylated 3′-end. This processing could occur post- or co-transcription-ally, and the order of trans-splicing versus polyadenylation is not known.

The model proposes that transcription begins at a promoter upstream of the cluster and that through a combination of trans-splicing and polyadenylation, mature, capped mRNAs are generated, each coding for a single polypeptide. By this pathway, each mRNA would acquire a cap, which is believed to be a prerequisite for stability and translatability of eukaryotic mRNAs. (Note that without trans-splieing, acquisition of a cap could not occur by any known pathway, because caps can be added only to transcription initiation sites, probably because of their unique 5′-triphosphate end) In addition to the tandem reiteration of many genes, the evidence for this pathway is fourfold.

1. Heat shock causes the accumulation of polycistronic transcripts.
2. Pulse-chase experiments suggest a precursor - product relationship between polycistronic transcripts and mRNAs.
3. Intergenic regions are transcribed in nuclear run-on assays.
4. Kinetics of UV inactivation indicate some genes are 50 kb downstream of their promoter, with multiple coding regions intervening.

The question of the function of the mini-exon sequence itself is intangible at present because we know of no exception to the rule that all trypanosome mRNAs include this sequence. That such a sequence will prove essential in the translation of these mRNAs seems very likely but tells us little about their function. It would simply confirm that the translational machinery and mini-exon sequence have co-evolved and are now functionally interdependent, just as the bacterial ribosome requires its sequence-specific binding site. However, in the absence of in vitro translation or DNA transformation, it has not been possible to test this idea.

Translation

As in all other eukaryotes, translation in Kinetoplastida is the responsibility of the ribosome, a complex organelle composed of many proteins and RNAs (rRNA). Most of the rRNA in trypanosomes is encoded by a tandemly repeated segment of DNA (about 22 kb per repeat in *T. brucei*. In most eukaryotes the rDNA is transcribed as a single long precursor that is subsequently processed by several cleavages into the 18S, 5.8S, and 28S rRNAs. The Kinetoplastida adhere to this basic pattern, except that the small subunit rRNA (18S-like) is in fact the largest of any known, at about 21S, and the large subunit rRNA (28S-like) is further processed into two medium and several small RNAs.

The machinery involved in this processing is not vyet identified, but all the cleavage sites have been precisely located. Interestingly, there is no apparent commonality in either primary or predicted secondary structure between all sites. The signal for cleavage therefore remains cryptic. Cleavage of the 28S rRNA into two halves is known to occur in a few other eukaryotic systems (*e.g., Drosophila* and *Plasmodium berghei)* but in no other system does this result in the generation of several small, stable RNA fragments that remain associated with the mature ribosome. Eukaryotic ribosomes generally contain another small RNA with a sedimentation coefficient of about 5S (120 nucleotides). As in other eukaryotes, the trypanosome 5S RNA is encoded by its own tandemly repeated DNA rather than as part of the rDNA repeat described above. Ribosomal RNA is among the most conserved RNA and as such is frequently used to deduce evolutionary relatedness between organism.

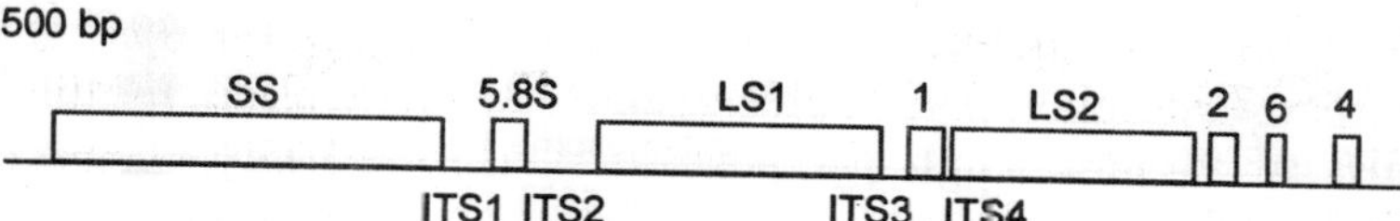

Fig. 7.8. Organization of ribosomal DNA. A portion of a single rDNA repeat of *T. brucei* is shown. The regions represented in stable RNAs are boxed and identified as small subunit RNA (SS), large subunit RNA (LSI and LS2), 5.8S RNA, and the small ribosomal RNAs (1, 2, 4, and 6). The intervening transcribed spaers (ITS), which are processed out and rapidly degraded, are also shown. The final component of the rRNA, 5S RNA, is encoded by its own DNA repeat.

The complete sequence of 5S rRN for two kinetoplastid species has been obtained. This shows that in addition to being the largest small subunit rRNA yet characterized, the order Kinetoplastida is one of the most distantly diverged (relative to the main eukaryotic line), apparently branching off before even the split into plants, fungi, and animals. As we progress further in our understanding of the basic molecular biology of the Kinetoplastida, information on their position on the phylogenetic tree will be increasingly important in enabling us to place new findings in an evolutionary context.

Antigenic Variation

This portion of the chapter deals with the extraordinary way in which the African trypanosomes successfully elude the immune system of their mammalian hosts. Antigenic variation describes the ability of the African trypanosomes to alter the antigenic nature of their surface coat. On an individual trypanosome, this coat is generally composed of a single type of abundant glycoprotein molecule known as the variant surface glycoprotein, or VSG. Each parasite contains many hundreds of genes coding for different VSGs. They effect antigenic variation by activating these genes one at a time, so that during the course of an infection initiated by a single parasite, the host is faced with a succession of antigenically distinct populations. No known stimulus triggers this switch; it appears to be spontaneous, occurring approximately once in every 10^6 divisions.

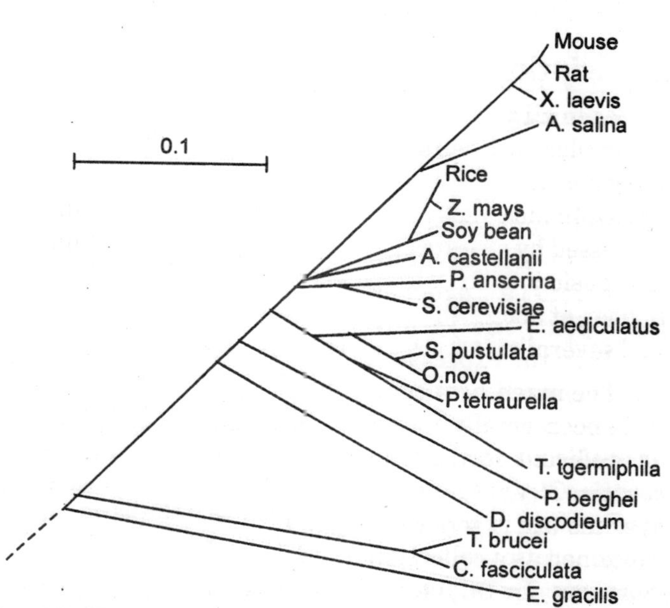

Fig. 7.9. Placement of trypanosomes on the evolutionary tree. Using the small subunit rRN A sequence as a measure of sequence variation, investigators have devised this phylogenetic tree, showing the placement of the *T. brucei* relative to other organisms, including plasmodia.

Antigenic variation is sustained through the appearance of new antigenic 'types before the preceding wave has been completely eliminated. It is slow enough, however, to prevent the full repertoire from being revealed to the host too quickly. This prolongment is further ensured by expressing the VSG genes in a nonrandom way, so that for a given strain, there is about a 40 percent chance of one antigenic type being followed by a second, defined type, with the remainder of the switched types largely limited to a small subset of the repertoire. The molecular basis of this phenomenon is now partially known. First, the expression of a gene is somehow controlled at the level of transcription, with only one VSG gene typically transcribed at any one time. In order for a VSG gene to be active, it must reside in a limited number of expression sites, which are invariably found very near the ends of the trypanosome chromosomes (the telomeres, discussed earlier).

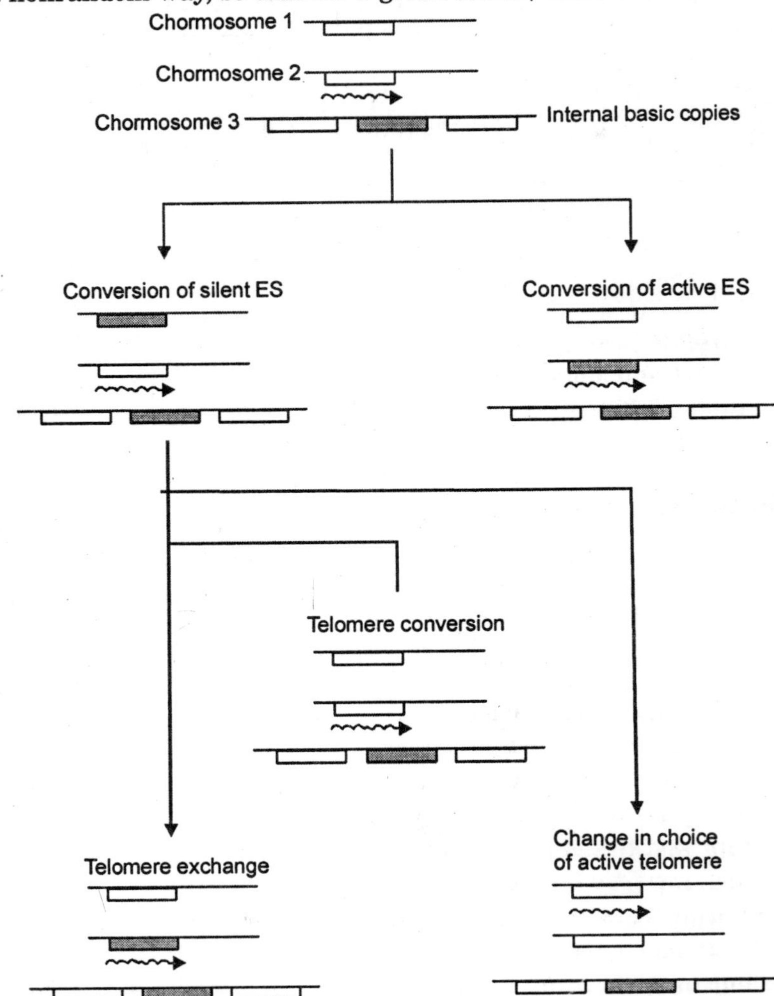

Fig. 7.10. Models for molecular genetics of antigenic variation. Activation of VSG genes requires that they first be duplicated and transposed into a telomeric expression site. In this drawing, several VSG genes (represented by the boxes) are shown on three chromosomes. The upper drawing represents a parasite expressing the "white" gene in expression site 1 (ES1), as indicated by the arrow. Activation can proceed by a one-step process if the basic copy of the "black" gene is copied into the active ES1, as shown by the right-hand pathway. Alternatively, the black gene could convert a silent ES (2), as shown in the left-hand pathway. This would be phenotypically silent with continued expression of the white gene. Following this (by any amount of time), the black gene could be activated by telomere exchange (such that the ES2 becomes activated), telomere conversion (such that a third copy of the black gene is generated, now in the active ES1) or a change in the choice of active telomere with no rearrangements.

A VSG gene enters an expression site by gene conversion a process whereby a gene in one locus (here the expression site) is replaced by a copy of a gene normally found in a different locus. The repertoire of VSG genes in trypanosomes is generally maintained as a collection of basic copy, or BC, genes, which are located at chromosome-internal positions. During gene conversion a duplicate of the basic-copy sequence replaces a gene in an expression site, the new copy being designated an expression linked copy, or ELC. The displaced ELC is permanently lost in this process, but this loss does not generally affect the total repertoire of the parasite, because usually there is a BC corresponding to this ELC at a stable, internal site within the genome. Once a telomere, the process of activation is unknown. However, several sorts of activation have been described. First, activation can occur directly as a result of the conversion if the target telomere is already active. Second, there can be telomere exchange such that the recently created gene is placed on an active chromosome and consequently is turned on. Third, the telomeric copy can in turn convert another, active telomeric locus, resulting in activation. Fourth, there can be no detectable rearrangements; in this case, the new telomeric copy is simply activated in situ, with corresponding switch-off of the previous expression site.

The way in which the trypanosome designates one or another of these expression sites for active transcription is not currently known. The only clue is that the nucleotides of a silent expression site are frequently modified. However, whether this modification is a cause or an effect of expression has not been determined. As discussed earlier, a given antigenic type is generally succeeded by a discrete subset of the repertoire. This subset seems likely to be partially defined as comprising genes already residing in expression sites. They are, it seems, more likely to be expressed than internal genes which have not yet entered an expression site. The tag identifying the single gene most likely to follow expression of any given gene is not yet known, but it could reside in adjacent sequences rather than in the VSG-coding portion itself.

Mitochondria and Respiration

Trypanosomes have a single mitochondrion that extends over the length of the cell. Like other eukaryotes, many of the mitochondrial proteins are encoded by circular DNA within this organelle. Trypanosomes are exceptional, however, in that their mitochondrial DNA is comprised of two sorts of circular DNA, the maxicircle and minicircle. Together, these two types of DNA comprise a catenated network that, for unknown reasons, is always found near the base of the flagellum. This DNA is easily visualized by basic stains, and before its composition was known, cytologists gave it the name *kinetoplast* for "body associated with motion." As it is found only in the trypanosomes and such closely related protozoa as *Leishmania* and *Crithidia,* it gives its name to the order comprising these genera, the Kinetoplastical.

It also gives its name to the mitochondrial DNA of trypanosomes, which is usually referred to as kinetoplast DNA or kPNA. Despite the unusual structure of its DNA, the fact remains that the mitochondrion must perform its essential function of generating ATP through oxidative respiration. In the trypanosomes, this process is highly regulated. When in the bloodstream of the mammalian host, there is an ample supply of readily converted energy in the form of glucose. The glycolysis necessary to convert the energy in glucose to ATP is handled by an unusual organelle, the glycosome, which will be discussed in greater detail later. Within the actively dividing, long-slender forms of the bloodstream, the mitochondrion is largely quiescent, with no detectable cytochromes or mitochondrial ATP generation.

Morphologically, the mitochondrion in this stage is a long, narrow tube with few cristae; biochemically, most of the enzymes and cytochromes necessary for respiration are absent. In the insect vector, the situation is very different, with proline and other amino acids being the major carbon source. In this stage of its life cycle, respiration is fully switched on and the mitochondrion is expanded with many cristae. Transcription of the mitochondrial genome is generally increased in this stage and there is a concomitant increase in the amount of the enzymes and cytochromes necessary for oxidative phosphorylation. The sorts of kDNA are very different in both structure and function.

The mini-circles are covalently closed circles of 0.7-2.5 kb, depending on the biological species. They have no known coding function and no substantial open-reading frame, but they may be transcribed to give one or more short transcripts. There are about 5000-10,000 minicircles in each kinetoplast network, but these are generally not homogeneous even within a single network. Instead, they are known to be highly variable in sequence, with only short, conserved regions likely to be involved in their replication. Notable exceptions to this variability are *Trypanosoma equiperdum* and *Crithidia fasciculata,* which have nearly homogeneous populations of minicircles. The other component of kDNA is the maxi-circle, which ranges in size from 20 kb to 38 kb. In other systems this component would be termed the mitochondrial genome. It contains genes for the ribosomal RNAs of the mitochondria (although no such ribosomes have yet been detected), several mitochondrial enzymes or subunits thereof, and some evolutionarily conserved open-reading frames of unknown coding function.

As with other mitochondrial genomes, the maxi-circle uses UGA to code for tryptophan (as opposed to using it as a stop codon in the "universal" code. Comparisons with the mitocondrial genomes of other organisms reveal several properties unique to the Kinetoplastida. First, the putative ribosomal RNAs encoded by the maxi-circle are substantially smaller than those found in any other organism, prokaryote or eukaryote (about 600 and 1150 nucleotides for the 9S and 12S RNAs, respectively). Second, no tRNA genes have yet been identified on any kinetoplastid maxicircle. The third and most unusual property of the maxi-circle is that several of its genes are "incomplete" yet functional.

The paradox describes the fact that transcripts from several of the genes are nonfunctional unless subjected to a remarkable process of uridine addition. This process of RNA editing, which has been observed in three disparate Kinetoplastida species, *T. brucei, Leishmania tarentolae,* and C. *fasciculata,* involves the addition of several (over 300 in some cases) uridine residues at precise positions within a transcript both in and out of the protein-coding region. Without such additions, the coding region is not a continuous open-reading frame and may even lack the start codon, AUG (*i.e.,* the corresponding gene encodes only AG at this position, so a U insertion is necessary for functionality). The mechanism of U addition, or even whether it is co- or posttranscriptional, is not yet known. Interestingly, the phenomenon is apparently developmentally regulated in *T. brucei,* so that in the bloodstream forms where respiration is quiescent, the transcripts lack the added Us and are thus nonfunctional. In the insect forms, on the other hand, the transcripts possess the Us and are fully functional, encoding essential mitochondrial enzymes. Transcription of the maxi-circle is not well understood.

No promoters have been definitively identified, and the primary transcription units have not been mapped. The ribosomal RNAs appear to be derived from longer primary transcripts (as is the case in mammalian mitochondria) based on pulse/chase experiments. The situation for the unedited precursors of the protein-encoding mRNAs is not yet known. The function of the maxi-circle is clear: it encodes several, but not all, of the mitochondrial enzymes, the remainder of these enzymes presumably being encoded in the nucleus. The function of the mini-circle is more enigmatic. Given its variability and lack of open-reading frames, it may be sewing as purely structural DNA, perhaps as a scaffolding on which to array the maxi-circle. The structure of such a scaffold could be a stack of parallel, elongated ovals (perhaps supercoiled), like baguettes in a baker's shop. This pattern is consistent with electron micrographs, which show a striated layer with a width equal to the predicted length of one or two mini-circles.

At least one of the "sharp turns" could be due to a periodicity in the sequence, wherein a run of adenosines is known to be present at about every 10 nucleotides. Such a nonrandom interval (approximately equivalent to the number of residues in one helix turn) can cause the DNA to bend a few degrees at each adenosine run; the cumulative effect is a substantial turn in a single plane. Replication of mini-circles first requires that they be freed from the complex by a topoisomerase and then replicated by a Cairns-type (theta) mechanism. The resulting two progeny molecules are reintegrated into the complex, where the nicks that remained after replication are sealed. Maxi-circles, on the other hand, replicate while still part of the network by a rolling-circle mechanism.

The linear extension of the rolling circle is cleaved at a discrete site into monomeric units which are not part of the network. The manner by which the linear units are circularized and reattached to the network is not yet known. The short-stumpy bloodstream forms, which are thought to be the forms infectious to the fly, also have increased levels of maxi-circle transcripts, consistent with the notion that they are preadapting to life in the fly.

Glycosomes

Trypanosomes rely on glycolysis for energy production when in the glucose-rich bloodstream of their mammalian hosts. This process enables them to aerobically metabolize glucose, producing two molecules of the three-carbon pyruvate. This process requires the consumption of four ATP molecules while generating six, resulting in a net gain of only two ATPs per glucose molecule metabolized. Because this process is the major source of ATP to these very motile parasites, the trypanosomes have evolved an unusually efficient system of glycolysis, which metabolizes glucose about 50 times faster than the system in mammals. This efficiency is attained through the enzymes involved being not only highly abundant but also concentrated into discrete, membrane-limited organelles called the glycosomes. These organelles, which are analogous to the micro-bodies found in other organisms, are about 03 microns (μm) across and number 200-300 per cell. They are bounded by a single membrane and constitute about 4 percent of the total volume of the trypanosome and 9 percent of the total cell protein.

The nine enzymes involved in glycolysis are the major components of the glycosome. In addition, however, the enzyme machinery involved in other metabolic processes, such as synthesis of ether lipids, beta-oxidation of fatty acids, and biosynthesis of pyrimidines, is also present, although in much lower amounts. One clear advantage to the compartmentalization of glycolysis is that as the intermediates in the glycolytic pathway are generated, they cannot diffuse too far before being acted on by the next enzyme in the pathway. The intriguing possibility of channeling of the intermediates from one enzyme directly to the next (without diffusion) has proved not to be the case. The genes coding for the different glycolytic enzymes in trypanosomes are nuclear-encoded (there is no detectable nucleic acid in the glycosome).

In general, the genes for the different enzymes are not physically linked. Interestingly, however, like many other nuclear genes in the trypanosomes, some glycolytic enzymes are encoded by two or more tandemly reiterated genes. In at least one case (phosphoglycerate kinase), the individual genes comprising a group are not absolutely identical, with their respective products being localized to the glycosome and cytoplasm, respectively. They are also present in different amounts in the various life stages. The glycosome enzymes have exceptionally high pis (9-10) compared with their homologues in the trypanosome cytoplasm and other organisms. For each enzyme, the extra, positively charged residues are found in two topological clusters spaced 40-50 angstroms (Å) apart. It has been suggested that these clusters are the tags that cause these enzymes to be transported into the glycosomes. This intriguing possibility is yet to be tested experimentally.

Coda

Clearly the application of molecular biology to the kinetoplastid organisms has revealed many unusual and unprecedented phenomena. But where do we go from here, apart from filling in the many missing steps in these unusual processes? Determining where trypanosomes fit into the larger picture of evolutionary biology is going to be just as exciting an area for future research. We know they are exceptional in so many processes, but how isolated are they as a group? The trypanosomes promise much in what they can tell us about the evolution of many complex processes whose origins are unclear from studies of more recently evolved systems. All this is in addition to the potential of applying a detailed knowledge of such fundamentally different pathways to effecting new means of controlling this critically important class of parasites. Exciting times await us.

BIOCHEMISTRY OF LEISHMANIA

The *Leishmania* organisms are single-celled members of the order Kinetoplastida. The leishmania has two distinct stages in its life cycle, the promastigote stage, in which the parasite is extracellular, flagellated, and motile, and the amastigote stage, in which the parasite is intracellular, nonmotile, and nonflagellated. The promastigote form is found in the insect vector, whereas the amastigote resides in macrophages of the mammalian host. The parasite multiplies in both stages of its life cycle. Human leishmaniasis is caused by at least 14 different species and subspecies of the genus *Leishmania*. Epidemiological studies over the last 25 years have demonstrated that in general leishmaniasis is a zoonotic disease; the parasite is transmitted to humans from a reservoir mammalian host by a phlebotomine sandfly vector during a blood meal. Only a subset of the numerous species of the genus *Leishmania* found to infect mammals also infects and causes disease in humans, presumably because of a combination of factors, including the intrinsic susceptibility of humans and the feeding habits of the sandfly vector. Enormous effort has been devoted to the isolation and characterization leishmanias that infect humans and to the identification of the principal mammalian reservoirs and species of phlebotomine vector.

The result of many such studies has been the correlation of particular clinical manifestations with certain species or subspecies of the parasite. *Leishmania* species identification is based on a variety of ecological, biological, biochemical, and immunological criteria. Previously, each cultured isolate of the parasite was analyzed by one or more of these criteria and grouped into species and subspecies. Controversies remain as to whether organisms isolated in distant geographic locations but sharing certain common properties belong to the same or distinct subspecies of the genus *Leishmania.* Whether these organisms represent strains of the same subspecies or distinct subspecies cannot be resolved because there is no single generally accepted method for species identification in the genus *Leishmania.* There is no defined sexual stage, so that traditional criteria for species identification cannot be applied. This problem complicates comparison of the disease epidemiology in distinct geographic locations and represents a potential limitation on the transfer of control measures from one geographic location to another. Recently there has been an interest in molecular mechanisms that might be unique to the parasite and therefore useful either in understanding the parasite itself or certain facets or the host-parasite interaction.

The obvious long-term practical goal of such work is the identification of potential new targets for chemotherapeutk attack of the parasite. This area should receive focus because although vaccine work is in progress, both the diversity of *Leishmania* sp. and the potential problems of antigenic change will significantly hamper those developments. Several questions are of fundamental interest to molecular biologists with regard to the parasite and its life cycle. These include organization, kinetoplast DNA structure and function, developmental control of gene expression, and novel biochemical and molecular processes. Recent work has defined the chromosomal organization of the parasite, but there is no information on the exchange of genetic information in these organisms and the establishment of genetic systems for analysis. This area is just beginning to develop. No sexual stage has been defined for these parasites, and controversy remains concerning even the most basic questions, such as whether the organism is haploid or diploid in either promastigote or amastigote stages.

The kinetoplast DNA of the parasite has stimulated particular interest; this fascinating set of DNA molecules is unique to the order Kinetoplastida, and the most extensive molecular characterization of any DNA molecule in the leishmania parasite has been of the kDNA maxi-circle and mini-circle. Control of gene expression in the various stages of the parasite's life cycle has been a focus of research since the parasite exists in vastly different environments within the insect vector and the mammalian host. One major interest for molecular biologists is to determine which adaptive mechanisms the parasite has developed to survive in two distinct environments: the sandfly gut and the host macrophage. Novel biochemical and molecular pathways that are specific for the parasite have attracted attention because their study could lead to new strategies for chemotherapeutic attack.

Genetic Organization and Function

The genome size of *Leishmania donovani* has been determined by Cot analysis and is estimated to be 5×10^7 base pairs (bp) per haploid genome. Recent work using pulse-field gel electrophoresis has demonstrated that the *Leishmania* species have a minimum of 16 chromosomes or chromosome-sized DNA molecules. Cot analysis of the ribosomal RNA demonstrates that there are approximately 150 copies of ribosomal RN A per haploid genome. The organization of the genomic DNA is in the very early stages of analysis. Separation of large chromosome-sized DNA molecules by the technique of alternating-field gel electrophoresis has allowed the identification of 16 to 23 chromosome-sized DNA molecules in the *Leishmania* species. In general, the chromosomes range in size from 150 kilobases (kb) to greater than 1000 kb. Before this work, it had not been possible to visualize the parasite's chromosomes because, unlike mammalian chromosomes, they are not visible as condensed chromosomes at mitosis.

Several laboratories have analyzed chromosomes from different leishmania species and have discovered that there is significant size polymorphism of certain chromosomes between species. This result is not unexpected; different yeast species also show chromosome size polymorphism. In addition, Southern blot analysis demonstrates species-specific variation in the size of specific chromosomes. Several structural genes have now been localized onto specific chromosomes-sized DNA molecules. In this way, a molecular karyotype of several *Leishmania* organisms has been derived. At least five chromosomes have been defined by the location of alpha and beta tubulin genes, rRNA genes, and the gene for the 5′-spliced leader sequence. The mapping of alpha tubulin genes and ribosomal genes demonstrated that each mapped to a distinct large chromosome in a variety of isolates. Both the beta tubulin gene and the 5′-spliced-leader sequence show evidence of size variation and variation in the number of chromosomes with homologous sequences. This pattern indicates a certain plasticity in the arrangement of genes on chromosomes and may have implications with regard to the evolution of the various strains and species.

Structure and Function of Kinetoplast DNA

In addition to the nuclear DNA, leishmanias have an unusual mitochondrial DNA, called the kinetoplast DNA, that comprises 10 percent to 15 percent of the total DNA of the cell. It is composed of two types of circular DNA molecules. The first, referred to as the maxi-circle (20,000 bp long), is the mitochondrial DNA of the parasite; the second type, the mini-circle (800-1200 bp), has no known coding function. The maxi-circle DNA is present in approximately ten copies per kinetoplast; the mini-circle DNA is present in 10,000-50,000 copies. Replication of both the kDNA maxi-circle and mini-circle has been extensively studied.

Elegant work has demonstrated that the maxi-circle replicates by the rolling-mechanism, whereas the mini-circle replicates as individual circles, using an RNA primer at the site of the putative origin of mini-circle replication. Several mitochondrial proteins have been identified by cross-hybridization with genes from yeast and mammalian cells. In addition, DNA sequence analysis of maxi-circle DNA has identified several open reading frames on the maxi-circle DNA. Transcripts of the maxi-circle genes have been identified in both the promastigote and amastigote stages. Developmental regulation of maxi-circle gene expression is less pronounced than in a related kinetoplastid organism, *Trypanosome brucei,* where several maxi-circle transcripts are developmentally regulated in their expression.

Kinetoplast DNA Mini-Circle

One of the most puzzling DNA molecules in nature is the kinetoplast DNA mini-circle. It has been studied extensively in both *Leishmania* and *Trypanosome* species, (both African and South American). It represents approximately 10 percent of the total cell DNA and is repeated 10,000 to 100,000 times per kinetoplast. Its function is unknown; no protein product or transcript has been demonstrated, although many mini-circle sequences contain a small open-reading frame of 30 - 50 amino acids. A unique 13-nucleotide sequence is contained in all mini-circles sequenced to date and is proposed to be the origin of DNA replication.

All mini-circles have a unique structural feature: a DNA sequence that contains a natural bend. This feature was first described because of anomalous migration of the mini-circle DNA in agarose and acrylamide gels. The DNA sequence itself is not conserved, but the structural feature of the bend is conserved. The function of this bend in the DNA is unknown, but one proposal is that it serves as a protein binding site important for the function or structure of the kDNA. DNA mini-circles appear to undergo relatively rapid DNA sequence divergence, presumably because of the noncoding nature of the DNA. This observation has led to a useful application of molecular biology to the differentiation of *Leishmania* species and subspecies and subsequently to the direct diagnosis of leishmaniasis in patients.

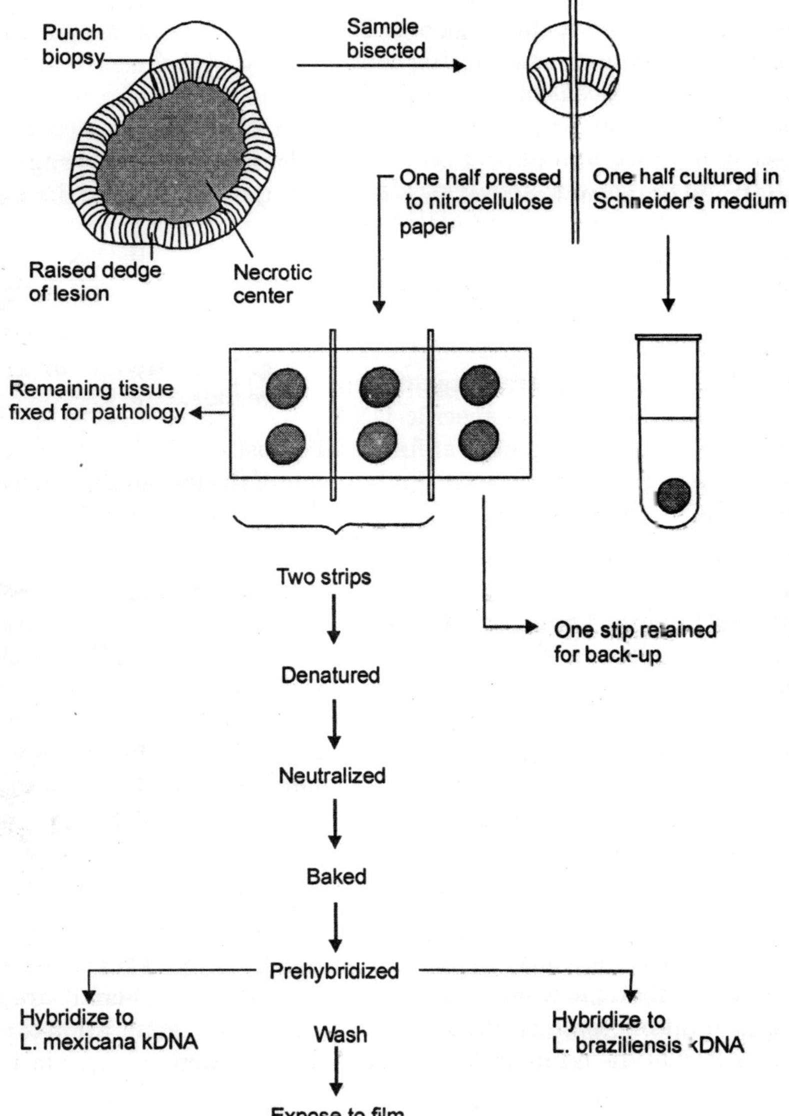

Fig. 7.11. A schematic representation of the method used for the DNA probe diagnosis of cutaneous leishmaniasis. A biopsy of the cutaneous lesion is taken and is divided in half; the tissue is then spotted directly onto a DNA-binding membrane. This membrane is then treated to denature the DNA and is hybridized with the specific probe for either *Leishmania mexicana* or *Leishmania braziliensis*. Specific hybridization is detected by x-ray film.

In the New World *Leishmania* organisms, the kDNA mini-circle isolated from L. *mexicana* does not share any sequence homology with those isolated from L. *braziliensis.* This difference in DNA sequence has provided the basis for a DNA probe that can distinguish these two *Leishmania* species directly when material obtained from a lesion is applied to nitrocellulose. This methodology has now been used to diagnose over 300 patients with cutaneous leishmaniasis offering both a more rapid and specific diagnosis than is currently available with any other methodology. Work to develop subspecies-specific diagnostic probes and molecular analysis of kDNA have shown species-related restriction site heterogeneity within the kDNA and several groups have cloned species- or isolate-specific kDNA fragments from visceral and Old World cutaneous strains of *Leishmania.* This work demonstrates that within the kDNA network, fragments of DNA exist with different taxonomic specificities. These cloned subspecies-specific DNA probes are important for the development of future diagnostic probes but are also interesting with regard to the kDNA sequence divergence. Perhaps one of the fascinating molecular questions in kDNA analysis is the mechanism of DNA sequence divergence within the kDNA mini-circle.

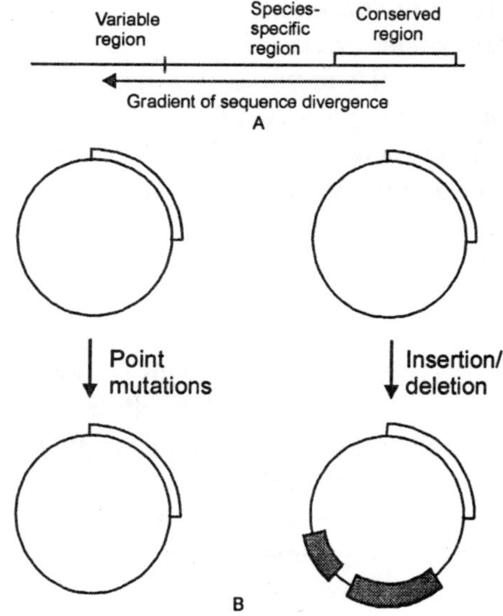

Fig. 7.12. Models of kDNA sequence divernce.

Within a single mini-circle, there are regions of conserved DNA sequence and regions of highly divergent DNA sequence. The 10,000 mini-circles within a single *Leishmania* organism are heterogeneous and fall into a number of different sequence classes. Within a given sequence class, the mini-circle sequences are identical. Sequence analysis of different classes of mini-circles in *Leishmania tarentole* and *T. brucei* has revealed that mini-circles of different classes share 150 bp of homology but are widely divergent throughout the rest of their length. In addition, in *L. mexicana amazonensis,* isolates. The maintenance of homogeneous sequences with different taxonomic specificities are present in a defined order along the length of a single cloned mini-circle. There appears to be a gradient of sequence divergence on the length of the mini-circle. There is a 230-bp region which is conserved across species boundaries, an adjacent 170-bp region that is found only in the *L. mexicana* complex and a 400 bp region which is highly divergent and found only in a single or small number of *L. mexicana amazonenesis* isolates.

The maintenance of homogeneous sequence classes in the face of rapid sequence variation presumably requires a mechanism whereby mutations in a single mini-circle are transmitted to other mini-circles. In addition, some mechanism must exist for the generation of rapid sequence divergence in one portion of the mini-circle molecule with maintenance of relatively conserved regions within the same molecule.

Developmentally Regulated Gene Expression

To study the developmental regulation of gene expression, it was important to choose the major proteins that are regulated during the transition from amastigote to promastigote. The major biosynthetic change is the increase in the synthesis of alpha and beta tubulin, the major structural proteins of the flagellum and subpellicular microtubules. These proteins can represent up to 25 percent of the total protein synthesized; they therefore represent the major transcriptional products of the parasite during the transition from amastigote to promastigote. *Leishmania* species have a novel cytoskeletal structure: organized arrays of microtubules, referred to as subpellicular microtubules, form the basis for the parasite structure. These microtubules are closely opposed to the inner surface of the plasma membrane and completely outline the parasite cytoplasm.

Both the amastigote and the promastigote stages contain subpellicular microtubules; the promastigote has in addition a flagellum with the standard 9 + 2 microtubule arrangement found in flagella from many different organisms. During transition from amastigote to promastigote, the flagellum is assembled, and this process presumably accounts for the large increase in synthesis of the alpha and beta tubulin proteins. The genes encoding both the alpha and beta tubulin proteins have been cloned and analyzed in *L. enriettii,* a rodent leishmania. Both genes are arranged in separate tandem arrays. The alpha tubulin gene is composed of 10 - 15 copies of a 2-kb identical repeat. All the copies of the alpha tubulin gene are contained on a single chromosome in all *Leishmania* species that have been analyzed.

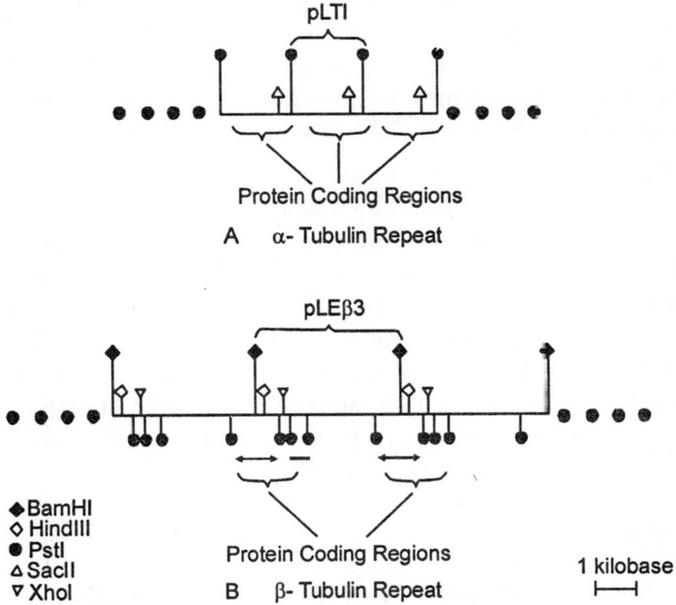

Fig. 7.13. Models for the arrangement of α- and β-tubulin genes in the genome of *L. enriettii.* A. The structure of tandemly repeated α-tubulin genes is indicated, along with selected restriction sites. The 2-kb *PstI* fragment that forms the insert for the α-tubulin genomic clone pLT1 is marked in brackets, and the α-tubulin protein-coding region is delineated by the bar. The direction of transcription is from left to right. B. The structure of the tandemly repeated β-tubulin genes is displayed, along with selected restriction sites. The 4-kb BamHI fragment that constitutes the insert of the *b*-tubulin genomic clone pLEB3 is indicated by the brackets, and the protein coding region is marked by the narrow bars. The direction of transcription is from right to left.

In contrast, the beta tubulin gene, which is approximately 4 kb long, is present both as a tandem array of 10-15 genes on a single chromosome and in separate copies on at least two other chromosomes. The exact function of the beta tubulin genes not contained in the repeat is unknown. They may be nonfunctional pseudogenes or they may be genes required for special functions, such as mitosis. Why the alpha tubulin gene complex does not have these genes outside the repeat is a curiosity.

The expression of alpha and beta tubulin in *L. enriettii* is controlled at the level of mRNA accumulation. Both of the tubulin mRNAs are present in 8-10-fold greater quantity in the promastigote than the amastigote. In another species, *L. mexicana,* translational control also may play a role in the regulation of tubulin protein expression. Approximate sizes of the alpha and beta tubulin mRNAs were determined by fractionating total RN A on denaturing gels and sequentially hybridizing the same Northern blot with the cloned probes for alpha and beta tubulin.

Both mRNAs appear to be slightly greater than 2 kb long, with the beta tubulin slightly larger that the alpha tubulin mRNA. It is interesting to note that although the beta tubulin gene is almost twice as large as the allpha tubulin gene, the final mRNA molecules are very similar in size. The sizes of the tubutin RNAs isolated from promastigotes are identical to those isolated from amastigotes. Detailed transcriptional analysis of the tubulin mRNAs has revealed that each mRNA is modified at its 5′-end to include the identical 35-nucleotide sequence. This 35-nucleotide sequence is not encoded upstream from the gene, but is spliced onto the structural gene RNA by a novel mechanism, described later, which results in the formation of a hybrid mRNA composed of transcripts from two distinct genes on separate chromosomes. DNA sequence and SI mapping analysis have led to the identification of putative gene control or RNA processing sequences.

The 35-nucleotide sequence is added onto the 5′-end of the structural gene mRNA at the position of a consensus splice site in both the alpha and beta tubulin upstream regions. There is a single splice site in the alpha tubulin and two putative splice sites in the beta tubulin. The most striking feature of the upstream region of both the alpha and beta tubulin genes is a tandem repeat of the dinucleotide AC just upstream from the RNA splicing site. The AC runs occur at different distances from the splice site in the two genes, but their position relative to the actual transcription initiation sites is unknown. Purine-pyrimidine runs of the type observed here may have the potential to form zDNA, but the role of DNA structure in gene expression in these parasites remains to be determined. In the alpha tubulin sequence, there is consensus polyadenylation site at the position of the 3′-end of the mRNA.

Other Protein Genes

Analysis of structural protein genes in *Leishmania* organisms has been limited mainly by the workers in the field. mRNA and DNA sequences homologous to the heat-shock proteins (HSP 70) have been identified both in African trypanosomes and *Leishmania* species. As described next, the genes encoding dihydrofolate reductase/thymidylate synthetase bifunctional enzyme have been isolated from drug-resistant parasites.

NOVEL BIOCHEMICAL AND MOLECULAR MECHANISMS

RNA Processing

The *Leishmania* species have an unusual mechanism of mRNA processing. Each mRNA is a hybrid RNA molecule containing RNA sequences derived from two distinct parts of the genome, in fact from two different chromosomes. Each mRNA contains both a unique sequence, which encodes a structural protein, and a common sequence, which is an identical 35-nucleotide cap at its 5′-end. In the *L. enriettii,* this cap, or spliced leader sequence (SLS), is encoded by a 438-bp gene arranged in a tandem array of approximately 200 copies. These spliced-leader genes are transcribed to give a 85-nucleotide precursor that has the 35-nucleotide cap sequence at its 5′-end. The genes for the SLS are located on a separate and distinct chromosome from that of the alpha and beta tubulins.

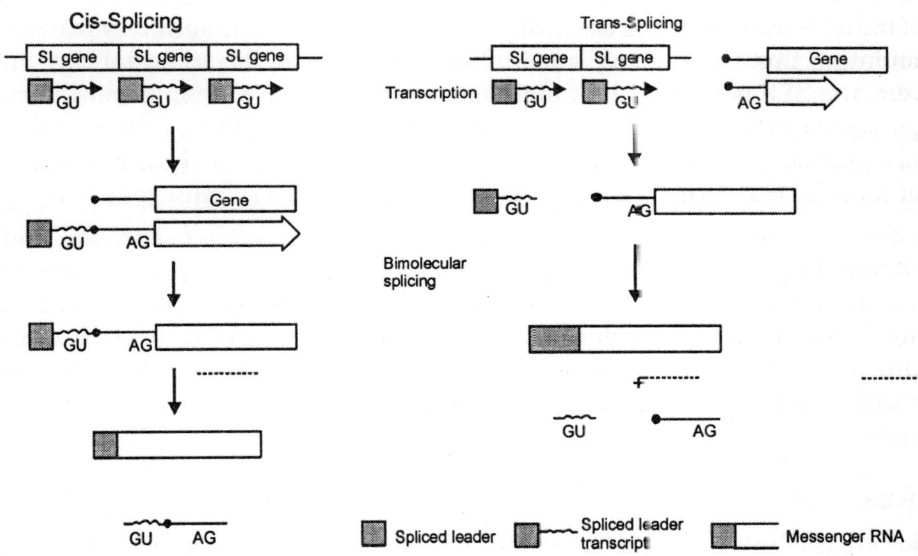

Fig. 7.14. Models of RNA processing after discontinuous transcription. In cis-splicing, the spliced-leader transcript could prime transcription of a gene, resulting in a precursor with an intron. Unimolecular splicing produces a single RNA by-product containing genomic sequence upstream to the AG consensus splice site and the remainder of the spliced-leader transcript. Alternatively, in cis-splicing, a ligation reaction after two unlinked transcription events could result in a unimolecular splicing event. In trans-splicing, independent transcription events followed by bimolecular splicing produce the two RNA by-products shown and mature mRNA. One RNA by-product contains a spliced-leader genomic sequence, and a separate by-product contains a transcribed genomic sequence from the gene processed to mRNA. Primed transcription after removal of the 3′-end of the spliced-leader transcript would also result in a RNA processing product comprising the free 3′-end of the spliced-leader transcript. GU represents the nucleotides of the consensus eukaryotic splice junction found in the spliced-leader transcript immediately 3′ to the spliced leader. AG represents the nucleotides of the consensus eukaryotic splice junction found immediately 5′ to the splice junction in all trypanosomatid mRNA genes examined to date.

An analogous gene was originally identified in *Trypanosoma brucei,* and work in that system has demonstrated that this 5 LS is added onto all mRNAs in the cell. The exact function of this SLS is not known. It may be a novel mechanism for the capping of mRNAs, or it may be important as a signal in the translation of proteins. Determination of function must await a transfection system that will allow site-specific mutagenesis of the gene. One of the major areas of current research is the elucidation of the mechanism of this novel hybrid mRNA formation. DNA sequence and primer extension analysis have demonstrated the presence of consensus splice sites both in the SLS transcript and in the upstream regions of the alpha and beta tubulin genes. There are two basic models for the addition of the SLS to mRNA precursors, the cis-splicing model and the trans-splicing model. In the cis-splicing model, the SLS is covalently attached to the primary transcript of the tubulin gene, either as a primer during transcription or posttranscriptionally by ligation, thereby creating a hybrid intron.

Subsequently unimolecular splicing occurs similar to that found in eukaryotic systems. This model predicts that there will be two products of splicing: a mature hybrid mRNA, with the SLS attached to the tubulin structural RNA, and the hybrid intron, consisting of the downstream region of the SLS transcript and the upstream region of the tubulin primary transcript. Evidence is accumulating that the splicing is more likely to be through a trans-splicing mechanism. In this model, the SLS gene and the tubulin structural gene are independently transcribed and splicing occurs between two separate transcripts. No covalent hybrid intron structure is formed. This model predicts that there will be three products of splicing: the hybrid mRNA, the downstream region of the SLS transcript, and the upstream region of the tubulin gene.

The downstream region of the SLS gene has been identified in *L. enriettii,* and recent work in *T. brucei* and *L. enriettii* has demonstrated that incubation with extracts known to debranch intron lariat structures results in the accumulation of the downstream region of the SLS transcript. What remains to be demonstrated is the presence of free upstream transcript from the tubulin or any structural gene. Future experiments in this area include the identification of splicing complexes and the isolation of enzymes responsible for this novel RNA processing mechanism.

Drug Resistance

Major areas of interest both for genetic analysis and for the more practical consideration of drug therapy are the mechanisms of drug resistance in the *Leishmania* organisms. Work in several laboratories has demonstrated that gene amplification plays a major role in the development of drug resistance. At least one mechanism of resistance is the overproduction of the target gene product in the resistant parasite, a pattern that is similar to the mechanism found in many drug-resistant mammalian cells and that may be an important in vivo step in the development of drug resistance. *Leishmania major* parasites have been selected for their resistance to the folic acid analogue methotrexate; in separate experiments, they have been selected for resistance to an inhibitor of thymidylate synthetase, 10-propargyl-5,8-dideazafolate (CB3717). These resistant parasites were selected after stepwise increases in drug concentration. *L. major* has a bifunctional enzyme that contains both DHFR and TS activity, and resistant parasites show an increased activity of the bifunctional enzyme. In mutants selected with methotrexate, this increased enzyme activity is accompanied by the amplification of two regions of the DNA: one of 30 kb, the R region, and the other of 80 kb, the H region.

In mutants selected with CB3717, the TS inhibitor, only the R-region DNA is amplified. These amplified DNA regions exist as circular extrachro-mosomal "factors" demonstrated by migration in chromosome gels and by visualization in the electron microscope. Initially, these circles were unstable in the absence of selective pressure; a decrease in the amount of amplified DNA was accompanied by a desrease in the bifunctional enzyme activity. After prolonged selective periods, stable mutants have been isolated. Analyasis of both stable and unstable resistant mutants by orthogonal field alternating gel electrophoriesis (OFAGE) demonstrates extrachromosomal circles homologous to the R and H regions. Thus, these circles must contain sequences for replication and stable inheritance. The structural gene for the bifunctional enzyme is located on the 30-kb amplified circle, and its protein sequence is derived from the cDNA sequence and genomic sequences.

It is interesting that the DHFR portion of the protein, located at the amino terminus has little primary sequence homology with mammalian or bacterial enzymes, whereas the TS or carboxyl portion of the enzyme has regions of significant homology with thymidylate synthetase isolated from other organisms. The DHFR portion of the protein does share secondary structure homology and known substrate binding sites with other DHFR proteins. Mutant parasites resistant to tunicamycin, an inhibitor of N-acetylglucosminyl-transferase in the dolichol pathyway of protein glycosylation, have been isolated in a similar stepwise manner. These resistant parasites have an increased level of the target enzyme; there is also evidence from a resistance-specific banding pattern of *Hind*III-digested, ethidium bromide-stained DNA that resistant parasites have amplified DNA.

The gene product of this amplified DNA has yet to be determined, however, it is interesting to note that these resistant mutants have increased virulence. Gene amplification of enzyme structural genes appears to be at least one important mechanism in the development of drug resistant in Leishmania species. An important question that remains to be addressed is the molecular mechanism of this circular amplification and its role in the development of drug resistance in the field.

Coda

The major development necessary to continue molecular genetic analysis of leishmanias is the development of a transfection system to allow the introduction and analysis of modified genes in these parasites. Initial attempts at this work in a related Kinetoplastida species, *Crithidia facisulata*, involved the introduction of a drug-resistant gene (G418) carried on a plasmid. Resistant parasites were isolated, but rearrangement of the plasmid NDA in these parasites made the interpretation of these results ambiguous. Several areas of research need to be pursued in order to develop a successful transfection system. First, some mechanism of introducing DNA into the cells is necessary.

Work in other eudaryotic systems suggests that electroporation, microinjection, or direct DNA addition should be tried. Second, a selectable marker, either drug resistance or an auxotrophic marker, is necessary. Finally, in order to achieve stable transformation, either an independent origin of replication or an integrating plasmid will be required. Highly repeated genes such as the SLS or tubulin genes may be good candidates for integration and the kDNA molecule or the amplified stable circles may provide the function of the origin of replication. One this transfection system is developed, of the structure and function relationships of the various putative controlling elements in gene expression can be directly assayed. This system also may help to define important regions of receptors or other virulence factors and could be the breakthrough that will take molecular work with this parasite from description to functional explanation.

BIOCHEMISTRY SCHISTOSOMES AND FILARIAE

Filariae and schistosomes rank among damaging helminth parasites infective to humans. Approximately 300 million people harbor schistosomes, 800,000 of whom die each year; filariae affect 300,000 individuals, and although they account for little mortality, they lead to blinding and disfigurement of significant numbers of people. Moreover, infections by these parasites tend to follow a chronic course that immeasurably saps the vitality and compromies the productivity of millions of children and adults. The study of the molecular biology of parasitic helminthes is a new discipline that arose from the opportunity to apply technologies offered by molecular genetics to questions raised by workers investigating the immunology, biochemistry, physiology, and biology of the organisms. Although it is clear that the use of recombinant DNA techniques in the study of schistosomes and filariae will primarily serve as an amplification system for providing quantities of proteins whose availability is otherwise limited, preliminary indications suggest that application of the new technologies also may aid in unraveling basic biological questions.

Vaccine Development

The goal of conferring sterilizing, or even partial, immunity to humans against schistosomes and filariae is obviously a desirable objective, and the search for safe, inexpensive vaccines that can impart this protection is a high priority. Such vaccines would complement current efforts to control these diseases with chemotherapy, vector control, and education. Although scientific progress to date is encouraging, scientific and economic considerations make if currently unclear whether such vaccines can be developed or if a large-scale vaccination program could be implemented. A major motivation to develop vaccines is the realization that in the absence of protective immunity, individuals treated with antihelminthic drugs tend to become reinfected. The life span of schistosomes and filariae within a vertebrate host is extraordinarily long; schistosomes have been documented to remain viable and fecund for up to 35 years. They stimulate a host immune response of variable intensity, and how they survive host attack has intrigued many scientists. While investigators for decades have attempted to identify antigens having immunoprohylactic potential, recent molecular and immunological methodologies have provided several powerful approaches to the problem.

Different laboratories have tackled the issue in different ways, with some degree of success. Convinced that target antigens should be routinely exposed to the immune system, some workers have proceeded to isolate and genetically express components of the parasites' surface or excretory/secretory antigens, and are testing their prophylactic efficacy in murine models. Others have produced and characterized monoclonal antibodies that were chosen for their ability either to kill worms in vitro or to confer resistance to infection after passive transfer into immunologically naïve recipients. Antibodies identified in this manner have then been used as reagents for isolating the protective antigen itself, usually from recombinant DNA sources. Others have concentrated on the identification of antigens recognized by antibody from hosts naturally resistant to infection but not recognized by antibodies of susceptible hosts.

The most thorough approach has involved fractionation of parasite extracts, assessment of the protective value of each fraction, and purification of active components. None of the candidate antigens characterized to date has proved capable of engendering sterile immunity, although this failure may reflect suboptimal stimulation of appropriate T-cell populations or of B cell of the right isotype rather than identification of an unsuitable antigen.

Schistosomes

The characterization of protective antigens with monoclonal or monospecific polyclonal antibodies that are cytotoxic to schistosomes has yielded a perplexing catalog of molecules. For example, preliminary data indicate that enzymes of the parasite's glycolytic pathway may represent important vaccine candidates. Adult schistosomes and filariae are anaerobes that use glucose as their sole source of energy. Their consumption of the sugar is prodigious: schistosomes consume the equivalent of 20 percent of their dry weight per hour. The importance of glycolysis to worm survival has been known for many years, and trivalent antimonials (once used as antischistosomal agents) appear to kill the organisms by interfering with the action of phosphofructokinase. Independent research groups have characterized three apparently protective antigens and identified them as aldolase, glyceraldehyde-3 phosphate dehydrogenase, and triose phosphate isomerase. Curiously, the enzymes catalyze sequential reactions that form the core of the glycolytic pathway.

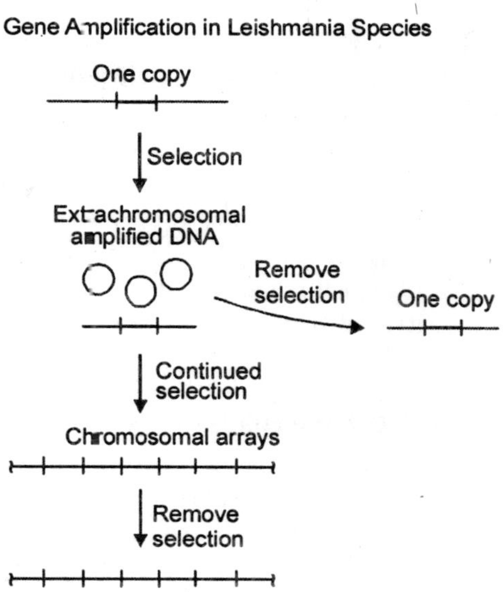

Fig. 7.15. Summary of the molecular structure of amplified DNA in unstable and stable *Leishmania* lines resistant to methotrexate.

It has recently been demonstrated that aldolase also is a potential vaccine target of *Plasmodium falciparum.* Should these proteins indeed impart protection when used as vaccines, their interplay with the immune system would raise several questions. For example, since these enzymes are not rate-limiting to the pathway (whereas membrane transport and the three kinases are), it is surprising and unclear why immune responses to them have antihclminthic consequences. Furthermore, the fact that the enzymes are normally found in the cytoplasm of eukaryotic cells raises the perplexing question of how the immune system can launch an attack on them in intact organisms. Finally, the amino acid sequences of the three proteins are very similar to host enzymes, which raises the concern that vaccines based on these proteins might elicit autoimmune responses. Another empirically determined class of schistosome vaccine candidates consists of glutathione metabolizing enzymes. One of these proteins was selected on the basis of the observation that mice resistant to *Schistosoma japonicum* infection mount a response to an antigenic enzyme, whereas mice that are susceptible to infection do not.

The antigen has been cloned and expressed in bacteria, and has been found to be a 26-kilodalton (kD) isoenzyme of glutathione S-transferase. Unlike the case for the glycolytic enzymes, the protein demonstrates only a modest degree of resemblance to the host homologue. Mice infected with *S. japonicum* produce antibodies that cross-react with the enzyme from *Schistosoma mansoni,* and sequencing of the S. *mansoni* homologue has revealed that they are nearly identical in structure differ significantly from a 20-kD isoenzyme isolated from *S. mansoni*. The similarity between the enzymes of the two parasite species is much greater than that noted for their heat-shock protein, HSP-70 (which ranks among the phylogenetically most conserved proteins), and is reflected in the sequence homology of 5′ untranslated portions of their RNAs, which begin with an unusual string of polyT.

Aldolase

```
Sm ald  RHAADESTATMGKRLQQIGVENNEENRRLYRQLLFSADHKLAENI SGVILFEETLHQ
RM ald  IL********GSIA****S**T**T*****F*****LTA*DRVNPC*G*****H***Y*
HL ald  IL*******VG **N***R*K***T*****QF**I***V*SSINQS*G*****H***Y*

Sm ald  KSDDGKTLPTLLAERNIIPGIKVDKGVVPLAGTDNETTTQGLDDLASRCAEYWRLGC
RM ald  *A***RPF*QVIKSKGGVV*************NG******G*SE***Q*KKD*A
HL ald  KDSQGKLFRNILKEKGIVV***L*Q*G*******NK***I****G*SE***Q*KKD*V

Sm ald  RFAKWRCVLKISSHTPSYLAMLEMLMYL
RM ald  D**********GE*****A**IM*NANV*
HL ald  D*G***A**R*ADQC**S**IQ*NANA*
```

Glyceraldehyde-3-Phosphate Dehydrogenase

```
        1                                                          60
SCH     GSRAKVGINGFRGIGRLVLRAAFLKNTVDVVSVNDPFIDLEYMVYMIKRDSTHGTFPGEV
HUM     MGKV***VN**GR*****T****NSGK**I*AI********N*****FQY*****KFH*T*

                                                                  120
SCH     STENGKLKVNGKLISVHCERDPANIPWDKDGAEYVVESTGVFTTIDKAQAHIKNNRAKKV
HUM     KA*****VI**NP*TIFQ****SK*K*GDA****************ME**G**LQGG-**RV

                                                                  180
SCH     IISAPSADAPMFVVGVNENSYEKSMSVVSNASCTTNCLAPLAKVIHDKFEIVEGLMTTVH
HUM     ***************M***HEK*DN*LKII*************************N*G************
```

Fig. 7.16. Inferred amino acid sequence of the amino termini of two putative schistosome vaccine target antigens. In the top sequence, the primary sequence of a clone identified as encoding the schistosome aldolase (sch ald) is compared to that of clones encoding rat muscle aldolase (RM ald) and human, liver aldolase (HL ald). In the bottom panel, the sequence of the schistosome glyceraldehyde-3-phosphate dehydrogenase is compared to an isoenzyme from human beings. Asterisks represent regions of identity.

A second isoenzyme of the glutathione S-transferase of *S. mansoni* also has prophylactic potential. The protein has a size of 28 kD, shows little sequence similarity to the 26-kD isoenzyme, and induces production of antibodies that do not cross-react with the 26-kD homologue or with homologues of the host. The enzymes are normally found in the cytoplasm or microsomal fraction of metazoan cells; however, there are indications that they may also be found on the surface of the parasite. The functional role of the schistosome isoenzymes is unknown, but their presence in the tegument suggests that they may play a role in protecting the organisms from oxygen-mediated damage. Protective responses stimulated by the antigen appear to involve the production of specific IgGZa antibodies that direct eosinophil-mediated cytotoxicity. Another unexpected schistosome vaccine candidate is paramyosin.

The peptide was identified by fractionation of parasite extracts and testing the degree of protection conferred by each fraction in a mouse model. A cDNA encoding part of the schistosome paramyosin has been used to isolate a filarial homologue, and this homologue has been shown to confer partial resistence to filariae in immunized animals. Paramyosin is found in invertebrate catch muscles, where, in association with myosin, it forms complexes that permit strong, protracted muscle contraction, as seen in the adductor muscles of clams. Immunocytochemical studies indicate that the protein is also found in nonfibrillar form within elongate membrane-bound structures within the worm's tegument, although it is apparently not exposed on the surface. Like schistosome myosin, sequenced portions of the parasite's paramyosin indicate that it consists of a series of seven amino acid repeats (a,b,c,d,e,f,g), in which residues a and d are generally hydrophobic, and which are part of a 28-amino-acid unit.

Table 7.2

Inferred amino the amino acid sequences of the amino termini of the 26-kilodalton (kD) glutathione S-transferase of *S-Japonicum* (Sj26), the rat yb isoenzyme-(RYb),	Sm26	MAPKLGYWKIKGLVQPTELLLEYL
	Sj26	TKL * I***********************
	RYB	* PMI ****NVR**TH*I******T
	SYC	* PGG*PVLHYFDGRGFMEP I RWL*
	Sm28	* AGEHIKVIFDGRGRAES I RMT
the rat yb the rat yc (Ryc), and a 28-kD isoenzyme from S.mansoni (Sm 28) compared to the 26(kD) protens Mansoni (Sm 26). Asterisks represent regions of idecutety	Sm26	(#68) DVKLTQSMA I LRY I ADKHNMLGGCPKERAEISMLEGA
	Sj26	(#68) **********I**-******************************
	RYb	(#67) SR * I ***N**M**L*R*HLC*ETEE***IRADIV*NQ
	RYc	(#62) GM **V*TR***N***T*YALY*KDM****L*D*YSEC
	Sm28	(#65) VKWMLE*L **A**M*K*H**M*ETDE*YYSVEK*I*Q

Acidic and basic periods within the 28 amino acid units are almost out of phase with each other, so that each segment can be subdivided into zones of alternating charge. In general, it appears that the molecule, like myosin, evolved from a 28-amino-acid peptide whose gene underwent a series of duplications and that the register between the molecule and myosin is mediated by ionic interactions. The amino acid sequence of paramyosin predicts that the molecule forms a classic a helix, where an antiparallel second strand can make periodic contacts at regularly spaced intervals where amino acids having a nonpolar side chain are positioned. Amino acids facing the aqueous environment of paramyosin dimers tend to have hydrophilic side chains. Once the complete amino acid sequences of schistosome myosin and paramyosin are determined, it will become possible to construct and test models on how the two proteins interact.

Unlike the case of glutathione S-transferases, paramyosin appears to stimulate T lymphocytes to produce interferon gamma, which activates macrophages to kill larval parasites. Notably, it seems that antibodies play no role in this cytotoxic reaction. The protein also may elicit an inflammatory reaction in sensitized hosts, which could wall off invading larvae. Another schistosome antigen — myosin — has been cloned and is recognized by antibodies generated by animals that are poorly resistant to infection by sera from susceptible animals. Antibodies against the antigen rise dramatically in patients after chemotherapeutic cure. The protein is unique in being encoded by a single-copy gene (all organisms studied to date have their myosins encoded by a multigene family).

Preliminary indications are that, unlike paramyosin, schistosome myosin is not a promising vaccine candidate. Taken as a whole, these findings add a new dimension to our understanding of schistosome antigens susceptible to cytotoxic arms of the immune response. Particularly puzzling is the mechanism by which host antigen processing cells manage to recognize cytoplasmic components of living schistosomes and subsequently translate this information into an immune response capable of homing in on molecules enclosed by a plasma membrane. Also surprising is the fact that nearly all potentially protective antigens sequenced to date share significant blocks of homology with host homologues, yet they apparently do not elicit an autoimmune response.

Whether this is attributable to special modes of presentation of the antigens by the parasites or to an inherent lack of immunogenicity of evolutionary conserved domains is unknown, but the question has obvious relevance to vaccine development. The observation that schistosome antigens contain regions whose structure is highly conserved suggests that these may have important functional purposes.

Filariae

As is the case with schistosomes, experimental evidence suggests that it is possible to develop protective immunity to filariae. Several workers have reported that immunization with irradiation-attenuated L_3 larvae can lead to a significant depression of adult worm burden and microfilaremia. Monoclonal antibodies against 70-kD and 75-kD stage-specific surface antigens of *Brugia malayi* have been developed that upon passive transfer lower the microfilaremia of infected animals. Immunization of gerbils with a soluble extract of *B. malayi* induces a 50 percent reduction in adult worm burden and a 80 percent to 100 percent decrease in parasitemia on challenge.

The development of resistance correlates with the appearance of IgG antibodies directed at proteins of 25 kD, 42 kD, 60 kD, and 112 kD. Immunization of animals with purified 25 kD and 60 kD proteins leads to their ability to lower the number of circulating microfilariae. A cDNA encoding the 60-kD antigen has recently been cloned and sequenced. The predicted size of the cDNA's translation product is 63 kD, and the antigen does not have any consensus glycosylation sequences or any hydrophobic domains suggestive of membrane-anchoring sites. Interestingly, the mRNA for the protein begins with a 5′ untranslated leader sequence that is identical to the trans-spliced leader of some actin genes of *Caenorhabditis elegans* (a free-living nematode). The 22-nucleotide spliced leader is present in several locations in the parasite's genome, including (as in *C. elegans)* within the 5s rRNA gene repeat.

The spliced leaders are located at the 5′-end of a 109-base pair (bp) transcriptional unit of nonpolyadenylated RNA. The role of the spliced leaders, and why some genes acquire this sequence (as opposed to the trans-spliced leaders in trypanosomes, which is a universal feature of the mRNAs of the organisms) is unclear and awaits the development of appropriate assay systems which can evaluate their significance.

Chemotherapy

The drugs of choice currently used in the treatment of schistosomiasis and filariasis are praziquantel and ivermectin, respectively. Praziquantel is effective in a single oral dose and appears to paralyze the organisms by perturbing their tegument and by affecting calcium transport; the immune system probably plays a role in killing and clearing organisms debilitated by the drug. Praziquantel is also effective against other trematode and cestode parasites but not against nematodes whose cuticle blocks the action of the drug. Verapamil, a slow calcium blocker, also disrupts the schistosome tegument and may prove a less expensive alternative to praziquantel. The semisynthetic antibiotic mixture, ivermectin, similarly affects calcium fluxes at neuromuscular junctions, presumably by increasing the secretion of gamma amino butyrate (GABA) and enhancing the binding of the neurotransmitterf to its receptor.

Although not a perfect drug, ivermectin is a welcome alternative to diethylcarbamazine, in that it kills microfilariae more slowly, thereby decreasing the possibility of inducing potentially life-threatening Mazzoti-like reactions. The latter are thought to be allergic responses triggered by a sudden appearance of high levels of antigen released by dying microfilariae. The perceived need to develop new antihelminthic drugs stems from three problems. First, researchers suspect that the parasites will, in time, develop resistance to currently available drugs. Second, current therapeutic regimens are too expensive to apply on a large scale, and reinfection is a constant threat. Finally, in the case of filariasis, available drugs have serious side effects and are not completely effective.

It has been recently noted that veterinary use of ivermectin may introduce new ecological problems because the compound is not biodegradable and is toxic to a wide variety of invertebrates. Combining methods in molecular biology with emerging ones in computer modeling can in theory strengthen and facilitate innovative approaches to the rational development of therapeutic drugs. Antihelminthic drugs currently in use are all products of empirical research, an inefficient method considering that on the average only 1 in 10,000 tested compounds shows any activity.

The concept behind rational drug design is to take advantage of biochemical idiosyncrasies that distinguish parasite from host metabolism. This might be accomplished in at least two ways. First, methods in molecular genetics can be used to clone and express active parasite enzymes in more manipulate hosts (yeast and bacteria). If enzymatically active, the enzyme can then be studied by conventional biochemical techniques. If the gene can genetically complement a yeast strain deficient in the enzyme, it is conceivable that the protein's properties can also be studied in vivo. In the other approach, the primary sequence of the protein is determined, and if the three-dimensional structure of a homologue from another organism is known, predictions may be made about the conformation of the parasite peptide.

After determination of the probable active site is made, the molecules can be modified by in vitro mutagenesis in order to test the validity of the predicted model. Several laboratories are currently developing algorithms to predict the structure of chemicals that might interfere with the function of active sites of known morphology and chemistry. One group, for example, has developed a model that correctly predicted that methotrexate should inhibit the activity of the dihydrofolate reductase-thymidylate synthetase of leishmania. One of the better examples of rational identification of drug targets has centered on how purine nucleoside pools are formed in *S. mansoni.* This organism is incapable of synthesizing these compounds de novo and must rely on salvage pathways involving acquisition of precursors from the host. Purines and their respective nucleosides gain entry into the parasites by carrier-mediated mechanisms involving at least five transporters.

The surface of the worm contains nucleotidases that may be coupled to nucleoside transport. Culture media preincubated with adult schistosomes (which regularly slough off their outer surface) are capable of converting adenosine to adenine, inosine, and hypoxanthine, suggesting the existence of a salvage pathway consistent with the scheme and noteworthy in three ways: (1) the organisms lack a nucleotide kinase, so that the initial substrate for nucleic acid synthesis is always a free base; (2) adenosine deaminase activity is developmentally regulated, being absent in larval worms, which therefore cannot interconvert adenosine and guanosine nucleosides; and (3) a single enzyme, hypoxanthine guanidyl phosphoribosyl transferase (HGPRTase), catalyzes the ribosylation of guanidine and hypoxanthine.

The proposed pathway indicates that the availability of guanosine nucleotides is strictly dependent on the activity of the parasite's HGPRTase. Studies on the schistosome enzyme

HGPRTase have shown that it is structurally different from mammalian homologues, which makes it a plausible drug target. A cDNA encoding a schistosome HGPRTase has recently been cloned and sequenced. The message for the protein is approximately 1.5 kb in size and hybridizes to a cDNA encoding a mammalian counterpart under conditions of moderate stringency. The predicted size of the protein is 26 kD, indicating that the active 105,000-dal-ton (D) protein is a tetramer.

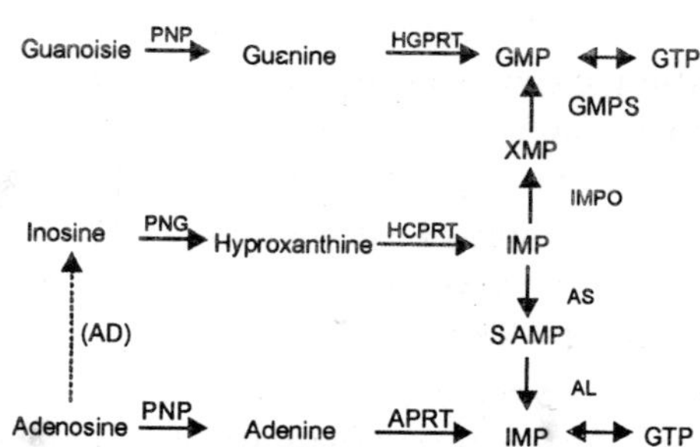

Fig. 7.17. Purine salvage pathways of schistosomes. Abbreviations: PNP, purine nucleoside phosphorylase; HGPRT, hypoxanthine-guanine phosphoribosyl transferase; GMPS, guanosine monophosphate transferase; IMPD, inosine monophosphate dehydrogenase; AS, adenylosuccinyl AMP synthetase; AL, adenylosuccinyl AMP lyase; AD, adenine deaminase. Solid lines represent general pathways; dashed lines indicate stage-specific ones.

Alignment encoding the predicted aminojacidseduence of the cDNA for the schistosome enzyme with that of humans and *Plasmodium falciparum* indicates that the enzyme is more closely related to that of humans (47.9 percent versus 36.9 percent). The areas of greatest identity between the human and schistosome enzyme correspond to the putative purine and phosphoribosyl binding sites. Interestingly, despite considerable differences in amino acid sequence the predicted α-β-α structure of the two enzymes is very similar and demonstrates considerable differences from that of malarial parasites. The 13 amino acid sequence stretch corresponding to the putative binding sites of the human and schistosome HGPRTase differs by one conservative and one nonconservative substitution, suggesting that development of inhibitors acting at this site will prove difficult. Several critical proteins of filariae that differ significantly in structure from those of mammals have been identified. These include their respiratory electron pathway complex III, tubulins, chitin, proteinases, amino acyl tRNA synthetases, and glutathione-folic acid metabolizing enzymes.

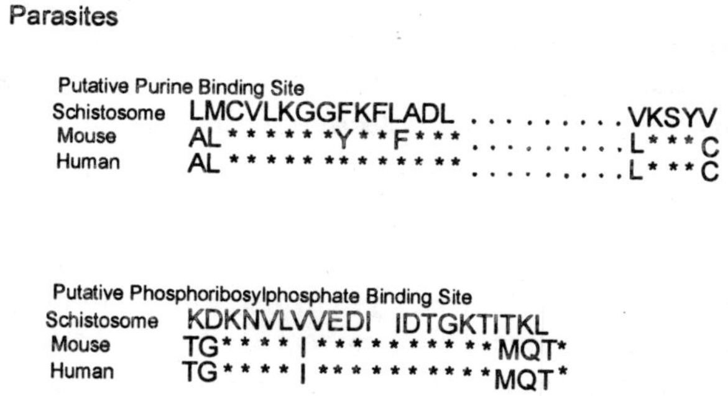

Parasites

```
Putative Purine Binding Site
Schistosome  LMCVLKGGFKFLADL . . . . . . . . . .VKSYV
Mouse        AL* * * * * *Y* *F* * * . . . . . . . . . .L* * *C
Human        AL* * * * * * * * * * * * * . . . . . . . . . .L* * *C

Putative Phosphoribosylphosphate Binding Site
Schistosome  KDKNVLVVEDI  IDTGKTITKL
Mouse        TG* * * * | * * * * * * * * * * *MQT*
Human        TG* * * * | * * * * * * * * * * * *MQT*
```

Fig. 7.18. Amino acid sequences of the putative purine and phosphoribosyl-phosphate-binding sites of the schistosome-XGPRTase are compared with those of the enzyme from mice and humans. Asterisks represent points of identity; periods denote the presence of amino acids that it do not appear to participate in binding.

There are indications that the filarial aldolase exerts a rate-limiting action on glycolysis since its activity is lower than that of phosphofructokinase, an enzyme that can catalyze 10 times more substrate than that present at steady-state levels. The filarial aldolase thus appears to be a nonequilibrium regulatory enzyme, and may prove a good chemotherapeutic target; even slight inhibition of such enzymes can greatly reduce the metabolic flow through the pathway.

Whether this is also true in malarial and schistosome parasites, where preliminary indications are that the enzyme may prove a vaccine target, is unclear. Derivatives of amoscanate, a noxious compound previously used in the treatment of hookworm and schistosome infections, have been shown to have microfilaricidal action. Part of the activity of these compounds appears to be attributable to inhibition of aminoacyl tRNA synthetases specific for arginine, lysine, and serine. The antimalarial compound mefloquine similarly inhibits filarial aminoacyl tRNA transferases, but not those of mammals, indicating that these enzymes represent a potential drug target.

Developmental Biology and Stage-Specific Gene Expression

Filaria and schistosome worms display a range of intricate developmental plans involving permutations of (1) life in aquatic, vertebrate, and invertebrate habitats; (2) shifts in temperature, ambient gas phases, and available nutrients; and (3) changes in the identity of hostile environmental pressures. The mechanisms whereby the organisms find it possible to transform from one stage to another are poorly understood, and how these are regulated at the level of gene expression is completely unknown. Several laboratories have recently begun to address the issue, primarily by analyzing stage- and tissue-specific transcripts. The parasites under discussion are genetically programmed to sequentially and specifically survive within widely different habitats. During these transitions, they display both non-stage-specific (glycolysis, DNA replication, transcription and translation, and gluconeogenesis) and stage-specific (respiratory pathways used by larval forms but not adults) metabolic strategies. While the inference of differential gene expression is easy to understand in the case of stage-specific gene expression, many workers have suggested that common metabolic pathways may be mediated by isoenzymes subject to stage-specific appearance.

Studies in comparative enzymology have revealed that protein-catalyzed reactions are subject to ionic, pH, and temperature defined modulation. Temperature is particularly important, as it influences rates of reactions, rates of protein denaturation, affinity toward substrate, and pH dependence. These considerations may critically affect the function of near-equilibrium, regulatory, or membrane-bound enzymes (changes in temperature affect membrane fluidity. How schistosomes and filariae have solved the problem of different stages adapting to different environments (for example, see Chapter 6) is unknown. The solution might involve the expression of separate genomic segments, the synthesis of enzymes that have broad physical optima, or synthesis of developmentally regulated factors that, depending on need, modulate enzyme function.

Schistosome glyceraldehyde 3-phos-phate dehydrogenase is encoded by a single-copy gene, indicating that the enzyme expressed during development in the snail host derives from the same genetic locus as that expressed during the parasitism of mammalian hosts. Some preliminary observations that deal with the developmental biology of the organisms are discussed next.

Developmentally Regulated Proteins of Unknown Function

Stage-specific gene expression by filariae and schistosomes requires a broad range of intrinsic genetic versatility, allowing the organisms to alternate from a poikilothermic to a homeothermic environment. Although investigations are still limited by technology (*e.g.,* development of systems for genetic analysis, DNA transformation, and in vitro culture systems), examination of the subject has in principle been made more accessible by recent advances in molecular biology. In one study, an expression cDNA library was screened with serum from schisto-some-infected patients. Of 21 cDNA clones expressing parasite antigens, three proved to be developmentally regulated. Northern analyses of adult worm, egg, and cercarial poly A^+ RNA revealed that two of the three stage-specific cDNAs hybridized to mRNAs which change in length during development.

While this observation in itself is not particularly unusual, the observation that the two RNAs appear to be transcribed from the same (single copy) gene is significant. How, and to what extent, schistosomes manage to produce multiple transcripts from a single gene is unknown; however, preliminary results indicate that the size variations noted above are due to the use of alternate start sites and alternative RNA processing.

Proteinases

One of the more surprising features of the schistosome proteins that stimulate a mammalian immune response is that most of these are synthesized by invertebrate as well as vertebrate hosts. A notable exception to this trend is the organism's proteinases, which appear to play specialized roles at different phases of development. Helminth proteinases are interesting for several reasons: in addition to playing a role in parasite nutrition, they may be essential for invasion of mammalian hosts, they may play a role in stimulating inflammatory reactions, they may be important for differentiation, and they may participate in the production of egg yolk.

It has been hypothesized that proteases secreted by the parasite or anchored onto its membranes may serve to evade the immune response by hydrolyzing bound immunoglobulin so that the Fc portion, necessary for recognition by effector cells, is removed. As many of these proteases are immunogenic, they may prove to be valuable diagnostic reagents and vaccine candidates. Two hydrolases, referred to as hemoglobinases, responsible for the digestion of hemoglobin by schistosomes have recently been cloned and expressed in *Escherichia coli.* One of the enzymes is a thiol proteinase that displays enzymatic similarities to vertebrate cathepsin B_2 and is synthesized by cells lining the parasite gut. Products of its activity are hematin, which is regurgitated, and a large number of oligopeptides, which result from digestion of hemoglobin.

The fate of the oligopeptides is unknown but it has been hypothesized that they are shuttled into the next anatomical compartment — the vitellaria, in the case of female worms, which consume the bulk of the hemoglobin — where they serve as egg yolk for the soon-to-develop miracidium. The enzyme is not essential to the parasites: they can survive for several months in protein-free, chemically defined media. On the other hand, worms maintained in culture quickly lose the ability to deposit eggs, and female schistosomula grown in media containing red blood cells reach sexual maturation, whereas those grown in the absence of the cells do not.

Moreover, unisexual female worms, or female worms grown in nonpermissive hosts, contain little pigment in their guts and do not achieve sexual maturity. The inferred ammino acid sequence of one of the cloned cDNAs encoding the enzyme is not homologous to any known protease. This is somewhat surprising as cystinyl proteases from slime molds to humans share recognizable blocks of homology, especially around regions forming the active site. The other shows clear homology to vertebrate cathepsins. Northern analysis indicates that female worms make more of the enzymes than do male worms and that only trace levels are produced during the intramolluscan stage of the life cycle.

The enzymes are predictable immunogenic in natural infections, induce a hypersensitivity response, and are one of the more promising substrates for development of an immunodiagnostic assay system. Another stage- and cell-specific schistosome proteinase has been cloned from a sporocyst cDNA library. The cDNA encodes the cercarial elastase which facilitates the schistosome breach of the human skin barrier, and is arguably the best-characterized macromolecule of a parasitic helminth.

It is a calcium-dependent serine protease that is exquisitely designed to fragment the major components of skin. In addition to elastase, the enzyme is capable of digesting keratin, types IV and VIII collagen, proteoglycan, fibronectin, and laminin. Molecular cloning of the cercarial elastase has provided new information about how the enzyme is regulated, how it functions, and where it is synthesized. Nucleotide-sequence analyses of the cDNA predicts that its mRNA untranslated leader, which is decidedly AT-rich, is unusually long (> 220 bp). The translation product of the longest open-reading frame indicates that the protein possesses an amino terminal extension of 27 amino acids not present in the mature enzyme. Because the extra amino acids are predominantly hydrophobic, they probably represent a peptide leader that maintains the enzyme in an inactive (zymogen) form. The length of the leader is also consistent with its potential functioning as a signal peptide that directs export of the elastase into secretion granules.

Notably, activation of the enzyme is predicted to occur at a site recognized by the enzyme itself, suggesting that once a few zymogen molecules are activated by autocatalysis, a massive burst of autocatalytic activation might occur. The preproenzyme consists of 244 amino acids, which, aside from regions flanking the active site histidine, aspartate, and serine, and regions corresponding to probable protein-folding areas, shares little homology with that of other serine proteases. The mature protein starts with the sequence IsoArgSerGly, which differs radically from the consensus ValValGlyGly, characteristic of serine proteases from invertebrate and vertebrate sources. Together these data suggest that the immune response stimulated by the enzyme should not cross-react with host tissues. In situ nucleic acid hybridization studies indicate that the cercarial protease mRNA is transcribed by the postacetabular gland cells (also known as the escape glands or penetration glands) during early stages of cercarial development, and studies involving monoclonal antibodies jndicate that the inactive form of the enzyme is stored within these cells in secretory vesicles.

Southern analyses indicate that the protein is encoded by one or a small number of genes that, because of the presence of intervening sequences, extend over several kilobase. Understanding the biosynthesis and activation of the schistosome penetration — elastase suggests that the potentiation of schistosome penetration of the host skin and its subsequent access to the circulatory system follows this course of events. At the 32-cell stage of cercarial development (when the acetabular cells, which are the largest cells in the organism, first become microscopically obvious) or shortly thereafter, the cercarial elastase gene is transcriptionally activated in acetabuler cells and an mRNA is expressed whose translational product is inactive and contains signals that direct it toward a subset of actabular secretion granules.

Elastase mRNA synthesis ceases sometime before full maturation of the cercaria. Upon contact with skin lipids and stimulation by heat, the cercaria secretes the storage granules and somehow a small number of preproelastase molecules become activated, perhaps by specific cleavage by another unidentified parasite protease. The remaining share of elastase zymogen would then be activated by autocatalytic processing. The observation that the active enzyme can be isolated from cercariae artificially exposed to linoleic acid suggests that host factors are not important in its activation. Preliminary data indicate that interference with the activity of cercarial elastase may be of practical value.

```
                                         -1
SE       MSNRWFVVVVTLFTYCLTFERVSTWL
REI       MLRFLVFDASLV*YGHSTQDFPETNAR
REII   MIRTLLLSALVGALSCGYPTYEVQHDVS

       1                   20                   40                  60
SE     IRSGEPVQHPAEFPF-----IAFLTTERTMCTGSLVSTRAVLTAGHCVCSPLPVIRVSFL
REI    VVG*AEARR-NSW*SQISLQYLSGGSWYHT*G*TIRRNW*M**A***S----------
REI    VVG*QEASP-NSW*WQVSLQYLSSGKWHHT*G****ANNW****A**IS---------

                           80                   100                 120
SE     TLRNGAQQGIHHQPSGVKVGYMPSCMSARQRRPIAQTLS--GFDIAIVMLAQMVNIQS
REI    SQMTF*VVVGD*NLSQNDGTEQ*VSVQKIMVHPTWNSNNVA-A*Y***LLR***S*T**NN
REII   NS*TYRVLLGR*SLSTSESGSLAVQVSKLVVHEKWN**K**NN*N***L*K**SP*A*T*

                           140                  160                 180
SE     GIRVISLPQPSDIPPPGTGVFIVGTGRDDDND---RDPSRLNGGILKKGRATIMECRHAT
REI    YVQLAV****EGT*:LANNNPCY*T*Q**TRT*GQLSQTLQQAYLPSVDYSICSSSSYWGS*
REII   K*QTAC**PAGT*LPNNYPCYVT*W**LQT*GAT*DVLQQGRLLVVDYATCSSASWWGSS

                           200                  220                 240
SE     NGNPICVKAGQNFGQLPAPGDSGGPLLPSLQGPVLGVVSHGVTLPNLPDIIVEYASVARM
REI    VKTTTM*C**GDGVRSGCQ*******HCL---VNGQYSV****SFVSSMGCNVSKKPTVF
REII   VKTTNM*C**GDGVTSSCN*******NCQ--ASNGQWQV**I*FFGSTLGCNYPRKPSVF

SE     VDFVRSNI
REI    TRVSAYISWMNNVIAYT
REII   TRVSNYIDWINSVIAKN
```

Fig. 7.19. The predicted schistosome elastase (SE) amino acid sequence is compared with that of rat pancreatic elastases I and II. Abbreviations: REI, rat pancreatic elastase. I; REII, rat pancreatic elastase II. Dashes indicate gaps and asterisks regions of identity. Components of the catalytic triad (histidine 46, asparate 106, and serine 202) are shown in bold lettering, and the numbering system is arbitrary. The leader sequences are shown on the top lines.

Monoclonal antibodies that both inhibit the enzyme activity and are cytotoxic to schistosomula have been developed, suggesting that the enzyme may represent a realistic vaccine target. Studies involving the use of chemicals that inhibit the enzyme indicate that the enzyme is essential for successful parasite invasion; parasites whose elastases are irreversibly inactivated do not reach the vasculature and instead die in the skin. Interestingly, the interstitial fluids of vertebrate skin contain proteins that are broad-spectrum inactivators of serine proteases. These antiproteases, which exist in multiple forms, are synthesized in the liver, from which they are subsequently secreted into the plasma.

The reactive center of the inhibitors lies near their carboxy terminus and acts as bait for foreign proteases that covalently bind to it and thereby inactivated. The finding that these proteins appear to have evolved in an accelerated Darwinian manner has led to the suggestion that they have been selected for their ability to recognize tissue-invasive parasites such as *Pseudomonas* organisms, schistosomes, and filariae. Whether helminth proteases secreted during tissue migration also have evolved in an accelerated manner is unknown, but it is noteworthy that the major protease of *Schistosomatium douthitti,* another digenetic trematode, is completely different from that of *mansoni*. Schistosomes possess two other proteases that appear to be stage-specific and that are thought to have a specialized function.

One is a surface-bound or secreted enzyme that degrades immunoglobulins so that the Fab fragment remains bound to the worm's tegument while the Fc fragment, which is the portion recognized by effector cells, is released into the circulation. The bound immunoglobulin therefore presents a hostlike barrier to the immune system. Some of the peptides generated by degradation of immunoglobulins appear to be immunosuppresive and may therefore further facilitate survival in the vertebrate host. The other enzyme is a protease synthesized within schistosome eggs. Although the function of this protease is unknown, some workers believe that it may play a role in assuring that eggs laid in the circulatory system manage to work their way into the lumen of the intestine. Less is known about the proteases of filariae.

Microfilariae of *Onchocerca volvulus* contain a collagenase that may facilitate their movement through interstitial spaces. The enzyme is immunogenic, shares epitopes with bacterial collagenases, and may act as a virulence factor by degrading host elastin and collagen fibers, thereby-leading to loss of elasticity of the skin and the "hanging groin" syndrome. The enzyme also may expose new epitopes of host collagen molecules to the immune system, and therefore lead to autoimmune manifestations. *O. cervicalis* contains a metaloprotease of 60 kD and a serine protease of 30 kD which display collagenase activity. Extracts of *Dirofilaria immmitis* microfilariae contain a 22-kD and a 76-kD protease. Proteolytic activity has also been noted in third-stage larvae of *Brugia pahangi* and B. *malayi.*

The role of the latter in skin invasion has not been determined. In general, the observation that tissue-invading helminths possess and secrete collagenolytic or elastinolytic enzymes is consistent with their migratory behavior. These classes of enzymes are notable for having a broad spectrum of substrate specificities; they therefore can degrade a large number of different proteins. The relative importance of these enzymes in the dissolution of tissue barriers, however, has proved difficult to assess. On an equally conjectural plane, it is possible that these enzymes play an additional role in preventing activation of host blood clotting mechanisms by interfering with proteolytic events in the activation of coagulation cascade.

Eggshell Proteins

The schistosome eggshell is a highly resilient, somewhat pliable biopolymer that appears to be composed of proteins cross-linked with each other by quinone bridges. The complex is highly resistant to proteases and contains pores that presumably allow the influx of nutrients required by developing miracidia and that permit the secretion of macromolecules, which lead to granuloma formation. Acid hydrolyzates of purified eggshells indicate that they are rich in glycine, aspartate, histidine, and lysine but contain unexpectedly low levels of tyrosine. The predicted amino acid sequence of a cDNA encoding a probable component of the schistosome eggshell shows areas of considerable similarity to keratin and silkworm chorion proteins.

The protein is unusual in that it contains regular repeats of GlyGlyGlyTyr, GlyGlyGlyCys, GlyGlyAsp, GlyGlyAsn, and GlyGlyLys, which indicates that it tends to form β-pleated sheets. The protein is tyrosine-rich, a feature consistent with the probability that the eggshells are cross-linked by quinone tanning, a process involving the hydroxylation of tyrosine, the oxidation of hydroxyl residues on the aromatic ring, and covalent binding of the latter to free amino groups of other amino acids. Production of the eggshell mRNA appears to be mediated by a small gene family that is under transcriptional control and that, unlike silkworm chorion genes, is not amplified during expression. The predicted amino acid sequence of another cDNA encoding a probable eggshell protein of 48 kD also demonstrates significant identity with chorion proteins. It consists of a series of pentapeptide repeats having a consensus sequence of GlyTyrAspLysTyr. The repeats would be predicted to lead toward a structure consisting of alternating beta turns and beta sheets, and the carboxy terminal portion of the peptide contains a histidine-rich region likely to form an alpha helix. The protein is remarkably rich in tyrosine, a feature that indicates that it is not an abundant component of the eggshell, as hydrolysis of the complex yields little of this amino acid. Treatment of the protein with phenol oxidase results in the formation of covalently linked polymers.

Epidemiology

Several laboratories have applied techniques of molecular biology toward the development of probes that might be useful in identifying infected mosquito, blackfly, or mollusc vectors or in assessing the occurrence, species, and distribution of larvae within in large areas. The Onchocerchiasis Control Program, for example, has, over the last 17 years, relied on vector-control measures to manage onchocerchiasis. It has assessed its success by assaying the prevalence of infective larvae in the blackfly population over a huge area of Africa. In recognition of the fact that light infections are difficult to detect and that onchocerchiasis has a relatively long prepatent period, the program has not relied on determination of the incidence of infection in the human population.

Repetitive Sequences

The major portion of the genome of eukaryotic cells consists of nontranscribed repetitive sequences whose function is poorly understood. Some workers theorize that these sequences play a structural role, somehow stabilizing the genome and perhaps even regulating gene expression, while others feel that these sequences provide a plasticity that may accelerate evolutionary processes. Repetitive sequences include duplicated pseudogenes, transposons, and satellite DNA. They can be clustered or dispersed throughout unique sequence DNA, and since they lack constraints imposed on functional genes, they are evolutionarily unstable. In most respects, the composition of helminth genomes has proven similar to that of other invertebrates. Their genome size is on the order of 10^8 bp, and their GC content ranges from 47 percent to 47 percent, compared with roughly 78 percent to 47 percent in mammals and 70 percent to 45 percent in invertebrates (gross exceptions exist).

Modified bases have not been found in parasitic helminths, suggesting that transcriptional regulation through methylation may not be of general importance. Although the genome size of parasitic helminths tends to be roughly twice that of free-living relatives (which posess some of the smallest genomes among the metazoa), the fraction consisting of repetitive DNA tends to be lower. Thermal denaturation studies indicate that the increase in genome size of these parasites is due to the introduction of low-copy-number sequences. Several workers have taken advantage of the inherent variability of repetitive sequences to develop nucleic acid probes that allow the identification of filariae in infected flies or mosquitoes, in blood samples, or in skin biopsies. Since these sequences exist in large copy number, these probes have the potential of being sensitive.

Although repetitive sequences have a high rate of drift, they may show considerable homology within a genus, thus limiting their usefulness. In one study, detailed later, this problem was circumvented by constructing oligonucleotide probes that spanned areas of divergence between different species, thus allowing species-specific identification of the parasites. In the case of *Onchocerca* species, an important question that might be clarified by the use of repetitive DNA probes is whether forest and savannah strains of the parasite have different biological characteristics. The ability to discriminate between the two forms of the parasite also may assist in determining whether outbreaks of infection are due to invasion of forest- or savannah-dwelling flies.

Preliminary indications are that the geographically separated strains may lead to different clinical manifestations and that they might be separable on the basis of differences in isoenzyme patterns. Early studies indicate that probes hybridizing to repetitive DNA may help to speciate members of the genus *Onchocerca* and to discriminate between forest and savannah types. The probes are sensitive enough to identify a single infective larva. In the case of *Brugia* spp., it is felt that development of species-specific probes may further our understanding of the parasite host range. *B. malayi* is thought to infect humans, nonhuman primates, and several other vertebrates, whereas *B. timori* is thought to infect only humans. *B. pahangi* is known to infect domestic animals and is a suspected human pathogen.

A genomic clone representing a repetitive sequence of *B. malayi* has been isolated. Twenty-five percent of the parasite's genome consists of repetitive DNA, and the isolated repeat, which consists of a tandemly arrayed 322-bp sequence repeated 30,000 times per haploid genome, accounts for approximately half of this DNA. Under stringent conditions, the probe can easily distinguish *Brugia* from other genera of filariae, but it differs from the major repeat of *B. pahangi* by only 11 percent. As differences between the two repeats are clustered in pockets, workers have designed oligonucleotides that can easily discriminate between the two species.

Ribosomal Genes and their RNAs

Eukaryotic ribosomal RNA genes are typically part of multigene families whose members are tandemly linked (mRNA) in the genome. In general the arrangement follows a plan wherein the gene encoding the 18S RNA subunit lies 5' to that encoding the 5,8S mRNA subunit, the latter being placed 5' to that encoding the 28s mRNA snbunit. The three subunits are transcribed by RNA polymerase I as a polycistronic unit whose spacer regions are subsequently removed by RNA splicing. Processed ribosomal RNAs are highly conserved in nucleotide sequence, a feature that has assisted in clarification of the taxonomic relationships between helminth parasites.

The intergenic regions, on other hand, vary in length and have evolved more freely. Since these spacer regions represent a homogeneous family of repeated sequences, they may present-species-specific differences that can be exploited toward the development of highly discriminating probes. Polymorphisms in RNA genes also have been exploited as a means of determining the species, sex, and strain of schistosomes within their molluscan host. There exist 17 known species of schistosomes, which are categorized into three groups on the basis of egg morphology. In addition, there appear to exist several strains of the parasite, a phenomenon that may account for differences in infectivity, virulence, isoenzyme patterns, and drug resistance among schistosomes.

Although these biological characteristics are unlikely to be related to polymorphisms in the structure of rRNA genes, their fixation within discrete subpopulations may nonetheless allow strain identification and thereby provide a marker that can be correlated later with differences in pathogenicity, longevity, and overall biological fitness. Coding sequences of helminth rRNA genes also have been used to determine phylogenetic distances between the organisms.

Sequence analyses of the 5′-end of the 28s rRNA of various nematodes and schistosomes indicate that the differences in size and sequence of these molecules occur within a few hotspots that have retained the same position relative to the core of all mRNAs examined to date. Preliminary data indicate that nematodes radiated a long time ago: the rRNA sequence of *Nematospiroides dubius* is as similar to that of *O. gibsoni* and *B. pahangi* as it is to that of *S. mansoni.*

Immunodiagnosis

Definitive diagnosis of schistosomiasis rests on the demonstration of parasite ova in stool or urine specimens. Similarly, detection of filariasis depends on the visual detection of microfilariae in blood or skin biopsy. Available methods to detect larval parasites are time consuming and insensitive, especially when applied in large epidemiological surveys — problems that have stimulated the development of immunologically based assay systems. Antigens that have rendered assays reproducible, sensitive, and specific have been identified: four of these have been cloned; two are proteolytic enzymes that are discussed elsewhere in this chapter.

Peliminary studies on the antibody response to schistosome peptides indicate that the response is erratic, demonstrates little correlation with clinical status, and persists for a long period after chemotherapeutic cure, thus limiting the information garnered by assay of specific antibodies. Nevertheless, the search for immunodiagnostic schistosome and filarial antigens has provided significant information relevant to understanding helminth biology. Strangely enough, it appears that immunodominant antigens of schistosomes include a family of macromolecules called heat-shock proteins (HSPs). HSPs are phylogenetically conserved proteins expressed universally by organisms in response to environmental stress. The function of HSPs is poorly understood, although they appear to protect cells from damage and at the same time rid them of denatured macromolecules.

Constitutively expressed HSP cognates are thought to play a role in mediating the assembly or disassembly of higher macromolecular structures; in surveying the cytoplasm for denatured, improperly modified, or foreign proteins; and in mediating the insertion of proteins into, or their translocation across, cell membranes. HSP cognates are subject to developmental regulation and are thought by some to be critical factors of differentiation. One of the major antigens of schistosome eggs is a 40-kD protein referred to as p40. The protein is also found in adult worms and elicits a strong immunological response in over 90 percent of infected humans. The antigen is actually a family of four nearly identical proteins that are differentially expressed during the schistosome life cycle.

One form of the peptide is produced by female worms, two in eggs, and four in adult worms. The sequence of p40, based on the nucleotide sequence of a cDNA, reveals that it is homologous to *Drosophila* and dogfish α-crystallins and a family of eukaryotic small heat-shock proteins. The primary sequence of the schistosome protein indicates that it is a descendant of a protein that was ancestral to the functionally different α-crystallins and small HSPs. The amino-terminal portion of the protein consists of a long stretch homologous to the α-crystallins, a central portion homologous to the carboxy terminus of small HSPs, an amino-proximal region homologous to α-crystallins, and a carboxy-terminal tail similar to that of some small HSPs. This sequence suggests that the protein is the result of an early intragenic duplication.

```
Schist P40      G E D G K V H F K V R F D A Q G F A P Q D I N V T S S E N - R V T V H A K K E - T T T D - - G
Sechist P40     G S G * * R L H V E V A V D P V Y K * E * L F * N V D S * - * * V V S G R H H K Q K S * Q H *
α-cryst 27      V R S D R D K * V I F L * V K H * S * E * L T * K V Q * D - F * E I H G * H N - E R Q * D H -
bHSP 22         I R H T A D R W R * S L * V N H * * * D E L T * K T K D G - V R E I T G * H * - E R Q * E H -
DHSP = 27       A T V N * D G Y * L T L * V K D * S - - E L K * K V L D E S V * L * E * * S * - Q Q E A E Q *
Schist p40 R    P A V * * D G * Q * C M * V S Q F K * N E L T * K V V D * - T * V * E G * H * - E R E * G H -

Schist p40      R - - - - - K C S R E F C R M V Q L P K S I D D S Q L K C R M T D D G V L M L E A P V K V - - -
Schist p40      * - - - - - S S S F A * * S Q S * A I * E T V * P L S V S A Q V - V G N T * V * * * * L E K - - -
α cryst A       - - - - - G Y I * * * * H * R * R * * S N V * Q * A * S * S L S A * * M * T F S G * K I P S G V
bhHSP 27        - - - - - G Y I * * * * T * K * T * * P G V * * P T * V S S S L S P E * T * T V * * * M P K L A T
dHSP = 22       - - - - - G Y S * * * H * L G R * V * * D G Y E A D K V S S S L S * * * * * T I S V * - N P P G V
dHSP = 27       - - - - - G M I Q * H * V * K * T * * * G F * P N E V V S T V S S * * * * T * K * * - P P P S K

Schist p 40     - D Q N Q S L T L N E S G Q V A V R P K S D N Q I K A V P A S Q A L V A K G V H G L S Y V D D
Schist p 40     - - - - - - - - - - - - - . Q H A I T H
α cryst         D A G H S E R A I P V * R E E K P - - S S A P S S
bHSP = 27       Q - S * - E I * I P V T F E S R A Q L G G R S C K I R
d HSP = 22      O E T L K E R E V T I E Q T G E P A K * * A E E P * D K T * * *
dHSP = 27       E Q A K S E R I V Q I Q Q T G - P A H L S V K A P A P E A G D G K A E N G S G E K M E T S K
```

Fig. 7.20. Regions of homology between schistosome egg antigen p40, bovine α-crystallin A2, *Drosophila* small heat-shock proteins (dHSP) 22 and 27, and human heat-shock protein 27 (hHSP 27). Sequences are aligned.

The function of the protein is unknown, although its similarity to the α-crystallins and small HSPs suggests that it probably forms large aggregates that somehow stabilize cellular structures. Background studies on the human immune response to *S. mamoni* indicate that although the response is erratic, virtually all infected individuals produce sera that react to the two schistosome proteins, one having a relative molecular weight (Mr) of 70 kD and the other an Mr of 38 kD. Antibodies to these schistosome proteins are not detectable in uninfected individuals or in persons infected with other parasitic helminths (species of *Brugia, Onchocerca, Chlonorchis, Heterobilharzia, Ascaris, Trichuris, Ancylostoma,* or Echiococcus), indicating that they may prove to be of diagnostic value.

The 70-kD schistosome antigen has been identified as an HSP-70 homologue; epitope mapping has localized the immunogenic portion(s) of the molecule at the carboxy terminus of the protein, the least conserved domain of HSP-70s in general. Similar properties are displayed by the *S. japonicum* HSP-70 cognate. Oddly, however, the 70-kD schistosome immunogens do not cross-react with each other. As immunization of rabbits with chicken HSP-70 leads to a response that cross-reacts with the homologue of fruit flies and humans, these findings indicate that schistosomes present the protein to the immune system in a manner that avoids an autoimmune response. Along similar lines, it has recently been noted that *P. falciparum* produces two different HSP-70s that are immunodominant and non-cross-reactive. These findings are of interest because they suggest that the parasite species-specific nature of the protective immune response generated by immunization with live-attenuated vaccines is not attributable to fundamentally different processes or targets, but rather to differences in the structure of homologous parasite antigens.

It is possible, however, that HSP-70s, molecules whose three-dimensional configuration is unknown, selectively present their hypervariable carboxy termini to the immune system, a property lost when animals are immunized with the denatured protein. The function of the schistosome HSP-70 is unknown. Curiously, HSP-70 also has been found to be a dominant immunogen of filariae, trypanosomes, plasmodia, and mycobacteria. In other eukaryotes, HSP-

70 is considered to be a molecular chaperone that escorts and influences the conformation of proteins destined to become part of, or which are going to be removed from, higher molecular assemblies. HSP-70 also plays a role in the translocation of proteins across cellular membranes. This process is thought to involve hydrophobic interactions between HSP-70 and the protein to be translocated, and it appears to be associated with conformational changes mediated by the binding and hydrolysis of ATP by HSP. HSP-70 has been implicated as playing a central role in the recognition of denatured proteins and in collaboration with ubiquitin, tagging these with signals recognized by lysozomal proteases.

Interestingly, synthesis of HSP-70 increases during senescence of fruit flies, perhaps as a result of increases in the production of improperly formed or glycosylated proteins. In light of the finding that the schistosome life expectancy is unusually long for an invertebrate, it would be interesting to see whether the same holds true for them. The schistosome HSP-70 is found in the tegument of adult worms and in the hatching fluid, surface cells, and nervous system of miracidia. Synthesis of the protein is constitutive in adult worms and sporocysts, although apparently it does not appear to be made by cercariae. The latter nonetheless maintain a previously synthesized pool of the protein. Transforming schistosomula synthesize massive amounts of HSP during a period in which production of nearly every other protein is repressed. The cue for HSP gene transcription in transforming larvae is not heat, as cercariae subjected to heat shock do not synthesize the uprotein, whereas cercariae placed in mammalian culture media do. Synthesis of other proteins slowly resumes 16-hour posttransformation, eventually plateauing to steady state levels.

Whatever the function of HSP-70, its mRNA is one of the more abundant species in adult worms, accounting for 0.1 percent to 1.0 percent of the total message. The observation that the protein is synthesized primarily by tegumental cells suggests that it may play a role in maintaining the integrity of this anatomical compartment. The observation that despite their similarities, the HSP-70s of S. *mansoni* and S. *japonicum* stimulate production of non-cross-reacting antibodies during the course of natural infection can be taken as a good omen for vaccine development.

```
     445
Sm   ERALTKDNNLLGKFELSGIPPAPRGTPQIEVTFDIDANGILNVSAVDKGTGKQNKITI
Sj   ** TM*************C********V***********************A***E*****
Hs   *** M**********R***************V*********************T*TKDS***A******

Sm   TNDKGRLSKEEIERMVADADKYKAEDEKQRDRVA SKNSLESYVYTMKQQVEGE-LKEK
Sj   *************D**INE*****S*****KN*IC************S***S***DEM*D*
Hs   **************QE*E********V**E*******A****AFN**AFN***SA**EG**G*

Sm   IPESDH QVIISKCEDTISWLDVHQSAEKHETESKREELEKVCAP IITKVYQ-A----
Sj   *S***RKNR*L****E**R*M*MN*L***E*F*E*KS*******M****AMNR-*----
Hs   *S*A*KKK* LD**QEV*****ANTL***D*F*H**K***Q**N***SCL**G*----

Sm   -----GGMPGGMHEAS--GAGGSG-------------------KGPTIEEVD
Sj   -GGVPS*******P------****G*-----------------**********
Hs   -----**-***-FG*Q-GPK****-----------------S*********
```

Fig. 7.21. Comparison of the predicted carboxy-terminal amino acid sequences of HSP 70 of *Schistosoma japonicum* (Sj), and *Homo sapiens* (*Hs*) to that of *Schistosoma mansoni* (Sm). Asterisks denote points of identity with the *S. mansoni* sequence, and dashes represent gaps.

It seems that in many experimental systems, immunity to the parasites is often specific to the species of sensitizing organism. The response to HSP-70 suggests that this may be due more to interspecific differences in the structure of parasite antigens than to different effector mechanisms or target antigens, and that once a protective antigen is found for one parasite, it may be possible to clone equally active homologues from species that are currently difficult or impossible to maintain in the laboratory. A case in point is the identification of paramyosin as a vaccine target in schistosomes, and subsequent demonstration of the prophylactic potential against filariasis of the nematode homologue.

Coda

Although studies on the molecular biology of schistosomes and filariae are at a stage of infancy, they have significantly contributed to the understanding of both the biology of the organisms and their interaction with their hosts. Many of the parasite antigens have now been assigned a biological function, a significant advance over their previous indexing in the literature on the basis of molecular weight. At the same time, the field has posed a number of new questions to researchers interested in biological and immunological aspects of parasite relationships with their hosts. The answers to these questions are likely to refine efforts to develop vaccines and immunodiagnostic systems. In conjunction with DNA probe technologies, such as those using the polymerase chain reaction, these studies offer the potential to develop highly sensitive probes that could detect and determine the species (and perhaps even strain) of parasites obtained in small biological samples. At a more basic level, the observation that parasitic nematodes possess both cis- and trans-splicing activities makes them a convenient system for studying the mechanics of eukaryotic RNA processing.

CHAPTER

8 Antibiotics

The term antibiotic was introduced by S. A. Waksman (Nobel Prize winner) in 1945, as the name of a class of substances of biological origin that are inimical to microorganisms. Strictly speaking, antibiotic means "against life." Since antibiotics, as commonly used, are inimical only to microorganisms they might logically be called "antimicrobiotics." Although substances of biological origin with antimicrobial properties were known before 1900, antibiotics as now defined first came into world-wide prominence about 1940. There followed a widespread search for new antibiotics. It was found that antibiotic substances are produced by a wide variety of living organisms, ranging from man to microbes.

The most important and widely used antibiotics at present are derived from microorganisms—mainly from various bacteria of the genera *Bacillus* and *Streptomyces* and certain common species of mold (*e.g., Penicillium*). Since the earliest studies with pure cultures, microbiologists had known that when certain airborne, saprophytic microorganisms contaminated their cultures they suppressed the growth of the desired species. The phenomenon was so commonplace and attention was so fixed on other problems that the antagonistic action of such contaminants was pushed aside as merely an inevitable nuisance. The true significance of the "nuisance" was largely overlooked until 1929, when Alexander Fleming (afterwards Sir Alexander Fleming, Nobel Prize winner with Florey and Chain) realized it, acted upon the basis of his idea, and discovered, named and described penicillin.

PENICILLIN

The antagonistic airborne saprophytic organism that first attracted Fleming's attention was a colony of the common mold, *Penicillium notatum.* When it contaminated one of his cultures in a Petri plate, Fleming experimented with its antimicrobial action. He passed broth cultures of the mold through filters, removed the mold filaments and was thus able to study the activity of the soluble growth products alone as they occurred in the clear broth. He found that the clarified, sterile broth contained a highly potent antimicrobial principle, the activity of which was readily demonstrated in contact with sensitive microorganisms, especially gram-positive bacteria. He called this principle *penicillin.* In a much purified and refined form it is still one of the forms of penicillin used therapeutically today. Fleming realized the practical possibility of his discovery and made use of it in his laboratory to eliminate gram-positive contaminants from cultures but was not, at that time, in a position to develop it more fully.

For some years the value of such antibiotic phenomena remained relatively unknown. Interest was reawakened when, realizing that many nonsporeforming pathogenic organisms in the soil are rapidly destroyed, presumably by antimicrobial substances produced by saprophytic microorganisms, Waksman suggested that the search for antibiotic-producing microorganisms be carried to the soil. In 1939 Dubos found, in bogs, organisms *(Bacillus brevis)* that produced two valuable antibacterial substances *(gramicidin* and *tyrocidin)*. These antibiotics, while of enormous value therapeutically for surface application and of great interest scientifically, were too toxic for internal use. Dubos's discoveries stimulated new interest in Fleming's observations on *Penicillium notatum;* work on penicillin was begun in England on a large scale by Florey, Chain, Abraham and others (the Oxford Group) in 1940, just after the outbreak of World War II. Because of the exigencies of the war, work on mass production methods was transferred largely to the United States.

Fig. 8.1. Sir Alexander Fleming, penicillin discoverer and Nobel Prize winner, points with his inoculating needle to a giant colony of *Penicillium notatum,* the organism that produces penicillin, on agar in a Petri dish.

Since that time penicillin has played a role in the greatest war in human history; in the most far-reaching piece of cooperative research ever organized prior to atomic research; in a great industrial development; in the most complete control over a variety of diseases ever achieved by man in a short time; and, because of its impact on venereal diseases, in a tremendous new social and moral trend, the possibilities of which are still only partly realized. The military importance of penicillin in World War II, then just beginning, can hardly be estimated. The development of American facilities for mass production of penicillin, not then available in Europe, was soon of great benefit to all concerned.

Production

Production of penicillin and other antibiotics is now a billion-dollar industry in the United States alone. In 1943 an ounce of penicillin cost around twenty thousand dollars. Now it costs about three dollars because of mass production methods.

Production of penicillin involves cultivation of *Penicillium notatum* under conditions most favorable to growth and penicillin production (*i.e.,* at about 24° C and at a pH between 7 and 8). As the mold is strictly aerobic, exposure to air is essential.

Commercially, a process using submerged growth in closed but vigorously aerated tanks holding thousands of gallons of medium most widely used today.

In all methods, suitable medium is first inoculated with suspensions of the conidia of *Penicillium notatum* or the closely related *P. chrysogenum,* which also produces penicillin. The entire procedure is a *pure-culture* process and necessitates costly aseptic technique at all stages.

During the 7 to 14 days of incubation in the large vat of aerated fluid medium the mold excretes at least three waste substances that are of importance: the yellow pigment *chrysogenin,* which must be removed by adsorption with charcoal or similar means; *penicillin;* and *notatin,* or *penicillin B.* The last occurs especially if the acidity of the medium is too great. It is removed during the final purification process. Other substances are also produced, some of them related to penicillin. In general, if a microorganism produces one antibiotic it produces others of the same nature, some valuable, others of little significance.

Medium

Different formulas are doubtless in use by various manufacturers, but basically they are similar and are mainly modifications of the following:

Corn-steep liquor	40.0 ml.
Lactose monohydrate	27.5 gm.
$NaNO_3$	3.0 gm.
$MgSO_4\ H_2O$	0.25 gm.
$KH2PO_4$	0.50 gm.
$ZnSO_4\ H_2O$	0.044 gm.
$MnSO_4\ H_2O$	0.020 gm.
Glucose monohydrate	3.0 gm.
Water ad	1000.0 ml.

After about seven days, growth is complete, pH rises to 8.0 or above, and penicillin production ceases.

When no more penicillin is being formed, the masses of mold growth are separated from the culture fluid by centrifugalization and filtration. The complex process of extracting the penicillin from the clear fluid then begins. The method involves various extractions with organic solvents and recrystallizations. These are chemical engineering problems and do not concern us at present.

Properties and Uses

The material extracted from the culture medium is in reality a mixture of six chemically related penicillins called X, G, two forms of F, dihydro F and K. The most important is penicillin G (benzyl penicillin), which is usually meant when the term penicillin is used. Penicillin is readily soluble in water and alcohol, but the latter inactivates it. It is quickly destroyed by acid, so that in the unmodified form it is ineffective if taken orally. In the crystalline form it is quite stable, even for several days at 100° C or for months refrigerated in the dark; however, in aqueous solution it is very unstable and quickly decomposes.

Penicillin acts mainly by inhibiting the enzymes concerned in the incorporation of muramic acid into the mucocomplex of the bacterial cell wall. Whether this action is of the nature of an antimetabolite is not known at present. Since synthesis of cell wall is carried on only by young, actively multiplying cells, the antibacterial action is manifested only during active growth. Because septum formation appears particularly vulnerable to penicillin, rod-shaped bacteria tend to form aseptate filamentous forms like mycelia in the presence of nonlethal concentrations of penicillin.

Among the most valuable properties of penicillin is its relatively low toxicity for man and animals, even in large doses, coupled with the fact that its antibacterial potency is such that a concentration of 0.000001 gm. per ml. will exert marked bactericidal effects. It may be administered subcutaneously, intravenously or locally. A particular disadvantage of penicillin is that it is a potent allergen in some persons and can cause fatal allergic reactions.

Chemistry of Penicillins

The naturally occurring penicillins are compounds of the strong, monobasic acid, *penicillanic acid.* The penicillins usually occur as salts of sodium and potassium. The general chemical structure assigned. They are unstable to heat and to strong acids and alkalis. Laboratory synthesis of penicillin was announced shortly after the formula was known. The substance artificially synthesized was penicillin G.

Biosynthesis of Penicillins

It has been found possible to induce *Penicillium* cultures to synthesize new forms of penicillin. The method consists of "feeding," to cultures of *Penicillium notatium,* synthetic substitutes for the normal "R's" (*i.e.,* the distinctive side chains) of the penicillin molecule. These substitutes, "prefabricated" molecules, are referred to as *precursors* of the new "R." In the presence of such abnormal precursors, the mold unwittingly conjugates them "as is" with penicillanic acid, thus forming new and more valuable penicillins that have been synthesized partly by man, partly by the mold. An important product is *phenoxymethyl penicillin (penicillin V),* produced by feeding the precursor phenoxyacetic acid to the mold. Penicillin V is fairly stable to gastric acidity and is less quickly excreted from the blood than other penicillins. It is widely used for oral administration.

This same method (*i.e.,* providing abnormal precursors) has been extended to other industrial processes in which a desired product is synthesized partly by man and partly by a living organism. The field of *biosynthesis* is clearly wide open for the ingenious, well-trained and energetic student.

Residue common to all penicillins	Various forms of R or acyl group (Natural Penicillins)	Common name. penicillin:	Chemical designation (-penicillin)
	$-\overset{H_2}{C}-C_6H_5$	G or II	Benzyl
	$-\overset{H_2}{C}-C_6H_4-OH$	X or III	Para-hydroxybenzyl
6-amino-penicillanic acid residue: H_3C, H_3C, S, COOH · C(H), N, C=O, H H H, $-N-\overset{O}{\overset{\Vert}{C}}-$	$-\overset{H_2}{C}-O-C_6H_5$	V	Phenoxymethyl*
	$-\overset{H_2}{C}-\overset{H}{C}=\overset{H}{C}-\overset{H_2}{C}-C\equiv H_3$	F or I	2-pentenyl
	$-\overset{H_2}{C}-\overset{H_2}{C}-\overset{H}{C}=\overset{H}{C}-C\equiv H_3$	F	3-pentenyl
	$-\overset{H_2}{C}-\overset{H_2}{C}-\overset{H_2}{C}-\overset{H_2}{C}-\overset{H_2}{C}-\overset{H_2}{C}-C\equiv H_3$	K or IV	n-heptyl

Fig. 8.2. Various forms of penicillin. They differ only in the form of *R* in the general formula. The different forms of *R* are shown in the second column.

Penicillinases

Several hydrolytic enzymes can attack and destroy penicillins. One, an inducible enzyme, is commonly called *penicillinase.* It is a type of penicillinacylase. This type of penicillinase is produced by a number of dangerous pathogenic bacteria, notably *Staphylococcus* and several species of gram-negative intestinal bacteria. By virtue of this enzyme these and other bacteria are penicillin-resistant and so defeat therapeutic uses of natural penicillins in many cases. These bacterial penicillinases attack the β-lactam bond of penicillins G, X and, to a lesser extent, V, and are often referred to as β-lactamases.

Another kind of penicillinase is produced by many common genera of fungi, including *Penicillium, Aspergillus* and *Mucor.* The penicillinases produced by these organisms differ from bacterial β-lactamase in that they attack the peptide bond between the 6-amino-penicillanic acid residue and the "R's". These enzymes are sometimes called *penicillin amidases.* They attack penicillins V and K most rapidly, G more slowly.

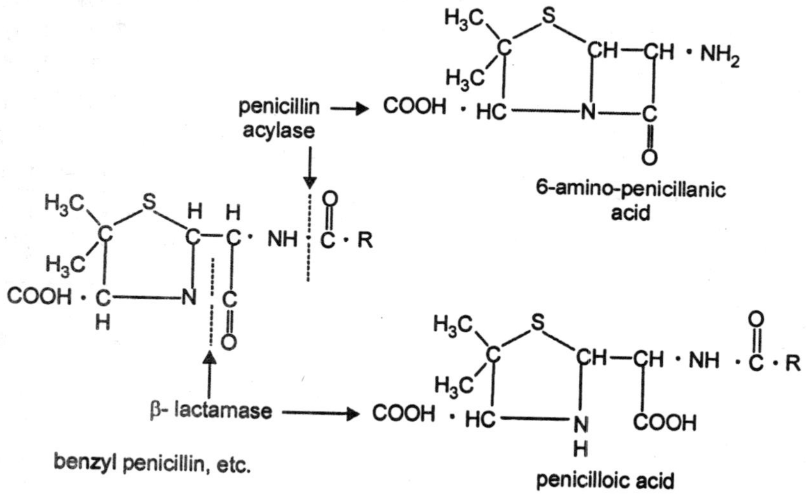

Fig. 8.3. Action of two major types of penicillinase: penicillinacylase (penicillin amidase) and β-lactamase (commonly called penicillinase). The dotted lines and arrows show which bonds are broken.

Semisynthetic Penicillins

When it was discovered about 1958, that 6-amino-penicillanic acid, the "nucleus" of all penicillins, could be produced in commercial quantities, it became possible to attach various artificially synthesized acyl groups to 6-amino-penicillanic acid on a large scale. Hundreds of such semi-synthetic penicillins have been made and tried. Some are extremely valuable, *e.g.,* phenethycillin, which is quickly absorbed into the blood and is as effective as penicillin G; methicillin, cloxacillin and nafcillin, which are not readily destroyed by gastric acidity and which are resistant to β-lactamase; and ampicillin, which is more effective against a range of bacteria, both gram-positive and gram-negative, than other penicillins, and is effective in oral administration. The acyl groups of some of these are shown in Fig. 8.5 comparison with natural penicillins.

The Cephalosporins

These are a group of antibiotics, the first described type of which, cephalosporin C, is produced by the mold, *Cephalosporium acremonium.* Although a relatively ineffective antibiotic, cephalosporin C is of great interest because it has a formula very similar to that of penicillin. Like penicillin, cephalosporin C lends itself to production of a large number of synthetic derivatives. Knowledge of these processes is based on the work of Woodward (Nobel Prize winner) and many other skillful chemists and biologists. The central residue of the cephalosporins is indicated in Fig. 8.6. Various acyl groups may be attached to this residue, with the production of much more valuable cephalosporins than cephalosporin C; *e.g., cephalothin.*

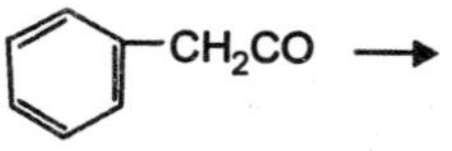

Penicillin and L Bodies

Numerous species of bacteria cultivated in the presence of penicillin in a medium of increased osmotic pressure (see Protoplasts) grow but produce no cell wall, *i.e.,* they produce protoplasts (L forms) or spheroplasts. In media of ordinary osmotic pressure these immediately burst because the strong, retaining cell wall is lacking. This is the basis of the lethal effect of penicillin; *i.e.,* removal of the cell wall followed by osmotic rupture. Animal cells (*e.g.,* our own body cells), viruses and the PPLO or *Mycoplasma* are completely resistant to penicillin because they do not synthesize cell walls. In general, the gram-positive bacteria are most susceptible to penicillin because their cell walls consist almost entirely of mucopolysaccharide.

$C_6H_5-CH_2CO \rightarrow$

Benzyl penicilin (G)

$C_6H_3(OCH_3)_2-CO \rightarrow$

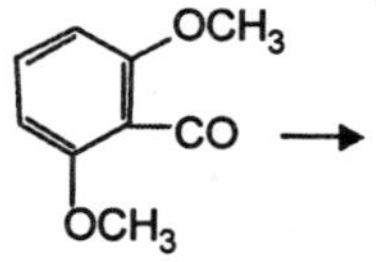

Methicillin
(Dimethoxyphenyl penicilline)

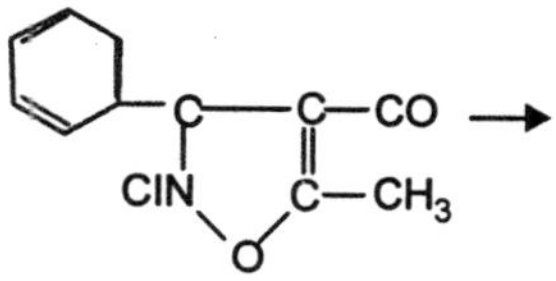

Cloxacillin
(5-methyl-3-phenyl-4-isoxyzolyl penicillin)

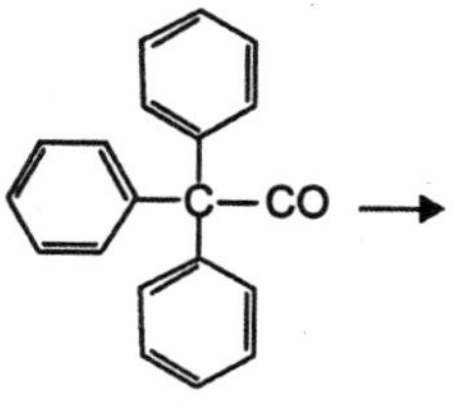

$(C_6H_5)_3C-CO \rightarrow$

Triphenyl methyl penicillin

Fig. 8.4. R groups of some representative bio-synthetic penicillins compared with that of a natural penicillin, benzylpenicillin (G).

Adaptive Penicillinase

Penicillinase (β-lactamase) production may appear in some bacteria as a result of induced enzyme formation. For example, strains of some kinds of bacteria are ordinarily sensitive to penicillin because they do not produce penicillinase. However, contact with nonlethal concentrations of penicillin appears to induce penicillinase production. They are then penicillin resistant.

Various Strains of Penicillin Notatum

Fleming's strain of *P. notatum* for a time was thought to be unique. It was later found that some "wild" strains of this organism were better producers of penicillin and that other closely related organisms, especially *P. chrysogenum,* also produce penicillin. These findings led to a very extensive search for better strains of *P. notatum* and for penicillin-producing strains of other organisms. Many thousands of cultures were tested, and it was found that the penicillin-producing property is widely distributed in the *P. notatum-chrysogenum* group and that some strains are better than others, especially for the various methods of production (*i.e.*, some were good in tanks, not so good on surfaces, and vice versa). Irradiation with ultraviolet light has furnished many thousands of mutant strains for testing, among them some very potent penicillin-producing mutants.

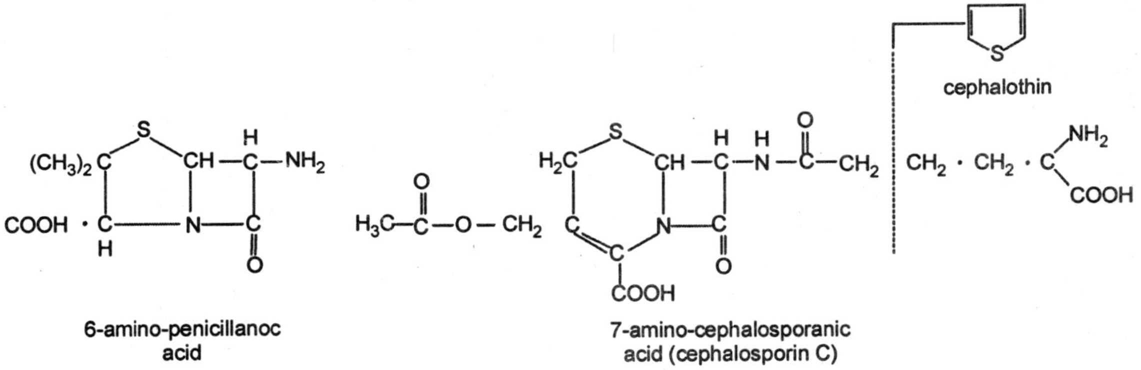

cephalothin

6-amino-penicillanoc acid

7-amino-cephalosporanic acid (cephalosporin C)

Fig. 8.5. Comparison of formulas of 6-amino-penicillanic acid, 7-amino-cephalosporanic acid, and cephalothin as a derivative of cephalosporin.

The same principle of hunting for new and more effective mutants for a variety of purposes can be applied in most fields of industrial microbiology, using yeasts, molds and bacteria.

Development of Resistance Drug-Fastness and Dependence

An undesirable aspect of the continued clinical use of antibiotics (as well as of sulfa drugs) is the emergence of mutants having *increased resistance* to the drugs *(drug-fast* strains) or being wholly *dependent* on the drugs. Drug-fastness and drug dependence have been described elsewhere.

Laboratory Uses of Penicillin

As mentioned, the value of penicillin as a selective bacteriostatic agent in laboratory bacteriology had been immediately recognized by Fleming and was adopted by others. It is a convenient aid to the isolation of penicillin-resistant organisms (such as the gram-negative whooping cough and influenza bacilli) from throat cultures, since it suppresses the growth of unwanted penicillin-sensitive (mainly gram-positive) organisms. It is also used for suppressing bacterial growth in animal tissue cultures used in studies of viruses, since viruses and animal cells are wholly resistant to most antibiotics.

Range of Activity

The clinical uses of penicillin are beyond the scope of this discussion, but some of the common infectious organisms against which natural penicillins are more or less effective are listed in Table 8.1.

Note that most of the susceptible species are gram-positive. Notable exceptions are *Treponema pallidum,* cause of syphilis, and gram-negative *Neisseria gonorrhoeae,* cause of gonorrhea; both very susceptible.

Natural penicillin is generally ineffective in infections caused by gram-negative bacilli, viruses, rickettsias or PPLO, and in infections with protozoa, yeasts and molds. As we have noted, several biosynthesized penicillins have much broader ranges of antimicrobial activity.

Table 8.1. Penicillin Sensitivity of Some Common Pathogens

Gram Stain	*Organisms**
	Highly Susceptible
+	*B. anthracis*
+	Various clostridia including Cl. *tetani* and *Cl welchii*
+	*D. pneumoniae*
–	*Neisseria spp.*, including gonococci and meningococci
+	Micrococci, *Staphylococcius aureus*
+	*Streptococcus pyogenets*
+	*C. diphtheriae* (in vitro)
±	*Treponema pallidum*
	Moderately Susceptible
–	*Borrelia* of relapsing fever
+	Various alpha (viridans) type hemolytic streptococci
–	*Leptospira spp.*
	Not Significantly Susceptible
–	Enterobacteriaceae
–	*Pseudomonas* species
–	Brucellaceae

• Except nonconforming mutants.

STANDARDIZATION OF ANTIBIOTICS

Standards of potency for all antibiotics are established by international agreement as to what shall constitute an *International Standard Sample* and an *International Unit.* Generally, the International Standard Sample of an antibiotic consists of a certain highly purified and tested lot of that antibiotic. For example, the 1960 International Standard Sample of the antibiotic tetracycline (see farther on) was a 500-gm. lot of highly purified tetracycline chloride furnished by an American manufacturer. The International Unit of tetracycline was defined as the antibiotic activity contained in 1.01 μg. of that particular standard sample; 1.01 μg. was equivalent in activity to 1 μg. of pure, crystalline tetracycline chloride.

ASSAY AND SENSITIVITY TESTING

Several types of procedure are in use for assaying the potency of antibiotic preparations for therapeutic purposes. This is commonly done in manufacturers' control laboratories under the supervision of the U. S. Food and Drug Administration. These methods, conversely modified, are also used for measuring sensitivity *(sensitivity testing)* of "unknown" organisms to antibiotics. This is commonly done in hospital laboratories and is essential to rational chemotherapy, although not always feasible. There are sources of error in all of these methods, and so they must be used and interpreted by experts.

ASSAY METHODS

Serial Dilution in Broth

Serial dilutions of the antibiotic being assayed are made in uniform amounts of standard broth in culture tubes. These are then inoculated with a uniform number of cells of the test organism, which is selected on the basis of its known degree of sensitivity. After incubation, turbidity (or its absence) is noted and may be measured by means of a photoelectric turbidimeter. The turbidities (amounts of growth) are compared with a dilution series made in the same way but with an antibiotic reference standard of carefully measured potency.

The method is time consuming and sometimes requires preliminary sterilization of the sample, which destroys some of its potency. However, the dilution method is an accurate procedure, and results can be stated in exact terms of units of antibiotic.

Sensitivity Testing by Broth Method

The serial dilution procedure—reversed—is often used to measure the sensitivity to various antibiotics of an organism freshly isolated from an infected patient. Graded dilutions of an antibiotic of known strength are placed in series of tubes. Each tube is inoculated with a drop of the organism whose sensitivity is being tested. If growth occurs in all tubes, the organism is resistant; if no growth occurs, the organism is sensitive. Growth in some of the tubes permits a measure of the sensitivity of the organism.

Serial Dilution (Streak Assay) in Agar

The streak assay method is rapid, convenient and much used for preliminary assays of crude material in the search for new antibiotics.

Graded dilutions of the substance to be tested (for example, a plant extract thought to have antibiotic properties) are placed in a series of Petri dishes. Into each of these is poured about 10 ml. of melted and cooled (45° C) agar, which is thoroughly mixed with the antibiotic dilution by tilting the plates back and forth. After the agar has hardened, the plates are marked into several sectors, each of which is streaked with a different test organism. After incubation, the highest dilution of the antibiotic that inhibits each organism is recorded.

This method—reversed—also is often used for sensitivity testing by using *known* strengths of known antibiotics and *unknown* organisms.

The Cylinder-Plate Method

A series of steel, glass or porcelain cylinders about 10 by 6 mm. in dimensions with perfectly square-cut ends (like a drainage tile) are heated and then sealed to the upper surface of already inoculated agar in a Petri plate by placing them, end up, on the solid agar. Dilutions of the single antibiotic to be tested are placed in the cylinders. The plate is then incubated. When they diffuse into the agar, the antibiotic dilutions reveal their potency by the width of the zone of inhibition of growth in the agar around each cylinder. The potency of different antibiotics cannot be compared on the basis of the width of their zones of inhibition, however, because a very potent antibiotic may not diffuse readily through the agar and will therefore produce relatively narrow zones, while a weak antibiotic may diffuse very well and produce deceptively wide zones. In such measurements the test organism, agar density, time, temperature and pH must be very exactly standardized. This is true of all the methods described.

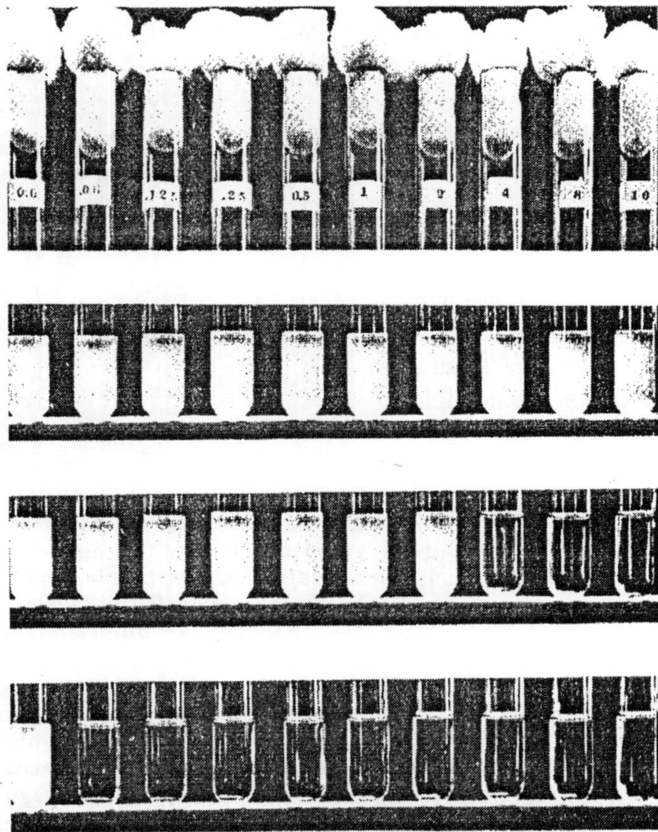

Fig. 8.6. Testing sensitivity of a bacterium to antibiotics or other chemotherapeutic agents by the "tube dilution" method. The top row of tubes of broth culture contain agent "A" in increasing amounts as shown by the figures on the tubes. The second row contains agent "B" in the same amounts, while the third row contains agent "C" in the same dilutions. All tubes were inoculated with organism "X" from the same culture at the same time and incubated together. Luxuriant growth, as shown by white turbidity, as occurred even in the highest concentration of agent "A," showing that organism "X" (perhaps from a patient very ill with this infection) is not in the least affected by this drug. The drug of choice will be agent "C" which prevents growth of organism "X" even in the smallest concentration. Agent "B" is slightly effective. The tube at the left of each row is a control tube containing no inhibitory agent.

The Disk or Tablet Method

This is possibly a little less exact than the procedures just described but is reasonably accurate and is quick, convenient and economical. It is therefore a preferred and widely used method for many purposes not requiring high degrees of precision. The principle is the same in the cylinder-plate method, but the cylinders are replaced by disks or tablets.

Nutrient agar of appropriate composition is heavily inoculated with the desired organism all over the surface of the solidified agar or mixed with the agar while still fluid before pouring the plate. If an antibiotic solution of unknown potency is being assayed (*e.g.*, manufacturer's sample, blood, urine) the organism used is a stock strain of known sensitivity to standard doses of that antibiotic.

Measured strengths of the antibiotic solution to be tested are applied to the inoculated agar in disks of sterile filter paper or in small tablets, which are placed on the surface of the agar plate before incubating. Several devices are available for application of the disks. The antibiotic diffuses outward from each disk or tablet and inhibits growth in the agar around it.

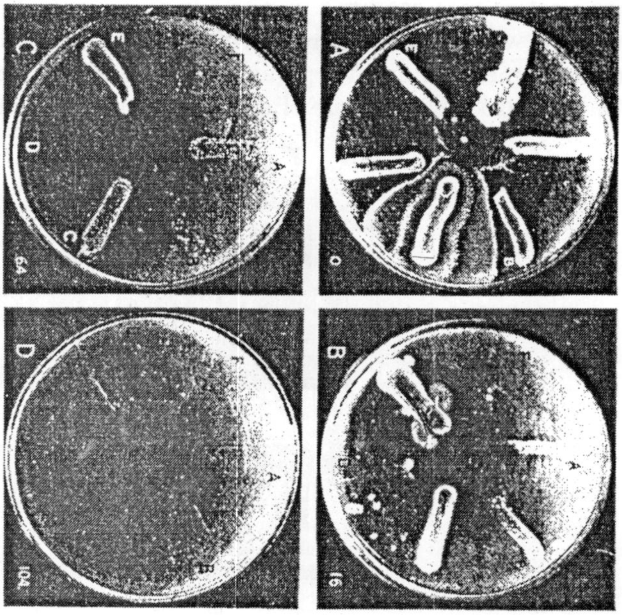

Fig. 8.7. Testing sensitivity of several species of bacteria to three different concentrations of streptomycin by the agar-dilution method. Plate A contains no antibiotic and is inoculated with *(A) Escherichia coli, (B) Salmonella typhi,* (C) *Proteus* sp., *(D) Klebsiella pneumoniae, (E) Pseudomonas aeruginosa, (F) Mycobacterium tuberculosis.* In Plate B, containing 1.6 mg. of streptomycin per ml. of agar, organisms *D* and *F* have been virtually eliminated. Larger amounts of the antibiotic, as shown in Plates C and D, successively eliminate all but a few *resistant* colonies of the other organisms. (What is the significance of these resistant colonies?) Note the great sensitivity of this strain of *M. tuberculosis* to streptomycin.

Growth-Indicator Methods

In these methods some early-perceptible indicator of growth is looked for rather than the often-delayed growth itself. In a representative procedure, sterile blood agar is poured into a Petri dish to a depth of about 3 mm. This is the base layer.

Similar fluid agar is heavily inoculated with the organism being tested. After this is thoroughly mixed, it is poured over the base layer to an equal depth (seed layer). When solid, antibiotic disks are placed on the surface, and the plate is placed in the incubator. The plate may be examined after two hours and then every hour thereafter. Diffuse growth of the organism, whether actual colonies are visible or not, produces a marked discoloration or hemolysis of the blood, except where inhibited under and around disks containing effective antibiotic concentrations. Here the blood remains unchanged.

Modifications of this method employ indicators of growth other than discolorations of hemoglobin; *i.e.*, pH indicators such as phenol red, and oxidation-reduction indicators such as resazurin. In all of these methods the change in color of whatever indicator is used reflects growth of the organism hours before growth itself is apparent.

ANTIBIOTICS FROM STREPTOMYCES

Streptomyces a genus of bacteria that has some of the physiological and gross morphological characteristics of molds like *Penicillium notatum. Streptomyces* species are, however, minute in size as compared with true molds and are entirely procaryotic in cellular structure.

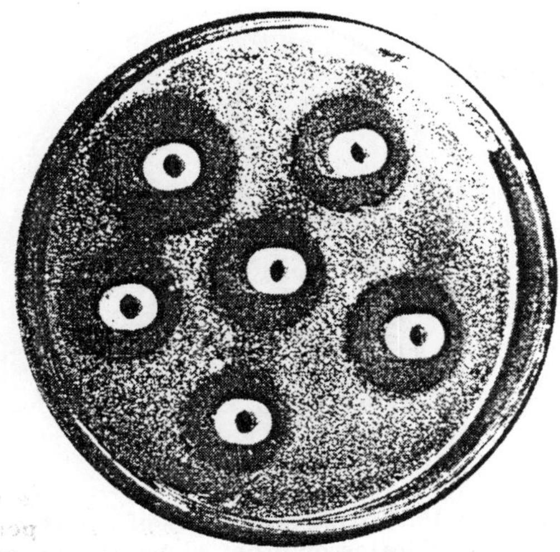

Fig. 8.8. Cylinder plate standardization of penicillin. The culture plate contains a very carefully standardized medium, uniformly and heavily seeded with staphylococci or spores of *B. subtilis.* In each cylinder was placed a measured quantity of penicillin-containing extract. Following incubation under standard conditions, growth of bacterial colonies gave the medium the pebbled, gray appearance seen everywhere except around the cylinders. Here, growth has been inhibited by penicillin. Measurement of these zones of inhibition permits standardization of penicillin in Oxford units by comparison with zones produced under identical conditions of test by standard solutions of known unit strength. The sensitivity of an unknown organism to any antibiotic is tested in exactly the same way; inoculation of the plate is made with the "unknown" in place of the staphylococci or *B. subtilis* routinely used. Known amounts of various antibiotics may be placed in the cylinders.

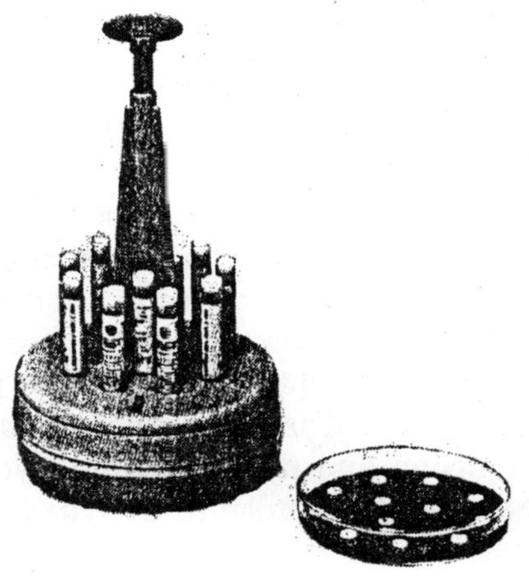

Fig. 8.9. Antibiotic solutions are released automatically onto an inoculated plate when plunger of Sensi-Disc dispenser is pushed.

A number of species of *Streptomyces* produce antibiotics of medical and commercial value. More are constantly being found. Some of the antibiotics produced by *Streptomyces* are said to be active against chlamydias, others against certain viral neoplasms and some against certain protozoa. A description of *streptomycin* will exemplify all the *Streptomyces*-derived antibiotics, although each has its special properties and values.

Streptomycin

The discoverv of streptomycin dates from 1944, and followed naturally in a long series of researches by Waksman and his collaborators, Schatz and Bugie into the numbers and kinds of antagonistic microorganisms in the soil. The source organism *Streptomyces griseus,* produces moldlike conidia and aerial mycelia suggestive of some of the filamentous Ascomycetes (*e.g., Aspiergillus)* but is one species of the bacterial order of Actinomycetales.

The principles underlying the production, purification, assay and standardization of streptomycin are analogous to those of penicillin, differing mainly in respect to cultural details and technical procedures appropriate to the organism and antibiotic involved.

Aeration of the culture is essential for rapid streptomycin production, which is at its maximum in two to three days. Various strains of *Streptomyces griseus* are more or less active producers of the drug.

The molecular formula of streptomycin is $C_{21}H_{39}O_{12}N_7$. It is structurally much more complex than penicillin. Numerous salts and derivatives have been prepared, some of them of great therapeutic value. Dihydrostreptomycin, in which two hydrogen atoms are added is of especial importance. Both streptomycin and dihydrostreptomycin are much more stable than penicillin, both in dry form and in solution. In the refrigerator the solutions retain potency for months; at 37° C for about two weeks. Boiling for 10 minutes produces about 50 per cent loss of potency. Both drugs, *in vitro,* are bactericidal in commonly used concentrations.

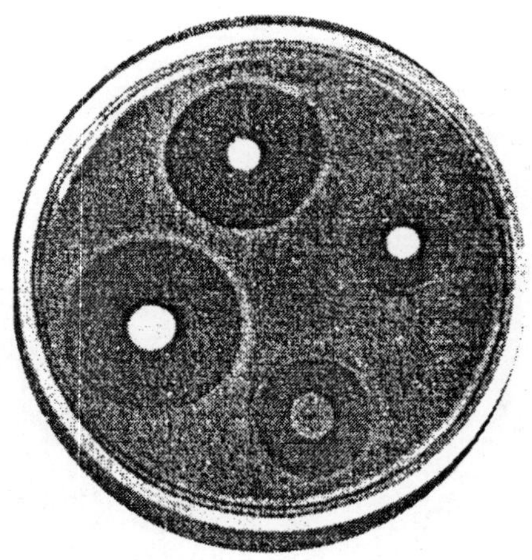

Fig. 8.10. Testing sensitivity of a bacterium to antibiotics or other chemotherapeutic agents by the "disk method." The entire surface of agar medium in a Petri plate is inoculated with the organism to be tested. Paper disks of uniform thickness containing graded amounts of the agent to be tested (or the same amount of different agents if a comparison is desired) are then placed on the surface of the agar. The agent diffuses into the agar and prevents growth of the bacterium in a zone around the disk. The width of the zone indicates, roughly, the sensitivity of the organism to the agent or agents being tested, though the *presence* or *absence* of a zone is of greater significance. (About 1/2 actual size.)

Scope of Action

The therapeutic range of streptomycin is in some respects broader than that of penicillin; it is active against a variety of gram-negative and several gram-positive organisms, some of which are listed in Table 8.3. It has been used in tuberculosis with striking results and in many infections with species of gram-negative rods that are not susceptible to penicillin or sulfonamides. Streptomycin is not so rapidly destroyed in or excreted from the body as penicillin. Some toxic effects (vertigo, deafness) have been described. It must be used only under medical supervision.

Mode of Action

Although the method is not entirely clear, streptomycin undoubtedly acts by interfering with enzyme action, notably enzymes concerned in early stages of the synthesis of proteins. Streptomycin has other, little-understood actions. Dihydrostreptomycin acts similarly. Both streptomycin and dihydrostreptomycin are much less active anaerobically than aerobically; *i.e.,* their action may therefore be related to aerobic oxidative metabolism that is dependent on the Krebs cycle. Like penicillin both are mainly bactericidal rather than bacteriostatic in their action.

No streptomycin-destroying enzyme has been reported, but resistant organisms appear and, unfortunately, often occur in patients during treatment. Strains of tubercle bacilli wholly dependent on streptomycin have also been found in treated patients.

Table 8.2. Partial list of antibiotics*

Antibiotics		*Characteristics***	*Source*
Common Name	*Trade Name*		
Penicillin***		Gram-positive bacteria; *Treponema*; *Neisseria*	*Penicillum notatum*
Fumagillin		*Entamoeba histolytica*	*Aspergillus fumigatus*
Paromomycin	Humatin	*E. histolytica*	*Streptomyces rimosus*
Streptomycin		Myco. Tuberculosis; Gram-negative bacteria	*Streptomyces griseus*
		Like streptomycin	Streptomycin; also some Species of *Streptomyces*
Tetracycline	Achromycin	Broad-spectrum	Chlortetracycline
Oxytetracyline	Terramycin	Broad-spectrum	*Streptomyces rimosus*
Chlortetracycline	Aureomycin	Broad-spectrum	*Streptomyces venezuelae*
Erythromycin	Ilotycin, Erythrocin	Braod-spectrum (not Enterobacteriaceae)	*Streptomyces erythreus*
Carbomycin	Magnamycin	Like erythromycin	*Streptomyces halstedii*
Oleandomycin	Matromycin	Broad-spectrum	*Streptomyces antibioticus*
Neomycin B***	Flavomycin	Mycobacteria	*Streptomyces fradiae*
Viomycin	Viochin	Like penicillin	Str. Floridae; *Str. funiceus*
Oligomycin		Fungi of plants	*Str. Diastatochromogenes*
Amphotericin B	Fungizone	*Candida* sp.	*Streptomyces nodosus*
Kanamycin	Kantrex	Broad-spectrum	*Streptomyces kanamyceticus*
Nystatin		Pathogenic fungi	*Streptomyces noursei*
Cycloheximide	Actidione	Saprophytic fungi	*Streptomyces griseus*
Griseofulvin	Grifulvin	Pathogenic fungi	*Streptomyces griseus*
Bacitracin		Like penicillin	*Bacillus subtilis*
Polymyxin B***		Gram-positive bacteria	*Bacillus polymyxa*
Pyocyanin****		Miscellaneous	*Pseudomonas aeruginosa*

* Several not listed here are valuable commercially, agriculturally and horticulturally.

** Not necessarily the only activity.

*** Several of these antibiotics are in reality mixtures consisting of related compounds such as the penicillins, polymyxin A, B, C, D, carbomycin A and B and so on.

**** Not used medicinally. One of the first known antibiotic substances.

Chloramphenicol

The organism producing chloramphenicol is much like other species of *Streptomyces*. It was found in 1947 by Paul R. Burkholder in soil collected near Caracas, Venezuela, and is called *S. venezuelae*. Like most other natural antibiotics, the drug is a by-product of growth. It may be produced in a manner analogous to the production of penicillin and streptomycin, but is now produced mainly synthetically.

Table 8.3. Some Bacterial Species Sensitive to Streptomycin

Gram-Positive
Mycobacterium tuberculosis
Staphylococcus aureus
Diplococus pneumoniae
Streptococcus species
Bacillus subtilis and related species
Gram-Negative
Salmonella typhi
Pasteurella tularensis
Klebsiella species
Brucella abortus
Proteus vulgaris
Salmonella paratyphi B
Bordetella pertussis
Haemophilus influenzae
Pseudomonas aeruginosa
Escherichia species

It affects many species of both gram-positive and gram-negative bacteria in high dilutions and may be given either orally or intravenously. Unlike either penicillin or streptomycin, it is also effective against the chlamydias and rickettsias. It is the first antibiotic discovered to have definite rickettsicidal properties. It has proved of inestimable value in the treatment of many infectious diseases, including especially typhoid fever. Because of its wide range of activity, it is classed as a *broad-spectrum* antibiotic.

Mode of Action

As in the case of streptomycin, although chloramphenicol does not with certainty compete with any known metabolite, it appears to block several important enzymes, possibly by *noncompetitive combination* (*i.e.,* at a site other than the functional center in the co-enzyme). Whatever its mechanism or mechanisms, the most obvious effect is to stop synthesis of proteins. There is also interference with oxidations at higher concentrations. Chloramphenicol does not stop the synthesis of nucleic acids.

The therapeutic excellence of chloramphenicol led to intensive chemical studies, as in the case of penicillin. The formula was soon learned and the drug synthesized *(Chloromycetin)* in 1949 in the laboratories of Parke, Davis and Company. It is the only antibiotic of major clinical importance that can be made synthetically on a profitable basis. It must be used only under medical supervision.

The Tetracyclines

This group includes three antibiotics: tetracycline *(Achromycin),* chlortetracycline *(Aureomycin)* and oxytetracycline *(Terramycin).* Their similar molecular structures are seen in Fig. 8.13. Chlortetracycline differs from tetracycline in having a chlorine atom attached to carbon atom No. 7 in place of hydrogen. It is produced by *Streptomyces aureofaciens.* Oxy-tetracycline has a hydroxyl group attached to carbon atom No. 5 in place of hydrogen (encircled). It is produced by *S. rimosus.* Tetracycline is an artificial derivative of chlortetracycline. All have similar antibiotic properties.

Unlike penicillin and streptomycin, tetra-cyclines are bacteriostatic in clinical concentrations. They are effective against many gram-negative and gram-positive species of bacteria, chlamydias and some rickettsias. For this reason, they are also classed as broad-spectrum antibiotics and have similar ranges of therapeutic activity. All are effective orally or hypoderm-ically. However, in any given patient, any one of these drugs may at times show surprising irregularities and give unexpectedly brilliant results—or fail. They are not identical.

Mode of Action

These antibiotics disrupt ribosomes and thus interfere with protein synthesis, especially in actively growing cells. They may also interfere with functioning of the Krebs cycle. Other noxious actions in the bacterial cell have been demonstrated but not yet explained.

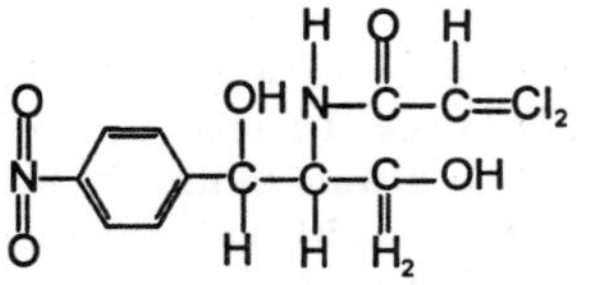

Chloramphenicol

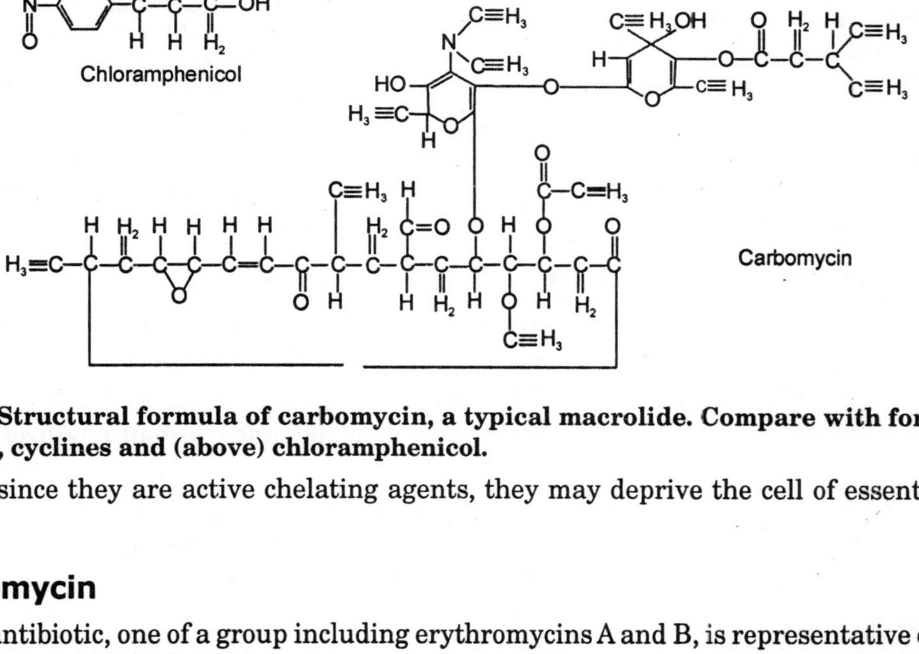

Carbomycin

Fig. 8.11. Structural formula of carbomycin, a typical macrolide. Compare with formulas of penicillin, cyclines and (above) chloramphenicol.

Also, since they are active chelating agents, they may deprive the cell of essential metal ions.

Erythromycin

This antibiotic, one of a group including erythromycins A and B, is representative of several valuable and widely used antibiotics, including *carbomycin* and *oleandomycin.* All are substances of a basic nature derived from *Streptomyces* species and having relatively high molecular weights. Because of their large, complex molecules, the group name *macrolides* has been given to them. They act principally against gram-positive bacteria, but also attack several species of gram-negative bacteria as well as other types of pathogens. They are therefore generally included with the broad-spectrum antibiotics. Their chief effect is to inhibit synthesis of proteins, possibly by blocking or destroying one or more types of RNA.

ANTIBIOTICS FROM BACILLUS

Several useful antibiotics have been derived from gram-positive, aerobic, sporeform-ing bacteria closely related to, or identical with, *Bacillus subtilis.* The first to be described, *tyrothricin or gramicidin,* from *B. brevis,* was discovered by Dubos, as previously mentioned. Another is *bacitracin,* discovered by F. L. Meleney.

Bacitracin

Bacitracin was first described in 1945. The source organism is a particular strain of *Bacillus subtilis,* a common and usually harmless organism widely distributed in dust. It was found contaminating a wound in a patient named Tracy, hence bacitraein. Bacitracin is a yellowish powder readily soluble in water. Like penicillin it is active clinically mainly against infections due to gram-positive bacteria. However, it also affects the gram-negative meningococcus *(Neisseria meningitidis;* cause of meningitis), gonococcus (cause of gonorrhea) and *Haemo-philus influenzae,* as well as *Treponema pallidum* (cause of syphilis). Unlike penicillin, streptomycin, chloramphenicol and the tetracyclines, bacitracin is quite toxic if given internally and is therefore used only for topical applications.

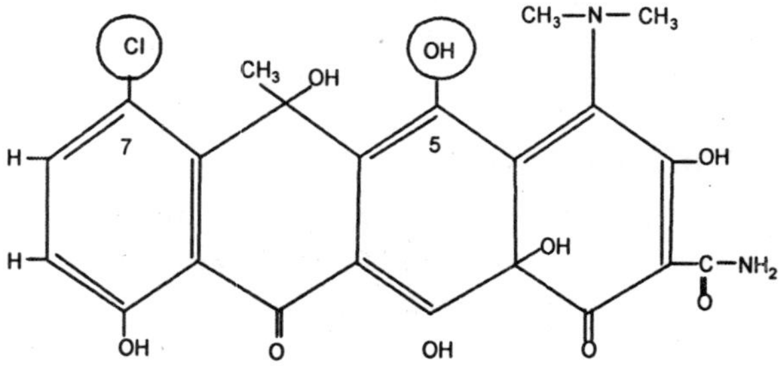

Fig. 8.12. Composite formula of the tetracycline group of antibiotics. In tetracycline each of the numbered carbon atoms (5 and 7) has an attached hydrogen atom. In chlortetracycline, chlorine (circled) replaces the hydrogen of the number 7 carbon atom; in oxytetracycline an OH group (circled) replaces the hydrogen of the number 5 carbon atom.

Other Polypeptide Antibiotics

Bacitracin is representative of several *polypeptidc* antibiotics with similar properties, derived from species of *Bacillus:* the polymyxins A, B, C, D, E *(B. polymyxa);* subtilin and bacitracin *(B. subtilis);* the gramicidins and tyrocidin *(B. brevis);* bio-cerin *(B. cereus);* and circulin *(B. circulans).* Commercial bacitracin, and probably these related polypeptides, are all mixtures of related substances.

These are curious chemically in that they contain rare types of peptide linkages and amino acids of the D-series which have not been found elsewhere in nature except in bacterial cell walls. They are alike in being surface tension reducers, or surfactants.

Mode of Action

Antibiotics such as bacitracin resemble other surfactants, such as the quaternaries and soap, in quickly destroying susceptible bacteria by attacking the cell membrane, possibly by acting as an emulsifying agent on the lipids in the membranes. They affect many gram-negative species, cell walls of which are rich in lipids. They act so quickly as to suggest that, unlike most other antibiotics, interference with enzymes is not involved. Unlike penicillin, these antibiotics leave the cell wall intact.

Other polypeptide antibiotics (*e.g.*, streptogramin, vernamycin and ostreogricin) produced by *Streptomyces* sp., act by inhibiting protein synthesis but not synthesis of nucleic acids or cell walls.

Lysostaphin

This antibiotic, discovered in 1964, is remarkably effective against coagulase-positive (pathogenic) *Staphylococcus aureus,* which is often resistant to other antibiotics. Instead of inhibiting synthesis of cell walls as does penicillin, lysostaphin destroys cell walls, regardless of penicillinase.

Nonmedical Uses of Antibiotics

Many antibiotics are extremely effective as antimicrobials but are too toxic for medicinal use. The industrial, agricultural and other uses of such antibiotics have grown enormously. Over three million pounds of antibiotics are used for such purposes annually. They are used in such ways as preventing spoilage of various products; supplementing stock feeds, greatly increasing growth; and controlling bacterial diseases of plants.

CHAPTER

9 Biochemical Regulation

The copying of genetic information from DNA into messenger RNA is the initial step in the chain of reactions leading to synthesis of the multitude of proteins and specialized RNA molecules needed by cells. The requirement for these macromolecules varies with conditions, and in eukaryotic cells, with the stage of differentiation. Therefore, it is not surprising that transcription is highly controlled. Cells make four principal kinds of RNA: ribosomal (rRNA), transfer (tRNA), messenger (mRNA), and a variety of small RNAs. The last, which range in length from a few up to several hundred nucleotide units, are designated variously as sRNAs, ncRNAs, miRNAs, siRNAs, snRNAs, and snoRNAs. The abbreviation *s, nc, mi, si, sn,* and *sno* stand for small, non-coding, micro, silencing, small nuclear, and small nucleolar, respectively.

All of the RNAs are synthesized as larger transcripts, which undergo cleavage and other modifications within the cell. Therefore, a second major topic in this chapter is the processing of RNA precursors. We will also consider the fact that cells may be hosts for RNA viruses, may occasionally harbor RNA plasmids, and must sometimes transcribe viral DNA. The absence of a nuclear membrane is a characteristic of bacteria that has a profound effect on transcription. Bacterial transcripts are processed rapidly, and their 5′ ends often enter ribosomes and are directing protein synthesis, while the 3′ ends of the genes are still being transcribed. In contrast, most eukaryotic RNA transcripts must be processed and transported out of the nucleus before they can function. As consequence, many aspects of the control of transcription differ between prokaryotes and eukaryotes.

TRANSCRIPTION AND PROCESSING OF RNA IN BACTERIA

Even after the existence of mRNA had been recognized, it was not obvious how formation of single-stranded (ss) RNA would be accomplished, using a double-stranded (ds) template. The fact that purified RNA polymerases can synthesize RNA from the four ribonucleoside triphosphates using ssDNA as the template suggested that transcription, like DNA replication, involves base pairing. In line with this conclusion was the fact that the ssDNA obtained from bacteriophage ϕX174 was converted by RNA polymerases into a dsRNA-DNA hybrid. However, when dsDNA served as the template, free ssRNA was formed. Thus, it appeared likely that at the site of the polymerase action the dsDNA was momentarily pulled apart into single strands and that one of these was copied by the polymerase. More recent experiments have confirmed this view.

The lac Operon

Much of the terminology used to describe the control of transcription originated with Jacob and Monod. Based on studies of the induction of enzymes in bacteria they proposed the *operon model*. An operon is a regulated cluster of genes, one of which is shown diagrammatically in Fig. 9.1. This is the *lac* operon of *E. colli.* Found at position 8 min on the genetic map, it is probably the most intensively studied group of *E. coli* genes. Three structural genes encode the amino acid sequences of β-galactosidase (*lacZ*), permease (*lacY*), and a transacetylase (acetyltransferase, *lacA),* which transfers acetyl groups from acetyl-CoA to β galactosides. To account for the apparently synchronous control of these three genes, Jacob and Monod proposed that they function as a *transcriptional unit* the operon, which encodes a single molecule of mRNA. They proposed that each operon is controlled by a segment of the DNA molecule located at the beginning of the operon, *i.e.,* at the 5′ end of the coding chain or 3′ end of the template chain. The first part of this *control region* they called the *promoter* (P). The promoter is the site of the initial binding of the RNA polymerase to the DNA, the binding constants for the association being very high. The rates of association and of initiation may be in fluenced strongly by various other proteins. One of these, the *catabolite gene activator protein* (CAP; also called cAMP receptor protein, CRP), is important to the *lac* operon. It also binds in the promoter region and stimulates transcription.

Repression and Induction

Immediately adjacent to the promoter is the *operator* (*O*), which is a binding site for a *represser* (R). When the operator is free, transcription is initiated and proceeds through the operator region and on to the genes coding for the three proteins. On the other hand, if the represser is bound to the operator, transcription is blocked. When the operon model was first proposed, the chemical nature of the represser was unknown. However, many repressers have been identified as oligomeric proteins able to undergo allosteric alteration. The *lac* represser is made up of four identical 360-residue subunits. Each subunit has a helix-turn-helix binding domain that is specific for the DNA sequence of the operator and an allosteric binding site for an effector. The *lac* operon is ordinarily subject to repression and is activated by the presence of an *inducer*, now known to be *allolactose*, D-Gal*p*-β1 → 6-D-Glc. However, in experimental work artificial inducers such as *isopropyl-β-D-thiogalactoside* (IPTG) are most often used. Jacob and Monod postulated that the free represser protein binds to the operator. In the presence of the inducer a conformational change takes place, destroying the affinity of the represser protein for the operator site. Thus, in the presence of inducer the operator is not blocked, and the transcription takes place. Such an operon is said to be *negatively controlled* and *inducible*. Important to the control of the operon is the *regulatory gene*, which codes for the synthesis of the repressor protein.

In the case of the *lac* operon, the regulatory gene (the *I* gene) is located immediately preceding the operon itself. However, for some operons the regulatory gene is located a considerable distance away. For example, the *gal* operon of *E. coli,* which codes for enzymes of galactose metabolism, is found at map position 17 min, while the regulatory gene is at 61 min.

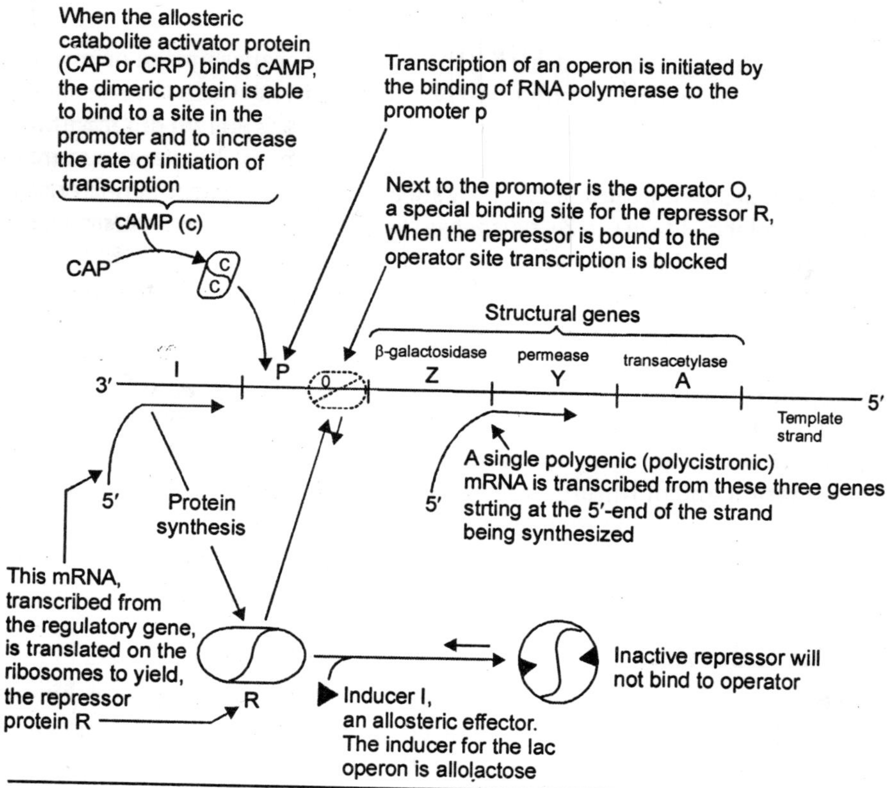

Fig. 9.1. Schematic representation of the *lac* operon of E. *coli* and of its control. Here only the template strand of the DNA is shown. However, the coding (nontranscribed) strand is usually the one labeled.

Regulatory genes are normally transcribed at a slow but steady rate, presumably because RNA polymerase initiates RNA chains infrequently at the promoter sites of regulatory genes. Thus, each cell of *E. coli* normally contains only about ten molecules of the protein. A mutation in a regulatory gene may lead to a defective repressor, which no longer binds at the operator. Then, transcription of the operon is uncontrolled, and mRNA is produced in greater amounts. In such a mutant strain (designated I^- in contrast to the normal I^+ strain) production of the enzyme representing the gene product becomes constitutive, just as is the formation of the enzymes of the central pathways of metabolism. The latter enzymes also appear to be produced regularly in large amounts without control by a repressor, establishing that *transcription* rather than translation regulates the expression of these genes. The operon can also become unregulated, *i.e., constitutively expressed,* if a mutation occurs in the operator site and the repressor can no longer bind because of the altered DNA sequence.

Nucleotide Sequence of the Lac Control Region

The sequence of the *E. coli* DNA representing the promotert-operator region of the *lac* operon is shown in Fig. 9.2. It includes the end of the *I* gene, at the left, and the beginning of the Z gene, at the right. The series of codons representing the peptide sequence Glu-Ser-Gly-Gln-Stop at the left-hand end corresponds to the known C-terminal sequence of the repressor, while the three codons at the right are those of formyl-Met-Thr-Met, the known N-terminal sequence of the Z gene product β-galactosidase. Detailed genetic mapping of the region in 1973 made it possible to assign operator and promoter regions with confidence as indicated. The mRNA transcript begins in the operator region as shown in the figure. The initiation codon for the Z gene is 39 bases from the end of the transcript. In this figure the original numbering of the nucleotides is printed in black. Now it is customary to number from the transcription initiation position (+1).

Green numbers are used in this way in Fig. 9.2. Position to the left and toward the 5′ end of the non-transcribed coding strand precede the initiation position. They are referred to as *upstream* and are numbered with negative integers. Positions to the right are *downstream* and numbered with positive integers. The operator region was located by digesting the DNA with deoxyribonuclease in the presence of the repressor protein. The bound repressor protected a region of 27 base pairs as indicated in the figure. The operator is centered on a region of local twofold rotational symmetry.

The symmetry is not perfect, the sequence being *quasipalindromic*. The following precisely symmetric synthetic sequence, which contains an 11-bp inverted repeat of the left half of the *lac* operator sequence, binds *lac* repressor 8-fold more tightly than does the natural *E. coli* operator.

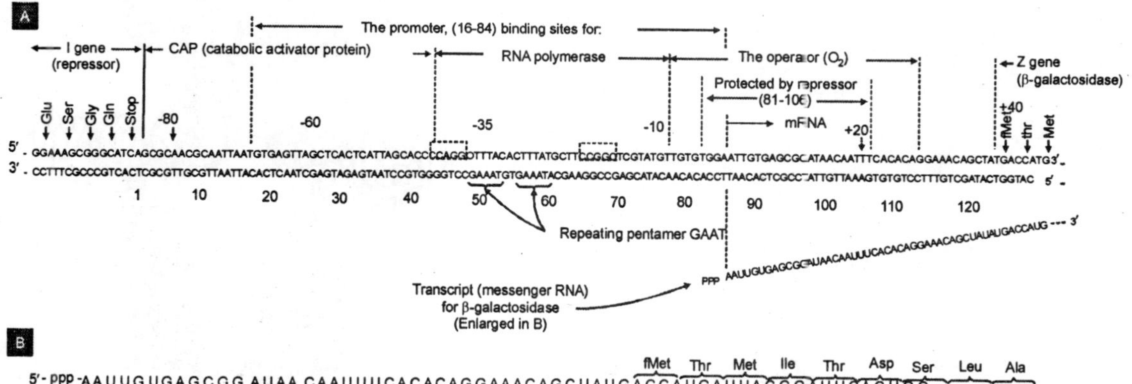

Fig. 9.2. (A) Nucleotide sequence of the *lac* promoter-operator region of the *E. coli* chromosome. The proposed locations of the *I* gene, the promoter (which contains CAP and RNA polymerase binding sites), the operator, and the beginning of the Z gene (P-galactosidase) are shown. Note the two regions of local 2-fold rotational symmetry, which are marked by bars and central dots and the repeating pentamer. Positions upstream (-) or downstream (+) from the +1 start position for transcription are marked in green. The -10 (Pribnow) and -35 promoter elements are boxed on the coding strand (top). Labels are usually applied to the coding strand. The mRNA is copied from the complementary template strand (bottom). (B) The sequence of an mRNA molecule initiated in the *lac* promoter-operator region of a mutant strain of *E. coli* with an altered promoter. The peptide initiation amino acid is identified by the symbol fMet, and the successive amino acids from the known N-terminal sequence of β-galactosidase have been matched with the codons.

5′-C-A-A-T-T-G-T-G-A-G-C-G-C-T-C-A-C-A-A-T-T-C

3′-C-T-T-A-A-C-A-C-T-C-G-C-G-A-G-T-G-T-T-A-A-G

The dyad symmetry of the operator sequence is probably important in providing tight binding to two subunits of the symmetric tetrameric protein. It is also possible that represser molecules move along DNA chains in a one-dimensional diffusion process, and that the symmetry of the operator site facilitates recognition by a protein moving from either direction.

The Repressor Structure

The *lac* repressor protein is a member of the *helix-turn-helix* family of DNA binding proteins. The first helix and the turn of this motif fit into the major groove of the DNA, the side chains from the helix interacting with specificity-determining groups in the major groove. The protein consists of three domains; the N-terminal DNA binding "head" (residues ~ 1– ~50), a core domain (residues 62-340), and a leucine heptad repeat domain (residues 340-360) that forms the dimer-dimer interface. The dimeric form of the repressor binds to the palindromic operator sequence.

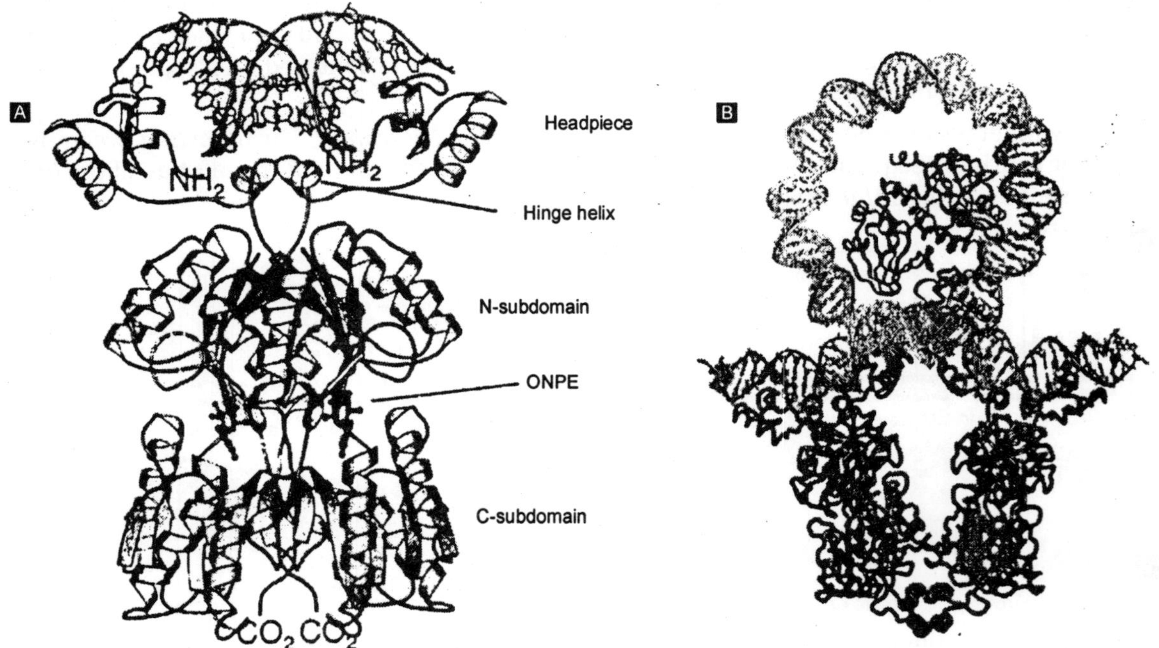

Fig. 9.3. (A) Ribbon view of the dimeric *lac* repressor bound to a natural operator and to the anti-inducer o-nitro-phenylfucoside (ONPF). The headpiece (residues 2-46) and the hinge helix (residues 50-58) form the DNA-binding domains. The core (residues 62-330), which is divided into N- and C-terminal subdomains, forms the binding site for ONPF. The C-terminal residues 334-360, which form a tetramerization domain, are absent from this MolScript drawing. Notice that the hinge helices bind to and widen the minor groove at the center of the operator. From Lewis *et al.* (B) Model of a 93-bp DNA loop corresponding to residues –82 to +11 of the *lac* operon bound to the tetrameric *lac* repressor. The active sites of the repressor are bound to the major operator O_1 and to the secondary operator O_3.

The tetramer can bind to two operator sequences. The inducer IPTG binds to the core domain near the ONPF site shown in Fig. 9.3. The sequence, structure, and binding site of the core domain resemble those of sugar transport proteins such as the galactose-binding chemoreceptor protein. Binding of the inducer causes the conformational change that appears to disrupt the interactions of the "hinge helices," seen in the center of Fig. 9.3A, with the DNA. This causes the repressor to dissociate from the operator allows RNA polymerase to bind and to initiate transcription. The *E. coli lac* repressor is one of the most investigated of all proteins. For example, 4000 single-amino-acid mutants have been prepared and studied.

Suppressor mutations were used to determine the function of various portions of the protein. Many of the mutant proteins were created using *amber* mutations that were induced in the gene at many positions. The mutated genes were transferred into plasmids for cloning. Each plasmid was used to infect five different strains of bacteria, each carrying a suppressor mutation that would introduce a different amino acid when the (termination) codon UAG was encountered. From these infected bacteria large quantities of the mutant forms of the *lac* repressor were isolated. It was found that many mutations near the N-terminal end interfered with binding of the repressor to DNA, whereas mutations near the center interfere with binding to the inducer. In addition to the main *lac* operator O_1, which is marked in Fig. 9.2, there are two weaker auxiliary operator sequences designated O_2 and O_3 located 401 bp downstream of O_1 in the lacZ coding region, and 92 bp upstream of O_1, respectively.

The DNA can bind to both O_1 and either O_2 or O_3 with a loop between them as in Fig. 9.3B. Binding of the 10-20 copies of the lac represser present in a cell of *E. coli* to the operator sequence is so tight that expression of the genes controlled is reduced 1000-fold. However, when placed in front of other operators or in different positions relative to the Pribnow sequence, its effectiveness varied greatly. The extent of repression appears to be affected both by thermodynamic factors and by relative rates of represser binding and of RNA polymerase movement.

Initiation of Transcription

The rate of RNA synthesis varies from one operon to another. Sequences of promoters, operators, and other control sequences as well as the state of repressors and activator proteins all affect these rates. However, in every instance the first steps in transcription involve the binding of RNA polymerase to DNA.

Bacterial RNA Polymerase

Most RNA polymerases (RNAPs) are large multisubunit proteins. However, bacterial viruses sometimes induce their own RNA polymerases, and these may be monomeric. For example, the 99-kDa (883-residue) phage T7-encoded polymerase is a single peptide chain with a structure and two-metal-ion active site resembling those of *E. coli* DNA polymerase I. It is able to carry out all of the steps of the transcription cycle of the virus. In contrast, the most studied bacterial RNA, that from *E. coli,* consists of five kinds of subunits with the composition $\alpha_2\beta\beta'\sigma\omega$. A similar composition has been found for RNAPs of other bacteria.

Functions of the five subunits can be correlated directly with components of archaeal and eukaryotic RNAPs. However, the latter contain additional subunits. The two α subunits in the *E. coli* enzyme have identical sequences, but their locations and interactions are different.

The three-dimensional structure of the *E. coli* RNAP bound to DNA in an initiation complex shows that the enzyme forms a groove into which the DNA can fit. It can then close to form a tunnel in which the template and nontemplate strands are separated. The polymerase is present in large amounts, -3000 molecules per cell in *E. coli*. The β and β' sub-units of the *E. coli* enzyme each contain an essential zinc ion. The Zn^{2+} in the β' subunit is present in a zinc finger motif near the N terminus.

Gene Symbol	*Subunit*	*Molecular mass (kDa)*	*Number of amino acid residues*
α	*rpo*A	36.5	329
β	*rpo*B	150.6	1342
β'	*rpo*C	155.2	1407
σ	*rpo*D	70.2	
ω	*rpo*Z	6	

It is thought to interact with the template strand of the DNA. The Zn^{2+} in the β strand is more loosely bound. The active site is largely in the P subunit. In the assembly of the RNAP complex a dimer of the small α subunit binds to β, after which β' is added. The sequences of the β and β' subunits have several highly conserved regions with homologous sequences from bacteria and from functionally equivalent regions in eukaryotic RNAPs. Three aspartate residues in the sequence NADFDGD may chelate two Mg^{2+} ions as in the active site of DNA polymerases. The basic chemistry of all of the polymerases may be similar, but the modular structure of the bacterial RNAP differs markedly from that of the DNA Pol I family. Of the RNA polymerase subunits σ (sigma) plays a unique role in initiation of transcription.

It is required for the recognition of promoter sites. However, it is not needed for elongation of an RNA chain and dissociates from the $\alpha_2\beta\beta'$ core complex soon after transcription is initiated. In a given bacterial species there is one predominant σ factor, but there are often smaller amounts of other σ factors with homologous sequences. In *E. coli* σ where the superscript number is the molecular mass in kDa) is predominant, but other specialized σ subunits recognize different groups of promoters. For example, $\sigma^N(\sigma^{54})$ binds to promoters that allow transcription of genes involved in assimilation of nitrogen as well as in aromatic catabolism. Synthesis of protein σ^s of *E. coli* is induced by stress such as carbon starvation. At high temperatures (*e.g.*, 40-49°C) σ^{32} is synthesized and permits transcription of genes for "heat shock proteins." Actively growing cells of *Bacillus subtilis* contain at least five different sigma factors. An additional four control gene expression during spore formation.

Promoter Seuences

In 1975, Pribnow pointed out that a series of six known promoters had a conserved 7-base sequence beginning six nucleotides upstream from the initiation site for transcription. Although this sequence varies somewhat from one promoter to another, it has been found in hundreds of *E. coli* promoters. This is called the –10 *region*, the *Pribnow sequence*, or Pribnow box (the last in recognition of the fact that people like to draw boxes around these special sequences). A typical 6-base consensus Pribnow sequence is 5′-TATAAT as written for the coding strand, whose sequence corresponds to that of the mRNA. Only three of these bases are highly conserved: 5′-TA_ _ _ T. For example, in the *lac* promoter the sequence is TATGTT.

The nucleotides of the –10 consensus sequence are present with the following frequencies (as percentages): T(80)A(95)T(45)A(60)A(50)T(96). The position of the –10 sequence is not exactly the same in all promoters but usually starts 5–9 bp upstream of the start position for transcription. About 16–19 bp upstream from the Pribnow sequence is another conserved sequence, the –35 *region*. A consensus sequence is TTGACA, the TTG sequence to the left being the most highly conserved. Both -10 and -35 regions are needed for efficient promoter activity. Lewin suggested the following "optimal" promoter sequence. The location of these two regions in transcription initiation complex is shown in Fig. 9.4B.

Naturally occurring promoters usually do not have the exact –10 and –35 consensus sequences, but artificially constructed promoters containing them are highly effective *in vivo*. The fact that most promoters depart from these "ideal" sequences is not surprising because cells need varying amounts of different proteins. Promoter strengths vary over a range of at least 10^4. Much of this variation comes from variations in the specific -10 and -35 sequences, which appear to be specifically recognized by RNA polymerase. This variation includes an upward extension of the –10 region for some promoters.

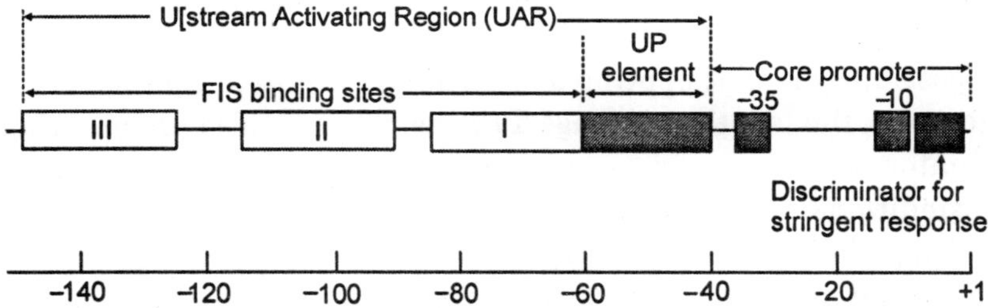

Activator-binding sites are also often present in an *upstream activating region* (UAR; see following diagram). This may extend from the –35 sequence through the remainder of the promoter region and as far upstream as –200 to –500 bp. For example, the CAP-binding site in the *lac* operon DNA is centered at –60. In spite of the variation it was possible to locate 2584 operators (of which only 392 were previously known) and to predict the location of 2405 promoters, when the complete *E. coli* genome sequence became known. The binding of the various RNAP subunits, repressors, and activators has been studied using mutant promoter sequences, antibiotic-binding sites, chemical crosslinking reagents, and a cysteine-tethered Fe-dependent DNA-cutting reagent. Other "footprinting" techniques, *e.g.,* observing cleavage of the DNA by hydroxyl radicals generated by reduction of H,O_2 by Fe(II), have also been employed. It was shown that RNA polymerase binds to both the –10 and –35 sequences and also to sequences further upstream.

The σ^{70} subunit associates with the DNA, principally the transcribed strand, along a region from about the –25 to the + 12 position relative to the transcription start site. The α subunits bind to an UP element from ~–40 to –60 via their C-terminal domains (CTDs).

Control of Stable RNA Synthesis

Whereas most mRNA has a relatively short lifetime, the stable ribosomal RNAs and transfer RNAs have much longer lives. Furthermore, in *E. coli* their transcription is coordinately controlled by seven *rrnB* P1 promoters. The genes for stable RNA have promoters with the usual –10 and –35 sequences, but they contain a complex upstream activating sequence that includes the UP element and three binding sites for protein FIS (*the factor for inversion stimulation*). This name reflects a second function, that of promoting inversion of a DNA segment in the Hin recombinase system. FIS is a dsDNA-bending protein.

A dimer of 11.2-kDa subunits, it is an abundant protein. Like HU, IHF, H-NS, and Dps, it coats a significant fraction of the DNA in the *E. coli* chromosome. It binds to the FIS sites using a helix–turn-helix motif. In addition, there is a GC-rich region at positions -7 to -1 with the consensus sequence 5′-GCGCC_C. It has been suggested that this *discriminator* is involved in the *stringent response*, the diminished rate of stable RNA synthesis observed during amino acid starvation. The very complex stringent involves ribosomes in the synthesis of guanosine 5′-triphosphate-3′-diphosphate (pppGpp) under some conditions. The regulator pppGpp, whose concentration may rise from ~ 50 μM to ~ 1 mM within minutes after deprivation of amino acids, may react directly with RNA polymerase at an allosteric site to inhibit transcription.

Another possibility is that pppGpp acts on another protein that binds to the discriminator sequence. However, the fact that a phage T7 gene, which is also under stringent control, lacks the discriminator sequence argues against this. pppGpp, whose concentration is usually high under conditions of slow growth, may be a major growth-controlling effector, which acts by inhibiting the replication of the rRNA needed for ribosome formation and protein synthesis when nutrients are scarce. However, some operons are activated by the same effector.

The Initiation Reaction

A promoter not only locates the site of initiation but also determines the direction of transcription and, therefore, the strand of the DNA duplex that is to serve as the template. The requirement for two specific recognition sequences ensures this directionality. The RNA polymerase may bind randomly to DNA, then move rapidly along the double helix until it locates a strong binding site where in binds to the recognition sequences of the promoter through specific interactions iivthe major groove of the DNA helix. A satisfactory mathematical model for initiation of transcription supposes that the polymerase and DNA bind reversibly to form a complex with formation constant K_f. This initial specific polymerase-promoter complex is referred to as a *closed complex* because it is thought that the bases in the DNA chain are all still paired. It is postulated that in a rate-determining step the closed complex is converted into an *open complex*, which is ready to initiate mRNA synthesis. In the open complex the hydrogen bonds holding together the base pairs have been broken, and the bases of the template chain are available for pairing with incoming ribonucleotide triphosphates.

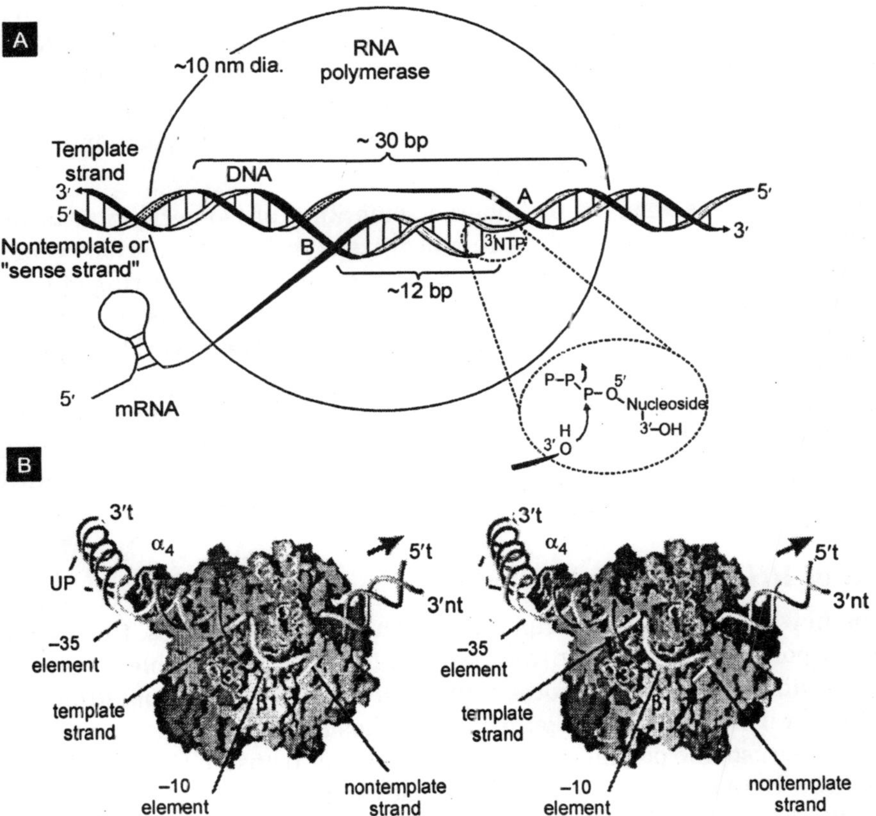

Fig. 9.4. (A) Hypothetical structure of a "transcription bubble" formed by an RNA polymerase. Shown is a double-stranded length of DNA with the unwound bubble in the center. This contains a short DNA-RNA hybrid helix formed by the growing mRNA. The DNA double helix is undergoing separation at point A as is the hybrid helix at point B. NTP is the ribonucleotide triphosphate substrate See Yager and von Hippel. (B) Stereoscopic view of the structure of RNA polymerase from *Thermus aquatieu* in a complex with a promoter DNA. Included are the αI, αII, ω, β, β′, and σ subunits. However, the α C-terminal domains have been omitted. The template (t) strand passes through σ tunnel, which is formed by the β and b′ subunits and two of the structural domains of the a subunit. The nontemplate (nt) strand follows a different path. The position of the -10, -35, and UP elements of the DNA are marked.

$$\text{E(Polymerase)} + \text{P(promoter)} \underset{}{\overset{K_f}{\rightleftharpoons}} \text{EP}_c\,\text{(closed)} \xrightarrow{k} \text{EP}_O\,\text{(open)} \rightarrow \cdots \rightarrow \text{Transcription} \qquad ...(9.1)$$

It is clear that the efficiency of initiation depends upon both the affinity K_f and the rate constant k for opening of the double helix. Notice that the Pribnow sequence is AT-rich; therefore, opening of the helix at this point would be easier than in a GC-rich region. Thus, the Pribnow sequence may represent a point of entry of RNA polymerase to form the open complex. Other upstream A•T tracts are often present frequently at about the -43 position in the UP element. They also seem to strengthen promoter activity.

The open complex is thought to undergo some kind of isomerization to form an *initial transcribing complex* in which the first chemical steps in RNA formation occur. These initial steps may involve rearrangements of subunit interactions and untwisting of DNA by torsional movements between subunits. Initiation of an RNA chain begins by reaction within the transcription bubble of either ATP or GTP with a second ribonucleotide triphosphate, form a dinucleotide still bearing a triphosphate at the 5′ end. Further addition of nucleotide units at the 3′ end by the same type of reaction leads to rapid transcription at a rate of ~50 nucleotides s^{-1} at 25°C. This is about one-thirtieth the rate of replication. The action of the RNA polymerase is apparently processive, a single molecule of the enzyme synthesizing the entire mRNA transcript.

$$\text{ATP(GTP)} + \text{NTP} \rightarrow \text{PP}_i + ppp\text{Pu}p\text{N} \qquad \text{...(9.2)}$$

Elongation of RNA Transcripts

After the newly initiated transcript has grown to 8-9 nucleotides, the sigma factor is lost from the RNA polymerase complex, the complex becomes very stable toward increased salt concentrations, and transcription proceeds processively in a *stably elongating mode*. As it does, the strands of the DNA duplex are pulled apart ahead of the polymerase and close up again behind the polymerase, the polymerase itself moving in a "transcription bubble". This is thought to lie within the ~ 10 nm diameter RNA polymerase complex and to encompass -30 bp of DNA. At the leading edge a "separator" opens the DNA, which then closes behind the bubble. Behind the polymerase active site the transcribed RNA forms a short hybrid helix with the DNA, presumably with an A type structure.

A severe topological problem is avoided if the transcribed RNA is separated at point B as the polymerase moves along the double helix. The driving force for the polymerization lies largely in the hydrolysis of the inorganic PP_i formed in the polymerization. Stabilization of the RNA transcript by formation of loops and other secondary structure may also be a factor. Although this picture seems clear and simple, many uncertainties remain.

Transcription does not proceed evenly but by pauses and spurts. This has suggested the possibility of an "inchworm" type of movement of RNA polymerase. However, the observations may also be explained by variations in the sequence. There are both pausing or stalling sites' and terminator sequences. The concentrations of the needed ribonucleotide triphosphate precursors will also affect the kinetics. In addition, defects in the DNA will be met. Transcriptionally linked repair may have to be called into play before transcription can continue.

Mispairing of bases can also occur in the growing RNA chain necessitating a pause during which the mismatch is recognized. Specialized proteins *GreA* and *GreB* participate in an editing step during which the RNAP backtracks for a few nucleotides, while a piece is hydrolyzed off from the 3′ end before transcription can continue. If the transcription complex moves straight along a DNA double helix, separation of the strands will create positive supercoils (overwound

repressors bind to only one side of the double helix. Addition of dimethyl sulfate to growing *E. coli* causes methylation at many sites in the DNA and has been used to obtain footprints *in vivo.*

Not only are certain guanosines in a promoter protected from methylation when active transcription is occurring but also guanosines on the opposite side of the double helix become unusually reactive. This suggested that the DNA helix is bent in the transcriptional initiation complex and that the resulting distortion makes the bases on the outside of the bend more reactive. There is also evidence that the repressor binds to the opposite face of the DNA that binds RNA polymerase in the –35 region of the promoter. Repressors may have similar recognition domains but may vary greatly both in size and in the functioning of their other domains, which may react both with small allosteric effectors and with other proteins.

The repressor *BirA* of the *E. coli* biotin synthesis operon is an enzyme. The 321-residue protein activates biotin to form biotinyl 5′-adenylate and transfers the biotinyl group to proteins such as acetyl-CoA carboxylase and also represses transcription.

Inducible Operons

The operon model as presented describes the negatively controlled inducible *lac* operon. There are many other examples of this type of control. Among them are control of the utilization of *N*-acetylglucosamine and xylose. Ten genes for catabolism of inositol are encoded in two negatively controlled operons in *Bacillus subtilis.* Similar controls are used by bacteria to protect against the antibiotic tetracycline. The *TetR* repressor controls a membrane associated protein TetA, which acts as an energy-dependent pump to remove tetracycline from the bacterial cell.

Synthesis of TetA is normally repressed, but tetracycline, if present, acts as an inducer to cause TetR to dissociate from its operator site, allowing transcription of the transporter gene. The gene order, gene content, and regulatory mechanisms in an operon are often poorly conserved among related species of bacteria. However, functional and regulatory relationships may be maintained.

Feedback Repression

A simple modification of the operon model accounts for *feedback repression* by end products of biosynthetic sequences. The product, *e.g.,* an amino acid, vitamin, purine, or pyrimidine base, acts as a corepressor binding to the aporepressor and causing an allosteric modification that inhibits transcription. It is not the aporepressor but the effector-repressor complex that binds to the operon and blocks transcription. An example is the tryptophan repressor complex, whose 3-dimensional structure is shown diagrammaticatty.

If free tryptophan accumulates within a bacterial cell, it binds to the allosteric site in the *TrpR* repressor, inducing a conformational change that permits the tryptophan-repressor complex to bind to at least three 21-bp operator sites.

Binding of the repressor-corepressor complex at the *aroH* operator locus represses the genes for the first three steps in the aromatic biosynthetic pathway. Binding at the *trpEDBCA* operator represses the genes for conversion of chorismate to tryptophan, and binding at the *trpR* locus lowers the rate of synthesis of the *trp* repressor, providing a counteracting effect that may be important in controlling the growth rate. The *trp* operon is also well known for another type of control called attenuation. Feedback repression controls methionine synthesis in *E. coli.*

The corepressor *S*-adenosylmethionine binds to an aporepressor that recognizes the following palindromic sequence in the DNA. This sequence occurs, with minor variation, in front of four different operons encoding enzymes of methionine biosynthesis. Together they form the methionine *regulon*. Other *E. coli* operons that are negatively regulated by operator -repressor interactions include those involved in biosynthesis of phenylalanine, tyrosine, arginine, threonine, and isoleucine. The *tyr* repressor modulates gene expression in at least eight operons, largely by repression. The *ile* repressor acts on both the *il* and *thr* operons. The purine repressor, 341-residue PurR, belongs to the LacI family of repressors. Its C-terminal domain also has some sequence similarity to that of periplasmic sugar-binding proteins. PurR represses several steps in the biosynthesis of IMP and in its conversion to GMP and AMP.

$$\begin{array}{c} \overrightarrow{\text{A G A C}}\,\text{G T C T} \\ \bullet \\ \text{T C T G}\,\underleftarrow{\text{C A G A}} \end{array}$$

In *E. coli* twelve biosynthetic genes, organized as nine transcriptional units, provide for the synthesis of arginine in eight enzymatic steps. As with the *trp* operons, synthesis of the represser ArgR is autoregulated. The hexameric represser has an N-terminal winged HTH DNA-binding domain. As is pointed out earlier acquisition of iron and control of its concentration is of crucial importance to bacteria. In *E. coli* the Fe^{2+}-binding protein *Fur* (ferric uptake regulator) represses promoters controlling siderophore biosynthesis as well as other responses. It is a *global regulator* that controls ~40 transcriptional units. Similar proteins repress synthesis of the diphtheria toxin by *Corynebacterium diphtheriae* uptake of iron in these bacteria and in *Mycobacterium tuberculosis,* and uptake of molybdate.

Positive Control by Activator Proteins

Cyclic AMP in bacterial cells mediates the phenomenon of *catabolite repression*. This is the inhibition of the transcription of genes for enzymes needed in catabolism of lactose and other energy-yielding substrates, when the more efficient energy source glucose is present. Glucose causes a decrease in the concentration of cAMP by a complex mechanism that may also cause a decrease in the concentration of inducer. When the glucose concentration decreases, the concentration of cAMP rises and stimulates the initiation of transcription in many operons. This is accomplished through the mediation of the 209-residue catabolite gene activator protein, CAP (also known as cyclic AMP receptor protein, CRP). This protein is a "global" regulator of gene expression that activates transcription at over 100 promoters in *E. coli.*

The CAP-cAMP complex binds to the *lac* promoter adjacent to the RNA polymerase site at the palindromic sequence in the DNA. The CAP molecule is a 45-kDa dimer of identical subunits, which resembles the repressors in having the HTH reading head that binds to the DNA. However, the HTH motif is at the C terminus. The cAMP binds to two sites in each monomer. Tightly bound cAMP molecules in an *anti* conformation bind in the center of the large N-terminal domains in the major allosteric sites. At higher cAMP concentration the second sites in the C-terminal DNA-binding domains are occupied. The CAP binds to DNA with the consensus sequence 5′-AAATGTGATCT/5′-AGATCACATTT, which maybe located at variable distances from the promoter.

How does binding of the CAP-cAMP complex increase the rate of initiation of mRNA transcription? The anwer evidently lies in direct interaction between CAP and the N-terminal domain of the RNAP α subunit. Binding of CAP induces a 90° bend in the DNA, which may facilitate the protein-protein interaction and may lead to looping. The galactose (*gal*) operon of *E. coli* is negatively controlled and inducible by D-galactose or D-fucose, which bind to the *gal* repressor. There are two overlapping promoter sites, one of which is stimulated by adjacent binding of a CAP molecule.

Fig. 9.6. MolScript ribbon drawing of the CAP dimer bound to DNA with two molecules of the coactivator cAMP bound per monomer. A *syn-cAMP* molecule is bound to the HTH domain and a loop from the N-terminal domain, while the second *anti-cAMP* is bound more tightly in the center of the larger N-terminal domain. The DNA sequence for each half site is 5'-ATGTCACATTAATTGCGTTGCGC-3¢.

A surprise came from the discovery that the operator was *upstream* from the promoter, that is, it comes before both the promoter and the structural genes to be transcribed. Later, a second operator sequence was found 90 bp away from the first and within the first structural gene. This suggested that the dimeric gal repressor binds the two operators to form a loop that blocks transcription. Another example of positive regulation by CAP is provided by the seven proteins required for uptake of maltose and its catabolism by *E. coli.* These are encoded in two operons that are controlled as a single regulon. An apo-activator protein becomes an activator when it binds maltose. An *E. coli* protein known as *FRN* (for fumarate nitrate reduction) is a global transcription regulator homologous to CAP. It is active only under anaerobic conditions in which it controls more than 100 genes. FRN contains an $[Fe_4S_4]^{2+}$ is required for dimerization and binding to DNA. Exposure to O_2 converts the cluster into an $[Fe_2S_2]^{2+}$ cluster with loss of activator activity.

The photosynthetic *Rhodospirillum rubrum* is able to adapt to growth on carbon monoxide as a carbon source. A CAP-like transcriptional activator *CooA* contains heme. It acts as a sensor for CO, which activates transcription, as does cyclic AMP with CAP. Sequence homologies suggest that several other bacterial activator proteins also have the HTH DNA-binding motif near their C termini. There are other types of transcriptional activators in bacteria. One is transcription factor 1 (TF1) encoded by a *Bacillus subtilis* phage. It is a member of the protein HU family.

However, unlike the nonspecific HU it binds to some sites specifically and activates transcription. The *E. coli* Ada protein is the acceptor protein in removal of methyl groups from DNA. The same protein is an inducer of transcription of DNA repair enzymes in the large ada regulon. Methylation of Cys 69 of the Ada protein itself converts it into a gene activator.

Control by Looping

The arabinose utilization operon of *E. coli*, *araBAD,* encodes proteins needed for uptake of arabinose and conversion to D-xylulose 5-*P*. The repressor AraC in the absence of arabinose binds at operator 1 (O_1) to prevent further synthesis of repressor (autorepression) and also at the *araI* region to repress transcription of operon *araBAD*. The operator 2 (O_2) site, which is 211 bp upstream from *oraI,* is also needed for this repression. A loop is apparently formed by repressor binding. Binding of arabinose to the repressor converts it into an activator, which stimulates initiation of transcription at the *BAD* promoter. Further stimulation is provided by the CAP-cAMP complex, which binds next at *araI*. Looping is a recognized control mechanism for a number of other operons as well.

Bacterial Enhancers

Positive regulatory DNA sequences that are distant from the genes controlled are often called enhancers. Their function is usually independent of position over a range of hundreds or more base pairs either upstream or downstream of the transcription initiation site. Quite common in eukaryotes, enhancers are less often found in bacteria. However, the binding sites for the *nitrogen regulatory protein C (NtrC* or Nr_I) of *E. coli* has the characteristics of an enhancer. It functions with the rather complex glutamine synthetase *(glnALG)* operon in a major control point for nitrogen metabolism. When the supply of nitrogen from NH_3 is low the NtrC protein, a product of gene *glnG,* binds to the enhancer, which is located over 100 bp upstream and is thought to contact the σ^N (σ^{54}) subunit of the RNAP by formation of a loop.

The process has been visualized by scanning force microscopy. Another interesting aspect of this control system is the activation of Ntrc by phosphorylation of a specific aspartate side chain (Asp 54). The NtrC-*P* form is the active enhancer-binding protein. NtrC is a member of the family of two-component sensor-response regulator pairs, which frequently control bacterial metabolism and behavior. The sensor protein is *NtrB* (NR_{II}), which is an auto-phosphorylating histidine kinase similar to the CheY protein of bacterial chemotaxis. NtrC-*P* dimerizes and binds to the enhancer sequence, where it appears to catalyze an ATP-dependent isomerization of the closed to open forms of the transcription initiation complex.

The isomerization may depend upon looping. Other operons that utilize the σ^N subunit of RNAP often also have upstream or downstream enhancers.

Other Two-component Control Systems

More than 60 different sensor-response regulator pairs have been discovered in bacteria. Many are associated with nutrition. For example, more than 30 genes of the phosphate (*pho*) regular are controlled by the sensor kinase ***phoR***, which detects low phosphate ion concentrations and phosphorylates the response regulator ***phoB***. Protein ArB senses changes in environmental O_2 levels, and response regulator *ArcA* regulates –30 operons in response. One effect is to activate ~30 genes needed for the conjugative transfer of DNA.

Transcription of rhizobial *nif and fix* genes controlled by the O_2 sensor *FixL* and its response regulator *FixJ*. FixL is a heme protein whose kinase is active only when the heme is deoxygenated. The *E. coli* proteins discussed elsewhere mediate transcriptional responses to accumulation of superoxide anions or hydrogen peroxide. In the absence of O_2 the *E. coli* FNR protein induces proteins of the anaerobic respiration pathways. Nitrate also has its own two-component system that senses nitrate availability and activates transcription of enzymes catalyzing nitrate respiration. An expanded two-component system induces sporulation in *Bacillus subtilis* in response to poor growth conditions. The crystal structure of one of two response regulators *(SpoOF)* has a structure closely related to that of CheY and the nitrate response regulator NarL.

Antisense RNA

Another mechanism of control of either transcription or of plasmid replication involves small molecules of RNA that are transcribed from the opposite strand than the template strand used for mRNA synthesis. These *antisense* RNA molecules have at least some part of their sequence complementary to that of the mRNA and to the corresponding sequence in DNA. A well-studied example is control of the copy number of the colicin E1 and other plasmids of *E. coli*. Two transcripts, RNAI and RNAII, are initiated upstream from the origin of replication. RNA II is a 555-nucleotide primer of replication. It is synthesized as a longer transcript that is cut by RNase H at *ori*. This cleavage is inhibited if the 108 nucleotide RNA I forms a duplex with the 5′ end of the RNA II, which has a complementary base sequence.

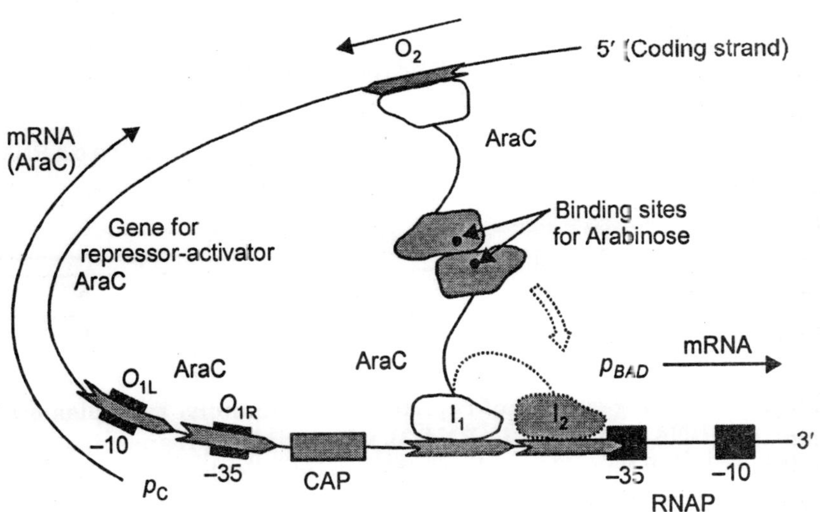

Fig. 9.7. Scheme showing regulatory region of the *araCBAD* operon of *E. coli*. In the absence of arabinose the protein AraC acts as a repressor, which binds to upstream region I of the promoter and also to operator O_2, forming a loop of ~210 bp of DNA. Binding of arabinose to the N-terminal domains of thes dimeric repressor allows the dimer to dissociate and the N-terminal domain of one subunit to bind to region I_2 of the promoter for genes *B,A,D* (dashed outline) activating their transcription. The AraC protein also binds to promoter P_c, repressing its own synthesis.

The process is more complex than this because both RNA II and RNA I have complex secondary structures and are brought together with the help of the small protein product of gene *gro,* which permits them to recognize each other prior to duplex formation. The major outer membrane porins of *E. coli* are encoded by genes *ompC* and *ompF.* A small 174-nucleotide RNA called *mRNA-interfering complementary RNA* (mic RNA), whose gene is upstream of the *ompC* promoter, is transcribed in the opposite direction from that of the porin gene. It is homologous with the 5′ end of *ompF* mRNA, and its function *is* evidently to inhibit translation of the *ompF* mRNA. Since *ompC* and the *mic*FRNA gene are apparently regulated coordinately, synthesis of the *ompF* product is inhibited if the *ompC* product is being synthesized, as happens when the bacteria are growing in a medium of high osmotic strength. Many other examples of regulation by antisense RNA are being discovered. Small noncoding RNA molecules (*ncRNAs*) often serve as templates or *guides* in processes ranging from synthesis of telomere ends to editing, modifying, and splicing mRNA. Both in bacteria and in eukaryotes dsRNAs are often formed and subsequently cleaved to single-stranded antisense RNAs that act as guides to initiate hydrolytic destruction of defective, toxic, or unwanted mRNAs. The recognition of these natural regulatory mechanisms has led to keen interest in artifical regulation by antisense RNA.

Synthetic antisense RNA injected into cells will inhibit expression of selected genes. There is the possibility of effective therapy against viruses, cancer, and inflammation if suitable antisense RNA could be generated within eukaryotic cells or introduced as drugs. Such drugs, which are typically ~15 nucleotides in length, are most satisfactory if they are stable, enter cells, and interact specifically with complementary sequences of DNA or RNA.

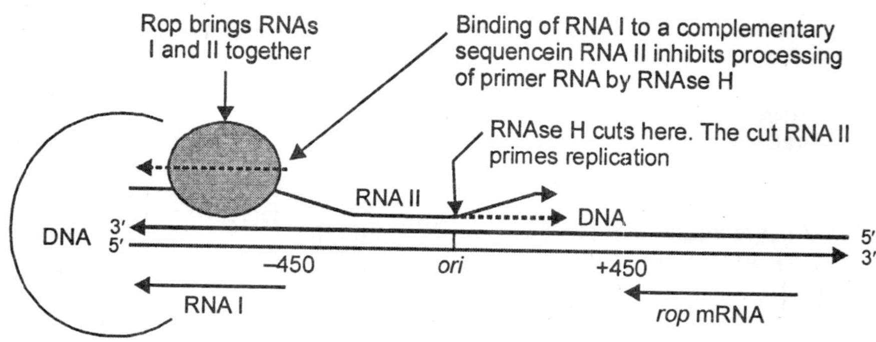

Fig. 9.8. Simplified scheme for control of replication of the ColE1 type plasmid by antisense RNA. The primer for DNA synthesis is RNA II whose initial transcript extends past the replication ori. It is cut by RNase H at ori and then primes replication of the upper strand as shown in the figure. The antisense RNA is RNA I. It binds to protein Rop whose gene location is also indicated in the figure. Rop assists RNA I and RNA II in undergoing a complementary interaction. However, both RNAs apparently maintain a folded tertiary structure, and only some segments interact. The interaction with the Rop protein evidently in some way prevents initiation of replication until the Rop concentration falls because of replication of the host cell.

Stability can be improved by use of linkages other than the natural phosphodiester. Phosphorothioates, in which sulfur replaces one nonbridging oxygen atom on phosphorus, is a favorite. Synthetic antisense oligonucleotide mimics may cause adverse reactions with proteins within cells. Nevertheless, future successes seem likely. See also RNA interference.

Termination of Transcription in Bacteria

Encoded in DNA are not only the initiation signals for transcription but also termination signals or *terminators*. Some of these are sensed by the bacterial RNA polymerase itself while the "reading" of others requires auxiliary proteins. Terminators can be either constitutive or regulatable. The simplest terminators result from GC-rich regions of dyad symmetry in the DNA. The RNA transcript is able to form a stable hairpin loop, possibly within the transcription bubble. If such a loop is followed closely by a series of uracils, the RNA and the polymerase will dissociate from the DNA template terminating transcription.

The low stability of AU base pairs may facilitate dissociation, but RNA polymerase may also recognize the terminator loop. Sometimes a terminator will have a series of adenines preceding the loop. This is often a bidirectional terminator; the transcript from the other strand of DNA will have a loop followed by a series of U's. There are many more complex termination mechanisms.

Attenuation

A major mechanism of feedback repression, known as attenuation, depends not upon a repressor protein but upon control of premature termination. It was first worked out in detail by Yanofsky *et al.* for the *trp*operon of *E. coli* and related bacteria. Accumulation of tryptophan in the cell represses the *trp* biosynthetic operon by the action of accumulating tryptophanyl-$tRNA^{Trp}$, which specifically induces termination in the *trp* operon. Other specific "charged" amino-acyl-tRNA molecules induce termination at other amino acid synthesis operons. The first structural gene in the *trp* operon, *trp*E, is preceded at the 5′ end by a 162-bp *leader sequence*, which is transcribed into mRNA. Within this RNA are two adjacent hairpin loops. The second of these loops has a GC-rich stem and is followed by eight consecutive U's. It is a typical efficient terminator.

An RNA polymerase, having just passed this sequence, will interact with the looped RNA formed behind it and will dissociate from the DNA to terminate transcription. However, if the terminator loop is prevented from forming, transcription will continue, and the structural genes of the operon will be expressed. The *trp* operon contains a short gene for a *leader peptide* preceding the terminator. Its RNA transcript is shaded, and its initiation codon AUG and termination codon AAU are boxed. In bacteria translation of mRNA begins while transcription is still in progress. The 5′ end of the mRNA may enter a ribosome before the RNA polymerase reaches the terminator loop, after which the leader peptide will be synthesized.

The ninth and tenth codons of the leader peptide gene are of special importance. They lie at the beginning of the 1:2 or "protector" loop code for tryptophan. If the level of tryptophan is high, tryptophanyl-$tRNA^{Trp}$ will be formed and the mRNA will move rapidly through the ribosome, these tryptophan codons will pass through, and tryptophan will be incorporated into the peptide. The 1 : 2 loop will be opened, but the terminator loop will remain intact. The result will be termination of transcription. However, if the tryptophan concentration is low, there will be a shortage of charged $tRNA^{Trp}$, and peptide synthesis will be "stalled" with these tryptophan codons in the active sites of the ribosome. This will allow time for the attentuator region to assume the alternative secondary structure shown in Fig. 9.9B. Here the 1 limb of the protector loop is stalled in the ribosome allowing the 2 limb to form the alternative 2 : 3 loop.

The terminator has been destroyed, and transcription continues through the rest of the operon. Thus a low level of tryptophan (and of tryptophanyl-tRNA) favors transcription of the *trp* synthetic operon. Cells of *Bacillus subtilis* also synthesize a *trp* operon transcript that can form either an antiterminator or a terminator loop. Tryptophan, when present in a high enough concentration, binds to a *trp* RNA-binding attenuation protein (*TRAP*). This is an 11-subunit protein, which has 11 tryptophanbinding pockets and also 11 binding sites for GAG or UAG RNA triplets. When tryptophan accumulates within the cell it binds to TRAP, which then wraps ~53 residues of RNA transcript containing 11 GAG or UAG triplets around its perimeter. This prevents formation of the antiterminator loop but allows the terminator loop to form.

At low tryptophan concentrations the antiterminator loop is formed, preventing formation of the terminator loop. Attenuation is also an important mechanism of control of transcription of biosynthetic operons for histidine, phenylalanine, leucine, isoleucine, and threonine. Like the *trp* attenuator region, attenuators for these operons contain codons for the amino acid whose synthesis is being regulated: seven Phe codons in the *phe* attenuator, seven His codons in the *his* attenuator, four Leu codons in the *leu* attenuator. The *thr* operon, which is sensitive to both threonine and isoleucine, has eight Thr and four Ile codons, while the *ilv* attenuator has four Leu, five Ile, and six Val codons permitting feedback repressor by three kinds of charged tRNA.

The pyrimidine synthesis operon *pyr* has three attenuator sequences, one right after the promoter and two others, one just before each of the two genes in the operon. This permits partially independent control of the two genes.

Rho and Other Termination Factors

Termination proteins can also react with specific regions of DNA or of an RNA transcript to terminate transcription. The best known termination factor is the rho protein; a hexamer of 45-kDa subunits. It interacts with transcripts at specific termination sequences, which are often C-rich, and in a process accompanied by hydrolysis of ATP causes release of both RNA and the polymerase from the DNA. Additional *E. coli* proteins, products of genes *nus A* and *nus G,* cooperate with the rho factor at some termination sequences.

The rho hexamer is a helicase that moves along the RNA transcript in the $5' \rightarrow 3'$ direction driven by ATP hydrolysis. If it locates an appropriate termination signal, it may utilize its helicase activity to uncoil the DNA-RNA hybrid segment within the transcription bubble. Cells also contain *antitermination proteins*, which prevent termination of transcription of rRNA or tRNA genes at the many loops of secondary structure that are possible with these transcripts. These antitermination factors are also important in regulating transcription during the lytic phase of growth of phage λ. Also important are rates of hydrolytic degradation of mRNA molecules.

Effects of Antibiotics

The antibiotic *rifamycin* appears to interfere with initiation by competing for the binding of the initial purine nucleoside 5′-triphosphate. The same bacterial RNAP that synthesizes mRNA also transcribes both rRNA and the tRNAs. Thus, the synthesis of all forms of RNA is inhibited by rifamycin. When a population of bacteria is subjected to this antibiotic, a few individuals survive. These rifamycin-resistant mutants are no longer sensitive to the antibiotic. Among them are some mutants that produce an RNAP with an altered β subunit. Since the mutant polymerases do not bind rifamycin, it was concluded that rifamycin binds to the β subunit and that the rifamycin-resistance gene *rpoB* or *rif* (which maps at 89 min) is the gene for the B subunits of RNA polymerase. *Streptolydigin* inhibits both initiation and elongation. *Actinomycin D,* by binding to DNA, inhibits DNA polymerases as well as RNA polymerases, the latter at a concentration of only 10^{-6} M.

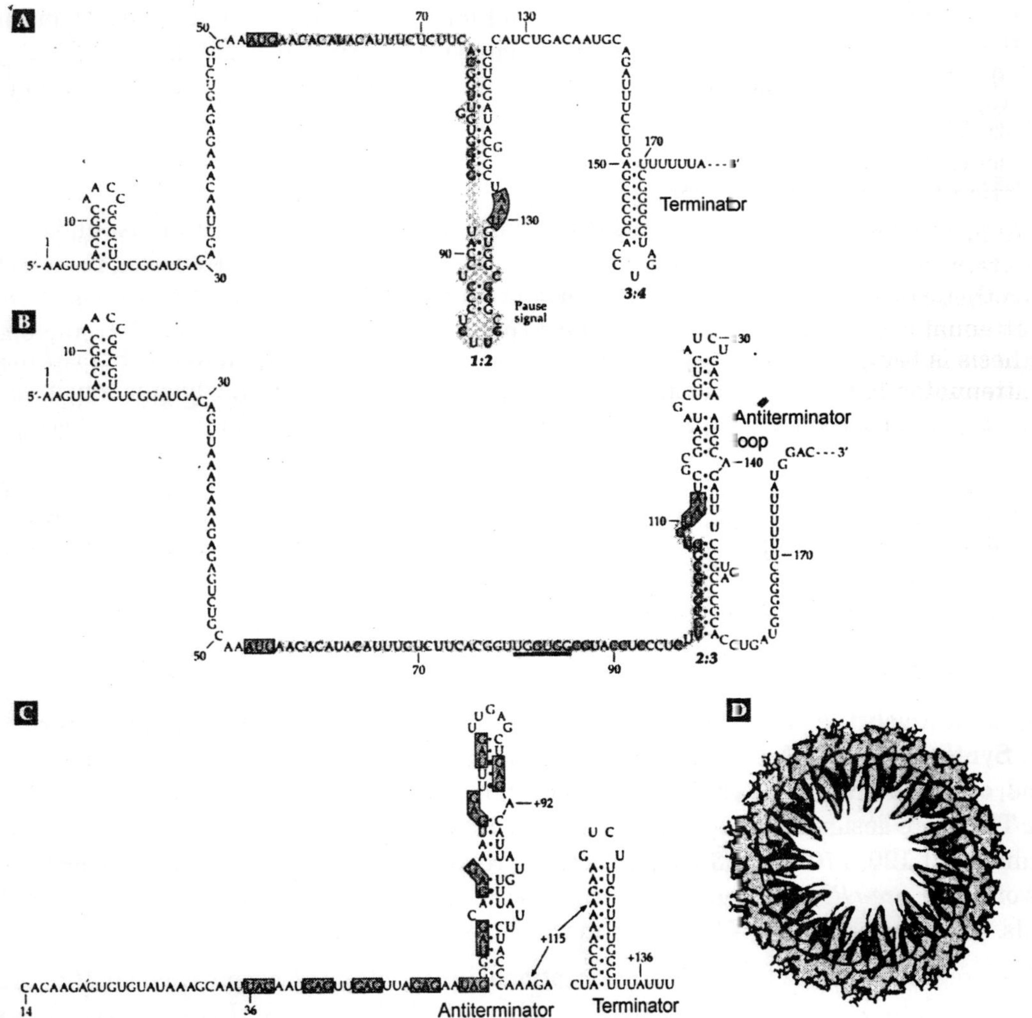

Fig. 9.9. (A,B) Alternative leader RNA structures that mediate control of attenuation in the *trp* operon of *Serratia marcescens*. The leader peptide initiation and termination codons are *boxed*. Tryptophan codons are indicated by *underlining* in B. Formation of the 5′ hairpin structure is predicted by computer analysis but is net implicated in control by attenuation. A structure 1:2 is thought to serve as a transcription pause signal. Structure predicted to form when there is an adequate supply of charged $tRNA^{Trp}$ and is thought to function as the transcription termination signal recognized by RNA polymerase. The 2:3 structure, or *antiterminator loop*, is predicted to form when charged $tRNA^{Trp}$ is unavailable. Its formation presumably precludes formation of structure 3:4, thereby allowing RNA polymerase to continue transcription into the structural genes of the operon. (C) Antiterminator and terminator loops, one of which may form from the leader sequence of the *Bacillus subtilis trp* operon mRNA. Numbering refers to the start point of transcription. The triplet repeats involved in attenuation are shaded. (D) Structure of the 11-subunit tryptophan RNA-binding attenuation protein (TRAP) as a ribbon diagram with 11 molecules of L-tryptophan shown as van der Waals spheres. The apparently circular RNA structure reflects the fact that the gap between the beginning and end of the bound RNA segment is averaged (randomized) over eleven binding sites in the crystal structure. The 53-residue RNA containing 11 triplet repeats of GAG or UAG is bound around the perimeter and is shown as a ball-and stick model.

The eukaryotic RNA polymerases are not inhibited by rifamycin, but RNA polymerases II and III are completely inhibited by the mushroom poison *α-amanitin.* Inhibitors of DNA gyrase also interfere with transcription as do chain terminators such as cordycepin (3′-deoxyadenosine) and related nucleosides.

Processing of Bacterial RNA

Newly formed rRNA and tRNA molecules are usually not functional but must undergo chain cleavage, methylation, and other alterations before they are "mature." However, most bacterial mRNA does not require processing. Bacteria produce a series of mRNA molecules of variable length, some corresponding to polycistronic (polygenic) and some to monocistronic operons. Most of the mRNA molecules produced are unstable with an average lifetime of about two min; however, some, such as those produced in bacteria about to undergo sporulation, survive much longer. Bacterial mRNA sometimes does undergo processing before it reaches the ribosomes.

For example, following infection of *E. coli* cells by phage 17, ribonuclease RNase III cleaves the large 7000 nucleotide "early RNA" transcript from the virus DNA into five defined fragments. Each fragment presumably carries the message for a single viral gene. Genes containing introns have been identified in several archaebacteria and in certain phage. The corresponding transcripts must be spliced, as are most eukaryotic transcripts.

Ribosomal RNA

Quantitatively the most important RNA, making up 90% of that present in cells, is ribosomal RNA. Synthesis of rRNA must be rapid, for an *E. coli* cell produces 5-10 new ribosomes per second, or 2×10^4 molecules of RNA per generation. Bacterial ribosomes contain three pieces of RNA. These are designated, according to their sedimentation constants, as 5S, 16S, and 23S and contain about 120, 1700, and 3300 nucleotides, respectively. All three pieces appear in cells as parts of larger *pre-rRNA* precursor molecules with extra nucleotide sequences at both the 3′ and 5′ ends. There are seven rRNA regions in the *E. coli* chromosome.

Each region consists of a single transcriptional unit containing a gene each for 16S, 23S, and 5S rRNA with interspersed tRNA genes as follows: 16S, tRNA, 23S, tRNA, 5S, tRNA. A single transcript (which can, in certain mutant strains, appear as a 30S molecule) is cut by the endonuclease RNAse III into the smaller pre-rRNA molecules. Other nucleases trim these to their final sizes, and methylases act to modify 24 residues in *E. coli* rRNAs. Most RNAs of all organisms contain *pseudouridine* (ψ), which is formed by isomerization of specific uridines present in the RNAs. In *E. coli* there is one pseudouridine in the 16S ribosome RNA and nine in the 23S RNA as well as one or more in most tRNAs.

The isomerization depends upon a carboxylate group of the enzyme, which evidently adds to the 6-position of the uracil ring to form a pivot around which the ring can rotate after it is eliminated from its attachment to the RNA and before it is reattached with a C-N linkage. Both bacteria and eukaryotes contain several pseudouridine synthases, which act to isomerize specific uridine residues in the precursor RNAs. In eukaryotes special *guide RNAs* direct the pseudouridine synthases to specific locations in their substrates. The same thing is true for 2′-*O*-methylases that modify selected ribose rings in precursor RNAs.

Transfer *RNA*

The genes for tRNA molecules in both bacteria and mammalian cells are grouped in clusters, which are transcribed as large precursors sometimes containing more than one kind of tRNA or containing tRNA fused to rRNA or mRNA sequences. At least three different nucleases are needed for cutting and trimming to form the mature tRNA. These enzymes may not always act in the same sequence. Thus, for some but not all tRNAs cleavage near the 3′ end is needed before cleavage can take place at the 5′ end. The best known processing nuclease is *RNase P*, which cleaves bacterial tRNA precursors to create the 5′ ends of the mature tRNAs. All of the 64 tRNA precursors present in *E. coli* are cleaved by this unusual enzyme, which contains an essential piece of RNA.

Cleavage of polycistronic tRNA precursors by RNase P or of the previously mentioned rRNA and tRNA pecursor by RNAses P and III releases mature tRNAs carrying extra nucleotides at the 3′ ends are removed by exoribonucleases, a surprising number of which are present in cells. In *E. coli* they include RNases II, D, BN, and T, as well as polynucleotide phosphorylase. The structure of the precursor to the minor *E. coli* tyrosine $tRNA_1$ are shown in the figure. This is encoded by the *amber* suppressor gene *SupF*. Transcription of its gene is initiated by GTP 43 bp upstream of the 5′ end of the mature tRNA and usually terminates at a ρ-dependent signal 225 bp beyond the CCA terminus of the tRNA.

Two steps
Tautomerization
...(9.3)

Uridine in an RNA chain
Pseudouridine in RNA

An endonuclease cuts the transcript a few nucleotides beyond the CCA end. It is then trimmed to an ~130-nucleotide piece still containing 2-3 extra nucleotides at the 3′ end. This intermediate is cut by RNase P at the 5′ end after which final trimming is done at the 3′ end. An important chemical achievement was the synthesis by H. G. Khorana and associates of the double-stranded DNA segment coding for this *E. coli* tyrosine tRNA and its precursor. This was one of the first synthetic genes. The synthesis was extended to include the gene termination region, which lies beyond the CCA end of the tRNA. Two noteworthy features appeared. There is a local center of dyad symmetry, which may serve as a termination signal. The operator is located in the 29-nucleotide sequence *preceding* the tyrosine tRNA gene. The 3′-terminal group of three nucleotides, CCA, is invariant among all tRNA molecules and is labile, undergoing active removal and resynthesis.

The rate of this turnover is sufficient to involve about 20% of the tRNA molecules of a cell per generation, but it is very much slower than the rate of participation of the tRNA molecules in protein synthesis. The physiological significance of end turnover is unknown. While this CCA sequence is encoded in bacterial tRNA genes, it is added in a separate reaction in eukaryotes. In addition to the cutting and trimming of precursors by nucleases, extensive modification of purine and pyrimidine bases is required to generate mature tRNAs.

In Fig. 9.10 the modifications in the mature tyrosine suppressor tRNA are shown, those in *E. coli* phenylalanine tRNA are indicated.

Modification usually begins with tRNA precursors. For example, the precursor in Fig. 9.10 methylated to form ribothymidine at position 63. Pseudouridine is then introduced at positions 40 and 64 of Fig. 9.10 by isomerization of the uridines present in the initial transcript. The Tψ pair at position 63 and 64 almost universally found in tRNA, but the positions are usually designated as 54 and 55. Position 8 in most tRNAs is occupied by 4-thiouridine. The sulfur atom is transferred from cysteine as S° using a PLP-dependent mechanism similar to that in Eq. 9.34 involving an *S-sulfonylcysteine (persulfide) intermediate.* Some modifications are completed on the mature tRNA. Some tRNAs require RNA splicing for maturation. The hehypermodified nucleoside *queuosine* is found in the first (wobble) position of anticodons of most eukaryotic tRNAs for Asn, Asp, His, and Tyr and also in most bacteria.

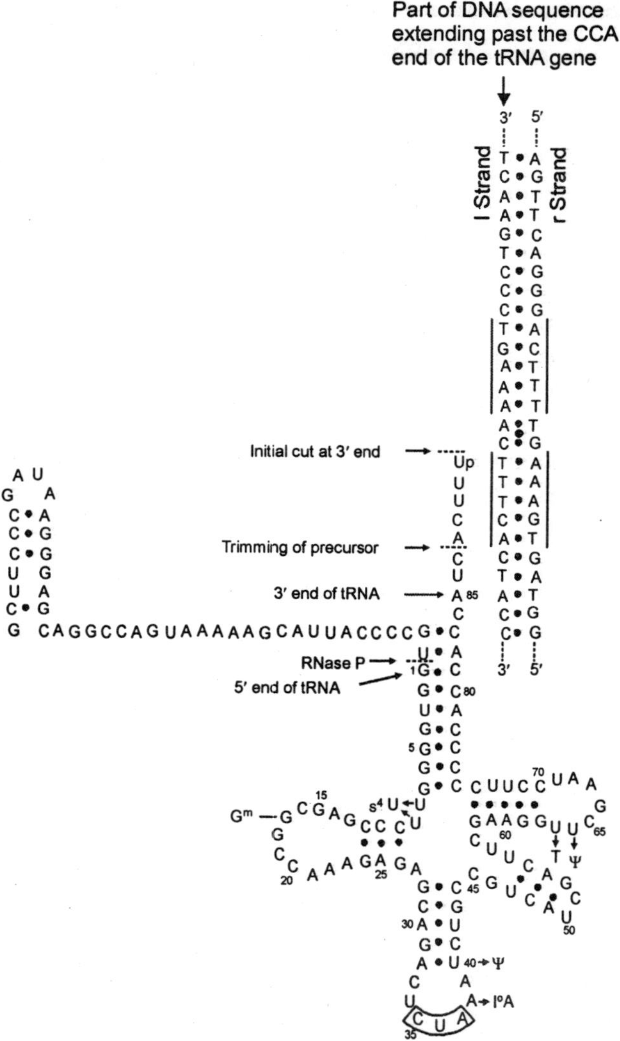

Fig. 9.10. Sequence of an *E. coli* tyrosine tRNA precursor drawn in a hypothetical secondary structure. Nucleotides found modified in the mature tRNA are indicated with their modifications (S^4, 4-thiouridine; G^m, 2′-*O*-methylguanosine; I^{10}, *N^6*-isopentenyladenosine; ψ, pseudouridine; T, ribothymidine; A partial sequence of the tRNA gene past the CCA end is also shown. Note the region of local 2-fold rotational symmetry (indicated by the bars and the dot). The anticodon 3′-CUA (shaded) of this suppressor tRNA pairs with termination codon 5′-UAG.

Bacteria apparently make aminomethyl-7-deazaguanine from guanine and transfer this compound into the appropriate position in tRNA (step *b)* by a tRNA-guanine transglycosylase. The incorporated nucleoside is then converted to queuine by incorporation of the 5-carbons of the ribosyl group in S-adenosylmethionine to form an intermediate epoxide. This is converted to queuine in the vitamin B_{12}-dependent step *d*. Eukaryotes are unable to form queuine and must obtain it as a nutrient factor or from intestinal flora. It is exchanged into tRNA by action of the transglycosylase. Queuine might be considered a vitamin. However, germ-free mice reared on a queuine deficient diet seemed normal. Therefore, the essentiality of queuine in the human diet is in doubt.

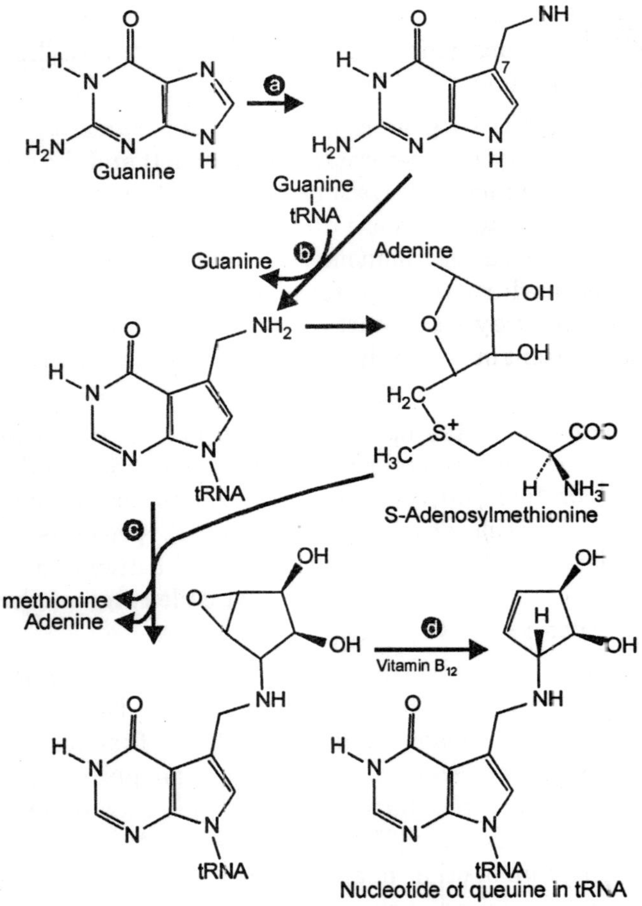

Guanine → (a) → (b) Guanine-tRNA, Guanine → (c) S-Adenosylmethionine, methionine, Adenine → (d) Vitamin B_{12}

Nucleotide ot queuine in tRNA

...(9.4)

VIRAL RNA IN PROKARYOTES

Bacteria not only transcribe their own genes but sometimes transcribe, or assist in transcribing, genes of invading DNA viruses or of integrated proviruses. In addition, they assist in replication of RNA viruses, another process that requires RNA synthesis. Viruses sometimes make use of host RNA polymerases but often synthesize their own catalytic subunits. Bacteriophage T4 uses the *E. coli* RNA polymerase and σ factors but modifies their action through the binding of several phage-encoded proteins. In contrast, phage T7 encodes its own relatively simple RNAP whose initiation complex and elongation complexes have been studied.

The Lysogenic State of Phage λ

The study of bacteriophage lambda has provided many insights into biological process. As we have seen, the DNA of phage λ, can become incorporated into the genomic DNA of *E. coli.* The resulting prophage contains many genes, but they remain largely unexpressed until the SOS signal is generated. Certain prophage genes are then expressed with the result that the λ DNA is excised as a replicating virus. How can the λ genes remain unexpressed in the prophage but be expressed rapidly at the time of excision? Part of the answer has been found in the *cl* and *Cro* repressors. The short L1 operon of the λ, prophage is transcribed continuously by the *E. coli* RNA polymerase. This operon contains genes *cl* and *rex,* which are transcribed from the *l* strands of the pro-phage DNA.

The protein C_1 (or CI) specified by gene *cl* is the *lambda repressor,* which binds to two operator sites in the prophage DNA. One operator (o_L) is to the left and the other (o_R) to the right of the *cl* gene. From a study of fragments of DNA protected by the repressor, it was concluded that each operator has three subsites, which are filled from left to right at o_L and from right to left at o_R successively by up to six repressor monomers. Each presumed subsite has a similar 17-bp quasipalindromic sequence to which a dimeric repressor can bind. The binding is cooperative, probably because the repressor molecules contact each other, apparently binding the DNA into a loopy. The right operator o_R controls not only the R1 operon but also the L1 operon, which encodes the λ, repressor *(cI* gene).

The first of the three subsites in the operator is adjacent to the L1 promoter P_{RM}, and binding of λ repressor activates that promoter at the same time that binding to the adjacent subsites blocks transcription of the R_1 operon. Thus, the λ repressor positively controls its own synthesis. At the same time, blocking of promoters P_R and P_L prevents synthesis of virally encoded enzymes that catalyze excision of the λ DNA and replication and transcription of the rest of the genes. The SOS signal causes rapid hydrolytic cleavage of the λ repressor and transcription of the other λ operons.

The matter is more complex than this. Gene products *cIII* and *cII,* from the *early left* and *early right operons,* respectively, stimulate the transcription of cl and are needed for establishing the lysogenic state initially. Once established these genes do not function since they are never transcribed. There are only a few molecules of the λ, repressor present in a cell, but this is ordinarily sufficient to maintain the prophage state. On the other hand, irradiation of the bacterium with ultraviolet light activates the SOS response and results in rapid hydrolytic cleavage of the λ repressor and transcription of other phage operons. Of special significance to the lytic cycle is the Cro repressor gene *ero,* found at the beginning of operon R1.

Although it binds to the same operator sequences as does the λ repressor, the Cro repressor has opposite effects. It represses transcription of operon L2 and hence synthesis of λ repressor, but it positively activates P_R and P_L. The earliest proteins synthesized during lytic development are the Cro protein and the product of the first gene *N* in the left operon L2. The N protein, an antiterminator that permits transcription to continue on past points t_L and t_R, is an unstable, short-lived molecule of t ½ ~2 min. Leftward transcription proceeds through genes *exo* and β, which are involved with recombination, and *xis,* which is required for excision. When the DNA is integrated into the *E. coli* chromosome, it is cut at points *aa′* inserted just to the right of the *gal* operon.

Prophage transcription can now continue past point a′ and into the genes of the bacterium. Translation of the mRNA formed from this early left operon generates the enzymes needed to free the prophage and to permit reformation of the circular replicative form of the phage DNA. The excision is also made near point a′, and it is easy to see how the nearby *gal* genes can sometimes be included in the excised λ genome. The product of gene *N* also permits rightward transcription through genes *O, P,* and *Q* and at a slower rate on along the rest of the chromosome to point a. Genes *O* and *P* code for proteins that permit the host replication system to initiate formation of new λ DNA molecules.

Replication begins at the point *ori* and occurs in both directions. Gene *Q* codes for a protein that activates transcription of the *late genes* beginning at promoter P_R. The chromosome can be divided into four major operons, the short one that produces repressor, and the early left, early right, and late operons. The early operons code largely for replication and recombination enzymes and control proteins. The late operon is concerned with production of proteins needed for assembly of the virus particles and must be transcribed at an even higher rate; hence the need for the product of gene *Q*. Within the late operon, genes *A* to F are involved in packaging of λ DNA and in formation of heads, while genes *S* to *J* are concerned with the production and assembly of tails. Genes *S* and *R* produce proteins that lead to destruction of the host membrane and to lysis of the cell. During the late stages of lytic growth the early genes are largely shut off by the Cro repressor. We can see that even in a virus the control of transcription can be a complex process.

Replication of RNA Bacteriophages

The small icosahedral RNA-containing bacteriophage are of interest because of the small number of genes and the possibility of obtaining a detailed understanding of their replication. The four genes in the 3569-nucleotide MS2 RNA code for the A protein (maturation protein), the coat protein, a replicase (RNA polymerase) subunit, and a protein needed for lysis of the host cell. The last gene overlaps both he replicase gene and that of the coat protein.

The somewhat larger phage Qβ contains a 4.2-kb RNA genome. One subunit of the Qβ replicase is encoded by the virus, but three bacterial proteins are needed to form the complete replicase. They are ribosomal protein SI and elongation factors EF-Tu and EF-Ts, proteins that normally function within *E. coli* in translation of mRNA. Their ability to associate with RNA has been exploited by the phage for a quite different purpose. Replication of a single-stranded virus must take place in two steps. From the (+) strand present in the virus a complementary (-) strand is first formed. Initiation of this step requires another bacterial host factor Hfq and GTP. The (-) strands formed do not associate with the (+) strands. They are apparently released from the replicase in a single-stranded form and presumably fold into highly structured molecules with many hairpin loops. The (-) strands are then copied (Hfq is not needed for this) to make a large number of new (+) strands for incorporation into the finished virus particles.

The Qβ replicase is able to synthesize *in vito* complete complementary strands to either (+) or (-) viral RNA molecules. However, the system is specific for the viral RNA and will not copy any arbitrary nucleotide sequence, certain sequences at the 3′ end being essential for initiation of replication. During replication in the test tube mistakes are made including premature termination and mispairing of bases. Thus mutation takes place, and it is possible to select RNA molecules much smaller than the original viral RNA that will be replicated readily by the Qβ replicase system. One such fragment contains only 114 nucleotides in a known sequence.

TRANSCRIPTION IN EUKARYOTIC CELLS AND IN ARCHAEA

There are three primary domains of life, represented by the bacteria, archaea, and eukaryotes. Some of the clearest evidence for the independent evolution of these three groups of organisms is found in the transcriptional apparatus. While the basic chemistry is the same, the details of initiation and control of transcription in bacteria and in eukaryotes are very different. The archaea share characteristics of both bacteria and eukaryotes. Archaeal RNA polymerases have a complexity similar to that of eukaryotes and also share a similar mechanism of initiation of transcription. Several of the protein transcription factors of archaea also resemble those of eukaryotes. However, in a comparison of DNA sequences from the complete genomes of four archaeal species, it was found that of 280 predicted transcription factors or transcription-associated proteins 168 were homologous to bacterial proteins and only 51 to eukaryotic proteins. This tends to confirm the ancient divergence of the three primary domains of life. In bacteria transcription and translation are closely linked.

Polyribosomes may assemble on single DNA strands. It has often been assumed that RNA synthesis occurs on loops of DNA that extend out into the cytosol. However, recent studies indicate that most transcription occurs in the dense nucleoid and that assembly of ribosomes takes place in the cytosol. In a similar way eukaryotic transcription occurs in the nucleus and protein synthesis in the cytosol. Nevertheless, some active ribosomes are present in the nucleus.

Eukaryotic Nuclei and Transcription

In cells with true membrane-enclosed nuclei the messenger RNA molecules are relatively long lived. They must move out from the nucleus to the sites of protein synthesis in the cytoplasm. In addition to the need for eukaryotic mRNA to travel further and to last longer than that of bacteria, a number of other differences are evident. Eukaryotic mRNAs are usually transcribed from single genes. Polygenic operons are uncommon in most animals but are numerous in C. *elegans*. Eukaryotic cells appear to rely less on negative control through specialized protein represors than do bacteria but use a greater variety of positive control mechanisms.

However, most genes are repressed by being held in a *silent state*. Another characteristic of eukaryotes is the extensive processing of transcripts. Most primary transcripts that give rise to mRNA appear first in the nucleus as *heterogeneous nuclear RNA (hnRNA)*. Like mRNA it has a base composition resembling that of DNA. The molecular size varies from 1.5 to 30 kb or more. It turns over rapidly, most of it having a half-life of ~10 min. However, some may last as long as 20 h. Only about 5% of the hnRNA ever leaves the nucleus as mRNA, most being degraded without export to the cytoplasm. The processing consists of capping at the 5′ ends, removal of introns (splicing), *cleavage* by nucleases, *polyadenylation* at the 3′ ends, *mthylation*, formation of pseudouridines, other covalent base modifications, and sometimes *editing*.

Because of the complexity of eukaryotic transcription there are many points at which control can be exerted during initiation of transcription, termination of transcription, splicing and methylation, transport of mRNA out of the nucleus, and the degradation of mRNA. However, the first points of control are found in the nucleosomes and in the structure of the chromatin. It has been recognized for many years that genes in heterochromatin are usually not transcribed. In most chromatin the great majority of the genes are silent most of the time. Genes are repressed by the folding of the DNA into nucleosomes and by the further folding into higher order folds or coils.

Various activating transcription factors as well as RNA polymerases must bind to the DNA, displacing it from the histones around which it is wrapped in the nucleosomes. The processes by which inactive nucleosomal DNA becomes active in response to external signals are beginning to be understood. Chemical alterations in the histones in the nuclear matrix and in other nuclear proteins and also in the DNA itself may be involved. As pointed out in the CpG "islands" that lie upstream of many genes are heavily methylated in the silent heterochromatin. Repressor proteins may bind to methyl-CpG groups. In recent years attention has been focused on the N-terminal "tails" of histones H3 and H4 in which lysine side chains undergo reversible acetylation and which may also be phosphorylated and methylated.

The resulting modifications in shape and electrical charge can affect the ways in which the histones interact with the DNA, with each other, and with other proteins in transcription complexes. The histone tails may interact with adjacent molecules either to pack the chromatin more tightly or to loosen it and allow transcription to take place. Active chromatin has long been known to be highly acetylated, while silent chromatin has a low degree of acetylation, but a high degree of methylation of both histones and DNA. Methylation of histone H3 on lysine 9 (H3-Lys^9 or H3-K9) is especially significant.

Several transcriptional activators form complexes with acetylating enzymes, the *histone acetyltransferases* (HATs), while transcriptional repressor proteins often associate with *histone deacetylases.*

The deacetylases are often found in very large complexes. For example, the mammalian complex *Sin3* contains two histone deacetylases plus at least five other subunits, some of which evidently bind to histones. Sodium butyrate in millimolar concentrations is a powerful inhibitor of these deacetylases. Special *silencer* sequences in the DNA are sometimes present and provide sites for binding of transcriptional repressors. Among these are *silent information repressors* (Sir proteins). They regulate regions of DNA that can be converted to a heterochromatin-like state. They have been located in silenced mating type loci in yeast, in telomeres, and in DNA containing ribosomal RNA genes. Sir proteins have an unusual histone deacetylase activity. The acetyl groups removed from histones are transferred by reaction with NAD^+ to ADP ribose. Regions of silenced DNA are often set apart by *insulator* or bounding regions. An important mechanism of silencing some genes is the use of an antisense RNA strand, as is illustrated for a bacterial system in Figure. This *RNA interference* is also used in animals and plants, often as a way of blocking replication of viruses. The small 20-25 nucleotide siRNAs that function in this way are abundant in C. *elegans* and in *Drosophila.* Histone acetyltransferases also form large complexes that acetylate not only histones but also other nuclear proteins.

The *SAGA HAT* complex of *S. cerevisiae* has a molecular mass of ~2.0 MDa and contains at least 14 subunits. Large acetylating complexes have also been identified in *Tetrahymena, Drosophila, Arabidopsis,* and mammalian species. Some subunits of these complexes have been identified as previously known transcription factors. Such multiprotein complexes are sometimes described as *cis-regulatory elements* (CREs). Changes in the structural properties of chromatin observed during silencing of genes or during their activation are often described as *chromatin remoling.* Large multisubunit complexes are involved. Their action is characteristically dependent upon ATP hydrolysis.

Complexes *SWI / SNF* and *RSC,* first found in yeast but also present in human cells, appear to participate in the disruption of nucleosomes needed for initiation of transcription. Among other distinctly different remodeling complexes are *ISWI* of *Drosophila,* its human homolog, and the human Williams syndrome transcription factor *WCRF.* The ATP-dependent component in these complexes has a conserved sequence that is shared with DNA helicases. RNA helicases are also required for all processes that form, modify, or utilize RNA. However, the chromatin remodeling complexes appear not to unwind DNA but to open the nucleosomal DNA for initiation of transcription. They may act in a processive fashion and be coupled to transcription. *Peptidyl-prolyl isomerases,* such as the cyclophilins may also be essential components of chromatin-remodeling complexes. Just as they participate in driving the cell cycle, ubiquitin and proteasomes also function in the control of transcription.

In many cases specific transcription factors are targeted for destruction after they are used to activate or repress a gene. However, a 19S regulatory complex, which consists of a base and a lid of the proteasome, may participate directly in control of transcription rather than in mediating proteolysis. In addition to ubiquitin a 97-residue relative designated SUMO-1 is linked to proteins by enzymes resembling E_1 and E_2 of the ubiquitin system. Conjugation with *SUOMO-1* regulates some transcription factors and has other functions, *e.g.,* participation in control of nuclear pores. The 81-residue protein *NEDD8,* which is 80% homologous with ubiquitin, controls some transcriptional processes in heart and skeletal muscle. Furthermore, ubiquitin-like sequences (UBX domain) are present in the C-terminal ends of a variety of specific proteins.

RNA Polymerases

Eukaryotic nuclei contain at least three RNA polymerases which have the following functions:

Polymerase I	Formation of large pre-rRNAs
Polymerase II	Transcription of most genes to give precursors to mRNA and most small nuclear RNAs (snRNAs and small nucleolar RNAs (snoRNAs)
Polymerase III	Formation of 5S rRNA, tRNAs, and small RNA U6

Polymerase I is localized in the nucleolus. Mitochondria contain a fourth RNA polymerase and chloroplasts a fifth. Like the bacterial polymerase, eukaryotic RNA polymerases are large 500-600 kDa aggregates of 9-14 subunits each. Yeast and human RNA Pol IIs each contain twelve subunits. There are two large nonidentical subunits, which in mammalian cells have masses of 214 kDa and 140 kDa and are homologous to the β' and β core subunits of the *E. coli* polymerase, respectively.

The active site contains one or two catalytic Mg^{2+} ions. The largest subunit has an unusual singly glycosylated C-terminal domain (known as the CTD). It contains the repeating sequence $(YSPTSPS)_n$. The number of repeats varies: n = 18 in plasmodia, 27 in yeast, 45 in *Drosophila,* and 52 in mammals. The numerous serine side chains in this tail domain undergo phosphorylation and dephosphorylation to varying extents during each catalytic cycle. This may be a way of easing the transcriptional complex through nucleosomes, but its most important function appears to be the linking of transcription to pre-mRNA processing. RNA polymerases I and III have properties similar to those of Pol II. Polymerases II and III are very sensitive to inhibition by the lethal mushroom poison *amanitin.*

However, both RNA polymerase I and RNA polymerases of mitochondria resemble the bacterial enzyme in being resistant. Genes transcribed by polymerase I, II, and III are often referred to as genes of classes I, II, and III, respectively. While mRNA is transcribed from class II genes, rRNAs, tRNAs, and some small RNAs, which must undergo processing but are not polyadenylated, are transcribed from genes of classes I and III. Each type of nuclear RNA functions in its own sites as independent "factories."

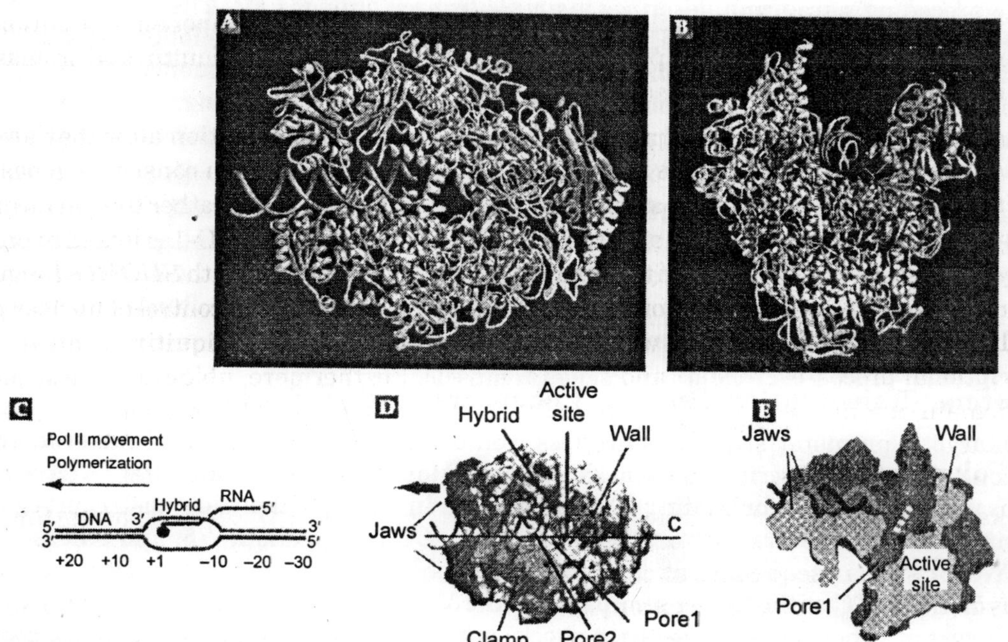

Fig. 9.11. Three-dimensional structure of yeast RNA polymerase II. (A) View from the "top," with backbones of ten subunits (of 12) in the 514-kDa protein shown as ribbon drawings. A 20-base-pair segment of B-DNA has been modeled in a location indicated by electron crystallography. (B) Side view looking toward the end of the DNA. Eight zinc atoms as well as an active-site magnesium (green) are visible. (C) Schematic drawing showing the transcription bubble as proposed for a transcribing polymerase. (D) Surface representation of the polymerase viewed as in (A). (E) Side view of a section cut along the line marked C in D. The dashed white line represents the axis of the DNA-RNA hybrid segment. The hybrid axis must lie at an angle with respect to the axis of the incoming DNA. Pore 1 may be a route for exit of RNA during "backtracking." The nucleotide triphosphate substrates may also enter via pore 1.

Transcriptional Units and Initiation of RNA Synthesis

Typical simple *transcriptional units* for class II eukaryotic genes contain the following elements: (a) site of initiation, (b) *TATA sequence* (Goldberg-Hogness sequence) at position ~–30 bp, (c) *upstream regulatory elements,* (d) *enhancers,* (e) a series of *coding sequences* or exons separated by introns, an *AATAAA sequence* that in the RNA transcript may establish the 3′ polyadenylation site, (f) a termination region.

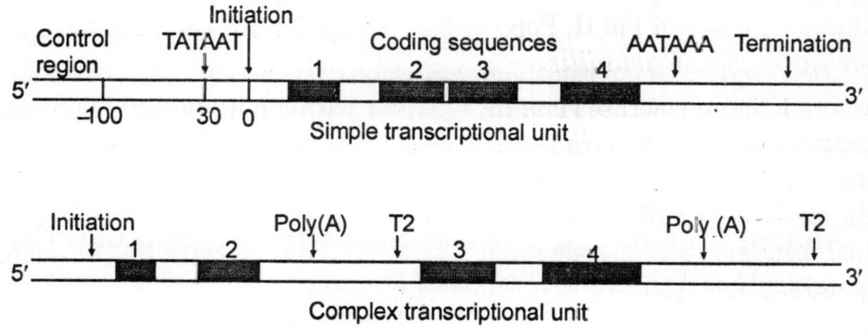

Fig. 9.12.

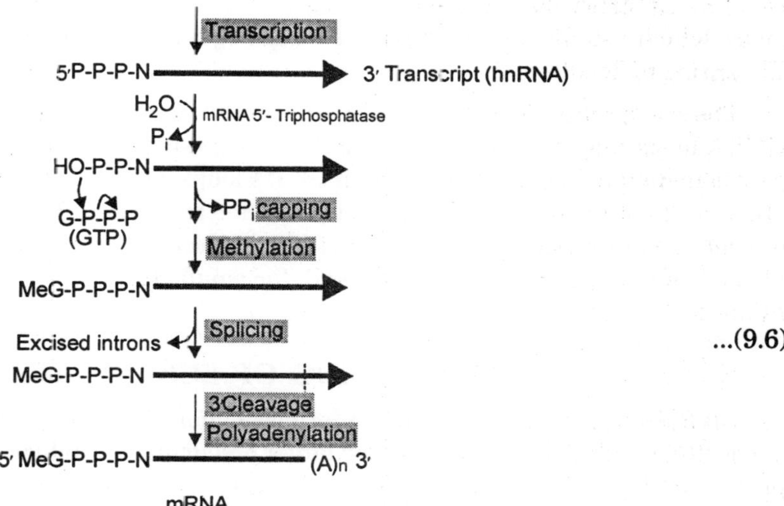

...(9.6)

Defects in elongation factors P-TEFb, ELL, and *CSB* (Cockayne syndrome complementation group) have been correlated with human diseases including cancer. Despite the complexity of the processes represented by Eq. 9.6 yeast cell is able to transcribe genes at rates of about one in every 6-8 s. This can be compared with a rate of about once in 2-3 s for RNA polymerase of *E. coli*. The least well defined step in eukaryotic transcription is termination, which follows the various steps of processing. The final 3′ end of processed transcripts of RNA pol II action in mammals is marked by the sequence AAUAAA, which is found about 10-30 nucleotides upstream from the end. This is usually followed by a polyU or GU-rich sequence.

In yeast the termination and polyadenylation signals are less clear. The initial transcript almost always continues beyond the AAUAAA signal, sometimes for hundreds of nucleotides. However, the excess RNA is rapidly degraded by a large complex of proteins. The precise 3′ end cleavage is energy-dependent, requiring creatine phosphate rather than ATP or GTP. Transcription termination by RNA polymerases I and III is more like that of bacteria. Terminator sequences are present in the DNA, and terminator proteins interact with them.

Conformational Properties of RNA

Newly formed RNA transcripts fold quickly into structures of complex shapes, the folding being influenced by interactions with proteins and with other RNA molecules. RNA chains are flexible, with many sterically allowed conformations. As with proteins folding probably begins with a nucleation event, perhaps involving monovalent or divalent metal ions and continues rapidly. Folding is affected by hydrogen bonding, base stacking, and binding of ions, and by formation of pseudoknots. As is apparent from the structures of tRNAs, the *Tetrahymena* self-splicing ribozyme, and ribosomal RNA structures, a large fraction of a folded RNA exists as hairpin or *stem-loop* structures. These are A-type structures with largely Watson-Crick base pairs.

However, mismatched pairs, triples, and quadruples of bases are also formed. Recently discovered RNA structural elements include *base platforms,* formed by pairing of adjacent bases, interdigitation of unpaired bases wobble pairs. Hydrogen bonding between riboses of consecutive nucleotides in two strands may help to form a ribose zipper. Guanine-rich tetraplexes, cytosine-rich i-motif structures, and water-mediated U•C base pairs also arise in RNAs. The terminal loops, which usually contain the consensus sequence *GNRA*, may constitute up to one-third of the entire molecule.

These loops interact with many binding proteins, such as those in the snRNA-protein particles. GNRA loops may also dock into the shallow groove of RNA helices. Adenosines that are not paired in double helices, *e.g.,* those in GNRA loops, are able to interact in a variety of ways with other parts of an RNA molecule or with other molecules. They are involved in helix packing interactions in virtually every RNA studied. Although examples are still rare, specific mRNA molecules may provide binding pockets for small regulatory molecules, *e.g.,* amino acids and thiamin.

PROCESSING OF EUKARYOTIC RNAs

All RNA found in eukaryotes undergoes major alterations prior to functioning. The cutting out of rRNA and tRNA molecules from larger precursors resembles that in bacteria, but subsequent processing is much more complex, as is that of mRNA.

Ribosomal RNA

Eukaryotic ribosomes contain four pieces of RNA, which are usually designated by their sedimentation coefficients. The 18S, 5.8S, and 28S RNAs are encoded as single transcriptional units with spacers separating the sequences that encode the mature RNAs. A typical animal cell contains several hundred copies of this transcriptional unit, all located in the DNA in the nucleolus, each having its own set of promoter sequences, enhancers, and transcription factors. The promoter sequences vary substantially among different species.

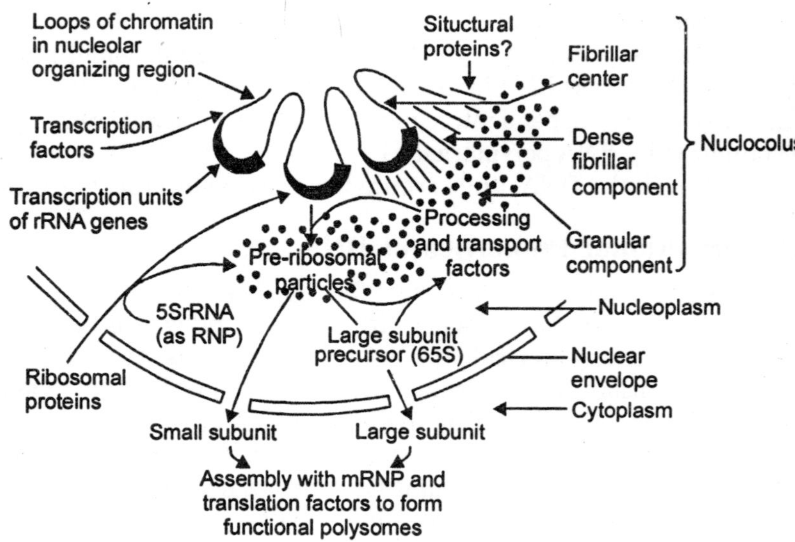

Fig. 9.15. Steps in ribosome formation in the nucleolus. Abbreviations are NOR, nucleolar organizing regions; RNP, ribonucleoproteins.

The primary transcripts from these units are the sole product of RNA polymerase I. Electron micrographs of portions of unwound cores of nucleoli have revealed fibrils of RNA coated with protein growing from the DNA strands of the pre-rRNA genes ~80-100 RNA chains of different length being transcribed concurrently from a single gene. The overall gene length in the electron microscope is 2.3 µm, only a little less than the calculated length for a fully extended DNA molecule in the B form. However, judging by the lengths of the transcripts formed, the pre-rRNA chains are folded extensively. The primary eukaryotic rRNA transcripts extend several hundred nucleotides past the 3′ termini of the mature 28S RNA molecules. As the primary transcripts, which are formed in the core region of the nucleolus, move away into the outer cortex of the nucleolus, cleavage occurs in a number of steps. Electron microscopy provided the first direct confirmation of the relationship of one pre-rRNA molecule to another. The 18S portion of the 45S RNA lies nearest to the 5′-end just as does the 16S rRNA in the large transcript of the prokaryotic rRNA genes.

Transcription is thought to occur from the loops of DNA that form the nucleolar organizing region. The 100-kDa **nucleolin,** the major protein of the nucleolus, binds to the non-transcribed spacer sequences in the DNA. It also binds to the newly formed transcripts, as do various proteins that enter the nucleus from the cytoplasm. More than 270 proteins, many of which participate in synthesis of ribosomes, have been detected in the nucleolus.

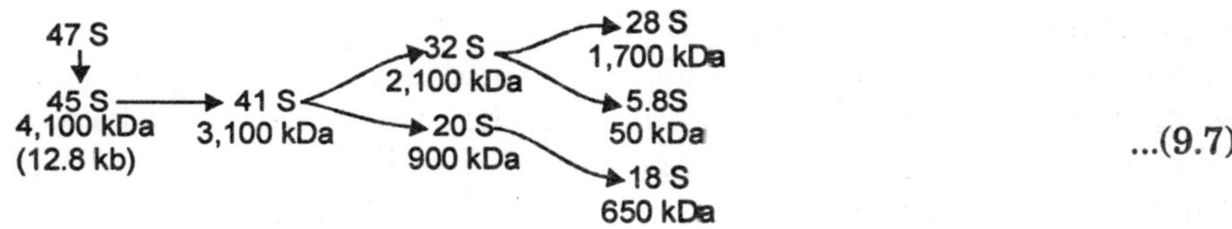

...(9.7)

Some of these proteins, acting together with the snoRNAs discussed in the next section, catalyze hydrolytic cleavage of the pre-rRNA molecules. For completion of pre-ribosomal particles additional protein molecules enter the nucleolus and associate with the pre-rRNA particles, then diffuse out of the nucleus.

Small Nuclear, and Nucleolar, and Cytoplasmic RNAs

The nuclei of all eukaryotic cells contain a group of small nuclear RNAs (snRNAs), some of which (snoRNAs) are localized to the nucleolus and function there. At least ten of these are always present yeast appears to contain more than 100. Some are present in small amounts, and it may be that a large number of snRNAs will also be found in other eukaryotes. The uridine-rich or U series of snRNA are especially abundant (10^4-10^6 molecules per nucleus). Several of them (U1,2,4-6,11,12,16,18) function in RNA splicing. Species U1-U10 contain from –60 to –215 nucleotides but some snRNAs from yeast have over 300 nucleotides and one over 1000. The most abundant snRNAs, the metabolically stable U1, U2, U4, U5, and U6, are exported from the nucleus. In the cytoplasm each snRNA associates with a complex of several proteins to form ribonucleoprotein particles *(snRNP particles).*

The proteins, known as *Sm proteins,* range in mass from 11 to 70 kDa and are designated B, B′, D1, D2, D3, E, F, and G. The proteins may associate to a complex B_2D_2EFG, which then binds the snRNAs. Whereas many proteins are found in most of the snRNP particles, some are associated with specific RNAs. Thus, mature U1 snRNP may have a partial stoichiometry $U1A_2B_2C_2D_4E$. The snRNA particle, known as U4/U6snRNP, contains two snRNAs. After returning to the nucleus the four snRNP complexes U1, U2, U5, and U4/U6, together with the pre-mRNA molecules, associate via an ordered assembly pathway that gives rise to the large *spliceosomes* in which the removal of introns takes place. Some of the snRNP complexes, *e.g.* the spliceosomal U6 snRNP and snRNPs involved in mRNA degradation, contain Sm-like (Lsm) proteins. More recently small 20-25 nucleotide antisense micro RNAs *(miRNAs)* have become recognized as important in control of the breakdown of mRNA, in formation of heterochromatin, and in control of specific stages in development in cellular defense mechanisms.

In animals the 21-23 nt miRNAs are formed by enzymatic cleavage of double-stranded dsRNA by a ribonuclease III-type enzyme called *Dicer.* In the resulting fragments, which have 5′-phosphate termini, one strand is antisense to a "target" sequence in mRNA. *Small temporal RNAs* (stRNAs) that guide development are cut by Dicer from RNA stem-loop structures. In other cases, *e.g.,* in formation of the siRNAs that silence individual genes, an RNA-dependent RNA polymerase acts on an RNA transcript to form a long dsRNA that can be cleaved by Dicer. This often happens with foreign RNAs, *e.g.,* from viruses or trans-genes.

Unwinding of the small fragments formed by Dicer, in a process that may require ATP, provides the single-stranded antisense siRNA molecules. Such antisense molecules bind to their target RNAs together with a group of proteins to form *RNA-induced silencing complexes* (RISCs). These complexes contain an RNAse (different from Dicer) that cuts the dsRNA of an siRNA-target complex, initiating destruction of the mRNA and silencing of its gene. This *RNA-interference* is widely utilized by plants and animals. It has also provided the basis for development of many practical tools for understanding gene sequences, for genetic engineering, and for design of new drugs.

Processing of Ribosomal RNA

Transcripts of rRNA genes vary in size from ~35 to 47S (6-15 kb) and often contain spacer regions at both ends as well as between the 18S, 5.8S, and 28S sequences. For example, human 47S transcripts have 414 extra nucleotides at the 5′ ends. One group of snoRNAs participates in the hydrolytic cleavage of pre-rRNA. RNAs U3, U14, snR10, snR30, as well as MRP RNA are always required. Also needed is U22, an intron-encoded RNA. The reactions represented by Eq. 9.7 are best known for *Saccharomyces cerevisiae*. Four spacers, the *5′-external transcribed* spacer (5′-ETS), the first and second *internal transcribed spacers* (ITS1 and ITS2), and the 3′-external transcribed sequence (3′-ETS), must be removed.

Removal of the 5′-ETS depends upon the snoRNA U3, which contains two highly conserved sequences able to form base-paired structures with the 5′ end of the 5.8S rRNA region of the pre-rRNA gene. Both U8 and U22 are also needed for cleavage of the pre-rRNA. Although the exact functions of snoRNAs and their associated proteins in the cleavage of pre-rRNA are still uncertain, these RNAs probably act as guide molecules for the cleavage reactions. They may have ribozyme activity and may perhaps be chaperones.

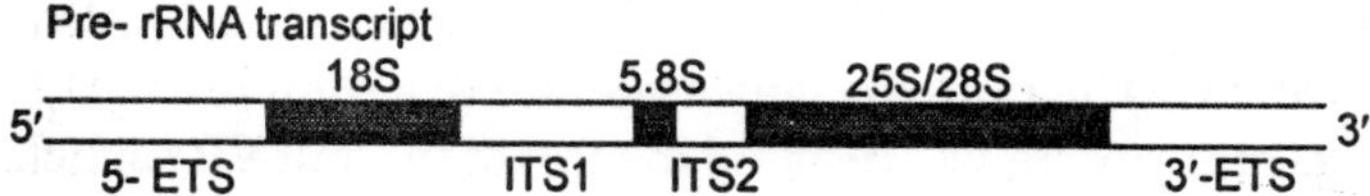

Cleavage at the 5′ end of the 5.8S region requires RNase MRP, a relative of the RNase P that cleaves at the 5′ ends of tRNAs. MRP *(mitochondrial processing protein)* also cleaves primers for mitochondrial DNA replication. The importance of the enzyme is emphasized by the existence of a hereditary defect in the MRP RNA that causes abnormalities in bone, cartilage, hair, and the immune system. Most bacterial rRNA genes have a tRNA gene in the position corresponding to that of 5.8S RNA in eukaryotes.

Table 9.3. Some Eukaryotic Small Nuclear and Nucleolar RNA Molecules

Designation	*Number of nucleotides*	
	Vertebrate	*Yeast*
U1[a]	164	568
U2[a]	188	1175
U3[b]	206-228	333
U4[a]	142-146	160
$U4_{atac}$[c]	131	
U5[a]	116-118	183 or 196
U6[a,d]	116	214
$U6_{atac}$[c]	125	
U7[e]	57-58	
U8[b]	136-140	
U9	130	
U10	60	
U11[c]	131-135	
U12[c]	150	
U13[b,f]	105	
U14[b,f]	87-96	125-128
U18[b,f]	67-70	102
X[b]	150	
U20[b,f]	80	
U21b,[f]	93	
U22(Y)[b]	125	
U24[b,f,g]	77	
U32-U40[b,f,h]		
SnR10[b]		245
SnR30[b]		605
SnR38[f]		93
SnR39[f]		85
SnR40[f]		96
SnR41[f]	s	
MRP RNA (RNA 7-2) 260-280	339	

[a] Major spliceosomal RNAs.
[b] Fibrillarin-associated SnoRNAs that function in pre-ribosomal RNA processing.
[c] Function in AT-AC spliceosomes.
[d] γ-Monomethyl cap.
[e] Required for 3′-end formation in histone mRNAs.
[f] SnoRNAs with long complementarities to rRNA. C and D sequences are present.
[g] Polyadenylated, noncoding.
[h] Participate in 2′-*O*-ribose methylation.

This provides an RNase P-dependent cleavage mechanism, which is alternative to action of other nucleases. The most abundant protein in the fibrillar regions of the nucleus, where the early stages of pre-rRNA processing occur, is *fibrillarin*. Many of the snoRNAs are closely associated with this protein. Fibrillarin is also well known as an autoantigen, which can induce formation of destructive antibodies that cause *scleroderma*.

Modification Guide RNAs

A second group of snoRNAs function in methylation, pseudouridine formation, and other RNA modifications. These snoRNAs have long sequences complementary to highly conserved regions of pre-rRNA, enabling them to form helical regions that may guide the docking with modification enzymes. Many of them also contain characteristic conserved sequences: C, 5′-UGAUGA; D, 5′-CUGA; H, 5′-AnAnnA; and 5′-ACA. Sequences C and D are present in snoRNAs that act as methylation guides, while the H and ACA sequences characterize guide RNAs for pseudouridine formation.

Transcription and Processing of snRNAs and snoRNAs

In higher organisms each of the snRNAs has several genes *e.g.*, there are 50 -60 U1 genes in the human haploid genome. However, in yeast there are often single copies. All of the snRNA genes, except for that of U6, transcribed by RNA polymerase II. The transcripts, which are capped at the 5′-end but are not polyadenylated, pass into the cytoplasm, where they undergo further processing and become associated with proteins. The 3′ ends are trimmed, the 7-methyl-guanosine of the cap is methylated further, and methylation may occur on other bases as well. SnoRNAs are not capped.

It was a great surprise to discover that many of the snoRNA genes lie within introns that occur in abundantly expressed genes for functionally unrelated proteins. Proteins with the autoimmune disease systemic *lupus erythematosus* make autoantibodies directed against the Sm proteins of snRNP particles. Antibodies from different patients vary in their specificities, making these antibodies a useful tool in the isolation and study of snRNAs and their protein complexes.

Processing of 5S RNA and tRNAs

The genes for 5S ribosomal RNA and all of the tRNAs are transcribed by RNA polymerase III. In the yeast genome the 5S RNA genes are located in the spacers between the transcriptional units containing the other rRNAs. However, in animals the 5S RNA genes are separate from the other rRNA genes and are not located in the nucleolus. In *Drosophila* ~500 copies of the 5S RNA gene are located in the right arm of chromosome 2. In *Xenopus* ~400 genes are active in somatic cells and another 24,000 only in oocytes. These are arranged as large transcriptional units, each containing several thousand copies of the 120-bp 5S RNA gene separated by 720-bp spacers.

Cleavage and trimming are required to form the final product, but there are usually no introns to be removed. Most eukaryotic tRNAs are formed from monomeric precursors, each gene acting as a transcriptional unit. Processing is similar to that in bacteria. Eukaryotic RNase

P usually cleaves the 5′ end, another enzyme cuts at the 3′ end. The 3′ CCA sequence of the mature tRNA is usually not present in the primary transcript but is added. As in bacteria extensive modification of bases also occurs in the tRNA precursors of eukaryotes. Many tRNA genes contain introns, which must be removed by splicing.

Messenger RNA, Caps, and Polyadenylate Tails

The first processing event most of the pre-mRNA and snRNA transcripts made by RNA polymerase II is addition to the 5′ end of a "cap," a terminal structure containing 7-methylguanosine from which a proton has dissociated to form a dipolar ion. The cap structure may be abbreviated 5′-m^7G(5′)pppNm—. The 5′ terminal ribose is often methylated on O2′. More complex caps are methylated at additional sites, *e.g.*, the guanine may be dimethylated on the 2-NH_2 group. Most snRNAs, including the U1-U5 and U7-U13 snRNAs, have such 2,2,7-trimethylguanosine caps. Many viral transcripts, including those of the much-studied vaccinia virus, have similar caps. However, U6 and some other snRNAs, which are transcribed by RNA polymerase III, have γ-monomethyl (me-ppp) caps or undergo a series of additions and deletions of uridylate residues at the 3′ ends. Cap synthesis occurs as follows. The 5′ end of an RNA transcript initially contains a triphosphate group arising from the fact that a nucleotide triphosphate serves as the primer in initiating transcription.

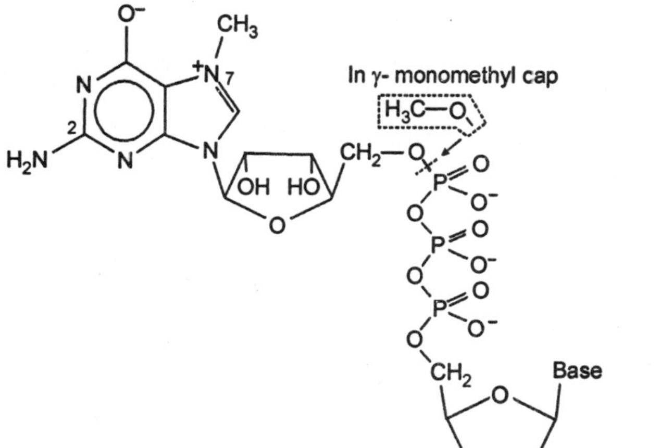

The terminal phospho group is removed by a triphosphatase leaving a diphosphate, which is then guanylated by GTP (Eq. 9.6). The capped transcripts are exported from the nucleus, after which additional methylation may follow. The cap structure affects several processes. A family of cap-binding proteins recognize the structure and may facilitate splicing as well as export from the nucleus. The cap is very important for ribosome binding and initiation of translation.

The trimethylated caps of snRNAs, on the other hand, may be signals for retention in the nucleus where they function. Following capping is the often elaborate process of splicing to remove introns. A poly (A) "tail" consisting of ~ 250 residues of adenylic acid is added next by poly(A) polymerase, a component of an enzyme complex that also cleaves the RNA chains. Most eukaryotic mRNA is polyadenylated with the exception of that encoding histones. The function of the poly(A) is unclear. It is needed for transport of mRNA out of the nucleus, but it does confer a greatly increased stability to the mRNA in the cytoplasm where the adenylate units are gradually removed.

In contrast, in chloroplasts and plant mitochondria polyadenylation is required for rapid degradation of mRNA. Polyadenylation may also increase the efficiency of translation. Polyadenylation occurs rapidly within ~1 min after transcription is completed. Mature mRNA molecules vary in lifetime. Some last for hours or days. Among the latter are mRNAs of maternal origin that accumulate in oocytes and are utilized during the early stages of embryonic development. Other mRNAs, *e.g.,* transcripts of the *c-fos* and *c-myc* proto-oncogene products, have half-lives of 30 min.

Some mRNA molecules are degraded while attached to ribosomes in response to recognition of the synthesized peptide. Longer lived mRNA molecules may be protected by RNase inhibitors. Hydrolytic removal of caps often initiates degradation, and Sm-like protein complexes participate.

Splicing

An essential modification of the precursor forms of large rRNA, tRNA, and mRNA molecules is the splicing out of intervening sequences. This occurs prior to polyadenylation of mRNA and is usually slow, the half-life of introns varying from a few seconds to 10-20 min. Splicing occurs by at least four distinctly different pathways.

Self-splicing RNA

The precursor to the 26S rRNA of *Tetrahymena* contains a 413-nucleotide intron, which was shown by Cech and coworkers to be self-splicing, *i.e.,* not to require a protein catalyst for maturation. This pre-rRNA is a ribozyme with true catalytic properties. It folds into a complex three-dimensional structure which provides a binding site for free guanosine whose 3′- OH attacks the phosphorus at the 5′ end of the intron.

The excised intron undergoes a third transesterification reaction, of uncertain significance (step *c*), to form a circular polynucleotide and a short displaced 15-residue oligonucleotide. The *Tetrahymena* pre-rRNA intron is a member of a group of similar *Group I introns,* many of which are found in fungal mitochondrial pre-mRNA and pre-rRNA. All are excised by a similar mechanism. Many are self-splicing, but others require a protein catalyst. A similar splicing sequence is involved in removal of a 1017-nucleotide intron from the thymidylate kinase gene of phage T4 and other introns in T-even phage. The later are among the relatively rare introns in prokaryotic systems. All Group I introns have several small conserved sequences, which suggest a common folded tertiary structure.

The conserved sequences are labeled A, B, 9L, 2, 9R, and 9R′. A is paired with B, 9R with 9R′, and 9L with 2. The sites of chain cleavage at the 3′ and 5′ ends of the intron are indicated by the heavy arrows. They are evidently selected by formation of the double-stranded regions.

Another type of intron (Group II) also undergoes self-splicing. The best known example is the last intron in the yeast mitochondrial pre-mRNA. The splicing pathway shown in Fig. 9.18B is similar chemically to that of the groups I introns. However, the initial attack is not by free guanosine, but by the 2′ OH of an internal adenosine, the intermediate product having a *lariat structure* with a loop at the end. Otherwise, processing is similar to that of group I introns. The same pathway with lariat formation is followed by the more widely used removal of introns from pre-mRNA in spliceosomes.

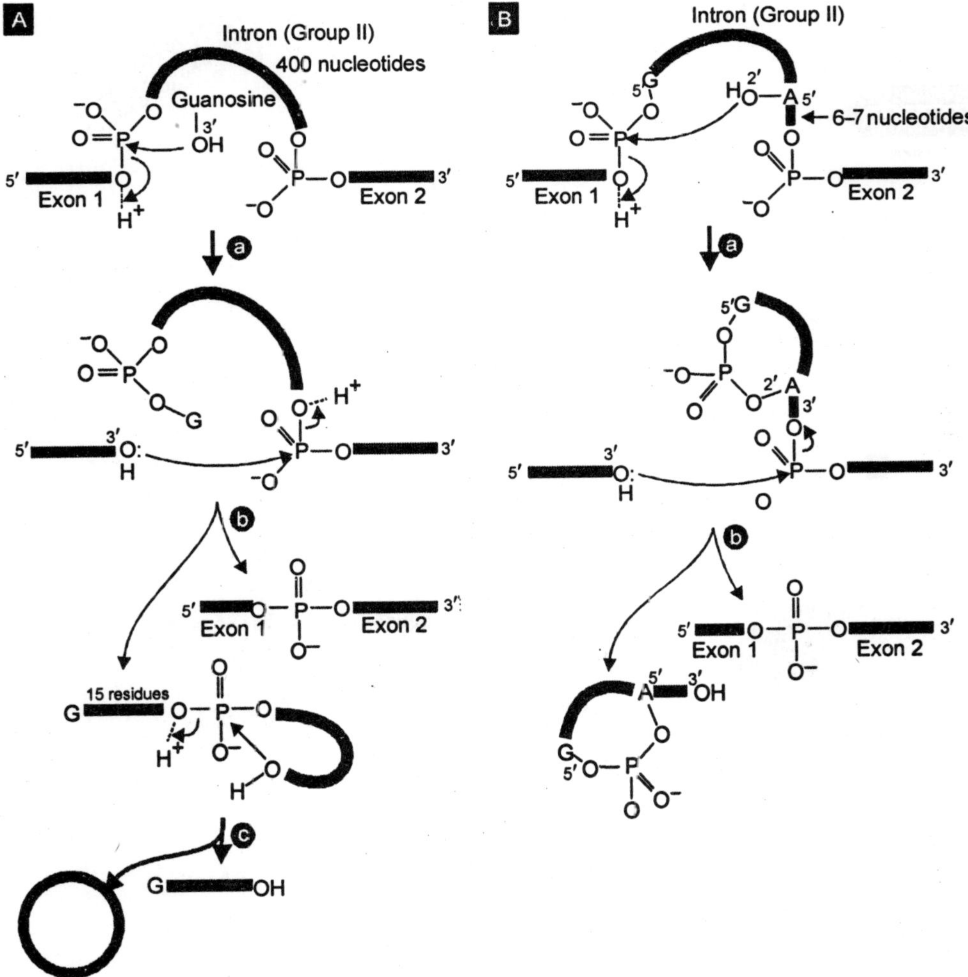

Fig. 9.16. Self-splicing by RNA containing introns (shaded) of groups I or II. (A) Excision of a group I intron to form a circular RNA and a small oligonucleotide. (B) Excision of a group II intron from yeast mitochondrial pre-mRNA as a circular RNA.

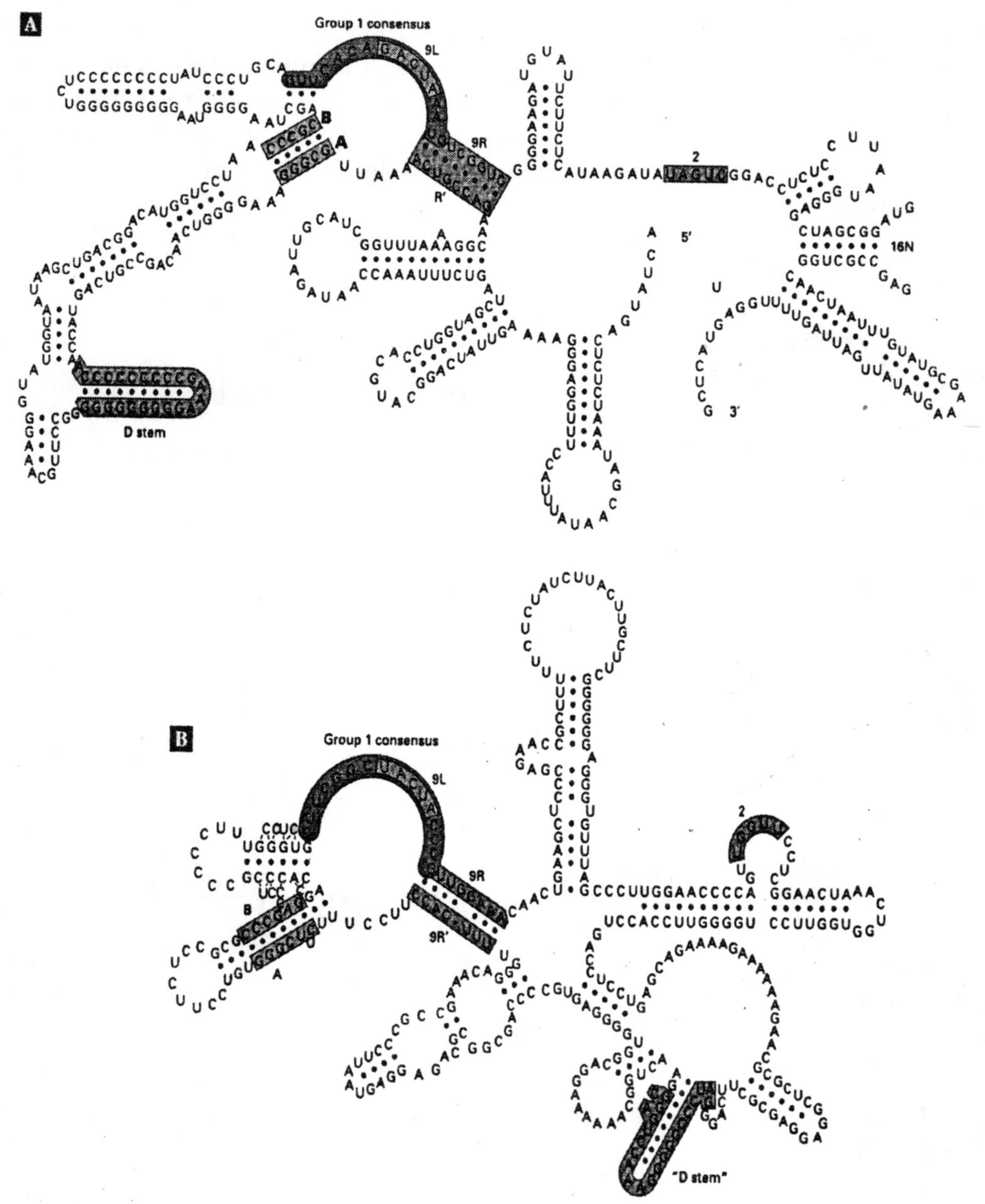

Fig. 9.17. (A) Proposed secondary structures of a self-splicing Group I intervening sequence from *Tetrahymena* according to Michel and Dujon. On the right, 16N indicates 16 nucleotides. (B) Potato spindle tuber viroid structure derived from pairing the conserved sequence elements that are also found in group I introns.

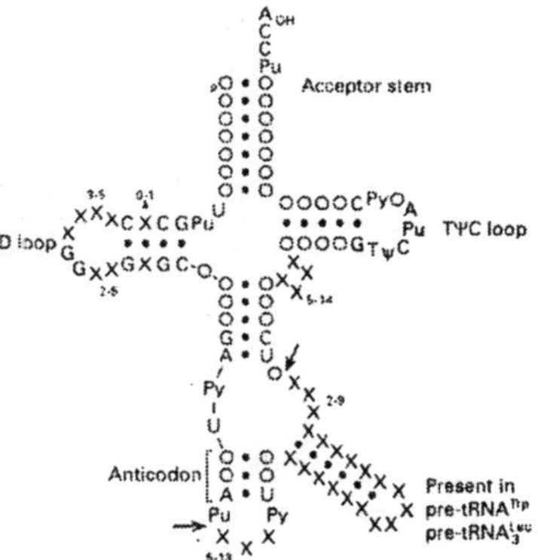

Fig. 9.18. Composite structure representing several tRNA precursors arranged in a similar secondary structure. The arrows indicate splice points. Variable positions are designated (O) for the mature tRNA and (X) for the intervening sequence and also in loops where insertions or deletion occur.

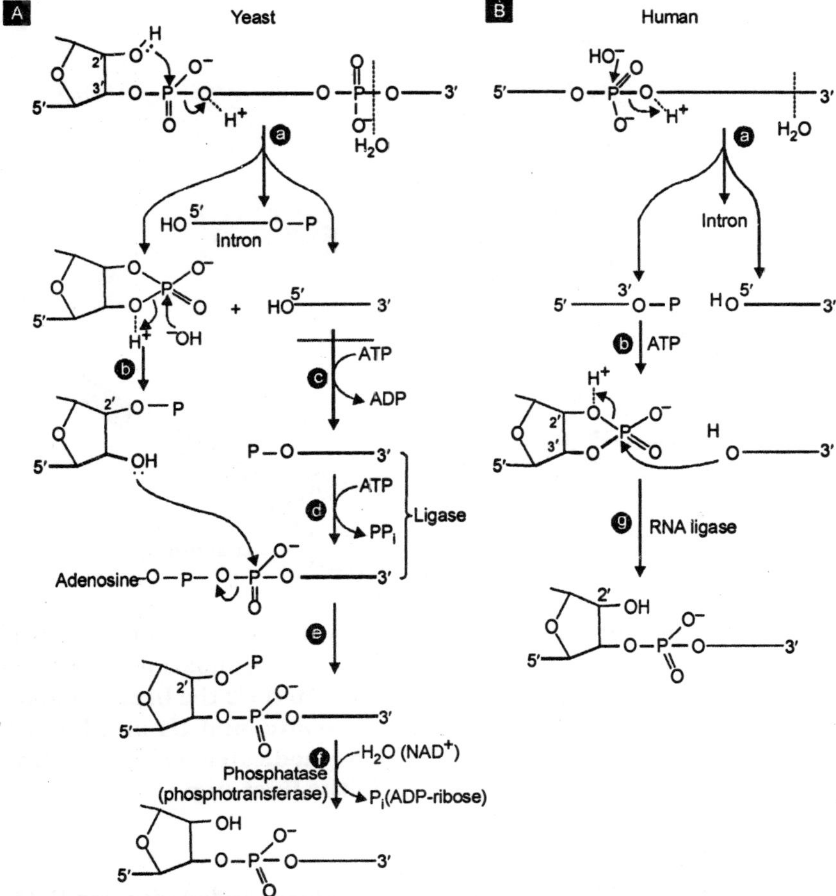

Fig. 9.19. Two related splicing pathways for tRNA precursors. (A) Pathway followed in yeast. (B) Related pathway used by HeLa cells.

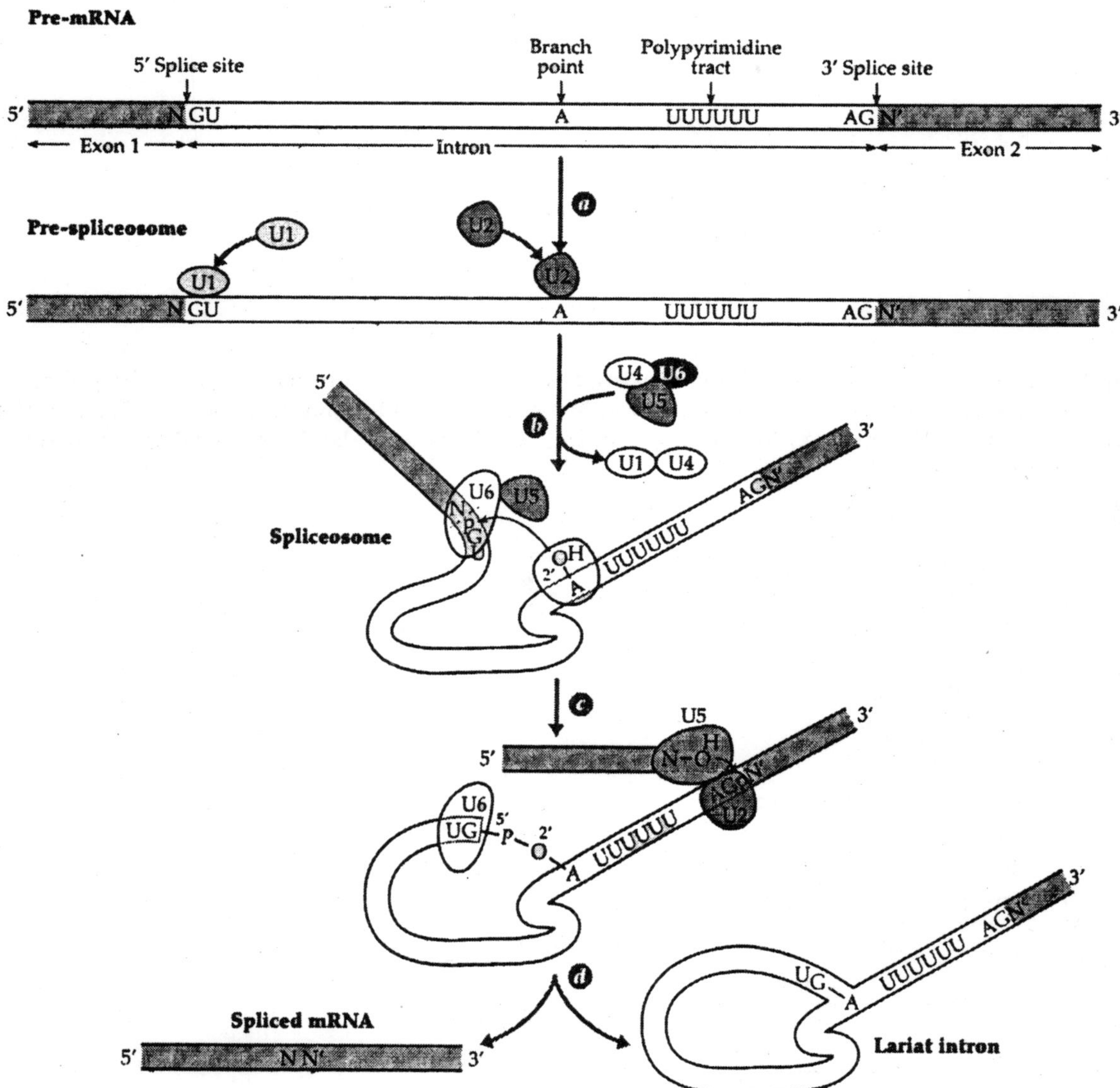

Fig. 9.20. Assembly and action of the spliceosomal complex. Four special sequence elements control the process: the 5′ and 3′ splice sites, the branch point (adenosine A), and a polypyrimidine tract. The snRNP particle U1 locates the 5′ splice site and U2 the branch point. The tri-snRNP U4•U6•U5 then binds, U6 recognizing the 5′ splice site, and U1 and U4 are released. The 2′-OH of the branch point adenosine attacks the phosphodiester linkage to form a lariet intermediate, which releases the intron in a lariet form in the final step.

Relationship to Viroids and Virusoids

The smallest of viruses are the naked ~250-460 nucleotide single-stranded circular RNA molecules called viroids. Viroids are closely related to group I introns. The conserved paired sequence characteristic of the group I introns are also present in viroids, and it looks as if viroids may be "escaped introns." Another group of "plant satellite RNAs" or virusoids replicate only with the help of larger RNA viruses. These satellite RNAs are replicated by a rolling-circle mechanism. The resulting long RNA molecules are self-cleaving, cutting themselves to form the unit length satellite RNAs. These are the simplest known self-cleaving RNA molecules.

Pre-tRNAs

In the removal of type I introns the formation of specific stem and loop structures directs the splicing reactions. Stems and loop structures already exist in tRNA precursors. Cleavage sites are usually located just to the 3' side of the anti-codon as is seen in the pre-tRNA molecule and in the composite structure. The chemistry of the splicing process is shown in the figure 9.21. Cleavage at the two splice sites excises the intron. Then the two ends are rejoined. In yeast cleavage at the 5′ end of the intron is by a multimeric enzyme with a pancreatic ribonuclease-type of action that leaves a 2′,3′-cyclic phosphodiester that is opened hydrolytically (step *b*). Another hydrolytic cleavage, catalyzed by a different subunit of the enzyme complex, occurs at the 3′ end of the intron. The resulting 5′ OH is phosphorylated by GTP (step *c*), the resulting 5′-phospho group reacts with ATP to displace PP_i and to form a transient adenosine-5′-diphosphate terminus (step *d*). This reacts with the 3′ OH formed at the 5′ splice site (step *e*) to create a phosphodiester linkage between the two pieces of RNA.

A phosphotransferase then removes the superfluous 2′ phosphate (step *f*). Steps *d* and *e* are catalyzed by an *RNA ligase*. In HeLa cells and presumably in normal mammalian nuclei, the initial cleavage at the 5′ end of the intron leaves a 3′ phospho group, which is cyclized, probably in an ATP-dependent process, to the 2′,3′-cyclic phosphate (step *b*). This is ligated to the other piece of RNA by a direct displacement on the cyclic phospho group (step c).

The Spliceosome

The hnRNA of nuclei, which includes all of the pre-mRNA, is associated with proteins, which sometimes form very large 200S particles. After limited cleavage with nucleases they tend to sediment in the 30S-40S range and to contain a variety of proteins. Some of the proteins may have been involved in control of transcription. Others participate in splicing. The smaller snRNP particles then appear to come into the nucleus and displace much, but not all, of the protein present in the pre-mRNA ribonucleoprotein particles. The 50-60S spliceosome complexes, with their protein and RNA components, are reminiscent of ribosomes.

Although smaller in size, they can be vizualized by electron microscopy. Each spliceosome is constructed from the four snRNP particles and additional proteins. Each spliceosome can accommodate ~500 nucleotides of pre-mRNA. As the pre-mRNA moves through the spliceosome the splice sites, which have only weakly conserved consensus sequences, must be located. The two ends of the introns, which may be much longer than the average 137-nucleotide exon, must be brought close together in the spliceosome. The exact splice sites are usually located by an invariant *GU* at the 5′ end and AG at the 3′ end. The sequence of the first 18 nucleotides of U1 snRNA is largely complementary to that of the 5′ splice site, which has the consensus sequence AG:**GU**RAGU, where the colon marks the junction. The **GU** is invariant. After the U1 snRNP binds to the pre-mRNA the U2 snRNP binds to another almost invariant sequence CURACU found 20 to 55 nucleotides upstream of the 3′ junction. The A in this sequence becomes a branch point.

It is brought close to the 5′ splice site with the aid of a preassembled complex of snRNPs U4, U6, and U5. In this complex U4 and U6 are tightly paired, additional proteins are also present, and enhancers may be located in adjacent exons. Upon binding of U6 to the 5′ splice site, the U1 and U4 snRNPs are released and the 2′-OH of the branch point adenosine attacks the backbone phosphorus atom (step *c*) at the 5′ splice junction forming a lariat intermediate. The 3′ end created at the 5′ junction must now be held and brought close to the 3′ splice junction, which is located with the aid of U5 snRNP. The 3′ splice junction, utilized in the second splicing step has the consensus sequence (T/C)N(C/T)AG:G. The first splicing step is dependent upon a divalent metal ion, but the second is not. Both steps appear to be in-line nucleophilic displacement reactions.

Additional *splicing factors* are needed for formation of the U4, U6, U5 complex and its function in the second splicing step, which also appears to require ATP. A small fraction of eukaryotic mRNA introns are characterized by AU and AC (rather than GU and AG) ends. The spliceosomes that act on these introns contain modified snRNAs U4 and U6, which are designated $U4_{atac}$ and $U6_{atac}$. They also require U11 and U12 snRNPs. Since pre-inRNAs usually contain many introns, a series of splicing events must occur. These apparently take place consecutively beginning at the 5′ end. Similar splicing pathways are followed in yeast, higher plants, insects, and mammals.

Alternative Splicing Pathways

RNA that contains many introns can undergo splicing in more than one way. Many examples of alternative splicing have been discovered. The mammalian isoenzyme forms of pyruvate kinase called M_1, M_2, L, and R are all tetramers of 60-kDa subunits. The M_1 and M_2 forms are encoded by a single gene. The two mRNAs contain 1593-nucleotide coding regions, which are identical except for a 160-nucleotide sequence that determines the amino acid sequence in a region responsible for intersubunit contact. The difference between the M_1 and M_2 forms involves a choice of two alternative exon regions, one or the other of which is omitted during splicing. Other examples have been found in human collagen in fibronectin, in neuropeptide formation among human proline-rich salivary proteins; cytoskeletal tropomyosin, platelet-derived growth factor, coagulation factor X, porphobilinogen deaminase.

Alternative splicing is very common in transcripts of viral DNA. Alternative splicing could have arisen accidentally, but it is controlled by proteins. Best known is the *alternative splicing factor* (ASF or SF2). It was first recognized by its function in *Drosophila melanogaster,* where the sex of individuals is determined by alternative splicing of an mRNA. In addition to ASF other serine- and arginine-rich *SR proteins* participate in alternative splice site selection.

Trans Splicing

Every mRNA in trypanosomes has, at its 5′ end, a short 35-nucleotide sequence that is not encoded in the transcribed gene. It was found that for each mRNA molecule two transcripts are formed and are spliced together, always with the 5′ piece from a short SL (spliced leader) gene being joined to each of the others. This trans splicing has since been observed in many plants, animals, and protists. Among these are the nematode *Caenorhabditis,* flatworms, yeast, plant chloroplasts and mitochondria, mammalian cells.

Modification and Editing of RNAs

Both mRNA and rRNA undergo rapid methylation of selected residues. About 1-3 internal adenylate residues per kb are methylated at their N^6 positions. These are usually the central adenylates in the sequences GAC or AAC. The methylated sites are not uniformly distributed but are clustered, sometimes in the 3′ untranslated ends of the RNA. Many more residues (55 in yeast and ~100 in vertebrates) in rRNA are methylated on selected ribose 2′-OH groups. About an equal number of uridine residues are converted to pseudouridines.

Methylation sites are apparently selected by the fibrillarin-associated snoRNAs U32-U40. These *methylation-guide* snoRNAs contain 10- to 14-nucleotide sequences that are complementary to segments of the rRNA that contain the methylation sites, and evidently provide rigid helical regions that are targets for the methylase action. The snoRNAs contain the previously mentioned C and D sequences. Methylation occurs on the ribose of the nucleotide that is base-paired with the fifth nucleotide that is upstream of the D sequence in the snoRNA. Each snoRNA directs methylation of a different ribose. In a similar manner the snoRNAs containing the ACA motif located three nucleotides upstream of their 3′ ends appear to direct the conversion of uridines to *pseudouridines.*

Editing of RNA Transcripts

Interpretation of the genetic code utilized by the mitochondria of trypanosomes and other kinetoplastid protozoa was confounded by the discovery that the DNA sequences of many genes, including the COIII cytochrome oxidase gene present in the kinetoplast maxi circle DNA, do not appear to encode the correct amino acid sequence. In fact, the RNA transcripts are extensively edited, mostly by insertion of multiple uridine residues at many positions and by occasional deletions at others. Editing of some mRNAs causes 45% of the message to be rewritten. Several additional kinds of editing were soon discovered, not only in protozoa but also in mammals, plants, and archaea. A frequent editing change is the ehydrolytic deamination of a C to a U residue.

For example, human apolipoprotein B is synthesized in two forms: apoB100, a full-length 512-kDa protein made in the liver and used for transport of cholesterol and triglycerides, and a shorter 241-kDa form, apoB48, used in absorption of dietary lipids. ApoB100 is synthesized from a full-length mRNA, but apoB48 is made according to a shortened mRNA in which a glutamine codon (CAA) has been converted by editing to the translation stop codon UAA.

A special enzyme deaminates only cytidine 6666 of the mRNA. C to U editing occurs in chloroplasts and mitochondria of plants. In *Arabidopsis* mitochondria 456 different C to U conversions have been identified in mRNAs. Deamination of adenosine residues produces inosine, which occurs in brain mRNA once in ~17,000 ribonucleotides. Some ionotropic glutamate receptors in the brain have subunits translated from inosine-containing mRNAs. A glutamine codon CAG is edited to CIG, an arginine codon. The arginine codon AGA is converted to the glycine codon IGA at another site, ATT is converted to ITT and TAG to TIC. All of these changes affect the properties of the glutamate-activated ion channel. The adenosine deaminases involved in these editing events are usually specific for double-stranded helical segments of RNA, *e.g.,* for the stems in stem-loop structures. At least one human adenosine deaminase not only binds to RNA but also contains a DNA-binding domain specific for Z-DNA. Returning to the trypanosomes and their relatives, mitochondrial RNAs undergo extensive insertion and deletion of U's. The editing site is located by a *guide RNA* (gRNA), which directs the hydrolytic cleavage of the chain and either the addition of U's to the 3′ cut end by transfer from UTP or hydrolytic deletion of U's from the 3′ cut end. The chain ends are then rejoined by an RNA ligase.

The functional significance of the editing of kinetoplast mRNA is uncertain. However, at least some of the edited mRNA is translated to give proteins that are presumably used. While trypanosomes usually insert only U's, the slime mold *Physarum polycephalum* may insert dinucleotides such as AA, AU, CU, or GU may also add nucleotides at the 3′-ends of RNAs. A to G editing occurs in RNA of *Drosophila.* Yet another type of editing has been observed in viral RNA from paramyxovirus.

The virally encoded RNA polymerase sometimes "stutters" reading the same template base two or more times, with a resulting insertion of a base. Editing of transcripts usually serves an essential biological function, creating the correct sequence for translation of the mRNA and often generating multiple isoforms of proteins.

Finishing the Transcripts

Additional modifications must be made to some mRNAs, and there will doubtless be many surprises as the details are worked out. One detail, which was discovered in the 1980s, is the specific function of snRNA U7 in recognition of the 3′ end of pre-mRNAs for histones. The U7 RNA apparently base-pairs with a sequence near the 3′ end cleavage site, acting as a cutting guide. Newly synthesized mRNA emerges from the nuclear pores as nucleoprotein complexes containing as many as ten different proteins.

A major component is a 78-kDa polypeptide thought to be associated with the poly(A) tail. These ribonucleoproteins are sometimes stored for long periods of time, for example, in mature seed embryos and in amphibian oocytes. They may also travel rapidly for long distances, *e.g.*, down nerve axons or from cell to cell in plants via transport in the phloem.

TRANSCRIPTION OF MITOCHONDRIAL, CHLOROPLAST, AND VIRAL GENES

In the compact 16-kb chromosomes of mammalian mitochondria, the genes are tightly packed against one another. Most genes are transcribed using heavy H strands as templates and specialized bacteriophage-type RNA polymerases encoded in nuclear DNA. A single promoter in the D loop region used to make a long transcript from the entire H strand. The transcript is then cut precisely by mitochondrial RNase P at the 5′ ends of the tRNAs. Similarly precise cleavage must occur at the 3′ ends because there are often no nucleotides or only one to a few nucleotides separating adjacent genes. The nucleases involved differ from those used in nuclear Trna 3′ processing.

In animal mitochondria tRNA genes sometimes overlap by one nucleotide. Polyadenylation of the transcripts provides a form of editing that is required to create a UAA translational termination signal to which termination proteins bind. The tRNA genes lack the 3′ terminal CCA, which must be added. The lighter L strand has its own promoter, also located in the D loop. Both promoters contain the conserved nonanucleotide sequence 5′-ATATAAGTA. The pre-mRNAs created by these cleavages are not capped but are polyadenylated. Since the promoters are simple and the transcription factors few, mitochondrial transcription is controlled largely by mRNA stability, translation, and posttranslational events. The 70-to 100-kb circular mitochondrial DNA molecules of yeast and of higher plants contain more genes than do animal mitochondria, but most of the increased size is accounted for by intergenic spacers and by a few long introns.

All yeast mitochondrial genes except for that of $tRNA^{Thr}$ are transcribed from a single strand. About 20 different primary transcripts have been identified in *Saccharomyces cerevisiae.* These originate at several points in the genome but always at the sequence 5-ATATAAGTA, the 3′A corresponding to the 5′ nucleotide of the transcript. One of these sequences is located at the origin of replication, suggesting the possibility that a normal RNA transcript provides the primer for DNA replication in yeast mitochondria. The 100- to 160-kb chloroplast genomes also have many prokaryotic features. They encode ~50 proteins as well as the tRNAs and rRNAs. Promoter and terminator sequences resemble those of bacteria and protein sequences are often homologous to those in bacteria. This applies, for example, to the a, β, and β′ subunits of RNA polymerase.

Viral Transcription and Replication

Because viruses contain small genomes, study of transcription of viral DNA and of replication of RNA viruses has played an important role in helping us to understand transcription in eukaryotes. An example is the discovery of the virus SV40 enhancer, which has been discussed else where. Study of viral life cycles is also essential to future progress in fighting viral diseases. Each of the many different viruses has its own often very complex life cycle. Only a few details can be given here. Eukaryotic DNA viral genomes, like that of phage λ usually contain early and late transcriptional units. The small papovaviruses, such as SV40 and polyoma virus, have 5.2 kb genomes.

Like the small RNA viruses, they make use of overlapping genes and alternative RNA processing. In SV40 DNA there are two overlapping promoters called early-early and late-early. The first of these contains a TATA sequence, and both promoters also depend upon a 21-bp repeat segment as well as the SV40 enhancer. At least two proteins, one that binds to the enhancer and one that binds to the 21-bp repeat, are needed for initiation of early transcription. The early-late promoter lacks the TATA sequence but requires the 21-bp repeat and enhancer. A 94-kDa encoded protein called the large T-antigen one of the regulators of transcription as well as of DNA replication. This protein is also sufficient to transform rodent cells in culture.

Although predominantly nuclear it is also inserted into the membrane where it acts as an antigen. The large icosahedral adenoviruses cause respiratory infections in humans and attack and may cause cancer in many other vertebrate species including birds and amphibians. The 35.9-kb genome of human adenovirus-2 encodes at least 30 proteins, 10 of which appear in the virion. One of these is covalently linked to the 5' end of the DNA. As with smaller DNA viruses extensive use is made of alternative splicing of the transcribed RNA. There are at least six early transcriptional units, each with its own promoter. A variety of mRNAs are created using the various coding segments and a complex array of regulatory sequences control transcription.

A 32-kDa phosphoprotein transcription factor designated E1A is encoded by a “pre-early” gene. It is required along with host-encoded transcription factors for viral transcription. The late region of the adenovirus genome encodes structural proteins for the virus coat. Most of its transcripts begin about 16.5% of the way along the 36.5-kb dsDNA.

However, the initial transcripts are cleaved at several different positions to yield a series of different 3′-poly(A)-terminated transcripts. In an exceedingly complex process the transcripts undergo splicing out of genes at their 5′-ends so that the final mRNAs typically code for single proteins. At the 5′ ends the cap is joined to short segments from the original mRNA 5′ end. Viruses SV40, polyoma, and some strains of adenoviruses are oncogenic in some species and cause transformations of cells in culture. Transformed cells always contain integrated viral DNA. That of SV40 can be incorporated at many different sites in the host genome. The integrated DNA does not always include the complete SV40 genome, and parts of the DNA may be inverted, deleted, or scrambled. Integration is not an essential part of the viral life cycle and has no effect on the infective properties of the viruses. Cells transformed by adenoviruses usually also contain only a fragment of the viral genome in their DNA. However, one small set of genes from early region 1A is present in all transformed cells. The encoded proteins appear to be modulators of transcription and may cause cancer by promoting uncontrolled transcription of certain genes.

Replication of RNA Viruses

The RNA (+) strands present in many RNA viruses often serve immediately after infection as a messenger RNA. However, replication requires formation of (-) strands of viral RNA from which new (+) strands can be transcribed for assembly into new virus particles. Other RNA viruses contain (-) strands of RNA or double-stranded RNA and, therefore, have significantly different life cycles.

Small RNA Viruses

The human polio virus, the common cold virus (rhinoviruses), and other picorna viruses have 7.2- to 7.5-kb genomes with considerable homology (50% between the polio and rhinoviruses) and similar overall structures. The polio genome encodes eight different proteins, one being a small 22-residue peptide that becomes covalently linked to the 5′-end of the RNA through a phosphodiester linkage to the side chain of a tyrosine. Cleavage of this linkage by a host enzyme allows the viral RNA (+) strand, which is polyadenylated at the 3′ end, to serve as an mRNA for synthesis of a single large 220-kDa polyprotein. This is cleaved by a host protease at several Gln-Gly bonds to form several proteins. These include the 22-residue RNA-linked peptide, two capsid proteins, and a capsid precursor protein. The latter is cleaved during capsid assembly by a viral protease at an Asn-Ser bond to give two more capsid proteins. These four proteins have masses of 7, 26, 29, and 32 kDa.

A viral protease, a large 58-kDa replicase, and a 37-kDa protein of unknown function are also cut from the polyprotein. A host protein initiates cleavage of the polyprotein, but the virally encoded protease later takes over this function. One of the best understood of the many viral pathogens of plants is the tobacco mosaic virus. Its 6.7-kb positive strand RNA encodes a replicase, coat protein, and at least one other protein.

Influenza Viruses

These negative-stranded viruses are classified into types A, B, and C, but it is only type A that infects nonhuman species including birds, horses, pigs, seals, mink, and whales. Type A influenza viruses have also caused the great pandemics such as those in 1918-1919 and in 1968. Influenza viruses are surrounded by a lipid bilayer in which the virally encoded *hemagglutinin* and a *neuraminidase* are embedded. The inside of the bilayer is coated with a matrix protein and within this coat eight pieces of RNA of total length 13.6 kb are coiled together with a basic nucleoprotein. Also present are ten molecules each of three other proteins. The eight pieces of RNA vary in length from 900 to 2500 nucleotides. Seven of them encode one each of the seven virion proteins. One encodes an additional nonstructural protein while the smallest piece, using overlapping nucleotide sequences, encodes two non-structural proteins. The existence of a fragmented genome evidently underlies the ability of influenza. A viruses to undergo rapid changes in antigenic behavior.

If a cell is coinfected with two strains of virus, the eight fragments act as independent chromosomes, which can be reasserted into new combinations in the progeny viruses. As a consequence, it is difficult to develop safe, live virus vaccines. A large reservoir of infection among migratory water birds and other animals facilitates the appearance of new strains and their rapid spread throughout the world. The first step in the replication of influenza viruses, which takes place in the cytoplasm, is the synthesis of (+) strands that can serve both as mRNA for synthesis of proteins and as templates for synthesis of new (-) strands. Three of the capsid proteins form the required RNA polymerase. This "transcriptase" is primed preferentially by 5′-capped 10- to 13-nucleotide segments of RNA that have been cut by a viral nuclease from host mRNAs.

The mRNAs made from viral RNA are polyadenylated and are translated by the host cell's ribosomes. However, some transcripts are used as templates to form viral (-) strands, which are not polyadenylated and which contain uncapped pppA at the 5′-ends.

HIV-1 and Other Retroviruses

Because of their association with viral oncogenes because of the *human immunodeficiency virus* (HIV-1) and the AIDS epidemic a great deal of attention is focused on retroviruses. Each retrovirus particle contains *two* identical single-stranded (+) RNA molecules, which may be as long as 10 kb. Their unique characteristic is that they induce synthesis of DNA, which must be integrated into the host genome before new viral (+) strands are transcribed. Retroviruses may sometimes cause cancer and may carry oncogenes. Study of the *Rous sarcoma virus* (RSV), which infects chickens, and of the related *avian myeloblastosis virus* (AMV) and of HIV has revealed a common structure and a complex life style that are largely shared by all known retroviruses. The organization of retroviruses always includes a sequence of genes designated *gag* (glycoprotein antigen core proteins), *pol* (polymerase), and *env* (envelope). These are often followed by an oncogene.

In RSV this is the *src* gene. At each end of the retrovirus gene sequence is a short direct repeat labeled R in Fig. 9.23. In RSV the R sequence is 21 nucleotides in length and in HIV it is 98 nucleotides long. The 5′ end of the viral RNA is capped, and the 3' end is polyadenylated. The dsDNA of the integrated form of the virus longer and at each end is bounded by *long terminal repeats* (LTRs). Each LTR consists of a sequence, designated U3, that is present next to R at the 3′ end of the viral RNA. In the LTR this is followed by sequence R and then by U5, a unique sequence that came from the 5′ end of the viral RNA. Each ds-LTR begins and ends with a short inversely repeated segment:

5′-TGT—ACA in RSV

5′-CTG—CAG in HIV

The integrated provirus is always bounded by a sequence of host DNA that is repeated without inversion at the opposite end. For RSV this is a 5-bp sequence. The LTRs in RSV are 569 bp in length, and those in HIV are 634 bp in length, 83 bp coming from U5, 98 from R, and 453 from U3. The LTRs themselves often contain promoter and other control elements and even entire genes. The organization of a retrovirus reflects the complex mode of replication, which is presented in simplified from. The key enzyme is the RNA-directed *reverse transcriptase.* The initial synthesis of DNA by this enzyme is primed by a tRNA. RSV uses $tRNA^{Trp}$ and HIV $tRNA^{Lys}$ for this purpose. The 3′ end of the tRNA, including the nucleotides forming both the acceptor stem and the stem of the TψC loop, must relinguish its normal base-pairing to form ~18 Watson-Crick base pairs with a *primer binding site* near the 5′ end of the retroviral DNA.

Because synthesis of the (-) strand of the retroviral DNA begins so close to the 5′ end of the template only a short piece of DNA, including sequences U5′ and R′ complementary to U5 and R, can be formed. For replication to continue RNA must be removed from hybrid regions. This is accomplished by the *RNaseH* activity of the reverse transcriptase. After removal of the RNA the primer tRNA must undergo a *strand transfer,* in which it shifts from the 5′ end of the viral RNA template to the 3′ end, utilizing pairing between the right-hand R sequence of the template and the R' sequence of the growing cDNA copy. This transfer is sometimes to the second of the pair of identical RNA molecules in the virus, providing a way of increasing diversity by recombination.

A second strand transfer of the growing (+) strand is needed to complete the dsDNA, which now contains the two identical LTRs. The 5′ and 3′ ends of the template RNA are doubtless held close together to facilitate strand transfer. Integration of the dsDNA into the host DNA can occur at many places. The mechanism of integration probably resembles that used by phage λ accounts for the duplication of host sequences at the two ends of the integrated virus. A virally encoded integrase catalyzes the process. It is the integrated virus that is transcribed to form new (+) viral RNA strands. Integrated retroviruses are usually transcribed as full-length RNA copies, which may or may not have introns spliced out.

The smaller spliced pieces encode the *env* and other genes, such as *src*. The *gag-pol* region is translated as a polyprotein that is cleaved into a number of pieces. These include four proteins of the virus core (encoded by *gag*), the reverse transcriptase with its associated RNaseH, and an *integrase* (all encoded by *pol*). There is also an *aspartic protease* only 99 residues in length within *pol*. The promoter and control region for transcription is located in the U3 region and is placed into a position where it can function only upon synthesis of the first LTR. The gene *env* encodes the major viral envelope protein and is translated from a spliced mRNA.

Accessory Regulatory Genes

HIV and some related retroviruses such as HTLV-1 (which causes rare T-cell leukemias) are distinguished from other retroviruses by a marked increase in the rate of DNA transcription within infected cells as compared with uninfected cells. This is thought to be a result of synthesis of virally encoded proteins that are trans-acting regulators of transcription. The HIV genome map the positions of the nine recognized genes marked in the three reading frames. In addition to the *gag*, *pol*, and *env* genes there are genes for six accessory proteins: Tat, Rev Vif, Vpr, Vpu, and Nef. Transcription is initiated at the promoter in the 3′LTR. This contains a TATAA sequence, an SP1 binding site, and an enhancer that binds transcription factor NF-κB. The full-length 9–kb transcript contains, according to Frankel and Young, the following essential regions, whose positions may vary somewhat from one isolate of the virus to another:

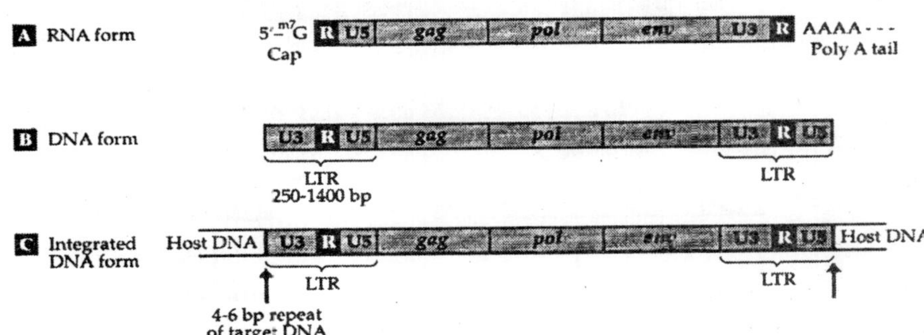

Fig. 9.21. comparison of the forms of a retrovirus. (A) Infective RNA (+) strand. (B) The double-stranded DNA form (C) The DNA form integrated into the host DNA. LTR, long terminal repeats.

Fig. 9.22. Simplified scheme for replication of the RNA genome of a retrovirus.

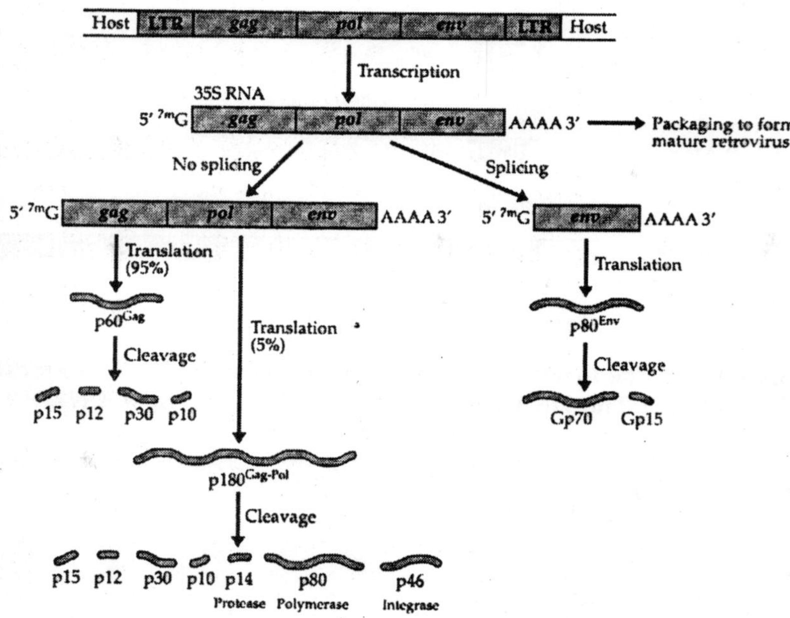

Fig. 9.23. Gene expression from a typical retrovirus that has been integrated into a host's genome. This figure illustrates how a variety of proteins are encoded by a single rather short piece of DNA.

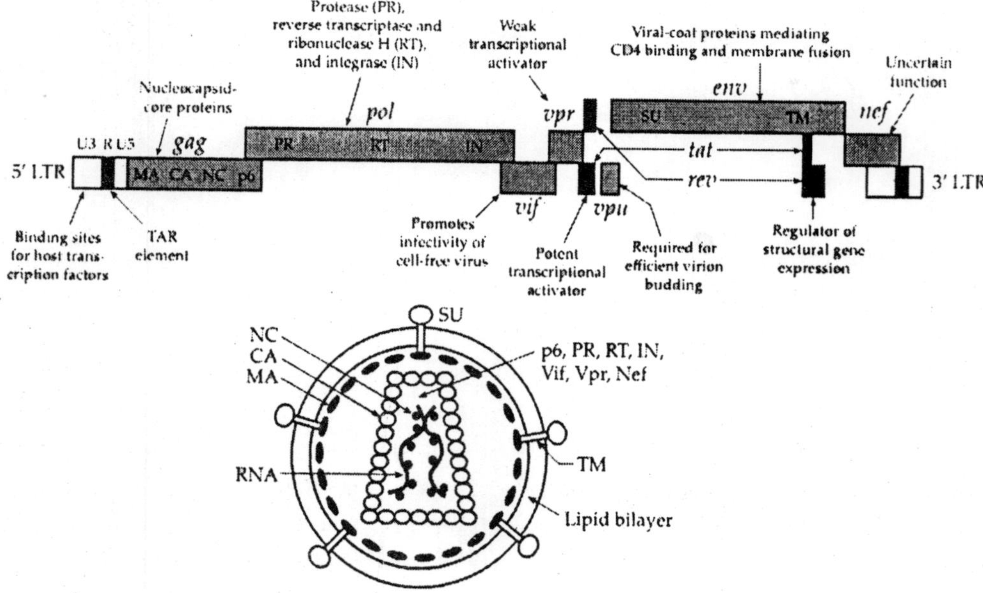

Fig. 9.24. Simplified genetic map of the AIDS virus HIV-1. All three reading frames are utilized to encode nine genes, which give rise to 15 proteins.

A complex pattern of splicing produces more than 30 mRNAs. When viral RNAs are first produced most are doubly spliced, allowing the split genes *tat* and *rev* to be expressed by synthesis of Tat and Rev. The *trans*-activator protein Tat is a small 86-residue cysteine-rich protein, which binds the Tar stem-loop structure and greatly stimulates transcription.

The 116-residue *rev* (regulator of expression of virion genes) gene product is also a transactivator, which is needed for rapid production of singly spliced (4 kb) or unspliced (9kb) *gag-pol* mRNA required for formation of virus structural proteins. The effect of Rev is probably on transport from the nucleus rather than on splicing.

Transcription is repressed by the 206-residue N-terminal myristoylated protein, Nef, a phosphoprotein that associates with cytoplasmic membranes. It has been difficult to learn its exact function, but it seems to be required for maintenance of the integrated provirus for long periods of time without extensive replication. Mutations in gene *nef do* not eliminate the ability of the virus to replicate in T lymphocytes and to kill them.

The 23-kDa protein Vif *(viral infectivity factor)* is not needed for growth but is essential for infectivity. Other genes in HIV are *vpu,* which encodes an 81-residue integral membrane protein (*virion protein* U), and *vpr,* which encodes the 96-kDa *virion protein R.* Several possible functions have been proposed for these small proteins.

HIV-1 is a member of the group of slow viruses or *lentiviruses.* Other lentiviruses include the human HIV-2, an immunodeficiency virus that attacks cats causing leukemia, and the human leukemia virus HTLV-1. Another lentivirus causes two diseases of sheep, *maedi,* a pulmonary disease, and *visna,* a paralytic condition somewhat similar to multiple sclerosis. Because of its slow development there has been doubt as to the cause of AIDS, but there is now little doubt that HIV-1 is the true culprit. Success in treating AIDS may depend upon better understanding of the complex life cycle of HIV-1. The cycle begins with the binding of the virion envelope protein to the immunoglobulin-like surface protein *CD4,* which is found principally on the type T4 helper T cells. Binding of CD4 to the HIV envelope proteins appears to activate the T cells to enter the cell cycle and to take up and integrate the virus. The virus infection destroys these CD4+ lymphocytes with a half-life of less than two days. A major effort is being made to devise a vaccine against HIV. However, rapid mutation of the *env* gene makes it difficult to accomplish. This high rate of mutation appears to be a result of a high frequency of errors by the HIV reverse transcriptase. There is hope from the fact that some individuals are naturally resistant to HIV infection.

Position from 5′end	***Description***
1 – 55	TAR, a 59-residue stem-loop structure; binding site for Tat
182–199	PB, primer binding site
240–350	Packaging signal. Binds envelope protein NC
248–271	Dimerization site with "kissing loop"
290	Major splice donor site, used to form all spliced mRNAs
1631–1673	Gag-Pol frameshifting region where –1 ribosomal frameshifting occurs to allow 5–10% synthesis of Gag-Pol polyprotein
7362–7596	Rev responce element. Binding site for Rev
5358 and 7971	Two major splice acceptor sites; other minor sites are also used
9205–9210	Polyadenylation signal

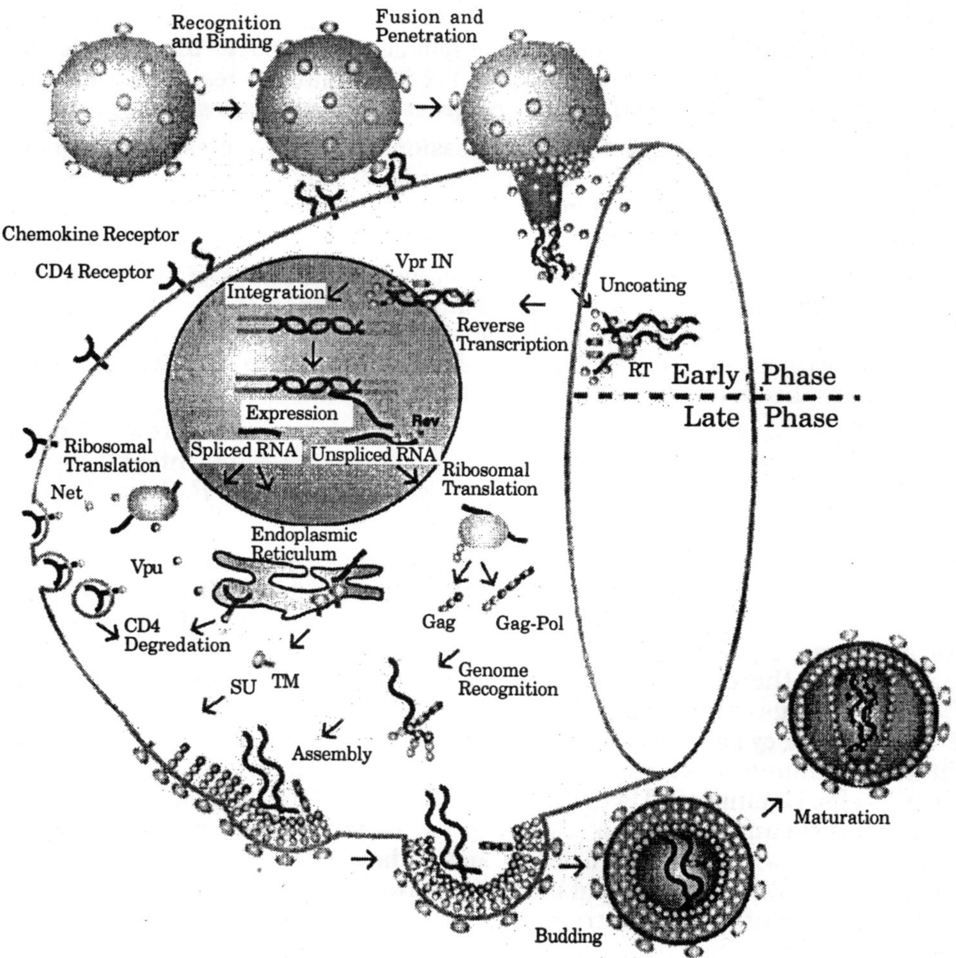

Fig. 9.25. General features of the HIV-1 replication cycle. The early phase (upper portion of the diagram) begins with CD4 recognition and involves events up to and including integration of the proviral DNA, and the late phase includes all events from transcription of the integrated DNA to virus budding and maturation.

Retrotransposons

Transposition of DNA, may seem to be a rare and relatively unimportant event in our body cells. However, trartsposon DNA accounts for 35% or more of the human genome and apparently plays a major role in evolution. Like other transposons, the DNA sequences known as retrotransposons also move about within DNA. However, they use an indirect mechanism that involves synthesis of mRNA and reverse transcription. The transcribed complementary DNA may be inserted back into the genome at new locations. The necessary chemical reactions parallel those involved in the replication of retroviruses.

Retrotransposens, truncated retrotransposons, and related sequences constitute as much as 16% of the human genome. There are two classes of retrotrotransposons: those with long terminal repeats (LTRs) and those without (LTRs). The first group is closely related to retroviruses, but its members lack genes for envelope proteins. They do carry *gag* and *pol* genes similar to those of retroviruses. Most retrotransposons are defective and do not move. Over evolutionary time they accumulate in the genome, sometimes to the extent that the genome size grows enormously. This has happened often during the evolution of plants, some of which (*e.g.*, certain lilies) have 40 times more DNA per cell than do humans.

Although most retrotransposons are inactive, some of them occasionally jump to new locations where they may mutate a gene and may sometimes cause disease. However, their major significance is probably in facilitating evolution, perhaps including the formation of new species. The non-LTR transposons are exemplified by the 6-7 kbp LINES the short 90-400 bp SINES. Mammalian genomes contain ~50,000 truncated members of the LINE-1 (L1) family and 3000-5000 full-length L1s. Only a few of these are active in our present population.

The RNA intermediates that participate in retroposition of LINES are generated by RNA polymerase II, while RNA polymerase III forms the RNA intermediates for propagation of SINES. Participation of these RNAs in trans splicing processes can modify existing genes, contributing to the remodeling of the genome. Yet another group of mobile elements in the genome are *intern genes,* which encode protein-splicing polypeptides. Many inteins also have *homing endonuclease* activity and cleave DNA at specific insertion sequences, initiating incorporation of intein DNA into new locations in the genome. Group II introns, which are found in bacteria and in organelles of fungi and plants, may also act as mobile DNA elements.

THE ANTIBIOTICS RIFAMYCIN, RIFAMPICIN, AND ACTINOMYCIN D

Rifamycin, produced by *Streptomyces mediterra-nei,* is of medical value because it affects *acid-fast* as well as gram-positive bacteria. The semisynthetic rifampicin has been especially useful in the treatment of tuberculosis. The ether linkage at the bottom of the ring at the right in the structural formula is cleaved, and the resulting hydroquinone is oxidized to a quinone within the bacterial At a concentration of 2×10^{-8} M rifampicin inhibits bacterial RNA polymerase 50%. It does not prevent the binding of polymerase to DNA but inhibits initiation of transcription. Mutants of *E. coli* resistant to rifampicin produce RNA polymerase whose f5 sub-unit has been altered, sometimes with a change in electrophoretic mobility. The related antibiotic *streptolydigin* also binds to the subunit of RNA polymerase and blocks elongation, resistant mutants mapping close to *rif* mutants.

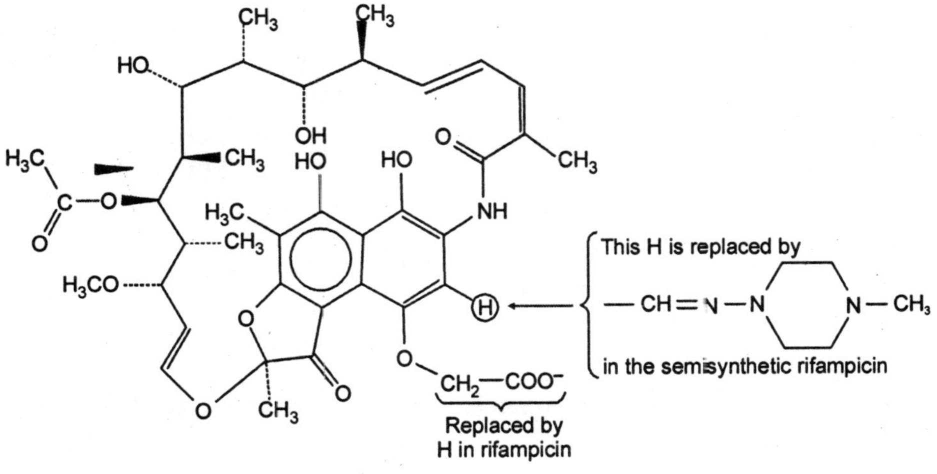

The actinomycins, which are also produced by *Streptomyces,* not only kill bacteria but also are among the most potent antitumor agents known. However, because of their extreme toxicity they are of little medicinal value. Actinomycin D, which is a specific inhibitor of RNA polymerase, contains a planar phenoxazone chromophore bearing two carboxyl groups, each one linked to an identical cyclic peptide made up of L-threonine, D-valine, L-proline, sarcosine (N-methylglycine), and L-methylvaline.

An ester linkage joins the methylvaline residue of the peptide to the side-chain hydroxyl of threonine. Two *cis* peptide linkages are present. Ignoring the obvious asymmetry of the phenoxazine ring, actinomycin possesses approximate twofold symmetry.

The antibiotic binds tightly to double-stranded DNA in regions containing guanine. A 2:1 deoxyguanosine-actinomycin complex has been crystallized, and the structure has been determined by X-ray diffraction. The phenoxazine ring lies at the center of the complex, one peptide loop extending above it. The twofold symmetry is present in the dideoxyguanosine complex as well as in actinomycin itself. The phenoxazine ring lies between the two flat guanosine rings in van der Waals contact. The two amino groups of the guanine rings form strong hydrogen bonds with the carbonyl groups of the threonine residues. There are also weaker nonlinear hydrogen bonds from the N-3 atoms of the guanines to the NH groups of the same threonines:

A symmetric pair of hydrogen bonds join the two carbonyl and NH groups of the D-valine residxies in the peptide loops.

Model building studies show that a similar complex can be formed with double-stranded DNA. While the amino groups of the guanine rings (see drawing) are hydrogen-bonded to the actinomycin, the other hydrogen atoms of the same amino groups as well as the N-1 hydrogen atoms and the carbonyl groups of the guanine ring are available for hydrogen bonding to form a GC base pair. Thus, the structure above can be modified readily into part of the double-stranded DNA molecule in which the phenoxazone ring of actinomycin is intercalated between two CG pairs. To do this the normal DNA structure has to be unwound by 18° at the point of insertion of the extra ring. Binding also occurs at other sites. Sobell suggested that actinomycin binds to a premelted region of the DNA helix within the transcription bubble and immobilizes it. This interferes with the elongation of growing RNA chains.

POWERFUL POISONS FROM MUSHROOMS

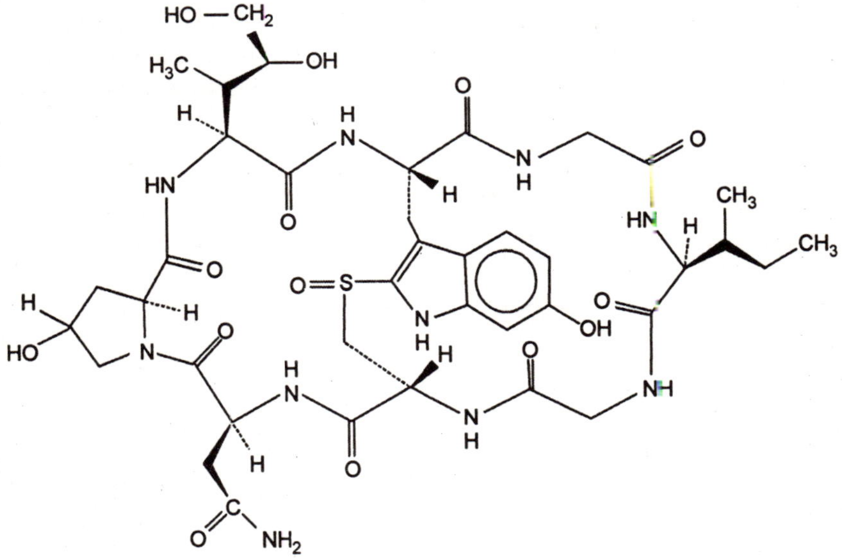

Several deadly species of the genus *Amanita* produce colorless toxic octapep tides, the *amanitins.* Two residues of glycine, one of L-isoleucine, one of the unusual L-dihydroxyisoleucine, one of L-asparagine, and one of L-hydroxyproline are present in a-amanitin. In the center a modified tryptophan residue has been combined oxidatively with an SH group of a cysteine residue. If the dihydroxyisoleucine residue of α-amanitin is replaced with unhydroxylated leucine, the resulting compound, known as amanullin, is nontoxic. The LD_{50} for mice is 0.3 mg kg^{-1} and 50 g of fresh *Amanita phalloides* may be sufficient to kill a person. Amanitins act slowly, and it is impossible to kill mice in less than 15 h, no matter how high the dose.

α-Amanitin completely blocks transcription by eukaryotic RNA polymerases II and III. Polymerase II is the major nuclear RNA polymerase, and its inhibition prevents almost all protein synthesis by the cell. Note that the amanitin molecule is semisymmetric overall, much as is a actinomycin with an aromatic group protruding from behind in the center. The same mushrooms contain several fast-acting toxic heptapeptides, the *phalloidins,* whose structures are similar to those of the amanitins. However, they contain a reduced sulfur atom (—S—) in the cross-bridge. They are specifically toxic to the liver. The same mushrooms also contain an antidote to the phalloidins, *antamanide.* This cyclic decapeptide, like the toxins, is made up entirely of L-amino acids, and it apparently competes for the binding site of the phalloidins. Unfortunately, it is of little value in treating cases of mushroom poisoning. Antamanide is a specific sodium-binding ionophore.

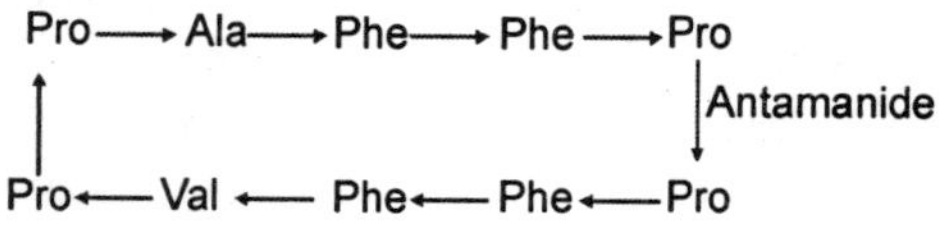

Pro ⟶ Ala ⟶ Phe ⟶ Phe ⟶ Pro
↑ Antamanide ↓
Pro ⟵ Val ⟵ Phe ⟵ Phe ⟵ Pro

Synthetic Antiviral Compounds

Most bacterial infections can be treated successfully with antibiotics, but the development of satisfactory antiviral agents has been slow. Yet we are susceptible to many dangerous virus diseases, and viruses also take a huge toll among domestic animals and plants. The first antiviral drug, 5-iodo-2′-deoxyuridine (idoxuridine), was introduced in 1962 and was used for 20 years by ophthalmologists to treat serious eye infections by the herpes simplex virus (HSV).

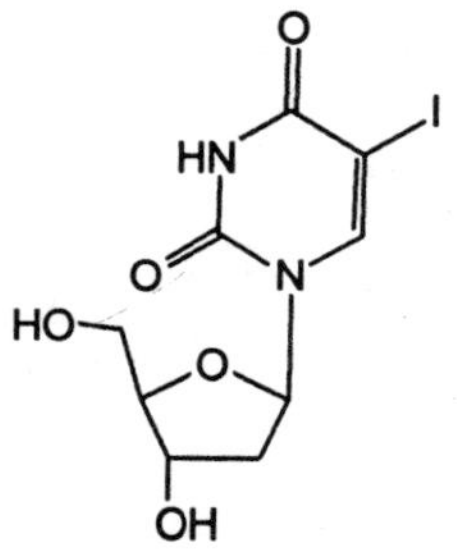

Idoxuridine

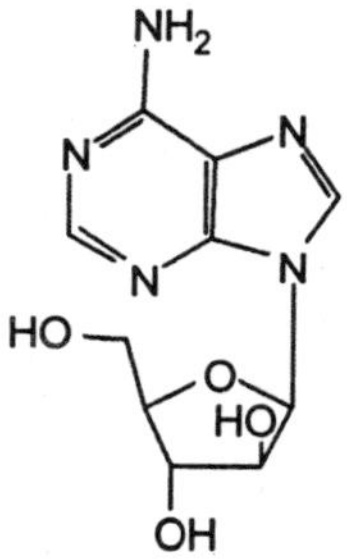

Ara-A

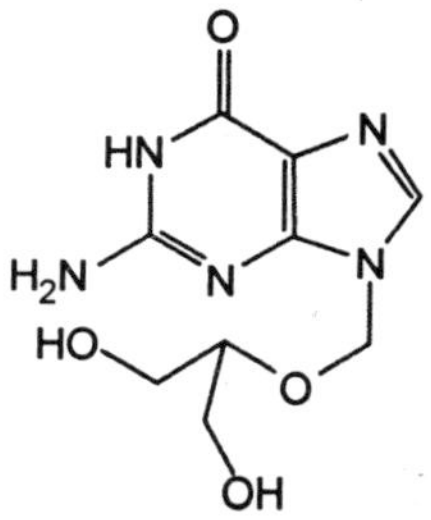

DHPG

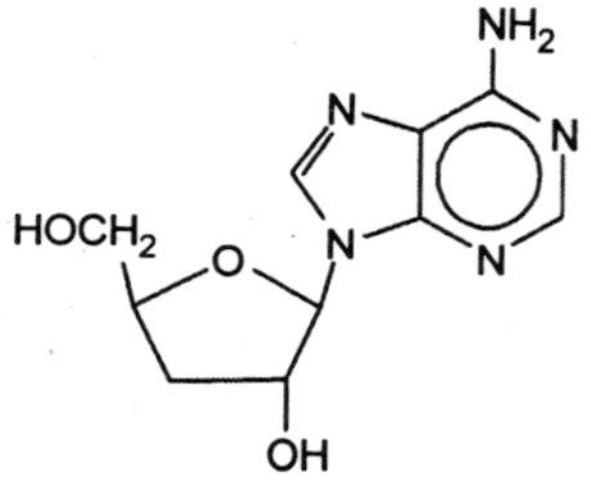

Cordycepin (3′-deoxyadenosine) preferentiallyinhibits synthesis of ribosomal and tRNA

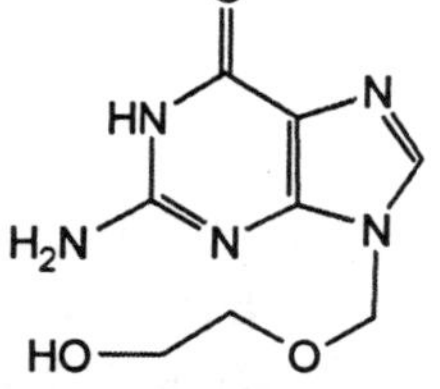

Acyclovir

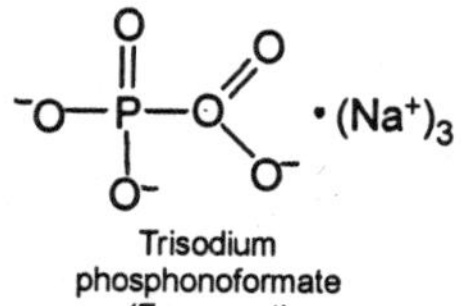

Trisodium phosphonoformate (Foscarnet)

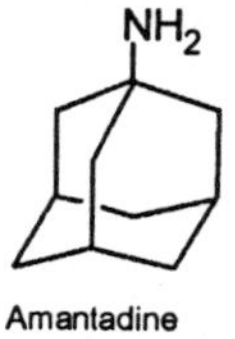

Amantadine

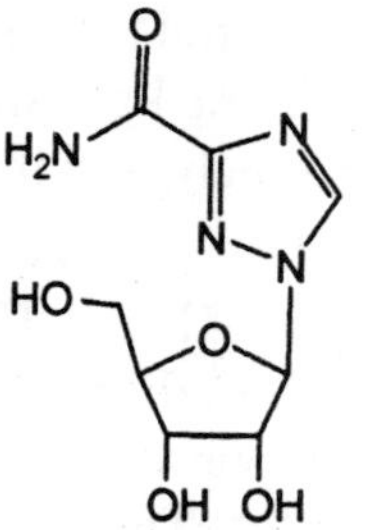

Ribavirin

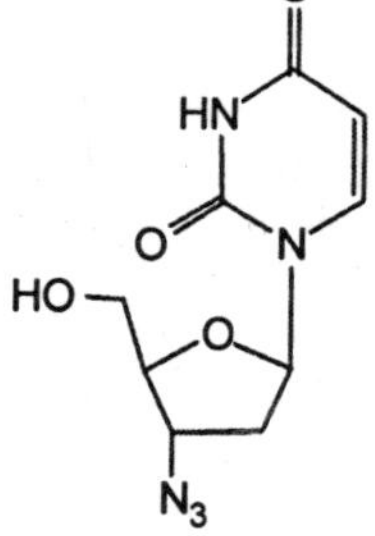

3′-Deoxy-3′-azidothymidine (AZT)

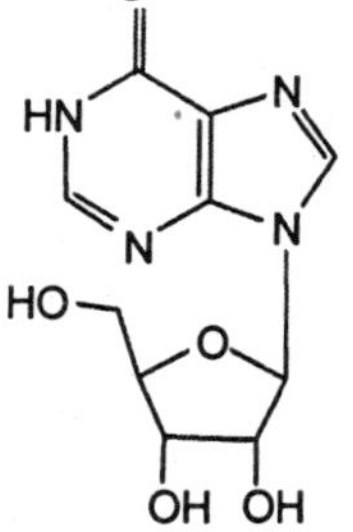

Dideoxyadenosine

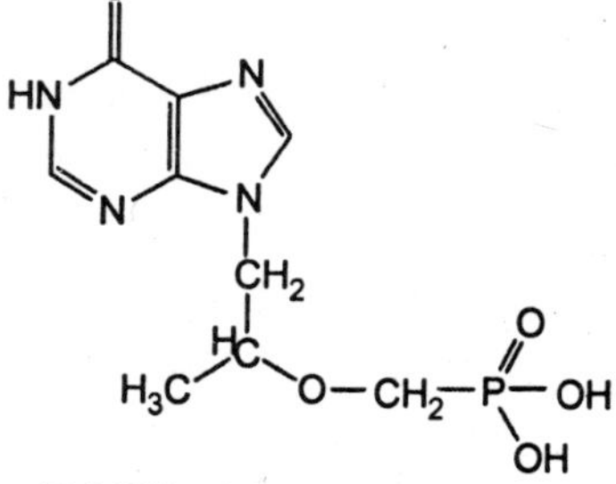

(R)-9-(2-Phosphonylmethoxypropyl) adenine

More recently 9-β-D-arabinofuranosyladenine (Ara-A) has become a preferred drug in treatment of ocular herpes infections. This compound, which is a naturally occurring antibiotic, can also be administered intravenously for life-threatening infections such as herpes encephalitis. Ara-A is quite toxic but the guanine derivative 9-(2-hydroxyethoxymethyl) guanine (acyclovir) is less so. Another acyclic 2′-deoxyguanine analog, 9-(1,3-dihydroxypropoxymethyl)-guanine (DHPG), is more soluble, more potent, and has a broader range of effectiveness.

One of the first effective drugs against RNA viruses was ribavirin (1-β-D-ribofuranosyl-1,2,4-triazole-3-carboxamide), an analog of guanosine. It has a broad range of action and is used to treat severe viral pneumonia and bronchitis caused by respiratory syncytial virus (RSV). This common virus usually is mild but may cause death in infants and children. Ribavirin may also be of some value in the treatment of AIDS.

How do these compounds exert their antiviral effects? The nucleoside analogs are usually phosphorylated to the corresponding mono-, di-, and tri-phosphate derivatives by cellular enzymes. Thus, Ara-A yields Ara-ATP, which inhibits the herpes virus-encoded DNA polymerase. Ara-A may also enter the viral DNA. In addition Ara-A may inhibit polyadenylation of virally induced mRNA. Some analogs, such as acyclovir, undergo little conversion to the monophosphate by cellular kinases but are efficiently phosphorylated by herpes virus-encoded thymidine kinase. Thus, acyclovir does little damage to uninfected cells. Ribavirin 3′-monophosphate may inhibit IMP dehydrogenase thereby interfering with GTP production. At the same time ribavirin triphosphate competes with GTP to inhibit virally encoded RNA polymerase.

Phosphonoformate is a pyrophosphate analog and inhibits both DNA polymerases and reverse transcriptase. However, toxicity may prevent long-term treatment of AIDS patients. Amantadine has a narrow antiviral specificity. It specifically inhibits initiation of the replication of influenza virus RNA of type A (but not of type B). Active only against retroviruses, 3′-azidothymidine is a reverse transcriptase inhibitor, which acts by a chain termination mechanism. It was synthesized in the early 1960s but only recently has been used in treatment of AIDS victims. More recently a series of 2′,3′-dideoxynucleosides, such as dideoxyinosine, have also been used. Acyclic phosphonates, such as phosphonylmethoxypropyladenine, avoid the need for metabolic phosphorylation of the drug.

Development of synthetic antiviral compounds is hardly beyond its infancy. Serious problems must be overcome with most of these compounds. Toxicity (sometimes carcinogenicity), development of resistance by viruses, and enzymatic destruction limit the utility of most drugs. For example, adenosine deaminase destroys Ara-A quite rapidly. With our rapidly advancing knowledge of viral life cycles and protein and nucleic acid structures many new drug targets have been identified. Among the targets for HIV are the reverse transcriptase, protease, and integrase. Computer-assisted design, as well as new techniques of synthesis and screening, have allowed development of many non-nucleoside inhibitors.

Oligonucleotide phosphoramidates and other triplex-forming compounds may be designed to bind to specific DNA targets.